Digital Synthetic Data and Outputs

Shalin Hai-Jew
https://orcid.org/0000-0002-8863-0175
Hutchinson Community College, USA

IGI Global
Scientific Publishing
Publishing Tomorrow's Research Today

Vice President of Editorial	Melissa Wagner
Director of Acquisitions	Mikaela Felty
Director of Book Development	Jocelynn Hessler
Production Manager	Mike Brehm
Cover Design	Jose Rosado

Published in the United States of America by
IGI Global Scientific Publishing
701 East Chocolate Avenue
Hershey, PA, 17033, USA
Tel: 717-533-8845 | Fax: 717-533-7115
Website: https://www.igi-global.com E-mail: cust@igi-global.com

Library of Congress Cataloging-in-Publication Data

LCCN: 2025017388 (CIP Data Pending)
ISBN13: 9798337322773
Isbn13Softcover: 9798337322780
EISBN13: 9798337322797
British Cataloguing in Publication Data
A Cataloguing in Publication record for this book is available from the British Library.

Table of Contents

Preface xvi

Chapter 1
Generative AI and Digital Synthetic Outputs 1
Aryan Vasa, Ajeenkya D.Y. Patil University, India
Aditya Shrivastav, Ajeenkya D.Y. Patil University, India
Sunil Sankathala, Ajeenkya D.Y. Patil University, India
Kashvi Chaturvedi, Ajeenkya D.Y. Patil University, India
Susanta Das, Ajeenkya D.Y. Patil University, India
Rizwan Shaikh, Dr. D.Y. Patil Institute of Management and Entrepreneur Development, India

Chapter 2
Exploring Synthetic Generation of Digital Data Through Data-Driven Learning 31
A. Prabhu Chakkaravarthy, Networking and Communications, SRM Institute of Science and Technology, India
Dhanalakshmi Jaganathan, SRM Institute of Science and Technology, India

Chapter 3
Digital Synthetic Generation From Learning and BigData 57
Ajay N. Upadhyaya, SAL Engineering and Technical Institute, India
S. Seenivasan, Rathinam Technical Campus, India
P. Selvakumar, Department of Science and Humanities, Nehru Institute of Technology, Coimbatore, India
Nilesh Anute, Sri Balaji University, India
S. Guru Basava Aradhya, Chandigarh School of Business, Chandigarh Group of Colleges Jhanjeri, Mohali, India
Mohit Sharma, Maharshi Dayanand University, Rohtak, India
T. C. Manjunath, Rajarajeswari College of Engineering, India

Chapter 4
Artificial Intelligence Agents Powered by Synthetic Data 87
Sachin Sharma, Amity University, Mohali, India

Chapter 5
Digital Synthetic Motion Outputs 123
Samrat Kundu, Brainware University, India
Kousik Dey, Brainware University, India
Rahul Bir, Brainware University, India
Sandip Pal, Brainware University, India

Chapter 6
Graph Neural Networks in Complex Image Recognition: From Theory to Real-World Applications.. 151
Minh Ly Duc, Faculty of Commerce, Van Lang University, Ho Chi Minh City, Vietnam
Sang Quang Nguyen, Faculty of Automotive Engineering, School of Engineering and Technology, Van Lang University, Ho Chi Minh City, Vietnam

Chapter 7
Productivity Enhancements from Digital Synthetic Data 183
S. Ida Evangeline, Government College of Engineering, Tirunelveli, India

Chapter 8
Spun Text 205
Shalin Hai-Jew, Hutchinson Community College, USA

Chapter 9
Generative Artificial Intelligence and Digital Synthetic Outputs 221
Wasswa Shafik, Dig Connectivity Research Laboratory (DCRLab), Uganda

Chapter 10
Physics Engines in Game Design for Digital Play 253
P. Selvakumar, Department of Science and Humanities, Nehru Institute of Technology, Coimbatore, India
S. Seenivasan, Rathinam Technical Campus, India
M. Jegadeeswaran, Rathinam Technical Campus, India
Vatsala Sharma, Chandigarh School of Business, Chandigarh Group of Colleges Jhanjeri, Mohali, India
S. Saravana Kumar, Vinayaka Mission's Kirupananda Variyar Engineering College, Vinayaka Mission's Research Foundation, India
Mohit Sharma, Maharshi Dayanand University, Rohtak, India
T. C. Manjunath, Rajarajeswari College of Engineering, India

Chapter 11
Word Art and Alphanumeric Images From Generative AI 283
Shalin Hai-Jew, Hutchinson Community College, USA

Chapter 12
Illustrating a Monthly Professional Newsletter With Synthetic Visuals 295
Shalin Hai-Jew, Hutchinson Community College, USA

Chapter 13
Human Consumption of Digital Synthetic Outputs: Cross Checking Real or Fake Information.. 337
Deepanshu Pachauri, Sharda University, India
Saurabh Chandra, Bennett University, Greater Noida, India
Arunima Shastri, Christ University, Bangalore, India
Prangya Parimita Behera, KIIT University, Bhubaneswar, India
Bhupinder Singh, Sharda University, India

Chapter 14
Navigating the Real and the Synthetic: Authentication and Validation in Digital Synthetic Outputs.. 355
Mohit Sharma, Maharshi Dayanand University, Rohtak, India
Nitika Malik, Maharshi Dayanand University, Rohtak, India

Chapter 15
A Comparative Analysis of Generative AI Traffic Across Continents 379
Kassim Kalinaki, Islamic University in Uganda, Uganda
Madinah Nabukeera, Kyambogo University, Uganda
Moses Matovu, Islamic University in Uganda, Uganda

Chapter 16
Inside the Black Box: How Generative AI Creates Digital Synthetics 409
Valisher O. Sapayev, Mamun University, Uzbekistan
Noor Hayat, University of Central Punjab, Pakistan
Zokir Mamadiyarov, Termez University of Economics and Service, Uzbekistan
Munir Ahmad, Survey of Pakistan, Pakistan

Conclusion 435

Compilation of References 437

About the Contributors 497

Index 511

Detailed Table of Contents

Preface..xvi

Chapter 1
Generative AI and Digital Synthetic Outputs 1
Aryan Vasa, Ajeenkya D.Y. Patil University, India
Aditya Shrivastav, Ajeenkya D.Y. Patil University, India
Sunil Sankathala, Ajeenkya D.Y. Patil University, India
Kashvi Chaturvedi, Ajeenkya D.Y. Patil University, India
Susanta Das, Ajeenkya D.Y. Patil University, India
Rizwan Shaikh, Dr. D.Y. Patil Institute of Management and Entrepreneur Development, India

GenAI, producing high-quality synthetic data, is transforming industries and tackling issues like algorithmic bias, data scarcity, and privacy concerns. Realistic images, audio, video, and structured datasets are produced using important generative models. But they also have to deal with issues like training stability, computational costs, and moral dilemmas. The chapter examines the potential and drawbacks of synthetic data applications in modern era of GenAI in various fields. Synthetic data produced by AI is considered revolutionary for resolving data-related issues. Large quantities can be produced, can be tailored to meet specific needs, and excludes personal information. Hence, it is very helpful for testing software, training AI, and evaluating effectiveness of algorithms. Ethical concerns do raise questions regarding data privacy, bias, false information, copyright concerns, transparency, and economic effects. Improvement of efficiency, privacy, and credibility of synthetic data should be the main goals of future research.

Chapter 2
Exploring Synthetic Generation of Digital Data Through Data-Driven Learning 31
A. Prabhu Chakkaravarthy, Networking and Communications, SRM Institute of Science and Technology, India
Dhanalakshmi Jaganathan, SRM Institute of Science and Technology, India

A revolutionary method for producing realistic, high-quality digital content, including pictures, movies, and audio, using computational models is digital synthetic generation, which is fueled by machine learning and big data. With the use of deep neural networks, sophisticated learning algorithms, and enormous datasets, this technology makes it possible to create synthetic media that remarkably accurately replicates real-world data. In this work, we investigate the use of digital synthetic generation in a number of domains, such as autonomous systems, design, healthcare, and entertainment. Generative methods such as Generative Adversarial Networks (GANs) and Variational Autoencoders (VAEs) can create fresh content that satisfies requirements without the need for human creativity or manual input by training models on large datasets.

Chapter 3
Digital Synthetic Generation From Learning and BigData 57
Ajay N. Upadhyaya, SAL Engineering and Technical Institute, India
S. Seenivasan, Rathinam Technical Campus, India
P. Selvakumar, Department of Science and Humanities, Nehru Institute of Technology, Coimbatore, India
Nilesh Anute, Sri Balaji University, India
S. Guru Basava Aradhya, Chandigarh School of Business, Chandigarh Group of Colleges Jhanjeri, Mohali, India
Mohit Sharma, Maharshi Dayanand University, Rohtak, India
T. C. Manjunath, Rajarajeswari College of Engineering, India

Digital synthetic generation is a rapidly evolving field that involves the creation of artificially generated data, media, and environments using computational techniques. It plays a crucial role in industries such as artificial intelligence (AI), gaming, film production, virtual reality (VR), and scientific simulations. By leveraging procedural algorithms, deep learning models, and physics-based simulations, digital synthetic generation enables the production of realistic and scalable content without relying on real-world data collection. This article provides an in-depth exploration of the concepts, methods, and applications of digital synthetic generation, highlighting its significance in various domains and its potential to reshape industries. Concepts of Digital Synthetic Generation: Digital synthetic generation refers to the process of artificially creating digital content using computational models rather than capturing it from the real world. This can include images, videos, 3D environments, physics simulations, and even synthetic datasets for AI training.

Chapter 4
Artificial Intelligence Agents Powered by Synthetic Data 87
Sachin Sharma, Amity University, Mohali, India

The merging of Artificial Intelligence (AI) agentry and progressed built yields talks to a transformative paradigm in how cleverly systems work and associated with the world. AI masters, characterized by their independence, considering, organizing, and learning capabilities, are dynamically being utilized to perform complex errands with irrelevant human intercession. These specialists as often as possible require endless sums of data for planning, endorsement, and operation. This collaboration addresses some fundamental challenges. Firstly, designed data can overcome the obstacles of real-world data, such as deficiency, inclination, assurance concerns, and the tall fetched of procurement and comment. By producing data that accurately reflects the genuine properties of veritable data, This capability opens unused streets for amusement, testing, and the creation of novel advanced substance.

Chapter 5
Digital Synthetic Motion Outputs 123
Samrat Kundu, Brainware University, India
Kousik Dey, Brainware University, India
Rahul Bir, Brainware University, India
Sandip Pal, Brainware University, India

Synthesizing motion data digitally employs advanced AI models like generative adversarial networks (GANs), and diffusion networks to create realistic motion data for various applications including visual effect, video games, robot motion, and simulation. These technologies aid in large-scale data augmentation as well as in rapid prototyping, thus helping get around the limitations of real-world data. Their major uses lie in film de-aging, animation for interactive games, and robotic training in simulated environments. However, concerning privacy, trust, and misinformation, deepfake technology adds to the worries. Human oversight, authentication methods, and validation benchmarks must ensure that these technologies are put to ethical use. On the responsible side, synthetic motion output constitutes a way to enhance real-world experience in education, design, therapy, and interactive multimedia.

Chapter 6
Graph Neural Networks in Complex Image Recognition: From Theory to Real-World Applications 151
Minh Ly Duc, Faculty of Commerce, Van Lang University, Ho Chi Minh City, Vietnam
Sang Quang Nguyen, Faculty of Automotive Engineering, School of Engineering and Technology, Van Lang University, Ho Chi Minh City, Vietnam

This study presents a comprehensive review of recent advancements in deep learning for object recognition and image classification, with a particular emphasis on the growing role of Graph Neural Networks (GNNs) in addressing complex image analysis challenges. Drawing on research from reputable databases including IEEE Xplore, Springer, Elsevier, and arXiv over the past five years, the study assesses various models using four key criteria: accuracy, generalization ability, computational efficiency, and practical applicability. The analysis reveals that GNNs offer notable advantages over traditional Convolutional Neural Networks (CNNs), especially in capturing spatial and structural relationships within image data. This makes GNNs particularly effective in scenarios that demand an understanding of topological complexity, such as classifying white blood cells, detecting faults in power systems, and tracking nano-robots.

Chapter 7
Productivity Enhancements from Digital Synthetic Data 183
S. Ida Evangeline, Government College of Engineering, Tirunelveli, India

Synthetic data—artificially generated datasets that replicate the statistical properties of real-world data—has emerged as a powerful enabler of productivity in data-driven industries. By providing privacy-preserving, on-demand, and task-specific datasets, synthetic data addresses persistent challenges related to data scarcity, regulatory restrictions, and the high costs of acquisition and annotation. This chapter examines its role as a strategic asset in accelerating workflows across artificial intelligence and machine learning development, software testing, system prototyping, and collaborative research. It addresses technological underpinnings of the generation of synthetic data and provides cross-industry case studies in the fields of healthcare, finance, manufacture, and autonomous systems as well as quantitative and qualitative measures of productivity improvement.

Chapter 8
Spun Text 205
Shalin Hai-Jew, Hutchinson Community College, USA

Multimodal generative AI (GAI) tools, built on large language models (LLMs), are eminently able to output texts of various sorts, both purporting to be nonfiction (fact-based) and fiction (imagination-based). Synthetic data is applied to research and model training in terms of specialist uses; it is applied to various personal uses in the "retail" application sense. These GAI tools are able to review and critique other texts, brainstorm lists and ideas, write in various genres (essays, novels, novellas, memos, letters, and others), converse with people, create synthetic data, and theoretically generate unlimited text outputs, among others. Based on several years of experiential and hands-on use, this chapter offers assessments of how well several such tools perform in the generation of a newish genre of "100-word novelas" based on a defined prompt. This experimental case asks whether the novelas are able to spin text into gold.

Chapter 9
Generative Artificial Intelligence and Digital Synthetic Outputs 221
Wasswa Shafik, Dig Connectivity Research Laboratory (DCRLab), Uganda

Generative Artificial Intelligence and digital synthetic outputs are transforming industries by enabling the creation of realistic and innovative digital content. Leveraging advanced machine learning techniques, generative AI produces diverse outputs, including text, images, videos, and 3D models, with applications spanning entertainment, healthcare, education, and environmental sustainability. This technology has revolutionized personalized experiences, adaptive learning, and creative design, while also enhancing decision-making in fields like disaster management and precision agriculture. The proliferation of digital synthetic outputs raises ethical concerns related to data privacy, misinformation, intellectual property, and societal impact. Addressing these challenges requires a balanced approach that fosters innovation while ensuring transparency, accountability, and inclusivity. This abstract explores the transformative potential of generative AI, emphasizing its implications for sustainable development and its role in advancing equitable and responsible technological adoption.

Chapter 10
Physics Engines in Game Design for Digital Play 253
P. Selvakumar, Department of Science and Humanities, Nehru Institute of Technology, Coimbatore, India
S. Seenivasan, Rathinam Technical Campus, India
M. Jegadeeswaran, Rathinam Technical Campus, India
Vatsala Sharma, Chandigarh School of Business, Chandigarh Group of Colleges Jhanjeri, Mohali, India
S. Saravana Kumar, Vinayaka Mission's Kirupananda Variyar Engineering College, Vinayaka Mission's Research Foundation, India
Mohit Sharma, Maharshi Dayanand University, Rohtak, India
T. C. Manjunath, Rajarajeswari College of Engineering, India

Game design has evolved tremendously over the years, with technological advancements pushing the boundaries of realism and immersion. One of the most critical components behind this evolution is the physics engine. Physics engines are specialized software frameworks that simulate real-world physics in digital environments, enabling objects to move, interact, and behave naturally. From simple 2D side-scrollers to complex open-world simulations, physics engines are at the core of making virtual worlds feel alive. They are responsible for everything from character movement and collisions to destructible environments and realistic fluid dynamics. Without them, modern games would lack the level of realism and interactivity that players expect today. At their core, physics engines rely on mathematical equations and algorithms to mimic the physical laws of the real world, such as Newtonian mechanics, gravity, and friction. These calculations determine how objects interact in a game environment.

Chapter 11
Word Art and Alphanumeric Images From Generative AI 283
Shalin Hai-Jew, Hutchinson Community College, USA

"Word art" comes in various forms: word clouds, word trees, digital graffiti, calligrams, typography, calligraphy, designed signs, and others. These may be generated from data, digital image editing programs, online word art generators, and other tools, now, including art-making generative AI ones. This latest category of tool has struggled with accurate spelling, even as generative AI tools create their own visual "fonts" and aesthetics. The challenges for such tools in creating word art involves visual aesthetics, font aesthetics; meaning capture and conveyance; alphabetics, numerics, alphanumerics;; human readability, machine readability; and other challenges simultaneously. This work involves a light exploration of the "word art" aspects of generative AIs in the contemporaneous moment.

Chapter 12
Illustrating a Monthly Professional Newsletter With Synthetic Visuals 295
Shalin Hai-Jew, Hutchinson Community College, USA

The success of generative AI tools will depend in part on how easily these may be harnessed for practical and work purposes. This chapter explores the uses of several multimodal generative AI tools in the creation of text, imagery, and word art for a monthly grant newsletter. The visuals are used for a variety of purposes in the newsletter: decorative, reader effects, teaching and learning, data sharing, human memory enhancement, and others. This use-case illuminates some of the capabilities of several image-making generative AI tools (Deep Dream Generator, Gemini, and Microsoft Copilot). This work identifies some areas to improve to enable regular usage of such tools to generate synthetic imagery for workplace applications.

Chapter 13
Human Consumption of Digital Synthetic Outputs: Cross Checking Real or Fake Information 337
Deepanshu Pachauri, Sharda University, India
Saurabh Chandra, Bennett University, Greater Noida, India
Arunima Shastri, Christ University, Bangalore, India
Prangya Parimita Behera, KIIT University, Bhubaneswar, India
Bhupinder Singh, Sharda University, India

AI-Generated content, deepfakes and synthetic media, people are increasingly wondering how they can discern the difference between genuine and fake. People may become confused, change their minds, and make unwise decisions that put their safety at danger when they see false information and manipulated pictures and videos, such AI-generated deepfakes. AI detection techniques, and teaching people how to read and write have all helped with these difficulties. Blockchain technology analyzes the digital history of anything to make sure it is authentic. Two significant steps toward making things more open are regulatory frameworks and ethical AI practices. Users should be vigilant and examine a lot of trustworthy sources, including reverse image search or deepfake detectors. The influence of incorrect information will be lessened if individuals are more informed and encouraged to utilize internet resources properly. To use another metaphor, let's make sure that people know what they're talking about when we speak about synthetic things.

Chapter 14
Navigating the Real and the Synthetic: Authentication and Validation in Digital Synthetic Outputs 355
Mohit Sharma, Maharshi Dayanand University, Rohtak, India
Nitika Malik, Maharshi Dayanand University, Rohtak, India

The rise of digital synthetic data and outputs has blurred the lines between reality and artificiality. With the growing sophistication of generative AI, synthetic outputs—ranging from images and video to text and voice—are increasingly integrated into various industries, including education, healthcare, entertainment, and cybersecurity. However, this surge in synthetic content necessitates robust mechanisms for authentication and validation. This chapter examines current methodologies and technologies for distinguishing synthetic outputs from authentic ones. It explores automated validation techniques, human-in-the-loop authentication processes, and hybrid models that leverage machine learning to detect digital forgeries. The discussion also extends to ethical considerations and societal implications, particularly in an era where synthetic deception poses real-world consequences.

Chapter 15
A Comparative Analysis of Generative AI Traffic Across Continents 379
Kassim Kalinaki, Islamic University in Uganda, Uganda
Madinah Nabukeera, Kyambogo University, Uganda
Moses Matovu, Islamic University in Uganda, Uganda

This study presents a comparative analysis of user engagement with generative AI, using ChatGPT traffic data for March 2024 as a case study of one widely adopted tool. Using World Bank data and statistical methods, including ANOVA, it compares usage across continents. While statistical tests showed no significant differences in total visits or global traffic share, descriptive results revealed meaningful disparities in engagement intensity. Chatbot-based platforms, particularly ChatGPT, have dominated other generative AI tools due to their versatility in tasks such as content creation, learning, coding, and communication. These findings suggest a trend toward broader access but also highlight persistent structural barriers, such as device access, digital skills, and infrastructure, that shape real-world usage. Therefore, promoting digital literacy, improving infrastructure, and localizing AI tools remain essential to achieving more equitable global engagement with generative AI.

Chapter 16
Inside the Black Box: How Generative AI Creates Digital Synthetics.............. 409
Valisher O. Sapayev, Mamun University, Uzbekistan
Noor Hayat, University of Central Punjab, Pakistan
Zokir Mamadiyarov, Termez University of Economics and Service, Uzbekistan
Munir Ahmad, Survey of Pakistan, Pakistan

This chapter unpacks the inner workings and outer implications of generative AI systems that produce synthetic text, images, audio, and video. While these models demonstrate remarkable creativity and utility, they also embody a core paradox: we increasingly rely on tools whose decision-making processes we do not fully understand. Exploring architectures such as transformers and diffusion models, we examine how outputs emerge from complex, probabilistic systems. The chapter also addresses the challenges of opacity, emergent behavior, and limited interpretability, highlighting the need for ethical oversight, technical safeguards, and societal trust in an age where synthetic content is reshaping digital reality.

Conclusion 435

Compilation of References 437

About the Contributors 497

Index 511

Preface

Computation is harnessed in many ways to enhance people's capabilities and lived experiences. One slice of that enhancement involves the generating of digital synthetic data for practical uses (training machine learning models, filling out incomplete datasets for analytics, modeling data for online surveys, work needs, among others) and human experiences (education, entertainment, storytelling, visualizations, artistic expression, and others).

Such machine-generated synthetic data comes in various forms. It can be structured data, as is often used in quantitative research and data analytics. It can be semi-structured (or some say "unstructured") data, such as that used in qualitative research and data analytics. The data may be in various modalities of form: text, data tables (matrices), imagery, audio, video, slideshows, animations, and multimedia formats.

Synthetic data may be partially derived from real-world data (as seed) or wholly auto-generated from nothing. Synthetic data may be used as a hybrid with real-world data to create a more complete dataset.

Digital synthetic data stands in various work gaps to enable bridging past various work challenges to the other side. It enables extensions to human expression and experience, with immersions in the digital and the illusory. Its generation is computationally negligible in terms of cost in many cases but expensive for some of the more recent technology systems. (Figure 1)

Figure 1. Synthetic Data

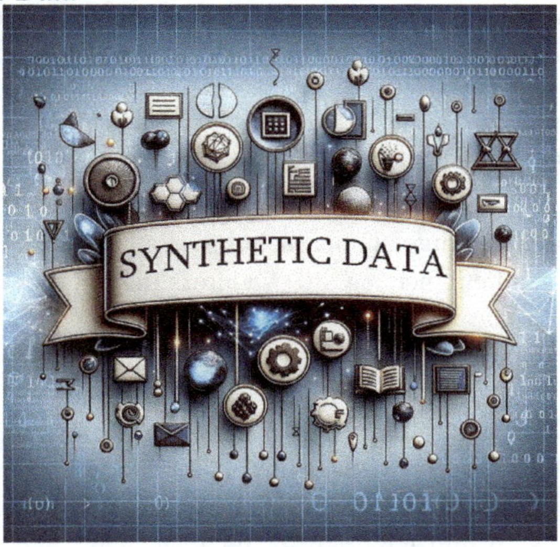

Digital Synthetic Data and Outputs is an edited collection that explores some of the advances in the area of synthetic data generation and applications.

Synthetic data has been created and used since the 1960s and 1970s. Researchers in computer vision created simplified artificial drawings computationally to train computers to recognize 3d shapes from 2d visuals. In the 1990s, researchers started using synthetic data to ensure data privacy while still enabling the sharing of derived datasets and real-world statistical observations (without any extractable personally identifiable information). Various procedural methods (based on code and algorithms and programs) were used for the generation of early visual synthetic data. The creation of derived datasets was based on statistical modeling, so the derived synthetic data would emulate the underlying statistical properties albeit with all-new row data. Researchers have used simulations and synthetic worlds to generate contents.

In the present day, generative AI models have been set up to create synthetic data in different ways, based on machine learning (from real and synthetic data). Some approaches are based on trying to ensure that the synthetic data mirrors real-world ones as closely as possible. Others define various inputs to enable the generation of the synthetic data to have particular features. Synthetic data may be integrated invisibly in various practical and experiential human practices. It may be used anywhere in a sequence of production.

As noted earlier, synthetic data may be structured data, as in quantitative data tables. It may be semi-structured (or unstructured) data, such as texts, visuals, music, audio, video, multimedia, slideshows, animations, and other mixes. The modalities of data may vary across a range of digital and digital-physical types. Also, there can be various human value-added along the way. One fairly recent innovation has been human-machine co-creation, such as of designs and artworks.

EVOLVING SYNTHETIC DATA

The works highlight a range of approaches to synthetic data. Some approach as computer scientists and engineers. Others address the topic as users. Some do so as thinkers and ethicists.

While synthetic data has been around for decades, in various modalities, the present day involves different outputs, different qualities of outputs, different methods to create synthetic data, and wide variations in their uses (for people, for computing machines). While a topic may have been around for decades, each generation offers something else. As such, this collection is released with a sense of occasion.

These works are early and exploratory. The types of synthetic data used are still within a kind of narrow band, with so much potential for more and for different. The space is fast evolving, with ever more applications across a broad spectrum of users.

Shalin Hai-Jew

2025

Chapter 1
Generative AI and Digital Synthetic Outputs

Aryan Vasa
Ajeenkya D.Y. Patil University, India

Aditya Shrivastav
https://orcid.org/0009-0000-8461-8171
Ajeenkya D.Y. Patil University, India

Sunil Sankathala
https://orcid.org/0009-0004-7809-7647
Ajeenkya D.Y. Patil University, India

Kashvi Chaturvedi
https://orcid.org/0009-0005-3192-3400
Ajeenkya D.Y. Patil University, India

Susanta Das
https://orcid.org/0000-0002-9314-3988
Ajeenkya D.Y. Patil University, India

Rizwan Shaikh
https://orcid.org/0009-0005-5689-7640
Dr. D.Y. Patil Institute of Management and Entrepreneur Development, India

ABSTRACT

GenAI, producing high-quality synthetic data, is transforming industries and tackling issues like algorithmic bias, data scarcity, and privacy concerns. Realistic images, audio, video, and structured datasets are produced using important generative models. But they also have to deal with issues like training stability, computational costs, and moral dilemmas. The chapter examines the potential and drawbacks of synthetic data applications in modern era of GenAI in various fields. Synthetic data produced by AI is considered revolutionary for resolving data-related issues. Large quantities can be produced, can be tailored to meet specific needs, and excludes personal information. Hence, it is very helpful for testing software, training AI, and evaluating effectiveness of algorithms. Ethical concerns do raise questions regarding data privacy, bias, false information, copyright concerns, transparency, and economic effects. Improvement of efficiency, privacy, and credibility of synthetic data should be the main goals of future research.

DOI: 10.4018/979-8-3373-2277-3.ch001

1. INTRODUCTION

With the rise of Generative Artificial Intelligence (Gen AI), artificial intelligence research and application have changed from traditional algorithmic models to systems that can produce new content on their own. By discovering patterns in large datasets, Gen AI creates original outputs like text, images, audio, and even video, in contrast to traditional AI models that concentrate on classification, prediction, and regression tasks (Takale et al., 2024). Generative AI is now used for more than just writing and art. Additional data addition, simulations, error correction, and model adaptation to new data types are some of the more technical tasks that it is currently being used for (Eigenschink et al., 2023). In the modern world, the technology has extended the boundaries of creativity and innovation in AI by not only replicating existing data but also comprehending patterns sufficiently to produce completely original concepts (Sengar et al., 2024). Such outputs mimic human creativity. Innovation has been sparked by this capacity to produce unique, realistic-looking synthetic data in a variety of fields, including healthcare, finance, the arts, and education. Deep generative models, which have undergone substantial development in the last ten years, are the cornerstone of generative AI. Earlier generative systems relied on probabilistic models and rule-based algorithms. However, as deep learning and processing power have increased, complex architectures like Diffusion Models, Transformers, Variational Autoencoders, and Generative Adversarial Networks (GANs) have appeared, each with unique uses and capabilities (Bandi et al., 2023; Sengar et al., 2024). The multidisciplinary technologies of machine learning, computer vision, and natural language processing form the foundation of generative artificial intelligence. From text and images to audio and 3D environments, these fundamental technologies enable systems to produce new content on their own. Architectures like Transformers and Generative Adversarial Networks (GANs) are essential. Examples of how models trained on large datasets can generate outputs that are both highly realistic and semantically coherent include autoregressive models like GPT and image-based networks like DALL-E (Lv, 2023; Park & Kim, 2022). Each model of generative AI has advantages and disadvantages. Despite their strength, GANs can be challenging to train and may introduce undesirable errors into the output. VAEs produce results that are often blurrier, but they are more stable. According to Sengar et al. (2024), diffusion models are more recent and generate synthetic content of very high quality, but they require more time to run. Transformer models, which are well-known for enabling ChatGPT and other similar tools, are now being utilized for purposes other than text, such as generating time-series data, synthetic medical records, and other structured data types (Eigenschink et al., 2023).

In generative AI, the goal of the specialised field of synthetic data generation is to create artificial datasets that closely resemble the statistical characteristics of real-world data (Kwatra & Torra, 2024; Ramzan et al., 2024). GPT-3, GPT-4, and StyleGAN are examples of advanced generative architectures that demonstrate AI's ability to replicate extremely realistic language and visual content (Bandi et al., 2023). These artificial results are now far more sophisticated than simple simulations, frequently reaching fidelity levels that render them identical to real data. Synthetic data goes beyond simple augmentation and is essential for resolving problems like data scarcity and class imbalance. For example, Gen AI can produce more samples of uncommon or under-represented cases to facilitate more equitable and precise model training, especially in delicate fields like law enforcement or finance (Figueira & Vaz, 2022). Synthetic data thus helps to improve model inclusivity and robustness in addition to increasing data volume. Because of its capacity to replicate intricate situations in highly risky or limited environments, it is also a fundamental component of privacy-preserving AI development and is being embraced by a growing number of industries (Ramzan et al., 2024; Takale et al., 2024).

Generative AI has gained significant traction as the focus in AI development moves from only creating complex models to improving data quality. As the need for trustworthy and representative datasets keeps increasing across a range of industries, synthetic data becomes even more crucial in tackling issues like imbalance, restricted access, and data privacy (Takale et al., 2024; Sengar et al., 2024). Especially in delicate fields like healthcare, synthetic data offers a way to get around privacy restrictions without sacrificing analytical usefulness. It allows, for example, the expansion of small clinical datasets and the simulation of uncommon medical situations without jeopardising patient privacy (Rujas et al., 2024). Additionally, this capability advances the objective of inclusive data representation, which is crucial for creating AI systems that are equitable. Researchers can replicate a variety of clinical information, including medical scans and records, by utilising generative models like GANs, VAEs, diffusion models, and language models. This allows them to avoid the risks that come with using real patient data (Ibrahim et al., 2025). To guarantee the reliability and security of these artificial datasets, however, strong frameworks are still needed in spite of these developments. Beyond the healthcare industry, the production of synthetic data is still used in more general contexts, like financial data simulation (Ramzan et al., 2024) and the development of dynamic, user-adaptive content that advances Web 3.0 (Bandi et al., 2023).

Although generative AI holds great promise for progress, it also presents urgent moral, legal, and societal issues that should not be disregarded. The possibility that artificial data will reinforce or even magnify preexisting biases in the original datasets is one major worry (Kwatra & Torra, 2024). Furthermore, there may be major privacy concerns if specific models—like GANs—inadvertently replicate traces of real data. As it gets harder to tell the difference between real and fake content, the growing realism of AI-generated media creates more issues and erodes public confidence in digital spaces (Takale et al., 2024). According to Kwatra and Torra (2024), these issues are particularly important in industries like healthcare and finance where synthetic data must faithfully depict uncommon or extreme situations. Generative models, in contrast to traditional AI models, need multifaceted evaluation criteria to assess the outputs' relevance and variability (Eigenschink et al., 2023). Standardised evaluation procedures, open-source benchmarks, and domain-specific solutions that put an emphasis on explainability, fairness, and transparency are among the strategies being pursued by researchers to address these issues (Sengar et al., 2024; Rujas et al., 2024). The use of techniques created for one kind of generative model, like language models, in other domains, like finance or transportation, is also gaining popularity (Eigenschink et al., 2023). Responsible AI practices, such as digital watermarking, deepfake detection, and legal compliance, are being emphasised as essential steps for the future (Sengar et al. 2024; Fisher et al. 2024).

The technology's operation, applications, and challenges are all examined in this review article to provide a comprehensive look at generative artificial intelligence and its synthetic outputs. It looks at the evolution of synthetic content and important generative models, as well as its practical applications in fields like healthcare, finance, education, and the arts. In particular, the article addresses privacy, equity, and regulation, as well as the benefits and drawbacks of using synthetic data. Moreover, it emphasizes the necessity of responsible innovation, cross-disciplinary cooperation, and guaranteeing accountability in AI systems as it looks to the future. The overall objective is to shed light on Gen AI and provide a road map for its moral and practical application in the digital age.

2. FOUNDATIONS OF GENERATIVE AI

Generative Artificial Intelligence (Gen AI) is a subset of AI that can simulate human creativity by producing completely original content, including text, images, audio, and synthetic data. According to Takale et al. (2024), Gen AI learns patterns from massive datasets and applies that knowledge to generate unique outputs, in contrast to traditional AI, which primarily analyses data to make predictions or classifications. In the early days of generative modelling, methods such as Markov Chains were used to predict sequences, such as speech or text, based solely on the sequence's previous step. Meanwhile, early artificial intelligence systems imitated intelligent behaviour using set rules. But according to Takale et al. (2024), neither of these approaches was able to scale or adjust to complex and evolving data. With the advent of deep learning, generative AI underwent a significant sea change. Generative Adversarial Networks (GANs) in 2014 and Variational Autoencoders (VAEs) in 2013 were two examples of technologies that enabled machines to recognize hidden patterns in data and produce realistic results (Figueira & Vaz, 2022). This advancement was furthered by the introduction of Diffusion Models for image generation and Transformers for language generation, which significantly raised the caliber and potential of generative AI systems (Sengar et al., 2024; Bandi et al., 2023). The foundational model architectures that underpin generative AI today are diverse, each with its own mechanisms and appropriateness for various forms of content production. In text, image, and other media formats, these models are the fundamental units that enable AI to produce realistic and practical synthetic outputs (Bandi et al., 2023).

One of the first deep generative models, Variational Autoencoders (VAEs), employ a decoding process to reconstruct input data after it has been compressed into a simplified representation called a latent space. With this structure, VAEs can produce new data that preserves the essential elements of the original input while allowing for flexibility. They excel at generating controlled and interpretable outputs, which makes them appropriate for applications that call for a harmony between creativity and accuracy (Goyal & Mahmoud, 2024). Generative Adversarial Networks (GANs) employ a different strategy by utilizing two neural networks in a continuous feedback loop: a discriminator that attempts to determine whether the data is real or fake, and a generator that generates synthetic data. Over a large number of cycles, both networks improve, the discriminator at identifying flaws and the generator at producing credible data. In domains such as image and video generation, in particular, this adversarial setup produces extremely realistic outputs (Ramzan et al., 2024; Singh et al., 2024). In order to comprehend and produce human language, Transformers and Large Language Models (LLMs), like those found in GPT models, were first created. These models generate content by taking into account the complete context of a sentence or document thanks to a mechanism known as self-attention. According to Bandi et al. (2023), transformers are particularly useful for tasks that call for coherence, fluency, and contextual relevance in both language and code because of their capacity to manage long-range dependencies in text. One of the more recent types of generative models is the diffusion model, which reverses a noise process. Iteratively, they transform random noise into structured, meaningful content, most notably high-quality images. This method has demonstrated exceptional efficacy in producing photorealistic images and intricate graphics, and it keeps pushing the boundaries of synthetic media generation performance (Kwatra & Torra, 2024; Sakirin et al. 2023).

A fundamental shift in the way digital platforms function, Generative Artificial Intelligence (Gen AI) allows for the production of unique, human-like outputs, whether they be text, code, images, or entire datasets, instead of just analysing preexisting data. Beyond the predictive power of conventional AI, this change opens up new possibilities for automation, creativity, and personalization. Gen AI's core components include large language models (LLMs), which can generate context-aware content by learning from large datasets, and deep learning models like Transformers. According to Wessel et al. (2025), these capabilities are not just technical triumphs but also revolutionary instruments that are changing the composition and operation of digital platforms. Intelligent automation is just one of the many ways that generative AI is transforming digital platforms. Gen AI enables platforms to react more intelligently and flexibly than they could in the past when they relied on inflexible systems with set rules. TikTok uses Gen AI to automatically and widely moderate content, and tools like GitHub Copilot can create code from plain text instructions. According to Wessel et al. (2025), this change enhances platform performance and user experience by making digital environments more interactive and effective. Gen AI is also making platform participation more accessible to all. Technical expertise was previously needed to create apps or digital content, which restricted who could participate. These days, however, people without any coding experience can create product listings or applications using common language thanks to tools like Shopify Magic and Microsoft AI Builder. Although this creates new opportunities for innovation and participation, it also poses new challenges for quality control and content authenticity. The third significant shift is hyper-personalization. Gen AI is capable of producing content that feels uniquely customized by analysing intricate user behaviour. Real-time user preference adaptation is provided by services like Netflix's personalized thumbnails and Spotify's AI DJ, which help platforms boost engagement. However, risks associated with such deep personalization include privacy issues and the potential for user manipulation through highly targeted content. Lastly, collaborative innovation, where users and AI systems work together to create, it is supported by Gen AI. Rather than merely issuing commands, users can collaborate with AI to improve designs or investigate innovative concepts, such as with Adobe Firefly or AI music tools. Human-machine cooperation is generating new types of innovation that were previously unattainable (Wessel et al., 2025).

In order to create digital synthetic outputs, in simple words, artificially generated data or content that appears and feels authentic, Gen AI is essential. Among these are 3D models, structured datasets, text, sounds, and realistic-looking but fake images. According to Goyal and Mahmoud (2024) and Figueira and Vaz (2022), many industries, including healthcare, finance, and robotics, use these outputs extensively to support tasks like training, analysis, and simulation. Synthetic data is especially helpful in situations where real data is expensive to collect, sensitive, or difficult to access. A privacy-friendly substitute, it preserves private information while simulating the structure and patterns of real data. In order to enhance model performance through augmentation, it additionally facilitates AI training by offering additional data (Endres et al., 2022). When applied in the real world, generative models can produce intricate simulations of challenging circumstances, such as atypical financial crises, customized marketing profiles, or individualized medical scans. These artificial results raise the performance of AI systems and enable their application in a greater variety of real-world and dynamic contexts (Ramzan et al., 2024). The ability to combine various skills across domains is making modern generative AI systems more adaptable. The integration of language and visual understanding is demonstrated by models such as DALL-E 2, which can convert text descriptions into images. This combination of skills, referred to as multi-modal generation, makes it possible to produce synthetic content that is more interactive and richer (Sakirin et al., 2023). Nevertheless, these potent models also have certain drawbacks. Mode collapse (where the model produces limited types of outputs), training instability (which is common in GANs), and how difficult it is to understand how models make decisions are still major concerns. Additionally, there is no guarantee that synthetic data is safe, even though it is frequently viewed as privacy friendly. If robust privacy measures like differential privacy are not implemented, some sophisticated attacks can still deduce actual information from synthetic datasets (Kwatra and Torra, 2024).

3. SYNTHETIC DATA: OUTPUT FEATURES AND CHARACTERISTICS

Synthetic data is data that has been artificially generated to mimic the structure and patterns of real-world data without including any records from real datasets. The data is generated by algorithms, frequently with the aid of Generative AI, rather than being gathered through surveys or sensors (Goyal & Mahmoud, 2024). The objective is to create data that can be used for training models and analysis while avoiding the privacy risks associated with using information that is difficult to access or sensitive (Endres et al., 2022). Since real data may be scarce, expensive, or subject to stringent privacy regulations, this kind of data has grown particularly valuable in industries like healthcare, finance, and AI development. Synthetic data can replicate the intricacy and diversity of real datasets, providing a trustworthy substitute that fosters innovation without sacrificing privacy (Ramzan et al. 2024).

The primary method for creating synthetic data is through deep learning models, which are a subset of generative artificial intelligence. These models are made to learn from actual data and then produce new, realistic data that replicates the patterns found in the original data. Generative Adversarial Networks (GANs) are a popular method using a generator and a discriminator working against each other. The discriminator attempts to determine whether the generated data is real or fake. Over the course of this competition, very realistic results are produced (Figueira & Vaz, 2022; Ramzan et al., 2024). Another popular model is the Variational Autoencoder (VAE), which creates new data while preserving the statistical properties of the original by compressing and reconstructing data into a lower-dimensional space (Sengar et al., 2024). For text generation, transformers and large language models (LLMs) are primarily utilized. According to Bandi et al. (2023), these models are helpful in applications such as chatbots, content creation, and text summarization because they comprehend context and anticipate sequences. For the creation of images, Diffusion Models excel. According to Kwatra and Torra (2024), they start with random noise and gradually transform it into a realistic image by using learned data patterns. The requirements of the task can determine which of these generative models are best suited for particular domains or kinds of synthetic data. There are many different synthetic outputs that generative AI can generate, each with its own unique uses. Assisting with tasks like content creation, document drafting, and chatbot conversations, Text-generation models such as GPT generate new text in response to user input (Goyal & Mahmoud, 2024). Often based on descriptive text, image-generation models like StyleGAN or Stable Diffusion can produce realistic images or creative illustrations (Sengar et al., 2024). Realistic animations, simulations, and even deepfake videos are made possible by Gen AI's ability to create consistent image sequences in the field of video synthesis (Takale et al., 2024). Applications for audio and speech synthesis include voice cloning and AI-generated music, which produce voices or music that imitate actual sound patterns (Sakirin et al., 2023). Finally, code generation helps developers and learners by converting natural language instructions into functional programming code using large language models like Codex. Gen AI is extremely versatile across domains because each type of output captures distinct qualities, such as visually precise images, tonal accuracy in audio, or semantic flow in text (Bandi et al., 2023).

There is no one right answer or "ground truth" to compare synthetic data to, making evaluation a challenging process. Instead, evaluation typically falls into three main categories and is contingent upon the objective. According to Ramzan et al. (2024), statistical similarity measures the degree to which the synthetic data resembles the real data using correlation or distribution overlap. Training machine learning models on synthetic data and then evaluating their performance on real-world data is the second method used to evaluate model performance (Endres et al., 2022). Using methods such as membership inference tests or attribute inference, third, privacy and security assessments examine whether synthetic data exposes private information (Kwatra & Torra, 2024). A significant obstacle is finding the ideal balance: if the data too closely resembles the original, it may reveal personal information; if it generalizes too much, it becomes useless. Particularly when it comes to language and image models, prompt engineering is a crucial component of synthetic data generation. The phrasing of a prompt, which is the instruction provided to the model, greatly affects the calibre and applicability of the content that is produced. Models are more successfully guided by techniques like few-shot or zero-shot prompting, prompt tuning, and others. Multimodal systems, where inputs include both text and images, are particularly in need of these strategies (Bandi et al., 2023; Goyal & Mahmoud, 2024).

A number of distinguishing features are also present in synthetic data. The degree to which it closely mimics actual data in terms of diversity and structure is known as its realism. Preserving privacy is a major benefit because it avoids exact duplicates of actual data. Additionally, it is adaptable, enabling users to produce data that is specific to requirements or circumstances. It can, however, either increase or decrease bias, depending on how it is produced; if done properly, it can aid in dataset balance. Last but not least, scalability is a key advantage since enormous amounts of synthetic data can be created to enhance AI training (Goyal & Mahmoud, 2024).

4. APPLICATIONS OF GENERATIVE AI IN PRODUCING SYNTHETIC DATA OUTPUTS

The use of deep generative models to produce synthetic sequential data in various fields is examined in this section. It looks at important studies in the field, assesses them using certain standards, and points out that the metrics and significance of these standards differ by domain. As generative artificial intelligence (AI) can produce new data that looks like pre-existing datasets, it is being used extensively in many different fields. Through the filling of conceptual gaps and the provision of real-time, tailored feedback, generative AI tools, in particular, massive language models like ChatGPT are revolutionizing STEM education. According to a quasi-experimental study conducted with Moroccan engineering students, ChatGPT considerably enhances conceptual knowledge of thermodynamics when included into an inquiry-based framework, particularly for abstract subjects like entropy and internal energy. According to El Fathi et al. (2025), there is still a need for hybrid teaching methodologies that combine AI with instructor guidance because of the limits in quantitative reasoning and complex problem-solving tasks. Using engineering design principles, it is possible to theoretically assess the generativity of such AI tools. Gen AI models exhibit several generative characteristics, including language-based synthesis, activation, and inversion. These viewpoints support Gen AI's capacity to produce synthetic outputs as well as to dynamically adjust to educational problems, revolutionizing the way knowledge is created and disseminated (Bordas et al., 2024). Although new technologies hold great promise, their implementation must be strategic and morally solid, with a strong emphasis on professional growth, contextual relevance, and trust in order to guarantee that educational goals are achieved effectively and fairly (El Fathi et al., 2025).

By making it possible to produce immersive digital spaces at low cost and in large quantities, generative AI is transforming the creation of content in the metaverse. From creating 3D architectural blueprints to creating real-time avatars and virtual interactions, artificial intelligence lowers the time and expense involved in creating a metaverse. This is especially noticeable when building complex digital environments, where AI helps with interior design and structure design (Castelli & Manzoni, 2022; Zhang et al., 2022). Even the fields of music production, design, healthcare, and natural language processing (such as text generation using Transformers) are some of the main ones where generative AI is used. In addition, key applications include creating human-like text content and creating and modifying images using methods such as Generative Adversarial Networks (GANs) (Takale et al., 2024).

4.1. Image Synthesis and Manipulation Using GANs

Generative Adversarial Networks (GANs) are among the most potent tools in image synthesis because of their capacity to create incredibly detailed and lifelike images. A discriminator determines whether the images are generated or real, while a generator generates images from random noise. These two components make up a GAN. More realistic results are produced as a result of this adversarial training process. GANs are widely used in image generation, where models can create realistic representations of objects, landscapes, people, and animals. These artificial images are utilized in domains where real data is scarce, like digital art production and machine learning model training. For example, transforming a photograph from day to night or applying an artist's painting style to a real-world image are two common uses for GANs. Furthermore, GANs have enhanced the accuracy of image editing and manipulation; users can now precisely change background settings, hairstyles, and facial expressions in photos. Users can generate image-type outputs in many different ways, using text, images, videos, etc. even to generate just images (Bandi et al., 2023; Eigenschink et al., 2023; Takale et al., 2024).

Text-to-image generation is an innovative as well as fast-growing method in which models generate detailed images from written descriptions. With the help of programs like DALL-E and Midjourney, which transform prompts into eye-catching images, this domain has grown in popularity. In contrast with previous methods that concentrated on image captioning (image-to-text), these systems work in reverse, producing images in response to narrative input. Various architectures, such as transformers, Variational Autoencoders (VAEs), diffusion models, and GANs, support this task. Diffusion-based models such as GLIDE and Imagen, for example, are renowned for generating remarkably realistic and cohesive images.

While AttnGAN adds multi-stage refinement with attention mechanisms, which greatly improves output quality, models such as StackGAN use a two-stage generation process to gradually improve image quality in GAN-based approaches. In applications like steganography, where hidden information is embedded in images, specialized models like SGAN (Steganographic Generative Adversarial Networks) and SSGAN (Secure Steganography Based on Generative Adversarial Networks) have also demonstrated promise. These models show how text-to-image synthesis has developed from simple captioning to creating rich, varied, and artistic visuals for a range of uses (Bandi et al., 2023; Eigenschink et al., 2023; Takale et al., 2024).

4.2. Text Generation

Machines now comprehend and produce human language using transformer-based models. A unique mechanism known as "self-attention" enables these models to comprehend word relationships over the course of a lengthy sentence or even a short paragraph. Their writing becomes more fluid and contextually accurate as a result. The ability to generate text is among transformers' most significant applications. Stories, poems, articles, and conversations can all be written by them in a human-like style. They are therefore frequently found in chatbots, virtual assistants, platforms for creative writing, and tools for creating content. Their prompt and precise responses make them perfect for real-time communication. In machine translation, transformers have supplanted more traditional statistical techniques by offering far more accurate and natural translations, facilitating inter-linguistic communication. These models are used to power a number of widely used translation services today. Transformers are also very good at summarizing lengthy texts by condensing them while retaining the main ideas and content. Applications like content curation, research tools, and news summaries greatly benefit from this (Bandi et al., 2023; Takale et al., 2024).

Additionally, significant advancements in translating, summarizing, answering queries, and rewriting content in a different style have been made possible by generative AI. The usual procedure for this process is to take user input (referred to as a "prompt"), transform it into internal representations, and then produce an output that is pertinent. For tasks like translation and paraphrasing, models such as T5 (Text-To-Text Transfer Transformer), ConvS2S, and Seq2Seq excel. By managing more complicated discussions, like responding to follow-up inquiries or declining to answer requests that are biased or unethical, InstructGPT goes above and beyond. There are models that are designed for more specialized tasks. For instance, TFix is made to find errors in code and recommend fixes, assisting programmers in creating better software. Adding GANs (Generative Adversarial Networks) to text-to-text generation expands its potential. These models bridge textual content with visual representation, as demonstrated by the fact that GlyphGAN can generate handwritten-style text in various fonts, while PassGAN is used for password generation and cracking (Bandi et al., 2023; Ferrari & McKelvey, 2023; Takale et al., 2024; Ramzan et al., 2024).

Transformer-based natural language processing (NLP) and text-to-text generation models together represent a major advancement in the way that machines interpret and produce human language. These models exhibit a remarkable array of abilities, ranging from poem and essay creation to language translation and code correction. Further extending these possibilities and allowing for more complex and imaginative outputs is the integration of transformers and GANs. This technology promises to change the way humans interact with machines as it develops further, resulting in more intelligent, efficient, and natural communication (Bandi et al., 2023; Takale et al., 2024).

4.3. Generative AI in Music and Creative Design

In creative domains like digital design and music composition, generative AI has had a big impact. By making it possible to produce unique and artistically complex content, these models are revolutionizing the way that designers, musicians, and artists approach their work. When it comes to music creation, AI models are able to create unique melodies, harmonies, and rhythms, surprisingly even imitating the compositional styles of well-known composers. These models are being incorporated into interactive platforms and software for music composition, helping professionals and amateurs alike produce high-caliber music. These days, a lot of people use these tools to create multimedia content, play games, and entertain. Generative AI is utilized in creative design to produce digital animations, graphics, and artwork. Artists and designers can quickly prototype or create several design variations, automate time-consuming tasks, and explore new creative directions with these tools. In addition to increasing productivity, this fosters creativity and creates new avenues for visual media innovation and personalization. In addition to music and design, generative AI has advanced text-to-speech (TTS) and text-to-audio techniques. Depending on the vocal style, these systems can transform written text into realistic speech or even entire audio compositions. GPS navigation systems, audiobooks, and voice assistants are a few examples of applications. A noteworthy development in this area is AdaSpeech, a customizable voice synthesis model that generates high-quality speech using two encoders: one for utterance-level features and another for phonemes. The mel-spectrogram decoder's incorporation of conditional normalization greatly enhances the generated voice's realism and fluidity. Diff-TTS, Grad-TTS, and ProDiff are a few diffusion-based models that have been created to make synthetic speech even more human-like. These models mimic the organic development of sound, resulting in more expressive and fluid outputs. DiffGAN-TTS also combines diffusion and GAN technologies to create speech that sounds even more natural and rich (Bandi et al., 2023; Eigenschink et al., 2023; Takale et al., 2024).

Beyond speech, generative AI can also generate audio, such as ambient noise and short musical segments. To effectively complete lyrics-to-singing (LTS) tasks, Jukebox, a VQ-VAE-based model, can create songs from lyrics, while GANSynth produces high-quality music directly from text inputs. Using text and audio encoders to recognize and categorize different sound types, like bird calls or crowd noises, effectively completing audio-to-text conversions, Microsoft's CLAP model is notable in terms of audio classification and captioning. In speech-to-speech conversion scenarios, SecGAN enhances noisy audio to improve speech quality, while MusCaps and other tools help with music genre tagging and captioning. The fields of speech, music, and design are rapidly changing as a result of generative AI, which provides tools for producing, improving, and interpreting audio-visual content accurately and creatively. These technologies continue to push the limits of what machines can produce, from crafting original music and realistic speech to categorizing background noise and automating design tasks. With increasing sophistication, the models create new and exciting opportunities for innovation in a variety of industries, from communication and accessibility to entertainment and design (Bandi et al., 2023; Takale et al., 2024).

4.4. Text-to-Code and Code-to-Text: Transforming Software Development

The development of software is being revolutionized by the powerful tools that generative AI has brought about, particularly the speed and ease with which code can now be written and understood. Writing and reviewing code has historically been a laborious and slow process. AI models can now translate plain English instructions into functional programming code thanks to text-to-code technology. This facilitates automating common coding patterns, writing new functions, and repairing bugs. On the other hand, code-to-text technology facilitates the conversion of programming code into readable documentation or explanations. This facilitates software understanding, review, and maintenance for developers. One well-known example is GitHub Copilot, which uses OpenAI's Codex. Copilot saves time and eliminates repetitive work by allowing developers to write comments in natural language and having it generate complete functions or even snippets of code directly within their code editor. CodeBERT is another important model in this field that can comprehend both natural language and programming. Writing documentation, examining code, and searching codebases are among the tasks it facilitates. CodeExecutor analyzes how code runs, GraphCodeBERT improves and translates code between programming languages, UniXcoder converts natural language to code and vice versa, and CodeReviewer automates code reviews. Text-to-code and code-to-text models are working together to revolutionize software development by decreasing manual labour, enhancing code quality, and increasing programming efficiency. Modern software engineering relies heavily on AI-driven systems, such as CodeBERT, CODEGEN, and AlphaCode, which enable developers to write code more quickly and intelligently (Bandi et al., 2023).

Applications of generative AI are quickly emerging in a number of new fields, including gaming, entertainment, and healthcare. These models are applied to tasks such as segmentation, anomaly detection, and medical image reconstruction in the healthcare industry. To enhance AI performance and medical training, they also assist in creating artificial medical images that are used to simulate clinical situations and train deep learning models. Procedural content generation in the gaming industry is frequently accomplished through generative techniques, particularly those based on genetic algorithms. The player experience is improved by the automatic generation of game environments, characters, levels, and plots, which gives games more realism, variety, and repeatability. Generative AI is utilized in the entertainment industry to produce immersive media for interactive storytelling, virtual reality (VR), and media production, including music videos, movies, and ads. Audiences can receive more individualized and captivating experiences thanks to these tools (Takale et al., 2024).

5. SOCIETAL IMPLICATIONS

The sociological lens provided by the term "hyperproduction" helps us comprehend how profound generative models are changing not only the content but also the infrastructure of cultural production itself. In the political economy of digital platforms, AI-generated media is becoming more and more integrated, impacting the creation and extraction of value. Deeper societal contemplation on media authenticity, governance, and power is required by this paradigm, as is regulatory innovation. The foundational dynamics of trust and authenticity, economic models, ethical standards, and cultural production are all being profoundly altered by generative artificial intelligence (Gen AI). As artificial intelligence (AI)-generated content and synthetic media proliferate, their societal ramifications necessitate careful examination from a number of angles (Ferrari & McKelvey, 2023).

5.1. Cultural and Economic Shifts

According to Ferrari and McKelvey (2023), Gen AI makes it possible for "hyperproduction" which is the mass production of artificial media that conflates simulation and reality. By integrating deep generative models into pipelines for cultural production, this change redefines traditional creative labour. When AI-generated representations take the place of human input in situations like virtual influencers or driverless cars, new forms of rentier capitalism are created where profit comes from managing access to platforms or model outputs rather than manufacturing goods (Ferrari & McKelvey, 2023).

5.2. Bias, Inequality, and Representation

Often, social biases present in the training data are reflected or amplified in synthetic data created. This can make underrepresentation and algorithmic discrimination worse, particularly in fields with low demographic diversity like facial emotion recognition (FER). Biased data can lead to incorrect classification, which can reinforce stereotypes and deny marginalized groups equitable treatment in AI applications like hiring or healthcare (Alipour & Gallegos, 2025).

5.3. Concerns about Surveillance and Privacy

The use of synthetic data is frequently marketed as a privacy-preserving measure. If models unintentionally memorize or divulge identifiable information from their training sets, the appearance of anonymity could be deceptive. Legal and ethical concerns are brought up by this, particularly when sensitive domains like financial transactions or medical records are used to create synthetic data (Eigenschink et al., 2023).

5.4. Misinformation and Trust Erosion

One of the biggest threats is the use of generative AI to produce realistic synthetic content, or deepfakes, which propagate false information. Gen AI tools have the ability to sway public opinion and foster mistrust of institutions and media through the use of fake news and edited videos. Synthetic media undermine traditional ideas of authenticity by making it harder for the general public to tell the difference between authentic and fake representations (Ferrari & McKelvey, 2023).

5.5. Labor Displacement and Creative Work

As Generative AI can automate content creation, it poses a threat to established creative industries. Artists and knowledge workers may lose their jobs or see the value of their human-authored work decline as AI tools write stories, create art, and compose music. While some contend that Gen AI can foster creativity, others see it as extractive, stealing content and style without giving credit or paying for it (Takale et al. 2024; Singh et al. 2024).

5.6. Ethical Governance and Responsible AI

To maintain openness, equity, and accountability in the development of generative AI, many researchers support responsible AI principles. This entails implementing explainable AI strategies, making sure datasets are inclusive, and creating explicit consent frameworks for data use. Nowadays, the main topics of regulatory debate centre on mandating watermarking of synthetic media, audit trails for AI outputs, and legal responsibility for abuse (Sengar et al. 2024).

5.7. Digital Inequality and Access Gaps

Gen AI models frequently demand large amounts of data and processing power, their implementation runs the risk of widening already-existing digital divides. The Global South may be excluded from the advantages and governance of these technologies since communities and researchers there might not have access to training infrastructure or foundational models. When a small number of actors control the development and application of Gen AI, this could result in technological hegemony (Bandi et al., 2023).

6. CASE STUDIES

6.1. Synthetic Data for Financial Market Modelling using GANs

Data scarcity and privacy concerns restrict access to comprehensive datasets in the financial sector, particularly for startups and less-traded assets. Using Generative Adversarial Networks (GANs) to aggregate financial data, such as stock prices, volumes, and market trends, Ramzan et al. (2024) tackled this problem. Notwithstanding errors such as duplicates and missing values, their GAN-generated datasets were able to accurately reproduce the original data's statistical characteristics. The modelling of more varied market scenarios was made possible by these artificial datasets, which improved the generalizability and resilience of financial prediction algorithms (Ramzan et al., 2024).

6.2. Healthcare Innovation with Synthetic Data for Privacy-Preserving AI

The development and validation of AI models in the healthcare industry are frequently impeded by stringent data privacy laws that restrict access to actual patient data. Due to these limitations, it is difficult to find the varied, high-quality datasets required for machine learning algorithm evaluation and training. In their thorough review, Rujas et al. (2024) addressed this problem and highlighted the expanding use of synthetic data as a potential workaround for these legal restrictions. One noteworthy example they cited in their study is the creation and use of artificial patient records, particularly in cardiology and oncology. In order to facilitate model training without disclosing private information, these synthetic datasets maintained the statistical and clinical realism of real-world data. Crucially, they made sure that data protection regulations were followed by removing the possibility of patient re-identification. In order to promote innovation and advancement in medical AI without jeopardizing patient privacy, this strategy has made research settings safer and more cooperative (Rujas et al., 2024).

6.3. Synthetic Cyberattack Data for Intrusion Detection Systems

Most of the time, real-world cybersecurity data is inaccessible because it could expose sensitive infrastructure. In order to create artificial, cyberattack datasets for intrusion detection system (IDS) training, Agrawal et al. (2024) investigated the use of GANs. They showed through their experiments on the NSL-KDD dataset that GAN-generated data could increase the robustness and accuracy of IDS models. These synthetic attacks helped create more robust defences without depending on sensitive operational data by covering a variety of patterns, such as spoofing and denial-of-service attacks (Agrawal et al., 2024).

6.4. Synthetic Medical Text and Time-Series Data for Clinical NLP and AI Models

A scoping review by Loni et al. (2025) focused on the synthesis of three forms of healthcare data: longitudinal records, time-series data (such as ECG signals), and clinical text using generative models. Large language models (LLMs) were used in one case to create synthetic medical notes in order to improve named entity recognition (NER) in clinical natural language processing (NLP) tasks. Concurrently, synthetic ECG signals were generated using adversarial networks and utilized to train diagnostic models. Hospital AI deployment became safer and more scalable as a result of these strategies' decreased reliance on actual patient data (Loni et al., 2025).

6.5. Diffusion Models in Digital Pathology for Synthetic Histopathological Images

In order to overcome data constraints and privacy concerns, digital pathology has embraced synthetic image generation. Using a diffusion model pipeline, Pozzi et al. (2024) created synthetic whole slide images (WSIs) from the GTEx dataset that included five different tissue types. Their comprehensive assessment, which included explainable AI techniques and pathologists' visual inspection, validated the high fidelity of these pictures. While maintaining the integrity of patient data, this method facilitates scalable model training and institutional data sharing (Pozzi et al., 2024).

7. LIMITATIONS OF GENERATIVE AI AND DIGITAL SYNTHETIC OUTPUTS

Even though generative AI has made remarkable strides, there are still many obstacles to overcome. The robustness of current models is frequently lacking in adversarial, compressed, or noisy environments. This is especially troublesome when it comes to deepfake detection, where developing synthesis methods put both AI-based and human perception to the test. Additionally, it is critical to prevent overfitting and diversify training datasets to improve generalization across real-world use cases. Notwithstanding the enormous potential of Generative AI (Gen AI), there are a number of significant obstacles to its practical application in the social, ethical, and technical spheres. Recognizing and addressing these limitations is necessary to guarantee the responsible and efficient use of synthetic outputs in a variety of fields (Cardenuto et al., 2023).

A major obstacle is the quality and realism of synthetic data, particularly in delicate domains like finance and healthcare. Statistical and contextual integrity must be maintained in addition to the generated data's realistic appearance. Studies reveal that Gen AI frequently has trouble with uncommon or edge cases, which can result in errors that jeopardize the dependability of AI systems that have been trained on such data (Ramzan et al., 2024; Ibrahim et al., 2025). In the medical field, for example, artificial medical records that don't represent diversity in the real world can lead to instruments that don't work well in real clinical settings (Loni et al., 2025).

At the moment, no uniform framework exists for assessing synthetic outputs in various fields. Generative approaches are hard to compare consistently because different models use different metrics, from coherence and utility to realism and novelty (Eigenschink et al., 2023; Figueira & Vaz, 2022). In industries like cybersecurity and healthcare, where precision and safety are essential, the absence of universal standards impedes scalability as well as trust (Ibrahim et al., 2025; Pozzi et al., 2024).

The use of synthetic data can reinforce preexisting biases if the training datasets are not representative or balanced. For instance, in fields like facial emotion recognition or diagnostics, models trained on sparse demographic data might not produce inclusive results, providing unfair or discriminatory results. If Gen AI is not properly managed, it may unintentionally make societal inequities worse rather than better (Alipour & Gallegos, 2025; Rujas et al., 2024).

A large number of Gen AI models, especially deep learning architectures like diffusion models and GANs, operate as "black boxes".Many times, users don't fully comprehend how or why particular outputs are created. They are not as useful in high-stakes domains like digital pathology or legal AI applications where interpretability and accountability are crucial because of their opacity (Pozzi et al., 2024; Agrawal et al., 2024; Bandi et al., 2023).

Frequently, generative models only work well in particular situations or with particular kinds of data. Without substantial retraining, a model developed to produce financial time-series data might not adjust to healthcare records (Ramzan et al., 2024; Agrawal et al., 2024). Additionally, models run the risk of overfitting, which produces redundant or less creative results that are unable to generalize to novel situations (Figueira & Vaz, 2022; Endres et al., 2022).

There are difficult moral and legal issues with the broad application of Gen AI. Misuse of synthetic content can undermine public communication and digital media trust, resulting in deepfakes and disinformation (Ferrari & McKelvey, 2023). Concerns about data misuse, authorship, and copyright are also growing more pressing. AI-generated scientific content, for instance, violates standards of accountability and responsibility because it is unable to claim ownership of the content it produces, hence it cannot be listed as an author (Chauhan, 2024; Singh et al., 2024).

The ability to precisely steer Gen AI outputs across modalities like text, image, and video is still challenging, even with advancements in prompt engineering. This unpredictability restricts the technology's use in situations where extreme precision is necessary, like legal writing or technical documentation (Sakirin et al., 2023).

A vast amount of computing power is needed to train and run Gen AI models. The high energy consumption and carbon footprint associated with large-scale AI model training create obstacles for smaller organizations and contribute to the growing environmental concerns (Takale et al. 2024; Bandi et al. 2023).

8. FUTURE DIRECTIONS AND CHALLENGES

Many industries are changing as a result of Generative AI's (Gen AI) increasing maturity and ability to produce high-quality synthetic outputs, from text and images to audio and simulation environments. However, a complex array of societal, ethical and technical issues accompanies this growth. For Gen AI to reach its full potential, future developments must be guided by responsible frameworks and well-defined goals (Figueira & Vaz, 2022; Eigenschink et al., 2023).

8.1. Standardized Evaluation and Benchmarking

The development of standardized frameworks for evaluating synthetic data is an important future direction. With fragmented metrics like realism, coherence, and privacy preservation being applied inconsistently, current practices differ greatly across domains (Ibrahim et al., 2025; Pozzi et al., 2024). In sensitive domains like healthcare and finance, domain-agnostic benchmarks and privacy-aware tools are essential to guaranteeing that synthetic outputs satisfy technical and ethical standards (Figueira & Vaz, 2022; Eigenschink et al., 2023).

8.2. Enhancing Domain-Specific Adaptability

As they are trained on generic datasets, the majority of Gen AI models are not as effective in specialized applications. Future research ought to concentrate on context-aware and personalized generation, where models can adjust to particular fields, like asset trends in finance or patient profiles in medicine. The improvement of conditional generation methods, which let users direct outputs depending on customized inputs, will also be part of this (Ramzan et al., 2024; Ibrahim et al., 2025).

8.3. Interpretable and Controllable Outputs

The increasing complexity of Gen AI, especially with the emergence of diffusion models and transformers, makes explainability and output generation control essential. Lack of transparency erodes confidence in AI systems, particularly in high-stakes fields like digital diagnostics and cybersecurity (Pozzi et al., 2024; Agrawal et al., 2024). Future research must prioritize explainable AI (XAI) and the creation of architectures that allow for interpretable and steerable outputs (Sakikin and Kusuma, 2023).

8.4. Multimodal and Cross-Domain Integration

Cross-modal synthesis is the next frontier in generative AI, where systems create content between various media types, such as converting text to video or medical reports into artificial images. Model versatility and application depth are increased by advances in cross-domain representation learning and synchronization algorithms, which are necessary to achieve semantic alignment across modalities (Loni et al., 2025; Sengar et al., 2024).

8.5. Mitigating Bias and Ensuring Fairness

The issue of bias is still very serious. Generative models run the risk of underrepresenting marginalized groups and perpetuating social injustices if they are trained on unbalanced data (Rujas et al., 2024; Alipour & Gallegos, 2025). For generative pipelines to produce synthetic outputs that support equity and inclusivity, bias detection and correction methods must be given top priority in future research (Agrawal et al., 2024).

8.6. Ethical Guidelines and Regulatory Development

Gen AI's ethical and legal frameworks are still developing. Concerns about authorship, responsibility, abuse, and digital rights become more complicated as it gets harder to tell synthetic content from authentic content (Chauhan, 2024; Ferrari & McKelvey, 2023). To combat misinformation, regulatory agencies must work with researchers and business leaders to create standards that can be enforced, such as incorporating watermarking or authenticity verification techniques. The creation of synthetic patient data by generative AI, which replicates real-world data without jeopardizing patient privacy, shows promise for medical research. Such information aids in the creation of medical training platforms, simulation tools, and predictive models. There are still ethical issues, though, like data integrity, possible biases, and the danger of relying too much on artificial datasets. These factors call for strong governance structures and ongoing model verification (Arora & Arora, 2022).

8.7. Addressing Scalability and Environmental Impact

High energy consumption and carbon emissions are caused by the massive computational resources needed to train large generative models (Takale et al., 2024). Future work must focus on energy-efficient architectures, like quantized learning techniques or sparse transformers, to guarantee sustainability and make Gen AI tools accessible to all (Bandi et al., 2023).

8.8. Simulating Rare or High-Risk Scenarios

A potential use case for synthetic data is the simulation of sensitive or uncommon occurrences, like financial crises, cyberattacks, or pandemics. In crucial industries, Gen AI can assist with training, forecasting, and stress testing by generating controlled scenarios in situations where real-world data is scarce or dangerous to collect (Endres et al., 2022).

8.9. Hybrid Generative Architectures

In the future of Gen AI, hybrid models that draw on the strengths of various architectures, like VAEs for latent control, transformers for semantic precision, and GANs for realism are likely to emerge in order to improve efficiency and adaptability. Such combinations could increase generalization, improve explainability, and increase adaptability to difficult real-world tasks (Sakirin & Kusuma, 2023; Sengar et al., 2024).

9. BIBLIOMETRIC ANALYSIS

In order to better understand the current state of research on digital synthetic outputs and Generative AI (GenAI), a bibliometric analysis (Figure 1) was carried out using VOSviewer. Key terms and topics taken from the pertinent literature are shown in the co-occurrence relationships in the generated visual map. The size of each node in the network indicates how frequently it appears, and each node represents a keyword or concept that is commonly found in the scholarly corpus. Co-occurrence link strength is indicated by the lines joining the nodes; thicker lines and closer proximity indicate stronger associations. Foundational terms like "artificial intelligence," "deep learning," "generative adversarial networks," "synthetic data," and "computer vision" are at the centre of the visualization's densely connected core, demonstrating their importance in the field. Subfields within the larger research landscape are represented by color-coded clusters, which are thematic groupings. As an example, one cluster pertains to medical applications (such as "radiology," "diagnostic imaging," and "precision medicine"), while others represent newer subjects like digital twins, bioinformatics, and privacy and AI ethics (Figure 1). The use of phrases like "large language models," "autoencoders," and "federated learning" indicates a growing interest in transformer-based architectures and privacy-preserving synthetic data generation.

Figure 1. Bibliometric Image using VOSviewer

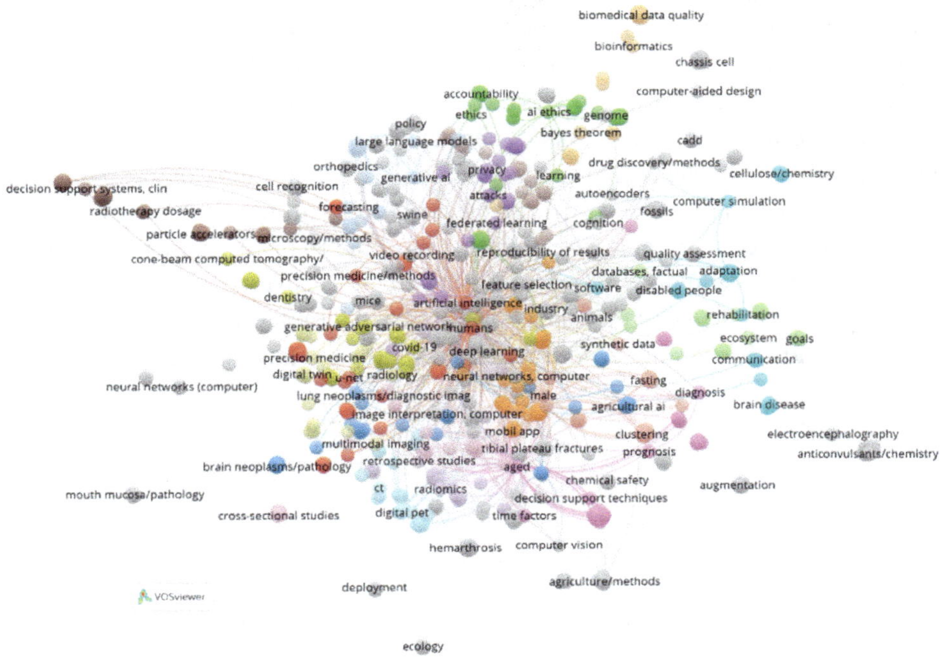

Additionally, terms like "agricultural AI," "digital pet," and "rehabilitation" are ancillary but related, demonstrating the multidisciplinary uses of GenAI in fields ranging from healthcare to the environment and society. This illustrates how artificial digital outputs are becoming more widely used in applications outside of machine learning.

In general, the bibliometric visualization highlights how multifaceted and quickly changing GenAI research is. Important thematic intersections are highlighted, and promising directions for further research are suggested, including domain-specific applications (e.g., education, ecology, and medicine), integration with decision-support systems, and the ethical governance of synthetic data.

10. CONCLUSION

Data synthesis, content production, and human-machine collaboration are just a few of the fields where generative AI is quickly becoming a game-changer. Digital banking, scientific communication, healthcare, and cybersecurity are just a few of the fields where the technology has demonstrated its capacity to replicate diverse, realistic, and contextually relevant synthetic outputs. Its powers are not limited to automation; they also encourage creativity, customisation, and data augmentation. In order to ensure responsible deployment, the field must simultaneously address several significant difficulties, such as model evaluation, ethical moderation, privacy concerns, and bias mitigation. As the corpus of research grows, it becomes clear that generative AI is changing social, educational, and epistemological paradigms in addition to technological infrastructures. To fully utilize generative AI while managing its inherent complexity, it will be essential to maintain interdisciplinary research, open development procedures, and flexible governance structures. The ideals ingrained in their creation and application will ultimately determine the future of digital synthetic outputs, in addition to technological advancements.

REFERENCES

Agrawal, G., Kaur, A., & Myneni, S. (2024). A review of generative models in generating synthetic attack data for cybersecurity. *Electronics (Basel)*, *13*(2), 322. DOI: 10.3390/electronics13020322

Alipour, P., & Gallegos, E. (2025). Leveraging generative AI synthetic and social media data for content generalizability to overcome data constraints in vision deep learning. *Artificial Intelligence Review*, *58*(5), 145. Advance online publication. DOI: 10.1007/s10462-025-11137-6

Arora, A., & Arora, A. (2022). Generative adversarial networks and synthetic patient data: Current challenges and future perspectives. *Future Healthcare Journal*, *9*(2), 190–193. DOI: 10.7861/fhj.2022-0013 PMID: 35928184

Bandi, A., Adapa, P. V. S. R., & Kuchi, Y. E. V. P. K. (2023). The power of Generative AI: A review of requirements, models, Input–Output formats, evaluation metrics, and challenges. *Future Internet*, *15*(8), 260. DOI: 10.3390/fi15080260

Bordas, A., Masson, P. L., Thomas, M., & Weil, B. (2024). What is generative in generative artificial intelligence? A design-based perspective. *Research in Engineering Design*, *35*(4), 427–443. DOI: 10.1007/s00163-024-00441-x

Cardenuto, J. P., Yang, J., Padilha, R., Wan, R., Moreira, D., Li, H., Wang, S., Andaló, F., Marcel, S., & Rocha, A. (2023). The Age of Synthetic Realities: Challenges and opportunities. *APSIPA Transactions on Signal and Information Processing*, *12*(1). Advance online publication. DOI: 10.1561/116.00000138

Chauhan, C. (2024). The impact of generative artificial intelligence in scientific content synthesis for authors. *American Journal of Pathology*, *194*(8), 1406–1408. DOI: 10.1016/j.ajpath.2024.06.002 PMID: 38879080

Eigenschink, P., Reutterer, T., Vamosi, S., Vamosi, R., Sun, C., & Kalcher, K. (2023). Deep Generative Models for Synthetic Data: A survey. *IEEE Access : Practical Innovations, Open Solutions*, *11*, 47304–47320. DOI: 10.1109/ACCESS.2023.3275134

El Fathi, T. E., Saad, A., Larhzil, H., Lamri, D., & Ibrahmi, E. M. A. (2025d). Integrating generative AI into STEM education: Enhancing conceptual understanding, addressing misconceptions, and assessing student acceptance. *Disciplinary and Interdisciplinary Science Education Research*, *7*(1), 6. Advance online publication. DOI: 10.1186/s43031-025-00125-z

Endres, M., Venugopal, A. M., & Tran, T. S. (2022). Synthetic Data Generation: A Comparative study. In *IDEAS '22: Proceedings of the 26th International Database Engineered Applications Symposium (IDEAS'22)* (p. 94-102). ACM Digital Library. DOI: 10.1145/3548785.3548793

Ferrari, F., & McKelvey, F. (2023). Hyperproduction: a social theory of deep generative models. *Distinktion: Journal of Social Theory*, 338–360. DOI: 10.1080/1600910X.2022.2137546

Figueira, A., & Vaz, B. (2022). Survey on synthetic data generation, evaluation methods and GANs. *Mathematics*, *10*(15), 2733. DOI: 10.3390/math10152733

Fisher, S. A., Howard, J. W., & Kira, B. (2024). Moderating Synthetic Content: The Challenge of Generative AI. *Philosophy & Technology*, *37*(4), 133. Advance online publication. DOI: 10.1007/s13347-024-00818-9 PMID: 39555113

Goyal, M., & Mahmoud, Q. H. (2024). A Systematic review of Synthetic data generation techniques using Generative AI. *Electronics (Basel)*, *13*(17), 3509. DOI: 10.3390/electronics13173509

Kwatra, S., & Torra, V. (2024). Empirical evaluation of synthetic data created by generative models via attribute inference attack. In *IFIP advances in information and communication technology* (pp. 282–291). DOI: 10.1007/978-3-031-57978-3_18

Lv, Z. (2023). Generative artificial intelligence in the metaverse era. *Cognitive Robotics*, *3*, 208–217. DOI: 10.1016/j.cogr.2023.06.001

Pozzi, M., Noei, S., Robbi, E., Cima, L., Moroni, M., Munari, E., Torresani, E., & Jurman, G. (2024). Generating and evaluating synthetic data in digital pathology through diffusion models. *Scientific Reports*, *14*(1), 28435. Advance online publication. DOI: 10.1038/s41598-024-79602-w PMID: 39557989

Ramzan, F., Sartori, C., Consoli, S., & Recupero, D. R. (2024). Generative Adversarial Networks for Synthetic data Generation in Finance: Evaluating statistical similarities and quality assessment. *AI*, *5*(2), 667–685. DOI: 10.3390/ai5020035

Rujas, M., Del Moral Herranz, R. M. G., Fico, G., & Merino-Barbancho, B. (2024). Synthetic Data Generation in Healthcare: A Scoping Review of reviews on domains, motivations, and future applications. medRxiv *(Cold Spring Harbor Laboratory)*. DOI: 10.1101/2024.08.09.24311338

Sakirin, T., & Kusuma, S. (2023c). A survey of Generative Artificial intelligence techniques. *Babylonian Journal of Artificial Intelligence*, *10–14*, 10–14. Advance online publication. DOI: 10.58496/BJAI/2023/003

Sengar, S. S., Hasan, A. B., Kumar, S., & Carroll, F. (2024). Generative artificial intelligence: A systematic review and applications. *Multimedia Tools and Applications*, *84*(21), 23661–23700. Advance online publication. DOI: 10.1007/s11042-024-20016-1

Takale, D. G., Mahalle, P. N., Sule, B., & Journals, M. A. T. (2024). Advancements and applications of generative artificial intelligence. *Journal of Information Technology and Sciences*, *10*(1), 20–27.

Wessel, M., Adam, M., Benlian, A., Majchrzak, A., & Thies, F. (2025). Generative AI and its Transformative Value for Digital Platforms. *Journal of Management Information Systems*. DOI: 10.1080/07421222.2025.2487315

ADDITIONAL READING

Bhuyan, S. S., Sateesh, V., Mukul, N., Galvankar, A., Mahmood, A., Nauman, M., Rai, A., Bordoloi, K., Basu, U., & Samuel, J. (2025). Generative Artificial intelligence use in Healthcare: Opportunities for clinical excellence and administrative efficiency. *Journal of Medical Systems*, *49*(1), 10. Advance online publication. DOI: 10.1007/s10916-024-02136-1 PMID: 39820845

Gašević, D., Siemens, G., & Sadiq, S. (2023). Empowering learners for the age of artificial intelligence. *Computers and Education Artificial Intelligence*, *4*, 100130. DOI: 10.1016/j.caeai.2023.100130

Joshi, S., Bhattacharya, S., Pathak, P., Natraj, N., Saini, J., & Goswami, S. (2024). Harnessing the potential of generative AI in digital marketing using the Behavioral Reasoning Theory approach. *International Journal of Information Management Data Insights*, *5*(1), 100317. DOI: 10.1016/j.jjimei.2024.100317

Khosrowi, D., Finn, F., & Clark, E. (2024). Engaging the many-hands problem of generative-AI outputs: A framework for attributing credit. *AI and Ethics*. Advance online publication. DOI: 10.1007/s43681-024-00440-7

Marvin, G., Hellen, N., Jjingo, D., & Nakatumba-Nabende, J. (2024). Prompt engineering in large language models. Data Intelligence and Cognitive Informatics (ICDICI 2023). In *Algorithms for intelligent systems* (pp. 387–402). DOI: 10.1007/978-981-99-7962-2_30

Mittal, U., Sai, S., Chamola, V., & Sangwan, D. (2024). A Comprehensive Review on Generative AI for Education. *IEEE Access : Practical Innovations, Open Solutions*, *12*, 142733–142759. DOI: 10.1109/ACCESS.2024.3468368

Pan, L., Zhao, Z., Lu, Y., Tang, K., Fu, L., Liang, Q., & Peng, S. (2024). Opportunities and challenges in the application of large artificial intelligence models in radiology. *Meta-Radiology*, *2*(2), 100080. DOI: 10.1016/j.metrad.2024.100080

Zhavoronkov, A., Mamoshina, P., Vanhaelen, Q., Scheibye-Knudsen, M., Moskalev, A., & Aliper, A. (2019). Artificial intelligence for aging and longevity research: Recent advances and perspectives. *Ageing Research Reviews*, *49*, 49–66. DOI: 10.1016/j.arr.2018.11.003 PMID: 30472217

KEY TERMS AND DEFINITIONS

Artificial Intelligence (AI): When machines, particularly computer systems, mimic human intelligence processes, this is referred to as artificial intelligence. These include self-correction, learning, and reasoning. AI allows systems in contemporary settings to analyze information, spot trends, and reach conclusions with little assistance from humans (Takale et al., 2024). Generative Artificial Intelligence: The goal of generative AI, a branch of artificial intelligence, is to create new data samples from learned data distributions. GANs, VAEs, transformers, and other models are used to produce artificial images, text, audio, video, and structured data that mimic real-world examples. Data scarcity, personalization, and creative automation issues are being resolved by GenAI, which is transforming various fields (Bandi et al., 2023; Goyal & Mahmoud, 2024). Digital Synthetic Outputs: Images, text, audio, video, and tabular data are examples of data artifacts that are produced by AI systems instead of being gathered from the real world. These are known as digital synthetic outputs. These results are essential for training AI models in situations where real data is scarce, delicate, or not available. (Figueira & Vaz, 2022; Cardenuto et al., 2023) Synthetic outputs reduce privacy and security risks while maintaining utility. Hallucinations: When artificial intelligence produces outputs that are believable in appearance but completely false or factually inaccurate, this is referred to as hallucination. This phenomenon is particularly troublesome in generative systems and large language models, where models may generate confident but deceptive information, raising questions about dependability and credibility in crucial applications (Fisher et al., 2024). Generative Adversarial Network (GAN): The generator and discriminator neural networks in a GAN compete in a zero-sum game, forming a deep learning framework. While the discriminator works to separate real data from fake, the generator produces artificial data. Highly realistic synthetic data can be generated by GANs using this adversarial process (Ramzan et al., 2024).

Diffusion Models: A class of generative models known as diffusion models generates data by reversing a slow noise process. They create realistic outputs by iteratively denoising samples, starting with random noise. Recently, these models have demonstrated state-of-the-art performance in digital pathology and image synthesis (Pozzi et al., 2024).

Transformers: These are models for deep learning that process input sequences in parallel instead of sequentially by using attention mechanisms. They have made advances in natural language processing, image generation, and multimodal learning possible and form the basis of contemporary large language models (LLMs) like GPT and BERT (Sengar et al., 2024).

Variational: Autoencoder (VAE): One kind of generative model is a VAE, which learns to encode input data into a probabilistic latent space and then decode it to reproduce comparable data. In contrast to GANs, VAEs concentrate on learning data distributions and provide superior training stability and interpretability in latent space. They are frequently employed in the imaging and healthcare fields (Eigenschink et al., 2023).

Chapter 2
Exploring Synthetic Generation of Digital Data Through Data-Driven Learning

A. Prabhu Chakkaravarthy
https://orcid.org/0009-0000-3180-8051
Networking and Communications, SRM Institute of Science and Technology, India

Dhanalakshmi Jaganathan
https://orcid.org/0000-0002-3919-4469
SRM Institute of Science and Technology, India

ABSTRACT

A revolutionary method for producing realistic, high-quality digital content, including pictures, movies, and audio, using computational models is digital synthetic generation, which is fueled by machine learning and big data. With the use of deep neural networks, sophisticated learning algorithms, and enormous datasets, this technology makes it possible to create synthetic media that remarkably accurately replicates real-world data. In this work, we investigate the use of digital synthetic generation in a number of domains, such as autonomous systems, design, healthcare, and entertainment. Generative methods such as Generative Adversarial Networks (GANs) and Variational Autoencoders (VAEs) can create fresh content that satisfies requirements without the need for human creativity or manual input by training models on large datasets.

DOI: 10.4018/979-8-3373-2277-3.ch002

1. INTRODUCTION

1.1 Background of Digital Synthetic Generation

The digital age has ushered in a paradigm shift in content creation, marked by the rise of **digital synthetic generation**—the automated creation of high-fidelity digital media such as images, audio, text, and videos through computational models. This phenomenon builds on decades of artificial intelligence (AI) research and leverages breakthroughs in machine learning (ML), especially deep learning. What began as a computational experiment in image manipulation has now evolved into a robust domain capable of producing hyper-realistic synthetic content that is often indistinguishable from real-world data (Karras et al., 2019).

The essence of digital synthetic generation lies in its **data-driven, algorithmic approach** to simulating human creativity. It departs from traditional content production by requiring minimal human intervention after initial model training. This innovation has significant implications across various industries including **healthcare, design, entertainment, and autonomous systems**. For instance, synthetic medical images are now used to train diagnostic AI systems without exposing patient data, mitigating ethical and legal concerns (Frid-Adar et al., 2018).

1.2 Role of Machine Learning and Big Data

Machine learning and big data are the twin engines driving synthetic content generation. **Machine learning**, particularly deep neural networks, enables systems to learn intricate patterns and generate novel outputs based on learned representations. Two major classes of generative models—**Generative Adversarial Networks (GANs)** and **Variational Autoencoders (VAEs)**—have revolutionized the space. GANs, introduced by Goodfellow et al. (2014), use a two-network system (generator and discriminator) to iteratively produce realistic data. VAEs, on the other hand, learn latent data representations to create similar data instances with high fidelity.

Meanwhile, **big data** provides the necessary fuel for these models to learn effectively. Massive datasets—from online images to sensor data—offer a rich source of variability and contextual information. **Big data analytics** enables the preprocessing, transformation, and real-time analysis of these datasets, further enhancing the quality and relevance of generated content (Chen & Zhang, 2014).

A key synergy emerges: **the more diverse and voluminous the data, the better the model performance** in terms of realism, diversity, and relevance of synthetic outputs. This intersection allows industries to simulate environments, test systems under diverse scenarios, and personalize content dynamically.

Figure 1. Interaction between Machine Learning, Big Data and Synthetic Generation

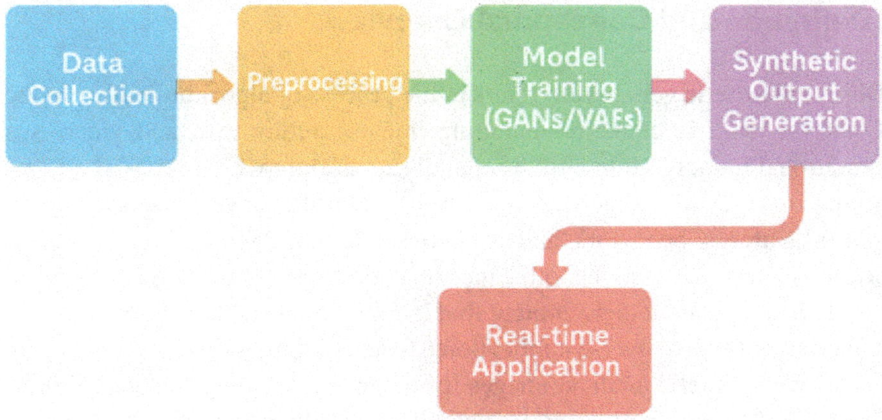

1.3 Objectives and Scope of the Chapter

The primary objective of this chapter is to explore how digital synthetic generation, powered by machine learning and big data, is transforming industries and information systems. It aims to provide a comprehensive overview of the underlying technologies, key generative models, applications in various sectors, and ethical and practical challenges associated with synthetic data.

This chapter will:

- Outline foundational concepts of synthetic generation;
- Analyze the role of big data in training and optimizing generative models;
- Discuss practical applications in domains such as healthcare, entertainment, and automation;
- Examine current limitations and ethical concerns;
- Propose future directions for responsible innovation in this field.

Through this examination, the chapter will offer valuable insights into how digital synthetic generation is not only a technological achievement but a **strategic asset** for data-driven innovation and decision-making in the digital era.

2. FOUNDATIONS OF DIGITAL SYNTHETIC GENERATION

2.1 Definition and Conceptual Overview

Digital synthetic generation refers to the process of algorithmically producing artificial yet realistic content—including images, audio, text, and video—using computational models. Unlike traditional digital design, which is based on manual input and rules, synthetic generation draws from **data-driven learning** to mimic real-world patterns, behaviors, and aesthetics. This capability arises from training algorithms on large datasets, enabling them to understand the structure and variability of real data and generate new content that is indistinguishable from it.

This concept spans diverse applications, from creating virtual avatars in gaming to generating synthetic medical images for diagnostic training. Its central promise lies in **scalability, efficiency, and privacy**—especially in contexts where acquiring real data is costly, time-consuming, or ethically sensitive (Gupta et al., 2019).

2.2 Historical Development and Milestones

The journey of synthetic generation has its roots in **computer graphics** and **simulation systems** from the mid-20th century. Early progress involved primitive shape modeling and animation tools. However, the field transformed dramatically with the advent of **machine learning** and **neural networks** in the 1980s and 1990s.

The 2010s marked a major milestone with the introduction of **Generative Adversarial Networks (GANs)** by Goodfellow et al. (2014). GANs opened the floodgates for high-quality, high-resolution image synthesis. Shortly after, **Variational Autoencoders (VAEs)** and **transformer models** (e.g., GPT series) further enhanced capabilities in generating text and multimodal content.

Notable developments include:

- **2014**: GANs enable realistic face generation.
- **2017**: DeepFakes use GANs for facial replacement in videos.
- **2020+**: Models like **DALL·E**, **StyleGAN2**, and **ChatGPT** showcase the integration of vision, language, and interactivity.

These milestones signal the shift from passive computation to **generative intelligence**, a hallmark of modern digital systems.

2.3 Key Technologies Involved (AI, ML, DL)

Three core technologies underpin digital synthetic generation:

Artificial Intelligence (AI):

AI serves as the overarching discipline enabling machines to simulate human-like cognition and behavior. It provides the foundational principles for pattern recognition, decision-making, and generative capabilities.

Machine Learning (ML):

ML is a subfield of AI focused on developing algorithms that learn from data. In synthetic generation, supervised and unsupervised learning methods are used to model relationships, classify inputs, and generate outputs. ML enables systems to generalize from training data to unseen scenarios, which is critical in generating new content (Jordan & Mitchell, 2015).

Deep Learning (DL):

DL, a subset of ML, uses **deep neural networks** to model complex data hierarchies. Convolutional Neural Networks (CNNs) are used in image generation, while Recurrent Neural Networks (RNNs) and Transformers dominate text and audio synthesis. GANs, VAEs, and diffusion models all rely on deep learning principles to perform their tasks (LeCun et al., 2015).

Figure 2. Technological Stack of Digital Synthetic Generation

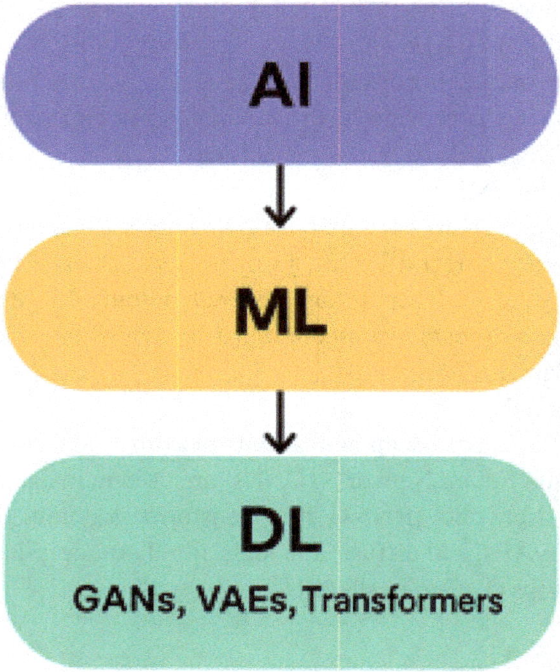

3. GENERATIVE MODELS IN SYNTHETIC MEDIA

3.1 Generative Adversarial Networks (GANs)

Generative Adversarial Networks (GANs), introduced by Goodfellow et al. (2014), represent a significant breakthrough in the field of synthetic media. A GAN consists of two neural networks—the generator and the discriminator—that are trained simultaneously in a competitive process. The generator attempts to produce realistic synthetic data, while the discriminator evaluates whether the data is real or generated. Over time, the generator improves its ability to create outputs that mimic real-world data, making GANs particularly powerful for generating images, videos, and even audio.

Applications of GANs are widespread, ranging from face generation (e.g., "This Person Does Not Exist") to art synthesis, deepfake creation, and data augmentation in healthcare and autonomous driving. However, training GANs is computationally intensive and often suffers from instability and mode collapse, where the generator produces limited varieties of outputs.

3.2 Variational Autoencoders (VAEs)

VAEs are another class of generative models that approach the problem differently. Proposed by Kingma and Welling (2013), VAEs use probabilistic frameworks to encode input data into a latent space and then decode it to generate new data points. Unlike GANs, which are adversarial, VAEs rely on reconstruction loss and a regularization term to guide training.

VAEs excel in generating smooth and continuous variations of data, making them useful in applications like facial expression morphing, medical image synthesis, and recommendation systems. They tend to be more stable during training than GANs but sometimes generate blurrier outputs due to the inherent probabilistic assumptions.

3.3 Diffusion Models and Emerging Techniques

Diffusion models represent a more recent innovation in generative modeling. These models work by gradually adding noise to data and learning to reverse the noise process to generate new samples. Notably, DALL·E 2 and Stable Diffusion have demonstrated the power of diffusion models in producing high-resolution, highly detailed images based on textual descriptions (Rombach et al., 2022).

Other emerging techniques include Transformer-based generative models, like GPT and Imagen, which extend generative capabilities into text-to-image and text-to-video domains. These models leverage attention mechanisms to understand complex patterns across modalities, enabling more nuanced and controllable content generation.

3.4 Model Training and Dataset Considerations

Training generative models requires large and diverse datasets to ensure quality, realism, and generalizability. Key considerations include:

- **Data Quality**: Noisy or biased data can propagate errors in generation.
- **Annotation and Labeling**: Structured datasets enhance model learning, especially for conditional generation.
- **Ethical Concerns**: Synthetic data must avoid reinforcing stereotypes or violating privacy.

Big data platforms and cloud computing infrastructures have become essential in handling the large-scale training requirements of these models. Model evaluation metrics like Frechet Inception Distance (FID) and Inception Score (IS) are commonly used to assess the quality of generative outputs.

Figure 3. Comparison of Generative Models in Synthetic Media

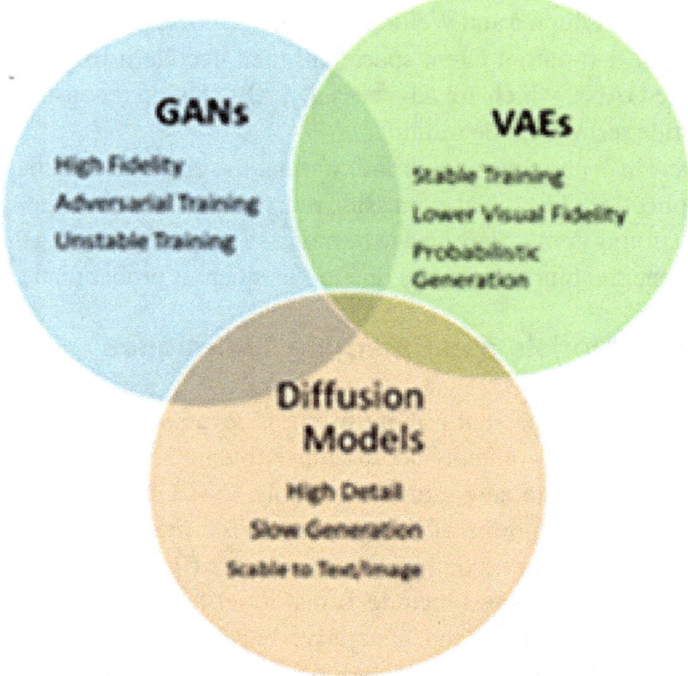

4. BIG DATA AND ITS ROLE IN SYNTHETIC GENERATION

Digital synthetic generation, an emerging field at the intersection of artificial intelligence and data science, relies heavily on big data as its foundational input. The ability to produce realistic and diverse synthetic content—whether visual, auditory, or textual—depends on training models using large, varied, and high-quality datasets. Big data enhances model performance, broadens generalization capabilities, and allows for dynamic content generation in real time. This section outlines the essential characteristics and sources of big data, details the processes involved in data preparation, and highlights the role of real-time data processing in synthetic generation systems.

4.1 Characteristics and Sources of Big Data

Big data is defined by its complexity and size, often summarized through the **5Vs** model: **Volume, Variety, Velocity, Veracity, and Value** (Gandomi & Haider, 2015).

- **Volume** refers to the vast amounts of data generated from sources like IoT devices, social media, medical imaging systems, and surveillance cameras.
- **Variety** includes different data types—structured (databases), semi-structured (XML, JSON), and unstructured (text, video, audio).
- **Velocity** is the speed at which data is generated and must be processed to remain relevant.
- **Veracity** relates to the reliability and accuracy of the data.
- **Value** is derived from actionable insights and synthetic outputs created using the data.

Common sources of big data include:

- **Medical imaging repositories** (e.g., X-rays, MRIs for generating synthetic patient data),
- **Traffic and surveillance systems** (for autonomous driving models),
- **Social media feeds** (for training generative language models),
- **Retail transactions and customer interactions** (for synthetic voice/chatbot systems),
- **Scientific databases and research papers** (used in AI-powered content summarization).

4.2 Data Preprocessing and Management

Before big data can be used for synthetic generation, it must undergo rigorous preprocessing to ensure quality and relevance. This step includes:

- **Cleaning**: Removing noise, correcting inconsistencies, and handling missing values.
- **Normalization and Transformation**: Scaling data and encoding formats suitable for machine learning algorithms (Kotsiantis et al., 2006).

- **Annotation and Labeling**: Particularly crucial for supervised learning and generative models like GANs or VAEs, where accurate labels improve model understanding.
- **Augmentation**: Artificially increasing dataset size through transformations (e.g., rotation, cropping in images; synonym replacement in text).

Efficient data management frameworks such as Hadoop and Apache Spark help process and store big data across distributed systems, allowing scalable synthetic content generation.

4.3 Real-Time Data Processing for Dynamic Generation

Modern applications increasingly demand **real-time or near-real-time synthetic generation**, especially in sectors like finance (automated reporting), gaming (dynamic environments), and healthcare (on-the-fly synthetic imaging).

Technologies like **Apache Kafka**, **Flink**, and **Storm** enable streaming analytics, where incoming data is processed instantaneously. This empowers generative systems to adapt dynamically to changing inputs. For example, a virtual assistant can generate context-aware responses using real-time user data, or a synthetic surveillance model can simulate evolving traffic conditions.

In reinforcement learning and online training setups, models continuously retrain using live data, enhancing adaptability and personalization (Kumar et al., 2020). These capabilities make real-time processing a cornerstone in the advancement of interactive, adaptive synthetic systems.

5. APPLICATIONS ACROSS INDUSTRIES

Digital synthetic generation has penetrated a multitude of industries, bringing forth unprecedented levels of personalization, efficiency, and innovation. Powered by machine learning and big data, generative models are enabling new modes of content creation, decision-making, and system automation. This section explores key sectors that are embracing synthetic generation technologies and outlines real-world applications with contemporary examples and research insights.

5.1 Healthcare and Biomedical Imaging

In healthcare, synthetic data generation is being leveraged to improve diagnostic systems, develop robust machine learning models, and protect patient privacy.

One of the major applications is the creation of **synthetic medical images** using GANs and VAEs. These models can generate realistic images of X-rays, MRIs, and CT scans, which are used for training and validating diagnostic algorithms. Since acquiring labeled medical images is expensive and privacy-sensitive, synthetic data provides a feasible solution.

For example, **Frid-Adar et al. (2018)** demonstrated that using GAN-generated liver lesion images improved the performance of classification networks. Similarly, **Shin et al. (2018)** used GANs to augment brain MRI datasets, boosting the performance of brain tumor segmentation tasks.

Another promising use is in **drug discovery**. Generative models like **Variational Autoencoders** and **Transformer-based models** have been used to generate novel molecular structures with desired properties (Zhavoronkov et al., 2019).

Synthetic data also enhances **predictive healthcare models** by providing balanced datasets that reduce biases and improve generalization.

5.2 Autonomous Systems and Robotics

Autonomous vehicles and robots rely heavily on synthetic data to simulate real-world conditions for training and testing purposes.

In autonomous driving, collecting data for all possible driving scenarios is impossible. Thus, **simulated environments** like NVIDIA's **DriveSim** or **Waymo's Simulation Engine** generate synthetic data of roads, weather, pedestrians, and vehicles to train perception and control systems. These simulations are particularly useful for rare or dangerous scenarios such as sudden pedestrian crossings or extreme weather conditions (Dosovitskiy et al., 2017).

In robotics, synthetic data enables **domain randomization**—where visual aspects of environments are varied synthetically to improve generalization when robots are deployed in real-world tasks. This approach has been used in robotic grasping, navigation, and manipulation tasks (Tobin et al., 2017).

GANs and 3D simulation engines have also been used to create virtual twins of physical robots, enabling rapid iteration in design and development.

5.3 Entertainment, Gaming, and Virtual Reality

Entertainment is among the most visibly transformed sectors through digital synthetic generation.

In **gaming**, generative algorithms are used to dynamically generate game content—levels, characters, narratives—providing infinite replayability. Tools like **Procedural Content Generation (PCG)** use neural networks to generate unique environments tailored to player behavior (Summerville et al., 2018).

Virtual reality (VR) and **augmented reality (AR)** systems benefit from synthetic environments where realistic avatars, textures, and immersive worlds are created using deep learning. **GAN-based avatar creation** can produce lifelike virtual humans from a single image, improving realism in metaverse applications.

In **film and media**, synthetic generation aids in:

- **De-aging actors** or creating digital doubles using GANs.
- **Voice cloning and dubbing** with AI-based speech synthesis tools like **Descript's Overdub**.
- **Script generation**, music composition, and storyboarding using transformer-based models like GPT-4 and MuseNet.

This application of synthetic generation has drastically reduced post-production costs and opened avenues for personalized storytelling and interactive films.

5.4 Marketing and Content Personalization

In marketing, synthetic data drives hyper-personalized advertising strategies and content curation.

Synthetic user profiles, generated from behavioral datasets, help brands simulate consumer interactions, enabling targeted ad campaigns and product design. This synthetic segmentation allows firms to run A/B tests and recommendation algorithms without compromising actual user data (Wang et al., 2020).

Text and image generation models, such as **DALL·E** and **Stable Diffusion**, can automatically produce visuals and copy tailored to individual preferences. Brands can use these models to generate millions of variations of an ad tailored for diverse demographic segments without manual design effort.

Chatbots and virtual assistants trained on synthetically generated conversational datasets enable 24/7 support and sales conversion, while **generative email personalization** tools write custom emails aligned with user tone and history.

Moreover, **emotion-driven content generation** allows real-time mood analysis from user inputs, enabling the dynamic generation of video, audio, or graphics to align with the user's emotional state.

5.5 Education and Training Simulations

Synthetic data plays a pivotal role in reshaping education, particularly in simulation-based and experiential learning environments.

Medical and military training programs use synthetic data and VR to simulate complex scenarios, such as emergency surgeries or battlefield environments. These simulations provide immersive, repeatable training without real-world risks (Laver et al., 2017).

Language learning applications now utilize synthetic conversations, generated through transformer-based models, for real-time practice. This not only ensures diversity of conversation topics but also allows learners to practice with AI characters possessing various accents and tones.

In **STEM education**, generative models are used to automatically create quizzes, lab simulations, and visualizations tailored to the learner's progress. For example, platforms like **Labster** use 3D synthetic environments to teach chemistry and biology through virtual labs.

Educators also use **synthetic classroom simulations** to train teachers on classroom management and instructional delivery by interacting with AI-based student avatars.

Figure 4. **Applications across Industries**

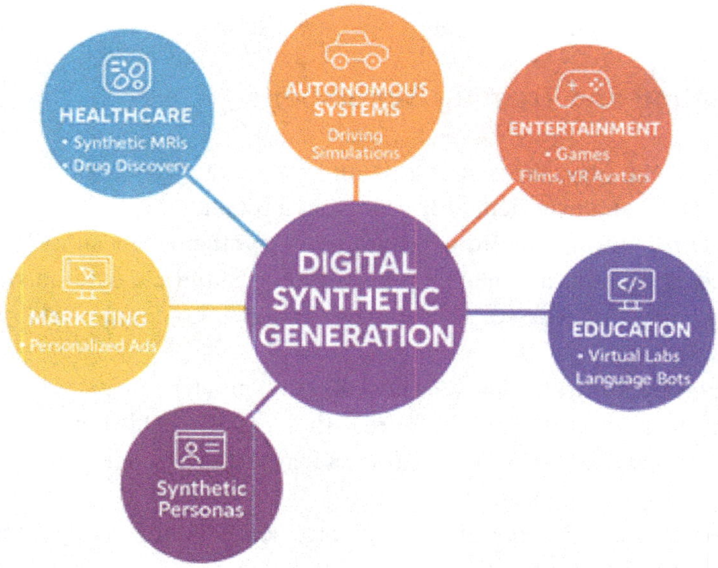

6. ETHICAL, LEGAL, AND SOCIAL IMPLICATIONS

The proliferation of digital synthetic generation technologies presents groundbreaking opportunities across sectors. However, this advancement also introduces profound ethical, legal, and social challenges that must be addressed to ensure responsible innovation. As synthetic content—generated through AI and big data—becomes indistinguishable from reality, concerns around privacy, misinformation, fairness, and governance are increasingly urgent.

6.1 Privacy Concerns and Data Protection

The use of large-scale datasets to train generative models raises significant privacy issues, particularly when these datasets include personally identifiable information (PII). Despite the rise of synthetic data as a privacy-preserving alternative to real-world datasets, it is not immune to risks. Poorly designed models can inadvertently memorize and reproduce sensitive data, compromising user privacy (Carlini et al., 2021).

Moreover, biometric data such as facial scans, voiceprints, and behavioral patterns are often used in generating deepfake content or training avatar systems in VR and gaming. Even synthetic versions of such data can lead to **identity misuse** if safeguards are not implemented. This is especially problematic in sectors like healthcare and finance, where confidential information forms the backbone of operational data.

Privacy-by-design frameworks are necessary to embed anonymization, federated learning, and differential privacy mechanisms into generative systems (Dwork & Roth, 2014). These ensure that the models learn patterns from data without replicating individual instances.

6.2 Misinformation and Deepfakes

One of the most contentious ethical challenges of synthetic media is its potential to spread **misinformation**. Deepfakes—highly realistic but fabricated video or audio content—can be weaponized for political manipulation, character assassination, or social unrest (Vaccari & Chadwick, 2020). The lines between real and fake are becoming increasingly blurred, and without transparency in content origin, trust in media and communications is eroding.

For example, synthetic speech generated to mimic a political leader's voice can be used to fabricate speeches or confessions. In one case, cybercriminals used voice deepfakes to impersonate a CEO and fraudulently transfer company funds (Westerlund, 2019).

To combat this, technologies such as **media provenance verification, digital watermarking**, and **AI-based detection algorithms** are being developed. However, these solutions must evolve in tandem with generative models, which are also becoming more sophisticated.

6.3 Bias and Fairness in Generative Systems

Generative models learn patterns from the datasets they are trained on. If these datasets contain historical biases or demographic imbalances, the outputs of these systems are likely to reproduce and even amplify those biases (Buolamwini & Gebru, 2018). For instance, a facial generation model trained predominantly on lighter skin tones may produce inaccurate or distorted images for people with darker skin.

Bias in generative systems can perpetuate stereotypes and result in discriminatory outcomes. In automated resume generation, facial animations, or even synthetic news reporting, gender, racial, and cultural biases can significantly impact fairness.

To mitigate these challenges, there is a growing call for **algorithmic audits**, **diversified training datasets**, and **inclusive design practices**. Incorporating fairness-aware learning mechanisms, such as adversarial de-biasing and fairness regularization, is becoming a research priority (Mehrabi et al., 2021).

6.4 Regulatory and Governance Frameworks

Despite growing adoption, regulatory responses to synthetic generation technologies remain fragmented and lag behind technical advancements. Laws surrounding data ownership, liability, and authenticity in AI-generated content are still under development in most jurisdictions.

The European Union's **AI Act**, for example, classifies synthetic content as high-risk when used in areas like recruitment or biometric surveillance, mandating stringent transparency and documentation requirements. Similarly, the U.S. **DEEPFAKES Accountability Act** aims to hold creators of malicious deepfakes legally responsible (European Commission, 2021; U.S. Congress, 2019).

Key questions remain unresolved:

- **Who is accountable** if a synthetic image causes harm—model creators, platform providers, or end users?
- Should **synthetic content be labeled** to inform users of its non-human origin?
- Can **intellectual property rights** be applied to AI-generated works?

Multi-stakeholder governance involving technologists, legal experts, policymakers, and civil society is essential to craft balanced and effective frameworks. These should protect public interest without stifling innovation.

Figure 5. Digital Synthetic Generation

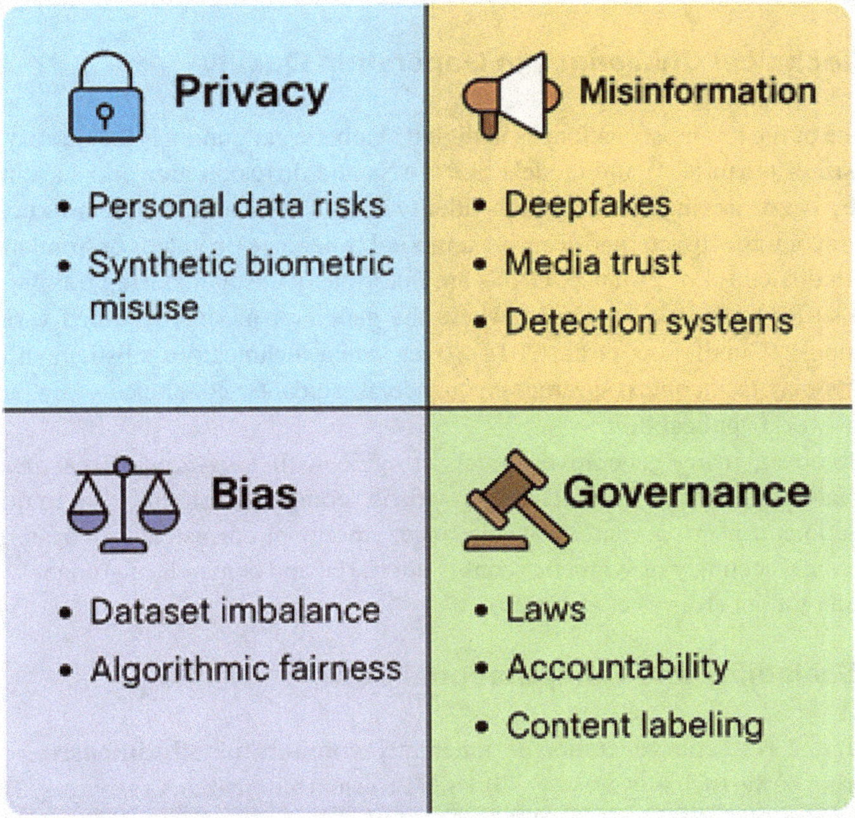

Digital synthetic generation is shaping the future of content creation, personalization, and automation. However, with great power comes great responsibility. Addressing privacy violations, combating misinformation, ensuring algorithmic fairness, and establishing robust governance structures are crucial to ensuring that synthetic media benefits society at large. Ethical and legal foresight, embedded from the design stage onward, is the only path to sustainable and trustworthy innovation.

7. CHALLENGES AND LIMITATIONS

The evolution of digital synthetic generation has unlocked immense opportunities across various domains. However, despite its transformative potential, several significant challenges and limitations persist. These span technical, infrastructural, and perceptual domains, hindering the full-scale deployment and acceptance of synthetic media systems.

7.1 Technical Challenges in Generation Quality

One of the foremost challenges in digital synthetic generation is the **quality and realism of outputs**. While models like GANs and diffusion networks have made impressive strides in generating high-fidelity images and audio, maintaining consistency, avoiding artifacts, and ensuring temporal coherence (in videos or simulations) remain difficult. For instance, GANs are notoriously unstable during training and often suffer from *mode collapse*, where the generator produces limited varieties of outputs (Goodfellow et al., 2014). Even when technically realistic, synthetic outputs may lack context-awareness, nuance, or semantic coherence—key factors in real-world applicability.

Moreover, many generative models struggle with *long-range dependencies*, especially in content like full-length videos, complex dialogues, or multistep simulations. Ensuring contextual relevance, emotional consistency in avatars, or biomedical accuracy in synthetic scans is not trivial and demands ongoing research and fine-tuning (Karras et al., 2020).

7.2 Scalability and Computational Cost

Digital synthetic generation is inherently **computationally intensive**, often requiring powerful hardware (e.g., GPUs, TPUs) and vast memory resources. Training large models, especially with high-resolution outputs, consumes considerable energy and time. For example, training a high-end diffusion model like DALL·E 2 or Imagen can involve *millions of GPU hours* (Ramesh et al., 2022).

Such requirements raise **accessibility barriers**, especially for smaller organizations or academic institutions. It also raises environmental concerns regarding **carbon footprints** associated with training and deploying large models (Strubell et al., 2019). Techniques such as *model distillation*, *quantization*, and *low-rank adaptation* (LoRA) have been proposed to mitigate these issues, but a balance between performance and efficiency remains elusive.

Furthermore, **scaling synthetic generation to real-time applications**, such as interactive simulations or autonomous systems, requires significant optimization of latency, storage, and compute distribution.

7.3 Interpretability and Trust in Synthetic Outputs

As synthetic content becomes indistinguishable from real data, **interpretability and user trust** become major concerns. Users and regulators need tools to understand **how and why certain content was generated**, especially in high-stakes areas like healthcare or finance. However, most deep generative models operate as "black boxes" with little transparency in decision-making.

The lack of explainability raises **ethical concerns**, particularly in determining accountability for misinformation or unintended harm. For instance, if a synthetic voice system outputs biased content, identifying the root cause (dataset bias, model architecture, or user prompt) is challenging (Bommasani et al., 2022).

Trust is also undermined by the proliferation of **deepfakes and media manipulation**, leading to skepticism of authentic content. Solutions like **watermarking**, **metadata embedding**, and **synthetic detection algorithms** are being developed, but they are not yet universally adopted or fail-safe.

Figure 6. Key Challenges in Digital Synthetic Generation

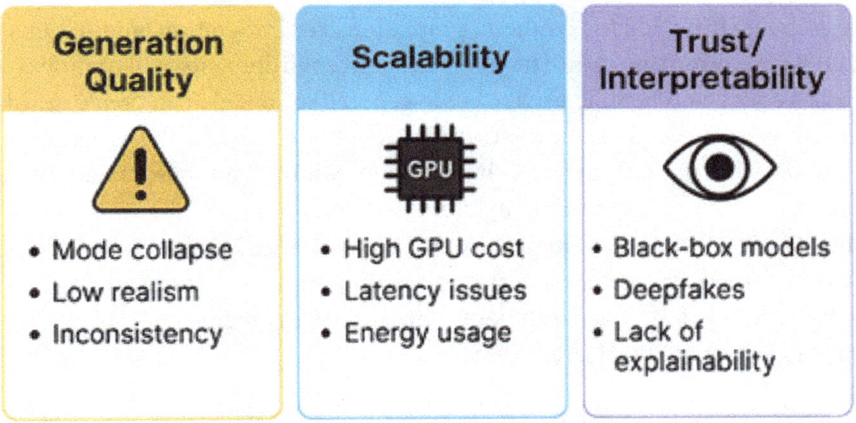

8. FUTURE DIRECTIONS AND INNOVATIONS

As digital synthetic generation matures, its future is being shaped by rapid advancements in generative AI architectures, convergence with emerging technologies like the Internet of Things (IoT) and edge computing, and a growing emphasis on ethical, sustainable, and collaborative frameworks. This section explores upcoming innovations that are likely to redefine how synthetic content is generated, managed, and integrated into daily life.

8.1 Advances in Generative AI Architectures

The field of generative AI is advancing beyond traditional models like GANs and VAEs to include **diffusion models**, **transformer-based architectures**, and **multimodal frameworks**. These next-generation models, such as Google's Imagen and OpenAI's DALL·E 3, offer significantly improved fidelity, coherence, and adaptability across text, image, and video domains (Saharia et al., 2022). Additionally, **foundation models**—large-scale pre-trained generative models—are increasingly capable of few-shot or zero-shot generation, reducing the need for task-specific datasets and training.

8.2 Integration with Internet of Things (IoT) and Edge Computing

A significant direction for synthetic generation is its integration with **IoT ecosystems** and **edge computing** infrastructures. In smart environments—like autonomous vehicles or connected healthcare devices—generative models can create real-time simulations and synthetic datasets to train systems continuously on-device (Shi et al., 2020). This not only reduces latency and bandwidth requirements but also improves data privacy by minimizing cloud transmission.

Edge AI chips, such as Google's Edge TPU and NVIDIA Jetson, now support on-device inference for generative tasks, enabling real-time synthetic outputs for applications like AR/VR environments, predictive maintenance, and intelligent assistants (Zhou et al., 2021).

8.3 Sustainable and Responsible AI for Synthetic Generation

As synthetic generation becomes more compute-intensive, **energy efficiency and sustainability** are gaining attention. Training a single large model like GPT-3 can emit hundreds of tons of CO_2, prompting research into **green AI**, model compression, and federated learning (Strubell et al., 2019). Innovations such as **efficient training algorithms**, **knowledge distillation**, and **quantization** aim to reduce the environmental footprint of generative models while preserving performance.

Furthermore, there's a concerted effort to embed **responsible AI principles**, including **fairness, accountability, and transparency (FAT)**, into synthetic generation pipelines. Organizations like the Partnership on AI advocate for clear labeling of synthetic content and robust audit trails to ensure traceability and ethical usage (PAI, 2023).

8.4 Human-AI Collaboration in Creative Work

Looking forward, one of the most exciting developments is the emergence of **human-AI collaboration** in creative domains. Artists, writers, and designers are leveraging generative tools to co-create content—using AI not just as a tool but as a creative partner. Platforms like Runway ML, Adobe Firefly, and Google's MusicLM exemplify this shift, offering intuitive interfaces for prompt-driven co-creation of images, video, and music.

Figure 7. Future Landscape of Digital Synthetic Generation

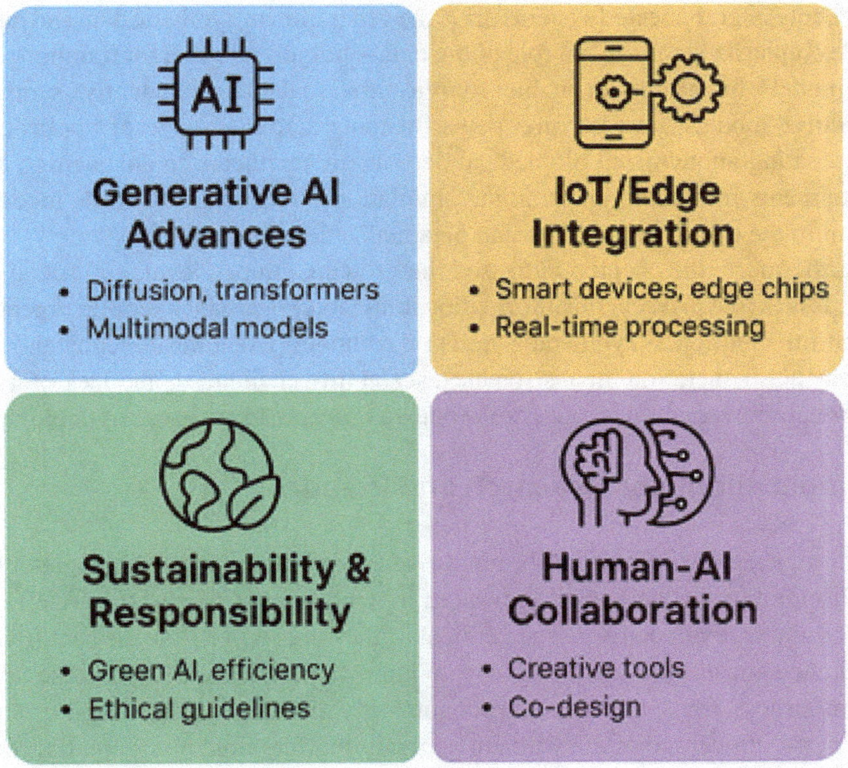

In the future, **interactive generative systems** will allow real-time refinement of content based on human feedback loops, aligning AI-generated outputs more closely with human intent. This symbiosis between human creativity and algorithmic precision may redefine workflows in design, journalism, education, and entertainment.

9. CONCLUSION

9.1 Summary of Key Insights

This chapter has explored the dynamic and evolving field of digital synthetic generation, underpinned by the convergence of machine learning, deep learning, and big data. We began by defining the foundational concepts, tracing the historical development from early statistical models to modern deep generative architectures like GANs, VAEs, and diffusion models. The technological stack comprising artificial intelligence (AI), machine learning (ML), and deep learning (DL) illustrates how deeply interconnected systems drive the generation of synthetic data and media.

We emphasized the crucial role of big data—not just as a fuel for training generative models, but also as an enabler for real-time, scalable, and adaptive synthesis. Generative models have become powerful tools across a range of applications: from creating anonymized biomedical datasets in healthcare to enhancing virtual environments in gaming and training simulations. The utility of these models is evident in the quality, versatility, and personalization they offer.

Furthermore, the chapter addressed the growing ethical, legal, and social concerns. Issues like privacy, misinformation, bias, and governance require urgent and thoughtful attention as synthetic generation technology matures. Technical limitations such as generation inconsistencies, scalability challenges, and lack of model interpretability remain as hurdles on the path to widespread, trustworthy deployment.

9.2 Implications for Research and Practice

From a research perspective, the rapid evolution of generative technologies invites interdisciplinary collaboration. Advances in neural architectures, energy-efficient model training, explainable AI, and ethical frameworks will define the next frontier. Researchers must explore not only how to improve model accuracy but also how to embed fairness, transparency, and accountability into systems by design.

For practitioners, digital synthetic generation offers immense potential to revolutionize industries. Healthcare professionals can safely innovate using synthetic patient data, marketers can tailor content for individual preferences, and educators can offer immersive learning experiences powered by virtual simulations. However, implementation must be accompanied by robust data governance, awareness of potential misuse, and continuous evaluation to prevent unintended harm.

Policymakers and organizations need to stay ahead by framing regulatory guidelines that balance innovation and protection. Investments in regulatory sandboxes, ethical review boards, and transparency toolkits will help create environments where responsible experimentation thrives.

9.3 Final Thoughts on the Future of Digital Synthetic Generation

Looking ahead, the future of digital synthetic generation is rich with promise and complexity. As models become more capable, the line between real and synthetic will blur further. This will demand new ways of thinking about digital content authenticity, intellectual property, and human creativity.

Human-AI collaboration is poised to redefine creative industries, where machines become co-designers, assisting humans in storytelling, design, and artistic expression. The democratization of these tools could empower a wider population to create, innovate, and solve problems like never before.

At the same time, sustainability and equity must guide future developments. Ensuring that synthetic generation is environmentally conscious, ethically grounded, and socially inclusive will determine whether this powerful technology becomes a force for collective good.

In sum, digital synthetic generation is not merely a technological advance—it's a paradigm shift in how we understand, interact with, and generate digital realities. Its trajectory will be shaped not just by algorithms, but by the values and visions of those who develop and use them.

REFERENCES

Bommasani, R., Hudson, D. A., Adeli, E., ... (2022). On the opportunities and risks of foundation models. arXiv preprint arXiv:2203.02155.

Buolamwini, J., & Gebru, T. (2018). Gender shades: Intersectional accuracy disparities in commercial gender classification. *Proceedings of Machine Learning Research*, *81*, 1–15.

Carlini, N., Tramer, F., Wallace, E., Jagielski, M., Herbert-Voss, A., Lee, K., . . . Erlingsson, Ú. (2021). Extracting training data from large language models. arXiv preprint arXiv:2012.07805.

Chen, M., & Zhang, Y. (2014). Big data analytics: A survey. *Journal of Big Data*, *1*(1), 1–32.

Dosovitskiy, A., Ros, G., Codevilla, F., López, A., & Koltun, V. (2017). CARLA: An open urban driving simulator. *Conference on Robot Learning*, 1–16.

. Dwork, C., & Roth, A. (2014). The algorithmic foundations of differential privacy. Foundations and Trends® in Theoretical Computer Science, 9(3–4), 211–407.

European Commission. (2021). *Proposal for a regulation laying down harmonised rules on artificial intelligence*. AI Act.

Frid-Adar, M., Klang, E., Amitai, M., Goldberger, J., & Greenspan, H. (2018). Synthetic data augmentation using GAN for improved liver lesion classification. *IEEE Transactions on Medical Imaging*, *38*(3), 915–925.

Frid-Adar, M., Klang, E., Amitai, M., Goldberger, J., & Greenspan, H. (2018). Synthetic data augmentation using GAN for improved liver lesion classification. *Neurocomputing*, *321*, 321–331. DOI: 10.1016/j.neucom.2018.09.013

Gandomi, A., & Haider, M. (2015). Beyond the hype: Big data concepts, methods, and analytics. *International Journal of Information Management*, *35*(2), 137–144. DOI: 10.1016/j.ijinfomgt.2014.10.007

Goodfellow, I., Pouget-Abadie, J., Mirza, M., Xu, B., Warde-Farley, D., Ozair, S., & Bengio, Y. (2014). Generative adversarial nets. *Advances in Neural Information Processing Systems*, *27*, 2672–2680.

Gupta, A., Mittal, A., & Jain, S. (2019). Synthetic data generation for machine learning. *International Journal of Computer Applications*, *178*(7), 12–18. DOI: 10.5120/ijca2016910248

Jordan, M. I., & Mitchell, T. M. (2015). Machine learning: Trends, perspectives, and prospects. *Science, 349*(6245), 255–260. DOI: 10.1126/science.aaa8415 PMID: 26185243

Karras, T., Laine, S., & Aila, T. (2019). A style-based generator architecture for generative adversarial networks. *IEEE/CVF Conference on Computer Vision and Pattern Recognition (CVPR)*. DOI: 10.1109/CVPR.2019.00453

Karras, T., Laine, S., & Aila, T. (2020). A style-based generator architecture for generative adversarial networks. *IEEE Transactions on Pattern Analysis and Machine Intelligence, 43*(12), 4217–4228. DOI: 10.1109/TPAMI.2020.2970919 PMID: 32012000

Kingma, D. P., & Welling, M. (2013). Auto-encoding variational Bayes. arXiv preprint arXiv:1312.6114.

Kotsiantis, S., Kanellopoulos, D., & Pintelas, P. (2006). Data preprocessing for supervised learning. *International Journal of Computational Science, 1*(2), 111–117.

Kumar, V., Raza, M. Q., & Rawat, D. B. (2020). Real-time big data analytics: Applications and challenges. *Computer Communications, 160*, 257–275.

Laver, K., George, S., Thomas, S., Deutsch, J. E., & Crotty, M. (2017). Virtual reality for stroke rehabilitation. *Cochrane Database of Systematic Reviews*, 11. PMID: 29156493

LeCun, Y., Bengio, Y., & Hinton, G. (2015). Deep learning. *Nature, 521*(7553), 436–444. DOI: 10.1038/nature14539 PMID: 26017442

Mehrabi, N., Morstatter, F., Saxena, N., Lerman, K., & Galstyan, A. (2021). A survey on bias and fairness in machine learning. *ACM Computing Surveys, 54*(6), 1–35. DOI: 10.1145/3457607

. Partnership on AI. (2023). Responsible practices for synthetic media.

Ramesh, A., Dhariwal, P., Nichol, A., ... (2022). Hierarchical text-conditional image generation with CLIP latents. arXiv preprint arXiv:2204.06125.

Rombach, R., Blattmann, A., Lorenz, D., Esser, P., & Ommer, B. (2022). High-resolution image synthesis with latent diffusion models. *Proceedings of the IEEE/CVF Conference on Computer Vision and Pattern Recognition*. DOI: 10.1109/CVPR52688.2022.01042

Saharia, C.. (2022). *Photorealistic text-to-image diffusion models with deep language understanding*. Google Research.

Shi, W., Cao, J., Zhang, Q., Li, Y., & Xu, L. (2020). Edge computing: Vision and challenges. *IEEE Internet of Things Journal.*

Shin, H. C., Tenenholtz, N. A., Rogers, J. K., Schwarz, C. G., Senjem, M. L., Gunter, J. L., & Andriole, K. P. (2018). Medical image synthesis for data augmentation and anonymization using generative adversarial networks. *International Workshop on Simulation and Synthesis in Medical Imaging*. DOI: 10.1007/978-3-030-00536-8_1

Strubell, E., Ganesh, A., & McCallum, A. (2019). Energy and policy considerations for deep learning in NLP. arXiv preprint arXiv:1906.02243. DOI: 10.18653/v1/P19-1355

Summerville, A., Snodgrass, S., Mateas, M., & Ontañón, S. (2018). Procedural content generation via machine learning. *IEEE Transactions on Games, 10*(3), 257–270. DOI: 10.1109/TG.2018.2846639

Tobin, J., Fong, R., Ray, A., Schneider, J., Zaremba, W., & Abbeel, P. (2017). Domain randomization for transferring deep neural networks from simulation to the real world. *IEEE/RSJ International Conference on Intelligent Robots and Systems (IROS)*. DOI: 10.1109/IROS.2017.8202133

. U.S. Congress. (2019). DEEPFAKES Accountability Act.

Vaccari, C., & Chadwick, A. (2020). Deepfakes and disinformation: Exploring the impact of synthetic political video on deception, uncertainty, and trust in news. *Social Media + Society, 6*(1), 2056305120903408. DOI: 10.1177/2056305120903408

Wang, W., Wong, R. C. W., & Wang, J. (2020). Generating and evaluating synthetic personalized data for recommender systems. [TIST]. *ACM Transactions on Intelligent Systems and Technology, 11*(1), 1–23.

Westerlund, M. (2019). The emergence of deepfake technology: A review. *Technology Innovation Management Review, 9*(11), 39–52. DOI: 10.22215/timreview/1282

Zhavoronkov, A., Ivanenkov, Y. A., Aliper, A., Veselov, M., Aladinskiy, V., Aladinskaya, A., Terentiev, V. A., Polykovskiy, D. A., Kuznetsov, M. D., Asadulaev, A., Volkov, Y., Zholus, A., Shayakhmetov, R. R., Zhebrak, A., Minaeva, L. I., Zagribelnyy, B. A., Lee, L. H., Soll, R., Madge, D., & Aspuru-Guzik, A. (2019). Deep learning enables rapid identification of potent DDR1 kinase inhibitors. *Nature Biotechnology, 37*(9), 1038–1040. DOI: 10.1038/s41587-019-0224-x PMID: 31477924

Zhou, Z., Chen, X., & Li, L. (2021). Edge intelligence: Paving the last mile of artificial intelligence with edge computing. *Proceedings of the IEEE.*

Chapter 3

Digital Synthetic Generation From Learning and BigData

Ajay N. Upadhyaya
https://orcid.org/0000-0002-7583-6430
SAL Engineering and Technical Institute, India

S. Seenivasan
https://orcid.org/0000-0002-2155-9781
Rathinam Technical Campus, India

P. Selvakumar
https://orcid.org/0000-0002-3650-4548
Department of Science and Humanities, Nehru Institute of Technology, Coimbatore, India

Nilesh Anute
https://orcid.org/0000-0001-6599-813X
Sri Balaji University, India

S. Guru Basava Aradhya
https://orcid.org/0000-0002-3335-2681
Chandigarh School of Business, Chandigarh Group of Colleges Jhanjeri, Mohali, India

Mohit Sharma
https://orcid.org/0009-0007-2280-8077
Maharshi Dayanand University, Rohtak, India

T. C. Manjunath
https://orcid.org/0000-0003-2545-9160
Rajarajeswari College of Engineering, India

ABSTRACT

Digital synthetic generation is a rapidly evolving field that involves the creation of artificially generated data, media, and environments using computational techniques. It plays a crucial role in industries such as artificial intelligence (AI), gaming, film production, virtual reality (VR), and scientific simulations. By leveraging procedural algorithms, deep learning models, and physics-based simulations, digital synthetic generation enables the production of realistic and scalable content without relying

DOI: 10.4018/979-8-3373-2277-3.ch003

on real-world data collection. This article provides an in-depth exploration of the concepts, methods, and applications of digital synthetic generation, highlighting its significance in various domains and its potential to reshape industries. Concepts of Digital Synthetic Generation: Digital synthetic generation refers to the process of artificially creating digital content using computational models rather than capturing it from the real world. This can include images, videos, 3D environments, physics simulations, and even synthetic datasets for AI training.

INTRODUCTION TO DIGITAL SYNTHETIC GENERATION

Digital synthetic generation is a rapidly evolving field that involves the creation of artificially generated data, media, and environments using computational techniques. It plays a crucial role in industries such as artificial intelligence (AI), gaming, film production, virtual reality (VR), and scientific simulations. By leveraging procedural algorithms, deep learning models, and physics-based simulations, digital synthetic generation enables the production of realistic and scalable content without relying on real-world data collection. This article provides an in-depth exploration of the concepts, methods, and applications of digital synthetic generation, highlighting its significance in various domains and its potential to reshape industries. Concepts of Digital Synthetic Generation: Digital synthetic generation refers to the process of artificially creating digital content using computational models rather than capturing it from the real world. This can include images, videos, 3D environments, physics simulations, and even synthetic datasets for AI training. Parameters can be adjusted to create variations tailored to specific needs. Realism vs. Abstraction: Content can be hyper-realistic (e.g., deepfake technology) or abstract (e.g., procedural art). Automation: AI and procedural techniques automate content creation, reducing human effort. Synthetic Media Generation – AI-generated images, videos, and audio (e.g., deepfakes, AI art). Procedural Content Generation (PCG) – Algorithm-driven creation of game worlds, textures, and assets. Synthetic Data for AI and Machine Learning – Artificial datasets for training AI models. Physics-Based Simulation – Virtual testing environments that replicate real-world physical interactions. These different types of synthetic generation have distinct applications, methods, and technological foundations, which will be explored further. Procedural Content Generation (PCG): PCG involves rule-based algorithms that generate content dynamically, commonly seen in Video Games: Algorithmic generation of levels, landscapes, and characters. Automated creation of textures, buildings, and terrains using fractals and noise functions. AI-generated compositions using procedural algorithms. PCG allows for infinite content generation with minimal manual intervention, making it a powerful tool for digital entertainment.

Physics-based methods use real-world physical principles to generate synthetic content, such as Fluid and Particle Simulations, which are used in CGI and gaming for realistic water, fire, and smoke effects. Soft-Body and Rigid-Body Dynamics: Simulating real-world physics interactions (e.g., destruction physics in AAA games). Autonomous Vehicle Simulation: AI models trained on synthetic driving environments (e.g., NVIDIA DRIVE Sim). Physics-based synthetic generation is essential for scientific modeling, gaming physics, and AI training environments. Hybrid Approaches: AI + Procedural + Physics-Based: Many modern applications combine multiple techniques. AI-enhanced procedural generation (PCG: AI optimizes procedural generation to create more believable environments. GANs + Physics Simulations: AI-generated characters behave naturally within physics-based worlds. Deep Learning for Texture Generation: AI generates ultra-realistic materials for 3D assets. These hybrid approaches push the boundaries of digital content creation, enabling hyper-realistic simulations and intelligent procedural generation. AI Training and Synthetic Data: Autonomous Vehicles: AI models are trained in synthetic driving environments before real-world deployment. Medical AI: Synthetic medical images improve AI diagnostic accuracy without privacy concerns. Facial Recognition & Biometrics: AI-generated faces diversify datasets to reduce bias in recognition systems. Film and CGI Production: AI-Assisted Animation: AI enhances animation workflows, reducing manual effort. Deepfake Technology: Used for digital de-aging and face replacements in films. Physics Simulations: Water, fire, and destruction effects are synthetically generated using simulation engines. Virtual and Augmented Reality (VR/AR): AI-Generated 3D Environments: VR worlds are created dynamically for immersive experiences. Physics-Based Interactions: Realistic object physics enhances user engagement. Synthetic Avatars: AI-generated avatars populate VR spaces for social interaction. Challenges and Ethical Considerations: Despite its advantages, digital synthetic generation presents challenges.

BACKGROUND OF LEARNING AND BIG DATA IN DIGITAL SYNTHETIC GENERATION

Digital synthetic generation relies on computational models to create realistic digital content, including images, videos, 3D models, and simulated environments. Two foundational pillars of this technology are machine learning (ML) and big data, which enable highly detailed and scalable synthetic content creation. Machine learning provides the intelligence necessary for generating, optimizing, and refining synthetic data, while big data supplies the vast amounts of structured and unstructured information required for training these models. Together, they power applications in AI-generated media, procedural content generation, virtual environments, and scientific simulations, revolutionizing industries such as gaming, film, artificial intelligence, and autonomous systems. This article explores the role of learning and big data in digital synthetic generation, focusing on key techniques, applications, challenges, and future advancements. Learning in Digital Synthetic Generation: Machine learning plays a central role in digital synthetic generation by automating content creation and improving realism through adaptive models. Several key ML techniques drive this field: Generative Models for Synthetic Content: Generative Adversarial Networks (GANs): GANs use two competing neural networks—a generator and a discriminator—to create realistic synthetic images, videos, and textures (De Melo C M., et al, 2022). Example: This Person Does Not Exist generates AI-created human faces indistinguishable from real ones. Applications: AI-assisted game asset generation, deepfake technology, and synthetic training data for AI models. Variational Autoencoders (VAEs): VAEs learn latent representations of data to create variations of realistic synthetic content. Example: AI-generated music and texture synthesis.

Recent advancements in AI-based synthetic generation, diffusion models iteratively refine noise into high-quality images and videos. Example: Stable Diffusion and DALL-E create photorealistic images from text prompts. Neural Radiance Fields (NeRFs) for 3D Content: NeRFs leverage deep learning to generate high-quality 3D reconstructions from 2D images, transforming static datasets into interactive synthetic environments. Example: AI-powered digital twins for VR and AR applications. Self-Supervised and Unsupervised Learning: Self-supervised learning allows AI models to improve synthetic content generation without needing large amounts of labeled data. Unsupervised clustering enables the discovery of patterns in big data, leading to more diverse and realistic synthetic media (Lu Y., et al, 2023). By utilizing these techniques, AI-powered synthetic generation models can create, optimize, and refine digital content at an unprecedented scale. The Role of Big Data in Digital Synthetic Generation Big data provides the foundational knowledge that fuels machine learning models, enabling them to generate highly detailed and context-aware synthetic content. The quality, diversity, and quantity of data significantly influence the realism and accuracy of AI-generated outputs. Real-World Data for AI Training: Large-scale datasets, such as ImageNet and Common Crawl, provide reference points for AI models generating synthetic content. Example: Autonomous driving simulations use vast datasets of real-world traffic footage to train in synthetic driving environments. Sensor and IoT Data: Real-world physics simulations and synthetic environments benefit from IoT sensor data collected from the real world. Example: Weather modeling simulations generate synthetic storm conditions based on real meteorological sensor data.

Crowdsourced and Open Data Repositories: Platforms like OpenAI and Kaggle provide massive datasets used to refine synthetic generation models. Example: AI-generated voices in virtual assistants improve by analyzing large datasets of human speech. Data-Driven Procedural Content Generation: Big data enhances procedural content generation (PCG) by training AI models on massive datasets to create highly detailed digital assets. Example: Gaming Industry: No Man's Sky uses procedural generation trained on planetary data to create billions of unique planets. AI-driven world-building in games like Minecraft allows for infinite, dynamic landscapes. Example: Film and CGI Industry: AI-powered big data analysis generates ultra-realistic CGI characters, such as deepfake-driven de-aging effects in Hollywood films. Applications of Learning and Big Data in Synthetic Generation: AI-Generated Media and Deepfake Technology: AI-generated images, videos, and voice synthesis revolutionize content creation in advertising, film, and gaming. Example: Deepfake technology enables realistic face-swapping in movies (Moreno Y., et al, 2023). Autonomous Systems and Simulated Environments: Self-driving cars train in synthetic simulations before real-world deployment (e.g., NVIDIA DRIVE Sim). AI-powered robotics use synthetic environments to learn real-world tasks. Virtual Reality (VR) and Augmented Reality (AR): Big data-driven AI creates hyper-realistic VR environments and interactive digital worlds. Challenges and Ethical Considerations: Ethical Concerns in AI-Generated Synthetic Content Deepfake Misuse: AI-generated media poses risks for misinformation and fraud. Bias in AI-Generated Data: Synthetic generation models must avoid reinforcing existing biases. Data Privacy and Security: Synthetic data generation must adhere to privacy regulations (e.g., GDPR, CCPA) to protect user identities. Computational Costs: Processing vast amounts of big data requires significant computational resources and energy-efficient AI models. Despite these challenges, ongoing advancements in ethical AI frameworks will help mitigate risks while ensuring responsible use of synthetic generation technologies (Rathore M M., et al, 2021). Machine learning and big data are the driving forces behind digital synthetic generation, enabling the creation of hyperrealistic, scalable, and customizable digital content.

THE EVOLUTION OF DIGITAL SYNTHETIC GENERATION

Digital synthetic generation has undergone a remarkable transformation over the past few decades, evolving from basic algorithmic techniques to sophisticated AI-driven systems capable of generating hyperrealistic media, simulations, and interactive environments. This evolution has been driven by advances in computational power, artificial intelligence (AI), procedural generation, and big data analytics, which have revolutionized how digital content is created, optimized, and utilized.

From early rule-based and mathematical procedural methods to the emergence of machine learning models like Generative Adversarial Networks (GANs) and diffusion models, synthetic generation has become an essential tool in gaming, film, artificial intelligence, virtual reality (VR), and scientific research (Figueira A., et al, 2022). This article explores the evolution of digital synthetic generation, tracing its journey from traditional techniques to modern AI-powered innovations and examining its impact on various industries. Traditional Methods of Digital Synthetic Generation Before the advent of AI-driven models, synthetic generation relied on rule-based, procedural, and physics-based simulations to create digital content. These methods, though limited by computational constraints, laid the groundwork for modern approaches. Procedural Content Generation (PCG): Procedural content generation (PCG) was one of the earliest approaches to synthetic generation, using mathematical algorithms and predefined rules to generate digital assets. Example: Early Computer Graphics (1970s-1980s): Graphics in early computer systems were generated using vector-based and raster-based algorithms (Bauer A., et al, 2024). Example: The iconic Mandelbrot Set (fractal graphics) showcased procedural generation's potential for infinite complexity. Example: Video Game PCG (1990s-2000s): Rogue (1980) and later games like Diablo (1996) used randomized dungeon generation to create unique playthroughs. Spore (2008) demonstrated procedural creature and world generation based on evolutionary algorithms. Minecraft (2009) popularized procedural world-building, allowing infinite terrain generation. Procedural generation remains widely used in gaming, where it allows for scalable content creation without manual asset design.

Rule-Based and Deterministic Simulations: Early digital synthetic generation methods relied on explicitly programmed rules to generate digital environments, animations, and physics-based effects. Physics Simulations in Early CGI (1990s-2000s): Films like Jurassic Park (1993) and The Lord of the Rings trilogy (2001-2003) used physics-based particle simulations for realistic water, fire, and crowd behavior. Early CGI relied on mathematical physics models to simulate cloth, hair, and smoke effects. Deterministic AI for Digital Agents: Before deep learning, AI in video games and simulations was rule-based, using finite state machines (FSMs) and decision trees to generate NPC behavior (Murtaza H., et al, 2023). These approaches were computationally efficient but lacked adaptability, requiring significant manual adjustments to improve realism. Transition to AI-Driven Synthetic Generation: As computing power increased, machine learning (ML) and data-driven techniques enabled a shift from static rule-based systems to adaptive, AI-generated synthetic content. Machine Learning in Synthetic Generation: With the rise of AI, digital synthetic generation began leveraging data-driven learning models, allowing for automated, scalable, and highly realistic content creation. AI-Assisted Content Generation (2010s- Present): AI algorithms learned from large datasets to generate synthetic images, text, and 3D assets. Natural Language Processing (NLP) enabled AI-driven story generation (GPT models) (Paulin G., et al, 2023). GANs and diffusion models have revolutionized image and video synthesis. Generative Adversarial Networks (GANs): The introduction of GANs (2014) marked a breakthrough in AI-driven synthetic content generation. GANs operate with two competing neural networks: A generator that creates synthetic data. A discriminator that evaluates its realism and improves the generator over time. Impact: AI-generated human faces (This Person Does Not Exist). AI-enhanced CGI characters and deepfake technology. Realistic synthetic datasets for AI training. GANs enabled photorealistic synthetic media generation, transforming the gaming, film, and AI industries.

MACHINE LEARNING FUNDAMENTALS FOR DIGITAL SYNTHETIC GENERATION

Machine learning (ML) has revolutionized digital synthetic generation by enabling the creation of realistic, scalable, and adaptable content across various industries, including gaming, film, artificial intelligence (AI), and scientific simulations. Synthetic generation involves producing AI-generated images, videos, 3D models, text, and virtual environments that mimic real-world data or create entirely new digital artifacts. At the core of ML-driven synthetic generation are three fundamental learning paradigms: Supervised Learning – Trains models using labeled datasets. Unsupervised Learning – Identifies patterns and structures in unlabeled data (Morales et al., 2022). Reinforcement Learning (RL) – Teaches models through trial and error in an interactive environment. Each approach plays a crucial role in synthetic generation, enabling AI-powered content creation, procedural world-building, and intelligent character behavior in digital environments. This article explores the fundamentals of supervised, unsupervised, and reinforcement learning and their applications in digital synthetic generation. Supervised Learning in Synthetic Generation: What is Supervised Learning? Supervised learning is a machine learning technique where an AI model learns from labeled training data to make predictions or generate content. The model is trained with input-output pairs, allowing it to recognize patterns and generalize them to new data (Johri et al., 2025). Example: A model trained on real-world human faces can generate synthetic faces by learning patterns of human features. Applications of Supervised Learning in Synthetic Generation: AI-Assisted Animation and CGI: AI-trained models enhance 3D character animation, motion synthesis, and facial expression generation. Example: AI-driven de-aging effects in movies (e.g., The Irishman). Synthetic Data for AI Training: AI models generate synthetic datasets for autonomous vehicles, medical imaging, and biometric recognition. Example: AI-generated synthetic driving environments for self-driving car training. Supervised learning excels in high-quality, controlled synthetic content generation, but it requires large labeled datasets, making it computationally expensive (Kapoor et al., 2022).

Applications of Unsupervised Learning in Synthetic Generation: Procedural Content Generation (PCG) in Gaming: AI autonomously generates terrain, textures, and game levels based on learned structures. For example, No Man's Sky uses procedural generation trained on real-world planetary data. AI-Generated Art and Music: Unsupervised learning enables AI to create abstract art, AI-composed music, and generative poetry. Example: Deep Dream and AI-powered musical compositions. Unsupervised AI for World Simulation: AI learns natural world structures and generates realistic synthetic environments. Example: AI-trained weather pattern simulations for climate research. Unsupervised learning enhances creativity and efficiency in synthetic generation but lacks precision compared to supervised models. Reinforcement Learning in Synthetic Generation: What is Reinforcement Learning? Reinforcement learning (RL) enables AI to learn through trial and error by interacting with an environment (Malik et al., 2022). The AI agent receives rewards for desired actions, gradually optimizing its behavior to achieve the best results. Example: An RL agent learns game-level design by testing different layouts and optimizing for player engagement. Reinforcement Learning Algorithms for Synthetic Generation: RL techniques used in synthetic generation include: Deep Q-Networks (DQN): Used for AI-driven game character behavior and physics-based simulations. Policy Gradient Methods: Optimize procedural generation and AI-controlled NPC behavior. Monte Carlo Tree Search (MCTS): Used in AI-powered game design and autonomous world-building. Applications of Reinforcement Learning in Synthetic Generation: AI-Driven Game World Generation: RL trains AI to dynamically generate adaptive game environments. Example: OpenAI's ProcGen enables AI-powered procedural game world creation (Sarker, 2021). AI-Controlled NPCs and Digital Avatars: RL optimizes intelligent AI characters for realistic interactions. Example: AI-powered NPCs with adaptive behavior in open-world games.

BIG DATA FUNDAMENTALS FOR DIGITAL SYNTHETIC GENERATION

Big data is the backbone of digital synthetic generation, enabling AI models to create realistic, diverse, and scalable digital content. Synthetic generation involves producing AI-generated images, videos, 3D models, virtual worlds, and interactive environments, all of which require vast amounts of high-quality data. For synthetic generation to be effective, three key big data processes must be optimized: Data Preprocessing – Cleaning, transforming, and structuring raw data for AI models. Data

Storage – Efficiently managing large-scale datasets for quick access and scalability. Data Retrieval – Enabling AI models to access, filter, and utilize data efficiently (Zhang, 2022). Each of these processes plays a crucial role in training AI-powered synthetic content generators, ensuring that data is clean, accessible, and optimally structured for real-time use in various applications, from gaming and CGI to AI-driven simulations and virtual reality (VR). This article explores the fundamentals of big data preprocessing, storage, and retrieval for digital synthetic generation, highlighting their importance, methodologies, and real-world applications. Data Preprocessing for Digital Synthetic Generation: What is Data Preprocessing? Data preprocessing is the first and most crucial step in big data workflows. It involves cleaning, structuring, and optimizing raw data before feeding it into AI models for synthetic generation (Shende et al., 2025). Example: A dataset of real-world human faces must be cleaned, normalized, and structured before training a generative adversarial network (GAN) to generate synthetic faces. Steps in Data Preprocessing: Data Cleaning: Removing Noise and Outliers: Eliminating corrupted, duplicate, or irrelevant data. Handling Missing Data: Using interpolation or synthetic data filling techniques. Filtering and De-duplication: Ensuring datasets contain unique and high-quality data. Data Transformation: Normalization and Standardization- Adjusting feature scales for consistency. Feature Engineering: Extracting key patterns for AI models (e.g., edge detection in images). Dimensionality Reduction: Using PCA (Principal Component Analysis) to optimize large datasets. Data Augmentation: Generating additional synthetic variations of existing data to improve model robustness (Saldanha et al., 2022).

AI-powered video and Image Synthesis: Preprocessing enables deepfake technology, AI-enhanced video editing, and motion synthesis. Data preprocessing ensures AI models receive high-quality, structured input, leading to more realistic, efficient, and scalable synthetic content generation. Efficient storage solutions are crucial for scalability, performance, and accessibility in AI-driven synthetic generation. Types of Data Storage Solutions: Cloud Storage Solutions: Amazon S3, Google Cloud Storage, and Microsoft Azure Blob Storage. Scalable, cost-effective, and ideal for distributed AI training. Distributed File Systems: Hadoop Distributed File System (HDFS) for big data processing. Apache Spark & Databricks for real-time synthetic data pipelines. Databases for AI Training: NoSQL Databases (MongoDB, Cassandra): Stores unstructured multimedia data. SQL Databases (PostgreSQL, MySQL): Manage structured AI metadata (Alghushairy & Ma, 2022). High-Performance Storage for Real-Time Synthetic Generation: GPUs & TPUs: Specialized hardware for AI model training and inference. Edge Computing Storage: Used in real-time AI applications like AR/VR and autonomous vehicles. Data Retrieval for Digital Synthetic Generation: What is Data Retrieval? Data retrieval involves fetching and filtering relevant data from large datasets to optimize AI model performance. Efficient retrieval ensures fast, accurate access to training data, real-time AI models, and generated synthetic content. Methods of AI-Driven Data Retrieval: Search and Indexing Techniques: Vector Databases (FAISS, Milvus): Optimize AI model searches for image and video synthesis (Sardar et al., 2024). Inverted Indexing for Text Data Retrieval: Used in AI-generated synthetic language models. Graph-Based Retrieval for Complex Relationships: Knowledge Graphs (Neo4j): Enhance AI-driven synthetic storytelling and virtual NPC interactions.

DATA QUALITY AND VALIDATION IN DIGITAL SYNTHETIC GENERATION

The success of digital synthetic generation—the process of using AI to create realistic images, videos, 3D models, text, and virtual environments—heavily depends on the quality and reliability of the data used. Poor data quality can lead to inaccurate AI-generated content, bias, and unpredictable results, making validation processes critical in synthetic generation workflows. Ensuring accuracy, consistency, and reliability in AI-driven synthetic generation requires: Data Quality Assurance – Checking for errors, noise, and inconsistencies. Data Validation – Ensuring data integrity through automated and manual checks. Bias Detection and Ethical Considerations – Reducing biases in AI-generated content. This article explores key data quality and validation techniques, their importance in synthetic generation, and real-world applications in AI-generated media, gaming, deepfake detection, and scientific simulations (Iantovics et al., 2022). Data Quality in Digital Synthetic Generation: Why is Data Quality Important? Synthetic generation models rely on large datasets to train AI-driven content generators, such as Generative Adversarial Networks (GANs) or Diffusion Models. If the training data contains errors, inconsistencies, or biases, the AI will produce low-quality or unrealistic synthetic outputs. Example: A facial recognition AI trained on low-resolution images will generate blurry or distorted synthetic faces. Key Data Quality Attributes: Ensuring high-quality data involves optimizing several attributes: Accuracy – Data must reflect real-world characteristics correctly. Completeness – No missing or incomplete records in the dataset. Consistency – Uniform format and structure across datasets. Timeliness – Updated and relevant data, especially in dynamic AI applications. Uniqueness – Avoiding duplicate or redundant information (Marwala, 2024). Common Data Quality Issues in Synthetic Generation: Noisy Data: Data with random errors, distortions, or irrelevant information. Data Bias: AI models trained on unbalanced datasets produce biased synthetic content. Overfitting to Training Data: AI generates overly similar content, reducing diversity.

Figure 1. Application of Digital Synthetic Data generation

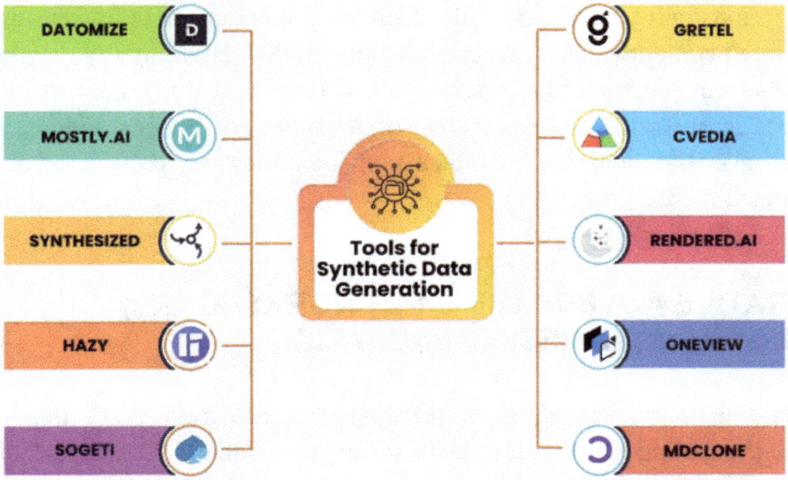

Data Validation in Digital Synthetic Generation: What is Data Validation? Data validation ensures that datasets meet predefined accuracy, completeness, and consistency standards before being used in AI-driven synthetic generation. Validation techniques are critical for preventing errors, filtering biased data, and improving AI performance. Types of Data Validation: Automated Validation Methods: Rule-Based Validation – Checking for missing values, outliers, and format inconsistencies. Statistical Validation – using statistical models to detect anomalies in datasets. Machine Learning-Based Validation – AI models detect fake or inconsistent data. Image and Video Quality Assessment – AI-based validation for AI-generated images, deepfakes, and CGI models. Procedural Game World Generation: AI validates procedural landscapes, terrain, and level designs for consistency (El Emam et al., 2022). Example: OpenAI's ProcGen system for testing game environments. Synthetic Data for AI Training: Validation ensures autonomous vehicle datasets accurately represent road conditions. Example: AI-generated synthetic driving scenarios for self-driving cars. By integrating automated and manual validation, AI-generated content becomes more trustworthy and aligned with real-world expectations. AI fairness tools (e.g., IBM's AI Fairness 360) detect bias in synthetic data (Debnath et al., 2025). Validation ensures AI-generated faces and animations look natural. Autonomous Vehicle Training with Synthetic Data: Self-driving cars use AI-generated road scenarios for training. Validation ensures realistic physics and lighting conditions in synthetic driving environments (Munappy et al, 2022). AI in Medical Imaging: AI-generated synthetic MRI and CT scans aid in training healthcare AI models. Data validation ensures that synthetic images accurately match real-world medical scans. Data quality and validation are fundamental pillars of digital synthetic generation, ensuring that AI-generated content is accurate, reliable, and ethical. High-quality data prevents AI models from generating inaccurate or biased synthetic content. Data validation techniques improve the trustworthiness of AI-generated images, videos, and 3D environments. Bias detection and ethical considerations help prevent the misuse of AI-generated media.

GENERATIVE ADVERSARIAL NETWORKS (GANs) FOR DIGITAL SYNTHETIC GENERATION

Generative Adversarial Networks (GANs) have revolutionized digital synthetic generation, enabling AI to create realistic images, videos, 3D models, and even synthetic text. First introduced by Ian Goodfellow in 2014, GANs use a dueling neural network architecture where two networks—the Generator and the Discriminator—compete against each other to produce highly realistic synthetic data. GANs have become essential in various fields, including AI-driven art, video game asset generation, deepfake creation, autonomous vehicle training, and medical imaging. Their ability to learn from large datasets and generate synthetic content that closely resembles real-world data makes them a powerful tool for AI-driven content creation. This article explores the theoretical foundations of GANs, how they work, and their applications in digital synthetic generation, including AI-generated media, procedural game content, deepfake technology, and scientific simulations (Aggarwal A., et al, 2021). The Discriminator is trained using a real dataset to distinguish real data from synthetic data. The Generator learns from the discriminator's feedback, improving its ability to create realistic outputs (Gui J., et al, 2021). Over many training iterations, the Generator produces synthetic data indistinguishable from real data. This adversarial training process enables GANs to generate photo-realistic images, human-like speech, and even synthetic environments. The Discriminator tries to maximize correct classification. The Generator tries to minimize the Discriminator's accuracy, improving its ability to create convincing synthetic content.

Applications of GANs in Digital Synthetic Generation: AI-Generated Art and Deepfake Technology: GANs have significantly impacted digital art and AI-generated media, allowing machines to create hyper-realistic synthetic content. Deepfake Creation: GANs power deepfake technology, generating fake human faces, videos, and voice clones. Example: AI-generated celebrity deepfakes in movies and media. AI-Generated Paintings and Artworks: Generative adversarial networks (GANs can mimic famous artists and create original artwork. Example: AI-generated paintings sold at auction (e.g., Christie's AI portrait "Edmond de Belamy") (Dash et al, 2023). Synthetic Driving Data for Autonomous Vehicles: AI generates virtual road conditions, pedestrians, and traffic for self-driving car training. Example: Tesla and Waymo use AI-generated synthetic driving data (Aragani et al. 2025). Smart City Simulations: GANs simulate urban environments for city planning and AI-driven surveillance. GANs provide a safe, cost-effective alternative to real-world data collection, accelerating AI model development for real-world applications. Ethical Concerns and Misinformation Risks: Deepfake Misinformation: GANs enable AI-generated fake videos used for misinformation and fraud. Privacy Violations: AI-generated synthetic identities raise privacy concerns. Bias in AI-Generated Data: GANs can reproduce biased content from training datasets. Ensuring responsible AI development and regulation is essential to prevent the misuse of GAN technology. Generative Adversarial Networks (GANs) are transforming digital synthetic generation, enabling AI to create realistic images, videos, 3D environments, and interactive content (Alqahtani et al, 2021). GANs power AI-generated art, deepfake technology, medical imaging, and game asset creation.

VARIATIONAL AUTOENCODERS (VAEs) FOR DIGITAL SYNTHETIC GENERATION

Variational Autoencoders (VAEs) are a powerful class of deep generative models used for digital synthetic generation, enabling AI to create realistic images, videos, 3D models, and structured data. First introduced by Kingma and Welling in 2013, VAEs leverage probabilistic modeling to learn meaningful latent representations of data, allowing them to generate new synthetic samples that resemble real-world data while maintaining smooth interpolation and controllable generation. VAEs are widely used in AI-generated media, game asset generation, text-to-image synthesis, healthcare imaging, and anomaly detection (An et al, 2023). What Are VAEs? VAEs are a probabilistic extension of traditional Autoencoders designed for unsupervised learning and synthetic data generation. They consist of two main components: The Encoder (Inference Model) Maps input data into a latent space representation (Ramakrishnan et al. 2022). Outputs a probability distribution rather than a single deterministic value. Decoder (Generative Model) Reconstructs data samples from latent variables. Learns to approximate the true data distribution. Advantages of VAEs over Other Generative Models: Probabilistic Latent Representations – VAEs learn a structured, continuous latent space (Dos Santos et al, 2024). Smooth Interpolations – Enables controllable data generation and modification. Better Training Stability – Unlike GANs, VAEs do not suffer from mode collapse. Compact Representation Learning – Useful for dimensionality reduction and feature extraction.

These properties make VAEs ideal for synthetic data generation in computer vision, gaming, and healthcare applications.

Applications of VAEs in Digital Synthetic Generation: AI-Generated Art and Image Synthesis: VAEs are widely used for image generation and AI-driven artistic creation, enabling AI-powered image Reconstruction – generating missing image parts from limited data. Latent Space Interpolation – Smoothly transforming one image into another. Text-to-Image Synthesis – Generating images based on textual descriptions. Example: DALL·E and VQ-VAE models use VAEs to create highly realistic AI-generated artwork. Future of VAEs in Digital Synthetic Generation: The evolution of VAEs is shaping the future of AI-driven content creation, including Hybrid VAE-GAN Models, which combine VAEs' structured latent space with GANs' high-quality generation. Neural Rendering with VAEs – AI-driven 3D scene reconstruction and animation. Multimodal AI Models – Combining VAEs with transformers for better text-to-image synthesis (Akkem Y., et al, 2024). As VAEs continue to improve, they will play a critical role in AI-generated digital content, personalized media, and interactive synthetic worlds. Variational Autoencoders (VAEs) are a cornerstone of digital synthetic generation, providing a structured, probabilistic approach to AI-driven content creation. VAEs generate realistic images, videos, game assets, and medical data. Their smooth latent spaces enable controllable and interpolated AI-generated content. Despite challenges, VAEs continue to evolve, influencing AI in gaming, healthcare, and multimedia. As AI-driven synthetic generation advances, VAEs will remain essential for building the next generation of AI-generated content.

DEEP LEARNING-BASED METHODS FOR DIGITAL SYNTHETIC GENERATION

Deep learning has revolutionized digital synthetic generation, enabling AI to produce realistic images, videos, text, and audio. Two of the most widely used architectures for synthetic data generation are Convolutional Neural Networks (CNNs) and Recurrent Neural Networks (RNNs). CNNs excel in spatial data processing, making them ideal for image and video generation. RNNs handle sequential data, making them well-suited for text synthesis, speech generation, and time-series prediction. Fully Connected Layers: Interpret extracted features for final predictions. By learning spatial hierarchies, CNNs can generate highly realistic synthetic content. CNNs are widely used for: AI-generated faces (e.g., StyleGAN for deepfake technology) Texture synthesis for video games Medical image augmentation CNN Applications in Digital Synthetic Generation AI-generated art and Deepfake Technology CNNs enable style transfer and deepfake generation, allowing AI to produce artistic images, AI-powered portraits, and realistic face synthesis (Gholamiankhah et al, 2021). Deep Dream uses CNNs to create AI-enhanced artistic images. Deepfakes: AI-powered video and image manipulations that swap faces or mimic real individuals. AI-Generated Paintings: CNNs replicate famous artistic styles and create new artwork. Procedural Content Generation in Games CNNs generate synthetic textures, game assets, and procedural environments in game design. Texture synthesis: AI-generated landscapes and photo-realistic game environments. 3D model generation: CNN-based neural rendering for realistic 3D characters. Scene generation: AI-generated backgrounds and level designs for gaming.

Hidden Layers with Recurrent Connections: Retains information from previous time steps. Output Layer: Generates sequential predictions. This makes RNNs perfect for synthetic text generation, automated speech synthesis, and AI-powered music composition. Types of RNNs for Digital Synthetic Generation: Vanilla RNNs: Basic form, useful for short-term sequential tasks. Limited by vanishing gradient problem, making it hard to retain long-term dependencies (Khalaf L I., et al, 2024). Long Short-Term Memory (LSTM) Networks: Introduces gates to retain long-term memory, solving gradient issues. Voice-mimicking AI for personalized soundtracks (Sreenivasulu et al. 2022). Example: AI-generated Bach and Beethoven compositions using RNN-based deep learning models. Challenges and Future Directions: Challenges in CNN and RNN-Based Synthetic Generation: CNNs require extensive computational power for high-resolution image synthesis. RNNs struggle with long-term dependencies, requiring improvements like transformers. Deepfake misuse raises ethical concerns regarding AI-generated media. Future of Deep Learning-Based Synthetic Generation: Hybrid CNN-RNN architectures for multimodal AI generation. Transformers replacing traditional RNNs in text generation (Anjaneyulu et al, 2024). As AI progresses, CNNs and RNNs will shape the future of synthetic media, bringing AI-generated content closer to human-like creativity.

DIGITAL SYNTHETIC GENERATION FOR IMAGE AND VIDEO APPLICATIONS

Digital synthetic generation has transformed image and video applications, enabling AI to create realistic synthetic images, enhance object detection, and improve tracking systems. Advances in deep learning, computer vision, and generative models allow AI to generate photorealistic visuals, detect objects with high precision, and track motion seamlessly. This article explores image synthesis techniques, AI-driven object detection, and tracking methodologies, highlighting their impact on gaming, healthcare, security, and entertainment. Image Synthesis: AI-Generated Visuals: Overview of Image Synthesis: Image synthesis involves generating realistic or stylized images using AI. Techniques include Generative Adversarial Networks (GANs), which are used for realistic face synthesis and deepfakes. Variational Autoencoders (VAEs) – Generate structured, smoothly transitioning images (Yang et al, 2022). VAEs create smoothly interpolated synthetic images, making them useful for AI-powered face morphing applications (Liu et al, 2021). Synthetic medical images for AI training. 3D object reconstruction from 2D images. Applications of Image Synthesis: AI-Generated Art and Creative Design: AI synthesizes new paintings, 3D models, and animations, enabling AI-powered artistic style transfer. Neural rendering for realistic visual effects in films. Procedural content generation for video games. Object Detection: AI-Powered Recognition Systems What Is Object Detection? Object detection enables AI to identify and classify objects in images and videos. It combines image processing, machine learning, and deep learning to detect faces, animals, vehicles, and human gestures. Enable autonomous vehicles to recognize obstacles. Improve medical diagnostics through AI-assisted image analysis.

Deep Learning Models for Object Detection: Convolutional Neural Networks (CNNs) for Object Detection: CNNs extract spatial features and identify patterns in images. AI-powered disease classification from medical scans. Early cancer detection using deep learning models. Object detection continues to evolve, enhancing AI-powered automation, healthcare, and security applications (Mahadev Madgule, et al. 2023). Object tracking enhances sports video analysis, enabling Player movement tracking for performance evaluation. AI-powered automated camera framing. Instant replay and predictive analytics for sports broadcasts (Aranjuelo et al, 2021). Augmented Reality (AR) and Virtual Reality (VR): AI-powered tracking enables real-time motion capture, enhancing Immersive gaming experiences with AI-driven avatar movements. Hand and body tracking for interactive AR applications. Virtual production in filmmaking with AI-based motion tracking. AI-powered object tracking continues to drive advancements in robotics, gaming, and smart surveillance. Digital synthetic generation is transforming image and video applications through Image synthesis techniques (GANs, VAEs) for AI-generated visuals, game design, and medical imaging. Object detection using CNNs and Transformers for autonomous vehicles, security, and healthcare. AI-powered object tracking for drones, AR/VR, and sports analytics. As AI advances, digital synthetic generation will continue to shape entertainment, automation, and interactive media, creating highly realistic AI-driven experiences.

DIGITAL SYNTHETIC GENERATION FOR AUDIO AND MUSIC APPLICATIONS

Digital synthetic generation has revolutionized audio and music applications, enabling AI to synthesize music, classify sounds, and generate new audio content with unprecedented realism. With advances in deep learning, neural networks, and digital signal processing (DSP), AI can now compose original music, enhance audio analysis, and create realistic speech and sound effects. This article explores key techniques and applications of music synthesis, AI-driven audio classification, and generative sound models, highlighting their impact on entertainment, gaming, healthcare, and interactive media. Music Synthesis: An Overview of AI-Based Music Synthesis: Music synthesis involves using AI to compose, arrange, and produce new musical pieces. AI-driven models learn from vast music datasets and generate compositions that mimic classical, jazz, pop, and electronic music (Civit et al, 2022). GANs enhance real-time music production, DJing, and virtual band simulations (Dwivedi P., et al, 2024). Applications of AI-Generated Music: AI-Generated Soundtracks for Films and Games: AI synthesizes customized music scores for Video game background music that adapts to gameplay. AI-powered film scoring for dynamic soundtracks. Adaptive music systems for interactive media. Therapeutic and Wellness Applications: AI-generated music is being used for Music therapy for mental health treatment. AI-driven relaxation soundscapes (meditation and sleep aids). AI-personalized playlists based on emotional analysis. AI-driven music synthesis is transforming composition, personalization, and interactive media, making AI-generated music widely accessible.

Audio Classification: AI-Driven Sound Analysis: What Is Audio Classification? Audio classification involves categorizing and recognizing sound patterns in audio signals using AI. This enables speech recognition, music genre classification, and environmental sound detection. Voice Assistants and Smart Devices: AI-driven audio classification powers: Speech-to-text applications (e.g., Siri, Alexa, Google Assistant). AI-powered voice biometrics for security authentication. Smart home automation with voice commands (Agchar et al, 2024). What Is AI-Generated Audio? AI-powered audio generation involves creating synthetic speech, environmental sounds, and music using deep learning. This enables Text-to-speech (TTS) synthesis for virtual assistants (Shi, 2021). AI-powered dubbing and voice acting for films. Text-to-speech (TTS) for accessibility applications. Digital synthetic generation is reshaping music, sound analysis, and AI-driven audio creation through AI-powered music synthesis for film scores, gaming, and personalized listening. Deep learning-based audio classification for voice assistants, security, and smart technology. AI-generated speech and sound effects for content creation, gaming, and accessibility.

As AI advances, synthetic audio applications will continue to enhance creativity, automation, and interactive digital experiences.

CASE STUDY: DIGITAL SYNTHETIC GENERATION FOR MEDICAL IMAGING APPLICATIONS—IMAGE SYNTHESIS, SEGMENTATION, AND ANALYSIS

Medical imaging is a critical field in healthcare, enabling early disease detection, diagnosis, and treatment planning. However, challenges such as limited labeled datasets, patient privacy concerns, and high costs of imaging data acquisition hinder progress. Digital synthetic generation, powered by artificial intelligence (AI) and deep learning, has emerged as a transformative solution for image synthesis, segmentation, and analysis in medical imaging. This case study explores how digital synthetic generation is revolutionizing radiology, oncology, pathology, and other medical fields by improving data augmentation, diagnostic accuracy, and automated medical analysis. Image Synthesis for Medical Imaging: What Is Medical Image Synthesis? Medical image synthesis involves using AI to generate realistic synthetic medical images that mimic real-world scans. This process is crucial for Augmenting training datasets for AI models. Overcoming data scarcity in rare diseases. Enhancing patient privacy by generating anonymized synthetic images (Thambawita et al, 2022). AI-driven enhancement of low-quality medical images. Variational Autoencoders (VAEs) for Medical Image Generation: VAEs produce structured, diverse synthetic images by learning latent space representations. Used for MRI synthesis in neurological studies (Andreini P., et al, 2022). Generates diverse synthetic tumor images for AI training. Improves AI-based anomaly detection in medical imaging. Applications of Image Synthesis in Healthcare: AI-Augmented Training Data for Medical AI Models: AI-generated images are used to train radiology AI models, improving lung disease detection from synthetic chest X-rays. AI-powered MRI segmentation for brain tumor classification. Synthetic CT scans for AI-driven organ segmentation. AI-Based Medical Image Segmentation: What Is Medical Image Segmentation? Medical image segmentation involves partitioning medical scans into meaningful regions to detect abnormalities (tumors, lesions, organ structures). Assist radiologists in AI-assisted diagnostics. Enable AI-powered automated contouring for treatment planning.

Enhancing AI-driven segmentation accuracy in complex structures. Improving robustness of AI-powered anomaly detection in medical scans. Deep learning-based segmentation is enhancing diagnostic precision and surgical planning (Osuala et al, 2023). Transformers improve medical analysis by enhancing AI-based histopathology image classification. AI-powered multimodal medical image fusion (MRI

+ PET). Deep learning-based prediction of disease progression from medical scans (Musalamadugu et al, 2023). Applications of AI-Powered Medical Image Analysis: AI for Radiology and Pathology Diagnosis: AI enhances diagnostic workflows by reducing radiologist workload with AI-assisted detection tools. AI-powered breast cancer screening from mammograms. AI-enhanced histopathology image analysis for tumor classification. Digital synthetic generation is transforming medical imaging applications through AI-powered image synthesis for data augmentation and privacy-preserving datasets. Deep learning-based segmentation for cancer detection, organ contouring, and AI-assisted surgeries. AI-enhanced medical image analysis for automated disease diagnosis and predictive healthcare. By integrating AI and digital synthetic generation, medical imaging is advancing toward faster diagnoses, improved patient outcomes, and enhanced AI-assisted medical workflows.

CHALLENGES IN DIGITAL SYNTHETIC GENERATION: DATA QUALITY, MODEL COMPLEXITY, AND EVALUATION METRICS

Digital synthetic generation, powered by artificial intelligence (AI) and deep learning, has revolutionized fields such as medical imaging, computer vision, gaming, and autonomous systems. However, despite its advancements, this technology faces several challenges that impact its reliability, scalability, and adoption. The key challenges in digital synthetic generation include: Data Quality – Ensuring that synthetic data is accurate, diverse, and free from biases. Model Complexity – Managing the computational and architectural challenges of deep learning models. Evaluation Metrics – Defining robust benchmarks for assessing the quality of synthetic outputs. This discussion explores these challenges in depth and examines how researchers and developers are addressing them (Williamson et al, 2024). Data Quality in Digital Synthetic Generation: Why Is Data Quality Important? Synthetic data is often used to train, augment, or replace real-world data in machine learning models. Poor data quality can lead to Biased AI models that fail in real-world applications. Inaccurate training data that affects model generalization. Unrealistic synthetic outputs that lack diversity.: Many generative models produce artifacts (visual distortions, noise) that reduce the quality of synthetic outputs. For example, AI-generated facial images may have inconsistent textures or unrealistic features (Xue B., et al, 2022). Solution: Researchers use adversarial training and perceptual loss functions to enhance synthetic realism. Ensuring high-quality, unbiased, and diverse synthetic data is critical for reliable AI applications. These models require large computational resources, careful hyperparameter tuning, and extensive training datasets.

High Computational Costs: Training state-of-the-art generative models requires massive GPU clusters and high memory usage. Example: Training DALL-E or

StyleGAN can take weeks on expensive hardware. Solution: Researchers explore model pruning, quantization, and efficient network architectures to reduce computational costs. Evaluation Metrics for Digital Synthetic Generation Why Are Evaluation Metrics Important? Assessing the quality of synthetic data is challenging because Traditional performance metrics (e.g., accuracy, F1-score) do not apply to generative models. Human evaluation is subjective and time-consuming. Generative models require domain-specific benchmarks to assess realism. Challenges in Evaluating Synthetic Data: Lack of Universal Evaluation Standards. Issue: There is no single metric that effectively measures synthetic data quality across different applications. Example: In image synthesis, researchers use Fréchet Inception Distance (FID), but in text generation, they use BLEU and ROUGE scores (Manduchi et al, 2024). Measures perceptual similarity in generated videos (Dou et al, 2023). Mel Cepstral Distortion (MCD): Assesses AI-generated speech quality. Developing standardized, task-specific evaluation metrics is essential for measuring progress in digital synthetic generation. While digital synthetic generation offers transformative applications across healthcare, gaming, and AI-driven media, it faces significant challenges related to Data Quality – Bias, lack of diversity, and artifacts reduce the reliability of synthetic data.

FUTURE DIRECTIONS IN DIGITAL SYNTHETIC GENERATION

Digital synthetic generation, powered by artificial intelligence (AI), deep learning, and big data, has rapidly evolved, transforming industries such as healthcare, gaming, autonomous systems, and creative media. As AI-driven generative models like Generative Adversarial Networks (GANs), Variational Autoencoders (VAEs), and Diffusion Models continue to advance, the future of digital synthetic generation holds exciting new opportunities. This discussion explores emerging trends, opportunities, and applications that will define the next generation of digital synthetic generation (Gill et al, 2022). Enables AI-driven storytelling, content creation, and automated media synthesis. Future of Multimodal AI: Cross-domain generation – AI models capable of generating text-to-video, speech-to-image, and image-to-3D conversions. Personalized AI synthesis – AI-generated content tailored to individual user preferences (Vijayakumar et al. 2024). AI-generated facial images often lack diversity across demographics. Solution: Fair AI training methodologies and bias-aware generative models. Future of AI in Art and Media: AI-driven film production and animation (Cardenuto et al, 2023). Personalized music synthesis based on mood and preferences.

Future of AI-Generated Data: Privacy-preserving synthetic datasets for medical and financial AI models. Automated synthetic data generation for AI research. AI-driven dataset augmentation for enhanced model performance. Synthetic data will become a cornerstone of AI development in healthcare, security, and autonomous systems. AI-Driven Personalization and Digital Assistants: Personalized AI Content Generation: AI-generated content is being tailored for individual users. Example: AI-driven news summarization and customized learning experiences. Future of AI Personalization: Hyper-personalized digital experiences powered by synthetic AI. Conversational AI capable of generating human-like interactions (Masood et al, 2023). AI can autonomously generate virtual cities, characters, and narratives (Pastuszak et al, 2022). Example: AI-driven game NPCs that evolve. Future of AI in the Metaverse: AI-driven personalized VR experiences. AI-generated real-time avatars for digital interaction. Autonomous AI-built game levels and immersive experiences. AI-generated virtual worlds will redefine gaming, education, and digital interaction. The future of digital synthetic generation is set to transform AI applications across industries. Key emerging trends include Multimodal AI integrating text, images, audio, and video. AI-generated 3D content revolutionizing gaming, VR, and architecture. Ethical AI models ensuring fairness and reducing bias in synthetic data.

THE ETHICS OF DIGITAL SYNTHETIC GENERATION

Digital synthetic generation, powered by artificial intelligence (AI), deep learning, and big data, has become a transformative force across industries such as healthcare, finance, gaming, and creative media. While synthetic data, AI-generated content, and automated decision-making systems offer immense potential, they also raise serious ethical concerns related to transparency, accountability, and fairness. Ensuring that AI-generated content is accurate, unbiased, and ethically sound is critical for building trust and preventing harm. This discussion explores the ethical challenges of digital synthetic generation and examines the principles and strategies for ensuring responsible AI development (Sindiramutty et al, 2025). Undisclosed AI-Generated Content: Issue: AI-generated images, videos, and text are sometimes presented as real without disclosure (Sugumaran et al. 2024). Example: Deepfake videos can be used for misinformation or fraud. Solution: Implementing watermarking and AI-generated content detection to ensure authenticity. Ethical Use of Synthetic Data in Research: Issue: Researchers using AI-generated data must be transparent about its origin and limitations. Example: AI-generated patient data used for medical research may not accurately represent real-world populations. Solution: Establishing ethical guidelines for synthetic data usage in research and industry applications. Accountability in Digital Synthetic Generation: What Is Accountability in AI? Accountability ensures that AI developers, organizations, and regulators take responsibility for the impacts of AI-generated content. It includes Clear ethical guidelines for AI model development. Legal and regulatory frameworks to prevent AI misuse. Auditing and monitoring AI-generated outputs for fairness and accuracy (Raza et al, 2025). AI-generated news articles spreading false information raise legal and ethical concerns. Solution: Implement AI governance structures to define liability and responsibility.

Ensuring Ethical AI Deployment: Issue: Some companies deploy AI-generated content without sufficient testing or ethical review. By holding developers and organizations accountable, AI can be used responsibly while minimizing harm (Memarian et al, 2023). Fairness in Digital Synthetic Generation: What Is Fairness in AI? Fairness ensures that AI-generated content and synthetic data do not reinforce biases or discrimination. AI models should be Equitable: Avoiding unfair advantages for specific groups. Unbiased: Ensuring training data does not reflect historical inequalities. Inclusive: Representing diverse populations in AI-generated media. Challenges to Fairness: Bias in AI-Generated Content: Issue: AI models trained on biased datasets inherit and amplify those biases. Example: AI-generated job candidate rankings may unfairly favor certain demographics. Solution: Using bias-mitigation techniques and diverse training datasets. Disparities in AI Access: Issue: AI-generated tools may benefit certain groups while excluding others. Example: AI-generated educational materials may not be accessible to non-English speakers or visually impaired users (Israel et al, 2023).

CONCLUSION

Digital synthetic generation, driven by machine learning (ML), deep learning, and big data, has emerged as a powerful tool across various industries, including healthcare, media, autonomous systems, and creative design. AI-driven generative models such as Generative Adversarial Networks (GANs), Variational Autoencoders (VAEs), and Diffusion Models have revolutionized how we create, manipulate, and analyze synthetic data. The integration of learning algorithms and vast datasets has enabled breakthroughs in image synthesis, audio generation, medical imaging, and realistic simulations. However, this new frontier also introduces ethical, technical, and regulatory challenges. This conclusion summarizes the key findings on digital synthetic generation, its impact, challenges, and the future potential of this evolving technology. Key Findings in Digital Synthetic Generation: Machine Learning as the Foundation of Synthetic Generation: Supervised, Unsupervised, and Reinforcement Learning Supervised Learning: Trains models on labeled datasets to improve accuracy in synthetic image classification, speech synthesis, and text generation. Unsupervised Learning: Identifies patterns in unlabeled data, enabling unsupervised image synthesis and anomaly detection. Reinforcement Learning: Optimizes AI-generated content by reward-based learning, applied in gaming AI, robotics, and adaptive synthetic environments. Deep Learning Architectures in Synthetic Generation: Generative Adversarial Networks (GANs): Enable photo-realistic image generation, deepfake synthesis, and virtual world creation. Challenges in Digital Synthetic Generation: Data Bias and Ethical Risks: AI-generated synthetic data may perpetuate biases present in real-world datasets. Addressing fairness in AI models requires bias detection and mitigation strategies. Scalability and Computation Costs: Training large-scale generative models demands high computational power, cloud storage, and energy efficiency improvements. Efficient AI model compression techniques are needed for real-time applications. Security Threats and Misinformation Risks: Deepfake technology can be misused for fraud, propaganda, and identity theft. AI-generated misinformation poses risks to media integrity and public trust. Future Trends and Opportunities: AI-Augmented Creativity and Human-AI Collaboration: AI-driven content co-creation in art, music, and storytelling will redefine creative industries. Generative AI in filmmaking and digital marketing will enhance visual effects, scriptwriting, and media automation. AI in Scientific Discovery and Research: AI-generated molecular structures will accelerate drug discovery. Synthetic medical data will drive advancements in disease modeling, personalized medicine, and AI-assisted healthcare. AI-Powered Virtual Reality and the Metaverse: AI-generated 3D worlds will shape immersive gaming, virtual training environments, and digital twins. AI-driven avatars, chatbots, and interactive experiences will enhance social interactions in the metaverse. However, as this technology advances, ethical considerations surrounding transparency, accountability, bias mitigation, and security must be addressed. The future of digital synthetic generation will depend on responsible AI governance, ensuring AI benefits society while minimizing risks. The digital synthetic generation stands at the frontier of AI-powered innovation, redefining how we create, analyze, and interact with digital content. With continued advancements in machine learning, data science, and ethical AI practices, the future promises unprecedented opportunities for synthetic intelligence in a data-driven world.

REFERENCES

Agchar I, Baumann I, Braun F, Pérez-Toro P A. (2024). "A Survey of Music Generation in the Context of Interaction". arXiv preprint arXiv.

Aggarwal, A., Mittal, M., & Battineni, G. (2021). Generative adversarial network: An overview of theory and applications. *International Journal of Information Management*. Advance online publication. DOI: 10.1016/j.ijinfomgt.2021.102363

Akkem, Y., Biswas, S. K., & Varanasi, A. (2024). A comprehensive review of synthetic data generation in smart farming by using variational autoencoder and generative adversarial network. *Engineering Applications of Artificial Intelligence*, *131*, 107881. Advance online publication. DOI: 10.1016/j.engappai.2024.107881

Alghushairy, O., & Ma, X. (2022). "Data storage". Encyclopedia of Big Data. Doi: DOI: 10.1007/978-3-319-32015-5_100047

Alqahtani, H., Kavakli-Thorne, M., & Kumar, G. (2021). Applications of generative adversarial networks (GANs): An updated review. *Archives of Computational Methods in Engineering*, *28*(2), 525–552. Advance online publication. DOI: 10.1007/s11831-019-09388-y

An, S., & Jeon, J. J. (2023). Distributional learning of variational autoencoder: Application to synthetic data generation. *Advances in Neural Information Processing Systems*.

Andreini, P., Bonechi, S., Ciano, G., & Graziani, C. (2022). *Multi-stage synthetic image generation for the semantic segmentation of medical images*. Machine Learning and Knowledge Extraction., DOI: 10.3390/make4020028

Anjaneyulu, B. P., & Reddy, C. P. K. (2024). "DeepFake Detection using Convolutional Neural Networks (CNN) and Recurrent Neural Network (RNN)". *International Conference on Data Science and Applications*. Doi: DOI: 10.1109/ICDICI62993.2024.10810970

Aragani, V. M., Anumolu, V. R., & Selvakumar, P. (2025). Democratization in the Age of Algorithms: Navigating Opportunities and Challenges. In Wongmahesak, K., Wekke, I., Seftyono, C., & Nurdin, N. (Eds.), *Democracy and Democratization in the Age of AI* (pp. 39–56). IGI Global Scientific Publishing., DOI: 10.4018/979-8-3693-8749-8.ch003

Aranjuelo, N., García, S., Loyo, E., Unzueta, L., & Otaegui, O. (2021). Key strategies for synthetic data generation for training intelligent systems based on people detection from omnidirectional cameras. *Computers & Electrical Engineering*, *92*, 107105. Advance online publication. DOI: 10.1016/j.compeleceng.2021.107105

Bauer A, Trapp S, Stenger M, Leppich R. (2024). "Comprehensive exploration of synthetic data generation A survey". arXiv preprint arXiv.

Cardenuto, J. P., Yang, J., & Padilha, R. (2023). *The age of synthetic realities: Challenges and opportunities*. Foundations and Trends in Signal Processing., DOI: 10.1561/2000000113

Civit, M., Civit-Masot, J., Cuadrado, F., & Escalona, M. J. (2022). A systematic review of artificial intelligence-based music generation: Scope, applications, and future trends. *Expert Systems with Applications*, *209*, 118190. Advance online publication. DOI: 10.1016/j.eswa.2022.118190

Dash, A., Ye, J., & Wang, G. (2023). *A review of generative adversarial networks (GANs) and its applications in a wide variety of disciplines: from medical to remote sensing*. IEEE., DOI: 10.1109/ACCESS.2023.3234567

De Melo, C. M., Torralba, A., Guibas, L., DiCarlo, J., Chellappa, R., & Hodgins, J. (2022). Next-generation deep learning based on simulators and synthetic data. *Trends in Cognitive Sciences*, *26*(2), 174–187. Advance online publication. DOI: 10.1016/j.tics.2021.11.008 PMID: 34955426

Debnath, P., Deshmukh, S. M., Laroia, M., Selvakumar, P., Manjunath, T. C., & Bhattacharya, S. (2025). Challenges Faced by Countries in Renewable Energy Adoption. In Özsungur, F., Chaychi Semsari, M., & Küçük Bayraktar, H. (Eds.), *Geopolitical Landscapes of Renewable Energy and Urban Growth* (pp. 295–320). IGI Global Scientific Publishing., DOI: 10.4018/979-8-3693-8814-3.ch010

Dos Santos, R., & Aguilar, J. (2024). *A synthetic data generation system based on the variational-autoencoder technique and the linked data paradigm*. Progress in Artificial Intelligence., DOI: 10.1007/s13748-024-00328-x

Dou, B., Zhu, Z., Merkurjev, E., Ke, L., & Chen, L. (2023). Machine learning methods for small data challenges in molecular science. *Chemical Science (Cambridge)*. Advance online publication. DOI: 10.1039/D2SC06141A PMID: 37384816

Dwivedi, P., & Islam, B. (2024). Generative Adversarial Networks Based Framework for Music Genre Classification. *SN Computer Science*, *5*(8), 1149. Advance online publication. DOI: 10.1007/s42979-024-03531-8

El Emam, K., Mosquera, L., Fang, X., & El-Hussuna, A. (2022). Utility metrics for evaluating synthetic health data generation methods: Validation study. *JMIR Medical Informatics*, *10*(4), e35734. Advance online publication. DOI: 10.2196/35734 PMID: 35389366

Figueira, A., & Vaz, B. (2022). Survey on synthetic data generation evaluation methods and GANs. *Mathematics*, *10*(15), 2733. Advance online publication. DOI: 10.3390/math10152733

Gholamiankhah F, Mostafapour S, Arabi H. (2021). "Deep learning-based synthetic CT generation from MR images: comparison of generative adversarial and residual neural networks". arXiv preprint arXiv.

Gill, S. S., Xu, M., Ottaviani, C., Patros, P., Bahsoon, R., Shaghaghi, A., Golec, M., Stankovski, V., Wu, H., Abraham, A., Singh, M., Mehta, H., Ghosh, S. K., Baker, T., Parlikad, A. K., Lutfiyya, H., Kanhere, S. S., Sakellariou, R., Dustdar, S., & Uhlig, S. (2022). AI for next generation computing: Emerging trends and future directions. *Internet of Things : Engineering Cyber Physical Human Systems*, *19*, 100514. Advance online publication. DOI: 10.1016/j.iot.2022.100514

Gui, J., Sun, Z., Wen, Y., Tao, D., & Ye, J. (2021). A review on generative adversarial networks: Algorithms, theory, and applications. *IEEE Transactions on Neural Networks and Learning Systems*. Advance online publication. DOI: 10.1109/TNNLS.2021.3088515

Iantovics, L. B., & Enăchescu, C. (2022). Method for data quality assessment of synthetic industrial data. *Sensors (Basel)*, *22*(4), 1608. Advance online publication. DOI: 10.3390/s22041608 PMID: 35214509

Israel, M. J., & Amer, A. (2023). Rethinking data infrastructure and its ethical implications in the face of automated digital content generation. *AI and Ethics*, *3*(2), 427–439. Advance online publication. DOI: 10.1007/s43681-022-00169-1

Johri, S., Selvakumar, P., Satyanarayana, P., Das, A., Gupta, S., & Bhattacharya, S. (2025). Urbanization and Energy: The Nexus of Global Challenges. In Özsungur, F., Chaychi Semsari, M., & Küçük Bayraktar, H. (Eds.), *Geopolitical Landscapes of Renewable Energy and Urban Growth* (pp. 453–478). IGI Global Scientific Publishing., DOI: 10.4018/979-8-3693-8814-3.ch016

Kapoor A, Gulli A, Pal S, Chollet F. (2022). "Deep Learning with TensorFlow and Keras Build and deploy supervised unsupervised deep and reinforcement learning models". Doi: DOI: 10.1007/978-1-4842-8246-5

Khalaf, L. I., Alhamadani, B., Ismael, O. A., & Radhi, A. A. (2024). "Deep learning-based anomaly detection in network traffic for cyber threat identification". *Proceedings of the ACM on Management of Data.* Doi: DOI: 10.1145/3660853.3660932

Liu, M. Y., Huang, X., Yu, J., & Wang, T. C. (2021). "Generative adversarial networks for image and video synthesis: Algorithms and applications". *Proceedings of the IEEE Conference on Computer Vision and Pattern Recognition.* Doi: DOI: 10.1109/JPROC.2021.3049196

Lu Y, Shen M, Wang H, Wang X, van Rechem C. (2023). "Machine learning for synthetic data generation a review". arXiv preprint arXiv.

Mahadev Madgule, N. (2023). Vinayaka, Yeshwant M. Sonkhaskar, Dhiren Ramanbhai Patel, R. Karthikeyan, P. Selvakumar, (2023). Mechanical properties and microstructure of activated TIG welded similar joints of Inconel alloys by desirability approaches. *Materials Today: Proceedings, 77*, 528–533. DOI: 10.1016/j.matpr.2022.12.250

Malik, S., Tyagi, A. K., & Mahajan, S. (2022). *Architecture generative model and deep reinforcement learning for IoT applications Deep learning perspective.* Artificial Intelligence-based Internet of Things., DOI: 10.1007/978-3-030-87059-1_9

Manduchi L, Pandey K, Meister C, Bamler R. (2024). "On the challenges and opportunities in generative AI". arXiv preprint arXiv.

Marwala, T. (2024). *Synthetic Versus Authentic Data.* The Balancing Problem in the Governance of Artificial Intelligence., DOI: 10.1007/978-981-97-9251-1_7

Masood, M., Nawaz, M., Malik, K. M., Javed, A., & Irtaza, A. (2023). Deepfakes generation and detection: State-of-the-art, open challenges, countermeasures, and way forward. *Applied Sciences (Basel, Switzerland).* Advance online publication. DOI: 10.3390/app13126941

Memarian B, Doleck T. (2023). "Fairness, Accountability, Transparency, and Ethics (FATE) in Artificial Intelligence (AI) and higher education: A systematic review". Computers and Education: Artificial Intelligence. Doi: DOI: 10.1016/j.caeai.2023.100141

Morales, E. F., & Escalante, H. J. (2022). *A brief introduction to supervised unsupervised and reinforcement learning.* Computational Learning and Intelligence., DOI: 10.1016/B978-0-12-820125-1.00017-8

Moreno, Y., Montero, A., Hidrobo, F., & Infante, S. (2023). "Synthetic Data Generator for an E-Learning Platform in a Big Data Environment". *International Conference on Technology and Education.* Doi: DOI: 10.1007/978-981-99-5414-8_39

Munappy, A. R., Bosch, J., Olsson, H. H., Arpteg, A., & Brinne, B. (2022). Data management for production quality deep learning models: Challenges and solutions. *Journal of Systems and Software*, *191*, 111359. Advance online publication. DOI: 10.1016/j.jss.2022.111359

Murtaza, H., Ahmed, M., Khan, N. F., Murtaza, G., Zafar, S., & Bano, A. (2023). Synthetic data generation State of the art in health care domain. *Computer Science Review*, *48*, 100546. Advance online publication. DOI: 10.1016/j.cosrev.2023.100546

Musalamadugu, T. S., & Kannan, H. (2023). *Generative AI for medical imaging analysis and applications*. Future Medicine AI. DOI: 10.2217/fmai-2023-0004

Osuala, R., Skorupko, G., Lazrak, N., Garrucho, L., García, E., Joshi, S., Jouide, S., Rutherford, M., Prior, F., Kushibar, K., Díaz, O., & Lekadir, K. (2023). Medigan: A Python library of pretrained generative models for medical image synthesis. *Journal of Medical Imaging (Bellingham, Wash.)*, *10*(6). Advance online publication. DOI: 10.1117/1.JMI.10.6.061403 PMID: 36814939

Pastuszak, J., & Węgierek, P. (2022). Photovoltaic cell generations and current research directions for their development. *Materials (Basel)*, *15*(16), 5542. Advance online publication. DOI: 10.3390/ma15165542 PMID: 36013679

Paulin, G., & Ivasic-Kos, M. (2023). Review and analysis of synthetic dataset generation methods and techniques for application in computer vision. *Artificial Intelligence Review*, *56*(9), 9221–9265. Advance online publication. DOI: 10.1007/s10462-022-10358-3

Ramakrishnan, T., Mohan Gift, M. D., Chitradevi, S., Jegan, R., & Subha Hency Jose, P. Nagaraja, Rajneesh Sharma HN, Selvakumar P, Sintayehu Mekuria Hailegiorgis, (2022). "Study of Numerous Resins Used in Polymer Matrix Composite Materials", Advances in Materials Science and Engineering, vol. 2022, Article ID 1088926, 8 pages, 2022. https://doi.org/DOI: 10.1155/2022/1088926

Rathore, M. M., Shah, S. A., Shukla, D., & Bentafat, E. (2021). The role of AI machine learning and big data in digital twinning A systematic literature review challenges and opportunities. *IEEE Transactions on Industrial Informatics*. Advance online publication. DOI: 10.1109/TII.2021.3084315

Raza S, Qureshi R, Zahid A, Fioresi J, Sadak F. (2025). "Who is responsible? The data, models, users or regulations? Responsible generative AI for a sustainable future". arXiv preprint arXiv.

Saldanha, D. M. F., Dias, C. N., & Guillaumon, S. (2022). Transparency and accountability in digital public services Learning from the Brazilian cases. *Government Information Quarterly*, *39*(2), 101680. Advance online publication. DOI: 10.1016/j.giq.2022.101680

Sardar T H, Pandey B K. (2024). "Big data computing advances in technologies methodologies and applications". Doi: DOI: 10.1007/978-3-031-28647-1

Sarker, I. H. (2021). Deep learning a comprehensive overview on techniques taxonomy applications and research directions. *SN Computer Science*, *2*(6), 420. Advance online publication. DOI: 10.1007/s42979-021-00815-1 PMID: 34426802

Shende, A. D., Tekade, S. A., Deshmukh, A. A., Tembhurkar, S. P., & Selvakumar, P. (2025). Water Quality Monitoring and Control With AI. In Ruiz-Vanoye, J., & Díaz-Parra, O. (Eds.), *Smart Water Technology for Sustainable Management in Modern Cities* (pp. 51–80). IGI Global Scientific Publishing., DOI: 10.4018/979-8-3693-8074-1.ch003

Shi, Z. (2021). "A survey on audio synthesis and audio-visual multimodal processing". arXiv preprint arXiv:2108.00443.

Sindiramutty, S. R., & Prabagaran, K. R. V. (2025). *Ethics and Transparency in Secure Web Model Generation*. Generative AI for Secure Web Applications., DOI: 10.4018/978-1-6684-9573-9.ch005

Sreenivasulu, A., & Manimozhi, T. Satheesh kumar P, Selvakumar P. and Singh P, (2022). "The Model of Smart Sensing Device For Sensitive Nanoclusters Modification in Sensing Properties," 2022 2nd International Conference on Advance Computing and Innovative Technologies in Engineering (ICACITE), Greater Noida, India, 2022, pp. 1043-1047, DOI: 10.1109/ICACITE53722.2022.9823846

Sugumaran K M D, Pandiyarajan V Y B S, Selvakumar P. and M. S, (2024). "Analysis of Electromyography Signals for Control of Mechanical Prosthesis using Machine Learning Techniques," 2024 2nd International Conference on Intelligent Data Communication Technologies and Internet of Things (IDCIoT), Bengaluru, India, 2024, pp. 1059-1067, .DOI: 10.1109/IDCIoT59759.2024.10467707

Thambawita, V., Salehi, P., Sheshkal, S. A., Hicks, S. A., Hammer, H. L., Parasa, S., Lange, T., Halvorsen, P., & Riegler, M. A. (2022). SinGAN-Seg: Synthetic training data generation for medical image segmentation. *PLoS One*, *17*(5), e0267976. Advance online publication. DOI: 10.1371/journal.pone.0267976 PMID: 35500005

Vijayakumar, G. (2024). Muniyandi Rajkumar, Rajiv Chandar N, Selvakumar P and Ramesh Duraisamy, (2024). Environmentally friendly TDS removal from wastewater by electrochemical ion exchange batch-type recirculation (EIR) technique. *Environmental Science. Water Research & Technology*, *10*, 826–835. DOI: 10.1039/D3EW00793F

Williamson, S. M., & Prybutok, V. (2024). Balancing privacy and progress: A review of privacy challenges, systemic oversight, and patient perceptions in AI-driven healthcare. *Applied Sciences (Basel, Switzerland)*, *14*(2), 675. Advance online publication. DOI: 10.3390/app14020675

Xue, B., & Zou, L. (2022). Knowledge graph quality management: A comprehensive survey. *IEEE Transactions on Knowledge and Data Engineering*, 1. Advance online publication. DOI: 10.1109/TKDE.2022.3150080

Yang, X., Fan, X., Wang, J., & Lee, K. (2022). Image translation based synthetic data generation for industrial object detection and pose estimation. *IEEE Robotics and Automation Letters*, *7*(3), 7201–7208. Advance online publication. DOI: 10.1109/LRA.2022.3180403

Zhang, T. (2022). "Data Synthesis". Encyclopedia of Big Data. Doi: DOI: 10.1007/978-3-319-32010-6_503

Chapter 4
Artificial Intelligence Agents Powered by Synthetic Data

Sachin Sharma
https://orcid.org/0000-0002-7821-3685
Amity University, Mohali, India

ABSTRACT

The merging of Artificial Intelligence (AI) agentry and progressed built yields talks to a transformative paradigm in how cleverly systems work and associated with the world. AI masters, characterized by their independence, considering, organizing, and learning capabilities, are dynamically being utilized to perform complex errands with irrelevant human intercession. These specialists as often as possible require endless sums of data for planning, endorsement, and operation. This collaboration addresses some fundamental challenges. Firstly, designed data can overcome the obstacles of real-world data, such as deficiency, inclination, assurance concerns, and the tall fetched of procurement and comment. By producing data that accurately reflects the genuine properties of veritable data, This capability opens unused streets for amusement, testing, and the creation of novel advanced substance.

I. INTRODUCTION

The relentless march of technological advance has introduced in a time where shrewdly frameworks are now not limited to the domain of science fiction. Artificial Intelligence (AI), once a theoretical construct, has penetrated different aspects of our lives, changing businesses, rethinking human-computer interaction, and opening uncommon conceivable outcomes (Pandey et al., 2024). At the bleeding

DOI: 10.4018/979-8-3373-2277-3.ch004

edge of this transformation lies the concept of Artificial Intelligence Agents (AI Agents) – independent substances able of seeing their environment, making decisions, and taking activities to realize particular objectives (Devliyal et al., 2024). These agents, imbued with the control of learning and adjustment, hold the key to opening a future where complex errands are robotized, complicated issues are illuminated with unparalleled productivity, and human potential is increased past current restrictions. In any case, the advancement and sending of strong and solid AI Operators confront a significant hurdle the insatiable request for high-quality, real-world information. Training advanced AI models, particularly those supporting independent operators, regularly requires endless datasets that are frequently costly, time-consuming to procure, privacy-sensitive, or basically inaccessible. This information shortage can extremely confine the capabilities and adaptability of AI Specialists, ruining their broad selection and constraining their potential affect. Enter Engineered Information – a game-changing worldview that guarantees to overcome these data limitations and quicken the advancement of AI Specialists (S. Sharma et al., 2025). Synthetic data, produced artificially through computational models, imitates the factual properties and designs of real-world information without containing any genuine touchy data. This imaginative approach offers a capable elective to conventional information procurement strategies, giving a essentially boundless supply of labeled, assorted, and controllable information that can be custom-made accurately to the preparing needs of AI Specialists. This comprehensive exploration delves into the transformative potential of engineered information within the setting of AI Agents (Bhatt & Sharma, 2025). We will unravel the elemental standards of AI Agents, look at the basic part of information in their development, and light up the bunch focal points that engineered information brings to the table. Moreover, we'll investigate the differing techniques utilized in generating engineered information, exhibit compelling applications of synthetic data-powered AI Agents over different spaces, address the characteristic challenges and limitations of this burgeoning field, and at last, offer a see into the energizing future that is standing by as AI Specialists and engineered information meet to reshape our world. At its core, an AI Agent is a cleverly framework that works independently inside an environment to achieve particular destinations. Drawing motivation from the concept of office in cognitive science, AI Agents are characterized by their capacity to see their environment through sensors, process data, make rational choices based on their information and learned designs, and act upon the environment through effectors.

II. LITERATURE REVIEW

Blythe and Tregubov's (2024) paper investigates the inventive utilize of cognitive agent simulations for generating synthetic data to train machine learning models. The authors propose that simulating the behavior and intuitive of cognitive specialists inside a virtual environment can deliver wealthy and relevantly pertinent synthetic datasets, especially useful when real-world information is scarce or sensitive. They argue that this approach permits for greater control over information characteristics, the incorporation of complex behavioral designs, and the era of edge cases that could be underrepresented in genuine information. The paper likely presents a system or strategy for planning such reenactments and illustrates its adequacy through particular utilize cases, highlighting the potential of cognitive specialist recreations to improve the preparing and vigor of AI models over different spaces. Meiser and Zinnikus (2024) present a survey analyzing the application of synthetic information to support key angles of dependable AI inside the vitality space. Their work highlights the noteworthy openings manufactured information offers in tending to challenges related to information shortage, security concerns, and the require for vigorous and dependable AI models in this basic segment. The study likely audits different synthetic data generation strategies pertinent to vitality applications, such as smart grids, renewable vitality determining, and vitality utilization optimization. Moreover, it dives into the challenges related with guaranteeing the devotion and representativeness of synthetic data, as well as the potential for inclination presentation. Eventually, the paper underscores the significant part of engineered information in cultivating transparency, fairness, and responsibility in AI arrangements over the vitality scene whereas exploring inborn information restrictions. Chakradhar, Vijey, and Tunk's (2024) chapter delves into the application of engineered information era for extricating profitable showcasing bits of knowledge. The authors investigate how artificially created datasets can overcome confinements related with real-world showcasing information, such as protection confinements, information shortage for specialty fragments, and predispositions. They likely talk about different procedures for producing synthetic customer information, counting statistic data, obtaining behavior, and online intuitive, whereas protecting key factual properties. The chapter likely highlights the benefits of utilizing synthetic data for tasks like showcase division, personalized showcasing campaign plan, and predicting consumer patterns, without compromising person protection. Besides, it may address the challenges of guaranteeing the quality and significance of engineered promoting information to derive exact and significant experiences for businesses. Antonucci et al.'s (2024) chapter focuses on the era and application of synthetic data particularly inside the setting of buildings. The creators likely address the challenges of getting comprehensive and differing real-world information related to building characteristics, vitality utilization, inhabitant behavior, and basic execution. They investigate different strategies for making synthetic datasets that can be utilized for preparing AI models in building-related applications, such as vitality effectiveness optimization, predictive maintenance, and smart building administration. The chapter likely examines the significance of capturing the complex interdependencies between diverse building parameters and recreating reasonable natural conditions. Moreover, it may dive into the approval of such synthetic data and its adequacy in improving the execution and reliability of AI frameworks outlined for the built environment, whereas moreover addressing potential protection concerns related with genuine building information. Mohammadabadi et al.'s (2024) paper investigates the application of generative artificial intelligence to upgrade smart network communication through disseminated learning. The authors likely address the challenges of information heterogeneity, protection concerns, and the require for strong communication in decentralized smart grid environments. They propose leveraging generative AI models to make manufactured information that can expand nearby datasets at diverse network hubs, encouraging more successful distributed learning without the have to be share sensitive real-world operational information. The paper likely examines how this engineered information can move forward the preparing of AI models for errands such as request estimating, blame location, and lattice optimization, whereas keeping up information security and improving the by and large versatility and productivity of shrewd network communication systems. The discoveries likely highlight the potential of generative AI in empowering collaborative learning and making strides AI execution in disseminated smart grid systems. Sharma, Khanna, and Gigras's (2025) chapter centers on the role of information analytics and synthetic data generation within the setting of IoT-based smart homes. The authors likely address the challenges of collecting and utilizing real-world information from smart home devices due to security concerns, information sparsity for certain events, and the heterogeneity of gadgets. They investigate how engineered data era methods can make practical and assorted datasets mirroring smart home sensor readings, client interactions, and vitality utilization designs. The chapter likely highlights the application of this engineered information for preparing AI models that empower progressed keen domestic functionalities, such as personalized automation, predictive support of appliances, and upgraded security. Besides, it may examine the strategies for guaranteeing the fidelity and utility of the produced synthetic data for successful information analytics and AI show preparing within the smart domestic space, whereas regarding client protection. Fragkouli et al.'s (2024) article investigates the potential applications of engineered information inside infectious disease inquire about. The authors highlight how the era of artificial datasets can address basic challenges such as limited persistent information, security concerns encompassing touchy health information, and the ought to show complex infection elements and episodes. They likely talk about how synthetic data, mimicking understanding socioeconomics, disease progression, treatment reactions, and indeed genomic data, can be utilized to prepare AI models for assignments like early disease discovery, expectation of epidemic spread, and the advancement of focused on intercessions. The paper likely emphasizes the significance of guaranteeing the statistical constancy and biological plausibility of the engineered information to abdicate important experiences and quicken headways in understanding and combating infectious illnesses, whereas too tending to ethical considerations and approval methodologies. Moreira, Marcel, and Rocha's (2024) article discusses the burgeoning intersection of engineered substances and AI-generated substance, with a specific center on security and protection suggestions. The authors likely investigate the advancements in AI's capacity to form increasingly reasonable engineered media, including images, recordings, and sound, and the challenges this postures for recognizing between true and artificial content. They likely dive into the potential for abuse, such as the creation of deepfakes for malicious purposes, and the coming about dangers to information integrity and person protection. The article likely emphasizes the urgent require for developing strong detection components and moral rules to explore this advancing scene, ensuring the capable creation and utilization of AI-generated synthetic realities whereas moderating potential security and security risks within the advanced age. Long et al.'s (2025) paper gives a comprehensive diagram of leveraging manufactured information to address machine learning challenges inside supply chains. The authors likely detail the different obstacles experienced in applying AI to supply chain administration, such as information shortage, information quality issues, and the complexity of supply arrange elements. They investigate distinctive strategies for creating synthetic supply chain information, counting simulating request variances, disturbances, and logistical processes. The paper likely highlights the applications of synthetic data in ranges like request estimating, risk administration, optimization of coordinations, and prescient support of equipment inside the supply chain. Besides, it distinguishes key challenges in guaranteeing the authenticity and significance of manufactured information and traces promising future inquire about openings in this pivotal space for upgrading supply chain flexibility and efficiency. Kyem et al.'s (2024) paper introduces a weather-adaptive engineered information era approach utilizing StarGAN to upgrade control line review. The creators address the challenges of constrained and imbalanced real-world information for preparing AI models to viably assess control lines beneath differing climate conditions. Their strategy leverages StarGAN to produce synthetic images of control lines beneath different climate scenarios (e.g., rain, fog, snow) by exchanging the fashion of real weather images onto existing control line symbolism. This increase with weather-adaptive engineered information points to move forward the vigor and exactness of AI models utilized for computerized control line review, empowering superior discovery of deficiencies and guaranteeing framework reliability over distinctive natural conditions. The paper likely illustrates the adequacy of this approach in moving forward the execution of review calculations compared to preparing on genuine information alone. De and Mitra's (2024) paper presents a novel strategy for generating synthetic data particularly for Asset Administration Shells (AAS) in Industry 4.0 scenarios. The authors address the challenges of information interoperability and the require for standardized, however adaptable, information models for speaking to mechanical resources. Their proposed procedure likely focuses on making engineered AAS information occurrences that follow to the Industry 4.0 guidelines whereas permitting for varieties and complexities experienced in real-world industrial situations. This engineered information can be significant for preparing AI models outlined to connected with and oversee industrial resources through their digital twins (AAS), encouraging assignments like prescient support, prepare optimization, and resource lifecycle administration. The paper likely highlights the benefits of their approach in terms of information standardization, adaptability, and the capacity to produce diverse and representative synthetic AAS information for progressing AI applications in keen manufacturing. Kshetri's (2024) article investigates the burgeoning part of generative artificial intelligence inside the budgetary administrations industry. The author likely examines different applications of generative AI, counting the creation of synthetic financial information for preparing extortion location models and risk appraisal devices, the generation of personalized financial advice and content for client engagement, and the automation of report era and compliance forms. The article likely talks about the potential benefits of generative AI in terms of efficiency picks up, upgraded client encounters, and improved risk management inside the financial sector. In any case, it also likely addresses the related challenges and dangers, such as information protection concerns, the potential for inclination in produced substance, and the require for strong administrative systems to guarantee the mindful and ethical deployment of these capable innovations within the financial landscape.

III. SYNTHETIC DATA

Synthetic data, in essence, is artificially generated information that imitates the measurable properties and designs of real-world information. It is created utilizing computational models, reenactments, or procedural era strategies, instead of being collected from real-world perceptions or intuitive. This crucial distinction offers a plenty of preferences for preparing AI Agents [Fig. 1]:

Figure 1. Synthetic data

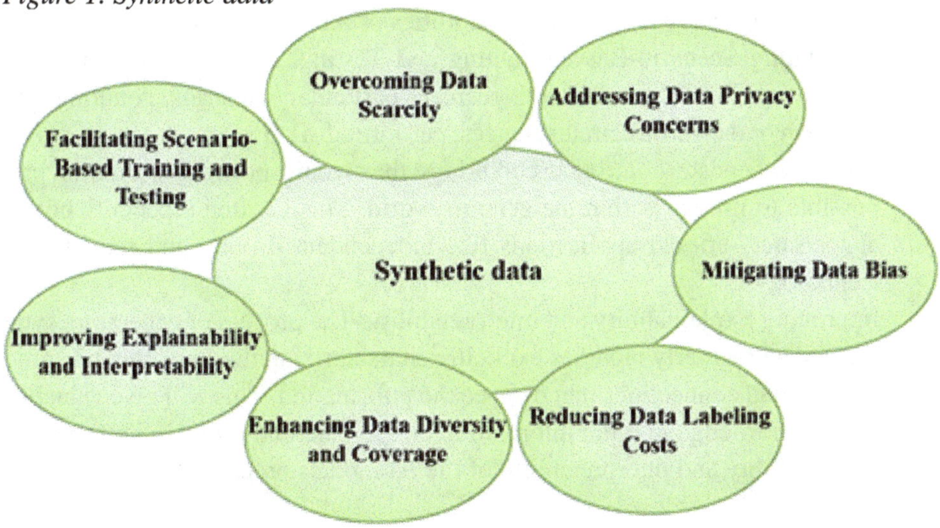

- Overcoming Data Scarcity: Synthetic data can be created in for all intents and purposes boundless amounts, giving a wealthy source of preparing information indeed when real-world information is rare or non-existent. This is often especially profitable for creating AI Operators for novel applications or for dealing with uncommon occasions.
- Addressing Data Privacy Concerns: Since engineered information does not contain any genuine individual data, it can be unreservedly shared and utilized for preparing AI models without raising protection concerns or abusing information security controls. This democratizes get to preparing information and encourages collaboration.
- Mitigating Data Bias: Synthetic data generation permits for exact control over the characteristics and dissemination of the information. This empowers the creation of adjusted datasets that moderate predispositions display in real-world information, driving to more pleasant and more impartial AI Operators.
- Reducing Data Labeling Costs: Engineered information can be created with culminate and steady names, disposing of the require for expensive and time-consuming manual comment. This altogether quickens the information planning prepare and diminishes the in general cost of preparing AI Specialists.
- Enhancing Data Diversity and Coverage: Synthetic data era strategies can be outlined to cover a wide extend of scenarios, varieties, and edge cases that

could be underrepresented or absent in real-world datasets. This progresses the strength and generalization capabilities of AI Specialists.

- Facilitating Scenario-Based Training and Testing: Synthetic environments can be meticulously outlined to recreate particular scenarios, counting uncommon or dangerous circumstances, permitting AI Operators to be prepared and tried beneath controlled conditions that would be troublesome or impossible to imitate within the genuine world. This is often especially pivotal for safety-critical applications like independent driving and mechanical technology.
- Improving Explainability and Interpretability: The process of generating synthetic data regularly includes explicitly characterizing the fundamental components and connections that oversee the information. This will give valuable bits of knowledge into the information era prepare and possibly improve the explainability and interpretability of the AI models prepared on it.

The era of synthetic data for preparing AI Agents includes a assorted extend of techniques, each with its possess qualities and reasonableness for distinctive sorts of information and applications:

- Simulation-Based Synthesis: This approach includes creating virtual situations that imitate the genuine world and producing information by reenacting the intelligent of AI Agents inside these situations. This is often broadly utilized in mechanical technology, independent driving, and gaming, where reasonable material science motors and sensor models can produce wealthy and changed information.
- Generative Adversarial Networks (GANs): GANs are a capable lesson of deep learning models that comprise of two competing neural systems: a generator and a discriminator. The generator learns to make manufactured data samples that are unclear from genuine information, whereas the discriminator tries to distinguish between genuine and manufactured information. This ill-disposed handle leads to the era of exceedingly reasonable engineered information.
- Variational Autoencoders (VAEs): VAEs are another sort of generative demonstrate that learns a probabilistic representation of the genuine information dissemination. By examining from this inactive space and translating back into the information space, VAEs can produce novel synthetic information tests that capture the fundamental structure of the genuine information.
- Statistical Modeling Techniques: Conventional factual strategies can moreover be utilized to produce engineered information by modeling the factual properties of real-world datasets and inspecting from the learned dissemina-

tions. Strategies like bootstrapping, Bayesian systems, and copula capacities can be utilized for this reason.

- Rule-Based and Procedural Generation: In some cases, synthetic data can be created based on predefined rules or procedural calculations. This approach is especially valuable for making organized information or for recreating particular designs and connections.
- Privacy-Preserving Synthesis: Techniques like differential security can be applied during the engineered information era prepare to guarantee that the coming about information does not uncover any delicate data from the initial genuine data. This permits for the creation of privacy-preserving engineered datasets that can be securely shared and utilized.
- Hybrid Approaches: Regularly, the foremost compelling approach includes combining numerous synthetic data era procedures to use their particular qualities and make more comprehensive and practical datasets. For illustration, recreation information can be increased with GAN-generated manufactured pictures to progress visual authenticity.

IV. ARTIFICIAL INTELLIGENCE AGENTS

The narrative of Artificial Intelligence Agents could be a wealthy embroidered artwork woven from strings of philosophical request, scientific formalism, and persistent designing endeavor. Its beginnings can be followed back to the mid-20th century, a period of burgeoning optimism encompassing the potential of computers to imitate human insights. The exceptionally idea of an independent substance competent of seeing, thinking, and acting inside its environment was profoundly interlaced with the early yearnings of AI as an entire. Whereas the term "AI Agent" as a distinct field set afterward, the foundational concepts and calculations that support it developed from these beginning investigations into typical thinking, problem-solving, and learning. The Dartmouth Workshop in 1956, broadly considered the birth of AI as a formal teach, laid the foundation for numerous of the mental interests that would afterward contribute to the improvement of AI Specialists. Early inquire about centered intensely on typical AI, pointing to represent knowledge explicitly and utilize consistent induction to illuminate issues. Algorithms just like the Logic Theorist (Newell and Simon, 1956), which seem demonstrate scientific hypotheses, and the General Issue Solver (GPS) (Newell, Shaw, and Simon, 1959), outlined as a all inclusive problem-solving engine, encapsulated this approach. These early frameworks, though constrained in their scope and flexibility, spoken to introductory steps towards making entities that could reason and act intellectuals. They utilized calculations established in look and typical control, such as depth-first look (DFS) and breadth-first search (BFS), to explore issue spaces and discover arrangements. The 1960s seen further progressions in symbolic AI and the development of frameworks that shown simple agent-like behavior in particular spaces. Terry Winograd's SHRDLU (1970), for occasion, was a characteristic dialect understanding program that seem connected with a simulated block world, reacting to commands and replying questions. Whereas not a fully independent specialist within the cutting edge sense, SHRDLU illustrated the potential for frameworks to see a simulated environment, process language, and take activities inside that environment based on logical thinking. The basic algorithms included advanced parsing strategies and information representation plans based on semantic systems. The late 1970s and 1980s saw the rise of expert frameworks, planned to emulate the decision-making capacities of human specialists in limit spaces. Systems like MYCIN (Shortliffe, 1976) for medical diagnosis and DENDRAL (Feigenbaum et al., 1965) for inducing molecular structure from mass spectrometry data joined rule-based thinking frameworks. These frameworks, whereas lacking true autonomy and learning capabilities, may be seen as early shapes of goal-oriented agents, applying a set of predefined rules to realize particular symptomatic or expository tasks. The core algorithms revolved around forward chaining and in reverse chaining induction instruments. A significant paradigm shift occurred within the late 1980s and 1990s with the resurgence of connectionism and the rise of machine learning. Neural systems, motivated by the structure of the human brain, advertised a distinctive approach to building cleverly frameworks, centering on learning designs from information instead of depending exclusively on unequivocal typical representations. Algorithms like backpropagation (Rumelhart, Hinton, and Williams, 1986) enabled the preparing of multi-layered neural systems, driving to critical breakthroughs in zones like design acknowledgment and expectation. This move laid the establishment for the improvement of more versatile and learning-oriented AI Agents. The concept of intelligent specialists as a unmistakable field started to cement in the late 1980s and early 1990s, driven by advancements in areas like conveyed AI, robotics, and program designing. Researchers begun to center on characterizing the key characteristics of specialists – independence, reactivity, pro-activeness, and social capacity – and developing models that may exemplify these properties. Designs just like the Subsumption Architecture (Brooks, 1986), which emphasized a bottom-up, layered approach to control in mechanical technology, and the Belief-Desire-Intention (BDI) model (Bratman, Israel, and Pollack, 1988), which given a system for thinking almost objectives and plans, risen amid this period. Calculations for way arranging (e.g., A* search), localization (e.g., Kalman filters), and control (e.g., PID controllers) became crucial for empowering operators to function viably in physical situations. The advent of the internet and the explosion of information within the late 1990s and early 2000s assist fueled the improvement of AI Agents. The accessibility of tremendous datasets empowered the preparing of more modern machine learning models, driving to critical advance in regions like natural language processing, computer vision, and data retrieval. Algorithms like support vector machines (SVMs) (Cortes and Vapnik, 1995) and various clustering algorithms (e.g., k-means) became important tools for building specialists that seem process and get it complex information. Reinforcement learning (RL) (Sutton and Barto, 1998), a paradigm where specialists learn ideal behavior through trial and error by getting rewards or punishments for their activities, also gained prominence as a powerful approach for preparing independent specialists in energetic situations. Algorithms like Q-learning and SARSA empowered operators to learn ideal approaches through interaction with their environment. The deep learning revolution of the 2010s checked another significant turning point within the history of AI Agents. The accessibility of massive datasets and capable computing assets empowered the preparing of profound neural systems with uncommon complexity, driving to breakthroughs in ranges like image recognition (e.g., AlexNet), natural language understanding (e.g., Transformer models), and diversion playing (e.g., AlphaGo). Deep reinforcement learning, combining the control of deep neural systems with reinforcement learning calculations (e.g., Deep Q-Networks), empowered the advancement of agents capable of acing complex assignments in simulated and indeed real-world situations. Algorithms like convolutional neural networks (CNNs) for handling visual information and recurrent neural networks (RNNs), including long short-term memory (LSTM) and gated recurrent units (GRUs), for preparing consecutive information became crucial building blocks for modern AI Agents. Nowadays, AI Agents are progressively modern and are being deployed in a wide run of applications, from virtual assistants and recommender frameworks to independent vehicles and mechanical frameworks. The field proceeds to advance quickly, with continuous investigate centering on regions like multi-agent frameworks, reasonable AI, ethical considerations, and the advancement of more common and versatile cleverly specialists. Algorithms like unified learning for preparing agents on decentralized information, consideration components for centering on pertinent data, and chart neural systems for preparing organized information are at the forefront of current research. The history of AI Agents is a confirmation to the persevering human interest with making intelligent substances and the tireless interest of calculations and structures that can bring this vision to reality. As we move forward, the combination of progressed machine learning procedures, advanced specialist models, and a deeper understanding of insights promises to open even more capable and transformative AI Agents within the years to come (S. Sharma, Gupta, et al., 2024). The journey from typical thinking to deep learning has been long and winding, but the goal – a world enabled by truly cleverly independent frameworks – remains a compelling and progressively achievable objective. Artificial Intelligence calculations that use the concept of "agents" speak to a paradigm move in how we plan and execute shrewdly frameworks. Rather than solid programs, the agent-based approach structures AI as a collection of independent substances – the operators – each with its own goals, capabilities, and discernment of the environment. These operators associated with their environment and with each other to collectively accomplish complex assignments that would be troublesome or outlandish for a single, centralized calculation. This modularity, flexibility, and ability to handle distributed information make agent-based AI algorithms especially well-suited for handling real-world issues characterized by complexity, instability, and energetic situations. At the heart of agent-based AI lies the basic concept of a cleverly specialist. As discussed previously, an agent is an independent substance that sees its environment through sensors and acts upon that environment through effectors to realize its design objectives. The "intelligence" of the specialist comes from the calculations it utilizes to handle data, reason around its state and the world, make choices, and learn from its encounters. When different such agents are outlined to participate or compete to solve a problem, we enter the realm of multi-agent systems (MAS). Several key AI algorithms and approaches are commonly utilized inside the agent-based framework:

- Classical Planning Algorithms: For deterministic and completely observable situations, calculations like STRIPS (Stanford Research Institute Problem Solver) and ADL (Action Description Language) are utilized to speak to states, activities, and objectives. Search algorithms such as A search* and heuristic look are at that point utilized to discover ideal or near-optimal plans. In multi-agent settings, arranging can ended up more complex, requiring calculations that consider the activities and objectives of other operators, such as partial-order arranging and hierarchical task network (HTN) arranging.
- Reinforcement Learning for Planning: In energetic and questionable situations, specialists can learn successful plans through fortification learning. Calculations like Deep Q-Networks (DQNs) and Policy Gradient methods (e.g., Reinforce, A2C, A3C) permit operators to learn optimal policies – mappings from states to activities – by collaboration with the environment and accepting rewards or punishments. In multi-agent RL, calculations have to be address the challenges of non-stationarity (where the environment changes from the point of view of one operator due to the learning of other operators) utilizing strategies like joint activity learning and adversary modelling (Sharma et al., 2025b).
- Supervised Learning: Agents can learn from labeled information to perform errands like classification, regression, and prediction. Algorithms like support vector machines (SVMs), decision trees, random forests, and various neural network architectures (e.g., multilayer perceptrons, convolutional neural net-

works, recurrent neural networks) can be utilized for this reason. In an agent context, supervised learning may be utilized to prepare a recognition module to recognize objects or a decision-making module to classify circumstances.

- Unsupervised Learning: Agents can find hidden patterns and structures in unlabeled data utilizing algorithms like clustering (e.g., k-means, DBSCAN), dimensionality reduction (e.g., PCA, t-SNE), and affiliation run the show mining. This may be valuable for operators to get it their environment superior or to recognize important features from tangible inputs.
- Reinforcement Learning: RL is a core learning paradigm for independent specialists, permitting them to learn ideal behaviors through trial and error. Single-agent RL calculations are broadly utilized for training agents in different spaces, from robotics control to amusement playing. Multi-agent RL amplifies these calculations to scenarios with multiple interacting specialists, requiring considerations of coordination, participation, and competition.
- Communication Protocols: Operators got to follow to particular communication conventions to trade data successfully. These conventions can run from basic message passing to more modern dialects and ontologies. Algorithms for natural language processing (NLP) can be utilized to empower operators to get it and produce human-like communication.
- Coordination Mechanisms: Different mechanisms encourage coordination among specialists, counting contract net conventions for errand assignment, voting components for collective decision-making, and arrangement conventions for settling clashes and coming to understandings. Game theory gives a hypothetical system for analyzing key intuitive among judicious specialists, and calculations based on game-theoretic concepts (e.g., Nash harmony finding calculations) can be utilized for planning operator techniques in competitive or agreeable settings.
- Swarm Intelligence Algorithms: Inspired by the collective behavior of social creepy crawlies like ants and bees, swarm insights calculations like Ant Colony Optimization (ACO) and Particle Swarm Optimization (PSO) empower a populace of simple operators to collectively illuminate complex optimization issues through decentralized interaction and self-organization.
- Computer Vision Algorithms: For operators that work within the physical world or associated with visual interfacing, computer vision calculations like object detection (e.g., YOLO, Faster R-CNN), image division, and include extraction (e.g., Sift, SURF) are significant for translating visual data.
- Natural Language Processing (NLP) Algorithms: For operators that connected with people or handle textual data, NLP calculations for errands like opinion examination, theme modeling, named substance acknowledgment, and machine interpretation are fundamental.

- Sensor Fusion Algorithms: Agents often get data from different sensors. Sensor combination calculations like Kalman sifting and Bayesian deduction are utilized to combine these boisterous and dubious inputs to get a more precise and vigorous understanding of the environment.
- Logical Reasoning: Agents can utilize formal logic (e.g., propositional logic, first-order logic) and induction rules to infer modern information and make consistent choices. Algorithms for theorem demonstrating and show checking can be utilized.
- Probabilistic Reasoning: In uncertain environments, specialists frequently got to reason probabilistically utilizing methods like Bayesian networks, Markov decision processes (MDPs), and partially observable Markov decision processes (POMDPs). Algorithms for conviction overhauling and esteem cycle are utilized in these systems.
- Case-Based Reasoning: Agents can solve unused issues by recovering and adapting solutions from comparative past experiences put away in a case base.

The choice of particular calculations inside an agent-based AI framework depends intensely on the application space, the complexity of the assignment, the nature of the environment (deterministic vs. stochastic, fully vs. partially observable), and the number of agents involved (S. Sharma et al., 2024). The power of the agent-based approach lies in its capacity to break down complex issues into smaller, more sensible sub-problems that can be handled by person specialists, cultivating seclusion, adaptability, and strength. As AI continues to development, the improvement and application of sophisticated algorithms inside operator models will without a doubt play a pivotal part in making progressively cleverly and independent systems able of addressing real-world challenges. The choice of synthetic data era technique depends on different variables, including the sort of information required, the complexity of the real-world conveyance, the required level of authenticity, and the computational assets accessible. The collaboration between manufactured information and AI Specialists is opening transformative applications over a wide run of businesses:

- Autonomous Driving: Synthetic data plays a significant part in preparing and approving self-driving car calculations. Realistic virtual situations can mimic different driving scenarios, counting dangerous conditions and uncommon occasions, permitting independent vehicles to be thoroughly tested and improved without the dangers and costs related with real-world testing. Manufactured information can too address the challenges of information shortage for particular edge cases and progress the vigor of recognition frameworks to different natural conditions.

- Robotics: Preparing robots for complex errands within the genuine world can be challenging and time-consuming. Synthetic data created from practical robotic simulations permits AI Agents controlling robots to learn control abilities, route procedures, and interaction conventions in a secure and controlled environment. This quickens the improvement and sending of robots for fabricating, coordinations, healthcare, and other applications.
- Healthcare: Synthetic medical information, such as simulated understanding records, medical images, and genomic arrangements, can be utilized to prepare AI Agents for assignments like disease conclusion, sedate revelation, and personalized treatment arranging without compromising patient security. Synthetic data can moreover offer assistance address the issue of information lopsidedness for rare infections and progress the execution of AI models for underrepresented quiet populaces (Rawat et al., 2025b).
- Finance: Synthetic financial data can be utilized to prepare AI Specialists for extortion location, chance appraisal, algorithmic exchanging, and client benefit. By generating reasonable but anonymized budgetary exchanges and showcase information, financial institutions can create test AI models without exposing delicate client data or affecting real-world markets.
- Cybersecurity: Synthetic network traffic data and simulated cyberattacks can be utilized to prepare AI Agents for interruption discovery, risk examination, and security vulnerability assessment. This permits cybersecurity experts to create more strong and proactive defense components against advancing cyber dangers in a secure and controlled environment.
- Retail and E-commerce: Synthetic customer data, including browsing history, purchase patterns, and demographic data, can be utilized to prepare AI Operators for personalized proposals, focused on publicizing, and inventory administration. Synthetic data can moreover offer assistance mimic diverse showcase conditions and client behaviors to optimize estimating and promoting methodologies.
- Education and Training: Synthetic data can be utilized to form practical and intuitively preparing reenactments for different callings, such as pilots, specialists, and crisis responders. AI Specialists inside these recreations can give personalized criticism and adjust the preparing scenarios based on the learner's advance.
- Gaming and Entertainment: Synthetic data is crucial to the creation of practical and locks in virtual universes and intelligent non-player characters (NPCs) in video recreations. AI Agents fueled by synthetic data can display more convincing behaviors, respond powerfully to player activities, and upgrade the generally gaming encounter.

These illustrations highlight the wide applicability of manufactured information in engaging AI Agents over differing spaces, driving advancement and making unused possibilities. In spite of its colossal potential, the field of engineered information for AI Specialists isn't without its challenges and impediments:

- Domain Gap: A key challenge is guaranteeing that the engineered information precisely reflects the complexities and subtleties of the real-world space. Discrepancies between the engineered and genuine information conveyances, frequently alluded to as the "domain gap," can lead to AI Agents that perform well in simulation but fail to generalize successfully to real-world scenarios. Bridging this hole requires modern modeling procedures and careful validation.
- Fidelity and Realism: Generating synthetic data that's adequately reasonable to prepare high-performing AI Agents can be computationally requesting and require significant expertise in both AI and the target space. Accomplishing a high level of devotion over different information modalities (e.g., visual, textual, temporal) remains an ongoing research challenge.
- Assessment and Approval: Assessing the quality and viability of synthetic data for preparing AI Specialists is pivotal but can be complex. Conventional measurements utilized for genuine information may not be directly appropriate to synthetic data, and modern strategies are required to survey its utility and affect on the execution of the prepared specialists.
- Bias Amplification: Whereas synthetic data offers the potential to moderate bias, ineffectively outlined era forms can incidentally present or indeed amplify biases show within the basic genuine information or the engineered information era demonstrate itself. Cautious consideration to decency and representation is basic.
- Computational Cost: Generating expansive and complex engineered datasets can be computationally costly, requiring critical handling control and capacity assets. Optimizing the productivity of engineered information era procedures is an imperative zone of inquire about.
- Standardization and Best Practices: The field of engineered information is still generally nascent, and there's a need of standardized techniques, assessment measurements, and best practices. Setting up these guidelines will be significant for cultivating believe, reproducibility, and more extensive appropriation.
- Trust and Acceptance: Building believe in AI Specialists prepared on synthetic data is basic for their widespread acknowledgment. Illustrating the reliability and robustness of these agents through thorough testing and approval is significant.

Addressing these challenges through continuous research, improvement of unused methods, and the foundation of best hones will be critical for realizing the complete potential of engineered information in enabling the next era of shrewdly independent specialists. The future of AI Agents is inseparably connected to the progressions in synthetic information era. As synthetic data becomes more practical, different, and promptly accessible, we will anticipate to see a significant acceleration within the development and arrangement of modern AI Agents over different domains.

- Enhanced Agent Capabilities: Synthetic data will empower the preparing of AI Agents with more vigorous recognition, thinking, and activity capabilities. Specialists will be better prepared to handle complex real-world scenarios, adjust to changing situations, and accomplish more yearning objectives.
- Democratization of AI Development: The capacity to generate high-quality preparing information on request will lower the boundary to section for AI advancement, permitting littler organizations and person analysts to construct and send effective AI Operators.
- Faster Innovation Cycles: The ease and speed of producing synthetic data will accelerate the iterative handle of AI Agent improvement, permitting for fast prototyping, experimentation, and refinement of operator plans and calculations.
- Addressing Data Deserts: Synthetic data will be especially transformative in spaces where real-world information is scarce, sensitive, or troublesome to get, opening the potential for AI-driven arrangements in zones like uncommon disease inquire about, disaster response, and space investigation.
- Personalized and Adaptive Agents: Synthetic data can be custom-made to particular people or settings, empowering the advancement of highly personalized and versatile AI Agents for applications like instruction, healthcare, and individual help.
- Seamless Human-Agent Collaboration: As AI Agents ended up more modern and reliable through preparing on different engineered information, we are able anticipate to see more consistent and natural collaboration between people and AI Agents in different angles of our lives.

The engineering of a typical AI Agent can be broadly categorized into a few key components:

- Perception: Agents possess instruments to sense and translate their environment. This will include preparing different shapes of input information, such as visual data from cameras, sound-related signals from microphones, textual information from archives, or sensor readings from physical devices (Dangi et al., 2025).

- Cognition: This is the "brain" of the specialist, mindful for preparing the seen data, thinking around the current state of the environment, arranging future activities, and making choices. Cognitive capabilities can run from basic rule-based frameworks to complex machine learning models, including deep neural systems.
- Action: Agents interact with their environment through effectors, which can be physical actuators (e.g., robotic arms, wheels) or virtual interfacing (e.g., software commands, API calls). The activities taken are based on the agent's internal reasoning and its goal-oriented behavior.
- Learning: A crucial aspect of intelligent agents is their capacity to memorize from encounter and adjust their behavior over time. This could include upgrading their inner information, refining their decision-making forms, and progressing their execution based on feedback from the environment.

AI Agents can be broadly classified based on their complexity, capabilities, and the nature of their environment:

- Simple Reflex Agents: These agents respond specifically to their current recognitions based on predefined rules. They need memory of past states and cannot reason almost future results.
- Model-Based Reflex Agents: These agents keep up an inner show of the environment, permitting them to reason approximately how their activities will influence the world. They can make more educated choices than straightforward reflex specialists.
- Goal-Based Agents: These agents have explicit objectives and endeavor to realize them by planning arrangements of activities. They can consider numerous conceivable ways and select the one that's most likely to lead to their craved result.
- Utility-Based Agents: These agents go past essentially accomplishing objectives; they moreover consider the "utility" or attractive quality of distinctive states and actions. They aim to maximize their overall happiness or performance based on a predefined utility work.
- Learning Agents: These agents can learn from their encounters and make strides their execution over time. They regularly consolidate a learning component that permits them to adjust their inside information and decision-making processes based on input from the environment.

The improvement of modern AI Agents capable of handling complex real-world issues hinges fundamentally on the accessibility of endless sums of high-quality information for preparing their fundamental machine learning models. This information empowers operators to learn intricate designs, generalize to unseen circumstances, and make exact and dependable choices. Information serves as the backbone of advanced AI, and its importance is especially articulated within the advancement of AI Agents. Machine learning algorithms, the center of most progressed AI Agents, learn by identifying patterns and connections inside huge datasets. The quality, amount, and differing qualities of this preparing information directly affect the execution, robustness, and generalization capabilities of the coming about AI Agent.

- Learning Complex Mappings: AI Agents often need to learn complex mappings between sensory inputs and suitable activities. For occurrence, a self-driving car must learn the intricate relationship between camera pictures, lidar information, and control commands for directing, speeding up, and braking. This requires enormous datasets of labeled driving scenarios.
- Generalization to Unseen Situations: A well-trained AI Agent ought to be able to perform viably not as it were on the information it was prepared on but moreover on novel, inconspicuous circumstances. This requires preparing information that's representative of the real-world environment the operator will work in and includes a wide run of varieties and edge cases.
- Robustness to Noise and Uncertainty: Real-world data is regularly noisy, fragmented, and subject to different shapes of instability. Preparing AI Agents on differing and reasonable information makes a difference them ended up more strong to these imperfections and make reliable decisions indeed in challenging conditions.
- Learning from Interactions: For many AI Agents, especially those working in dynamic situations, learning through interaction is significant. This includes collecting information from the agent's possess encounters because it interatomic with the world, permitting it to adjust and improve its behavior over time.

However, acquiring the necessary information for preparing high-performing AI Agents regularly presents significant challenges:

- Data Scarcity: In numerous spaces, particularly those including novel applications or uncommon occasions, adequate real-world information may basically not exist. Collecting sufficient information to prepare a vigorous AI Operator can be a time-consuming and costly endeavor.

- Data Privacy and Security: Real-world information regularly contains delicate individual data, raising critical protection concerns. Sharing and utilizing such information for preparing AI models can be legitimately limited and ethically risky.
- Data Bias: Real-world information can reflect existing societal inclinations, driving to AI Agents that sustain or indeed increase these inclinations. Guaranteeing decency and value requires cautious curation and increase of preparing information.
- Data Labeling Costs: Supervised learning, a common worldview for preparing AI Specialists, requires expansive sums of labeled information, where each information point is related with the proper yield or activity. Getting exact and reliable names can be a exorbitant and labor-intensive prepare.
- Data Diversity and Coverage: Real-world datasets may not enough cover the complete range of scenarios and varieties that an AI Agent might experience in its operational environment. This lack of differing qualities can lead to destitute generalization and unexpected failures.

V. COMPONENTS OF ARTIFICIAL INTELLIGENCE AGENTS

- Perception System: Usually how the specialist takes in data from its reenacted environment. It forms engineered "sensory" inputs. Think of it as the agent's "eyes" and "ears" in the virtual world.
- Synthetic Sensors: Mimicking real-world sensors like cameras (processing synthetic images/videos), lidar (processing synthetic point clouds), or indeed more unique information streams depending on the recreated environment.
- Data Preprocessing Modules: Cleaning, changing, and organizing the crude engineered information into a arrange the agent's learning calculations can get it.
- Cognitive Architecture: This is the "brain" of the specialist, mindful for decision-making, arranging, and thinking. It houses the AI models prepared on the engineered information.
- Learning Algorithms: These are the center AI models (e.g., neural networks, reinforcement learning algorithms, Bayesian networks) that learn designs and relationships from the engineered information.
- Knowledge Representation: How the operator stores and utilizes the "knowledge" it gains from the engineered information. This may include storing learned parameters in neural systems, making typical representations, or building inside maps of the reenacted environment.

- Reasoning and Planning Modules: Depending on the agent's complexity, it might have modules for logical deduction, goal-oriented arranging, and grouping era.
- Action/Actuation System: This can be how the agent interatomic with its recreated environment. It translates the agent's choices into activities inside the virtual world.
- Synthetic Actuators: Mimicking real-world actuators like automated joints (controlling development in a recreated robot), software commands (interacting with a reenacted program interface), or indeed unique activities inside the virtual environment.
- Activity Execution Modules: Translating the high-level plans or decisions into low-level commands that the synthetic actuators can get it and execute.
- Environment Simulator: This can be the virtual world where the specialist learns and interacts. It ought to be outlined to precisely speak to the key perspectives of the real-world environment the operator will inevitably work in. This incorporates:
- Physics Engine: Reenacting the laws of physics (gravity, collisions, etc.) on the off chance that the errand includes physical interaction.
- Object Models: Realistic representations of objects the specialist will experience, counting their geometry, appearance, and behavior.
- Agent Model (Self): A representation of the AI operator itself inside the reenactment, counting its sensors and actuators.
- Scenario Generation: The ability to create differing and challenging scenarios that the specialist needs to learn from. This might involve randomizing parameters, introducing variations, and creating edge cases.
- Data Generation Pipeline: This component coordinates the method of making the engineered information. It characterizes how the reenacted environment is inspected and how the "sensory" information is captured (Devgan et al., 2025b).
- Sensor Simulation: Accurately mirroring the yield of real-world sensors based on the simulated environment.
- Annotation and Labeling: Automatically producing ground truth names and comments for the engineered information (e.g., object bounding boxes, semantic segmentation, action labels). This is a major advantage of synthetic data.
- Data Augmentation (within the simulator): Applying changes and varieties to the recreated information to extend its differing qualities and strength.
- Data Management and Storage: Efficiently storing and overseeing the possibly expansive volumes of synthetic data produced.

- Domain Adaptation Techniques: Algorithms and procedures to diminish the "domain gap" between the synthetic and real data.
- Feature Alignment: Learning representations that are invariant to the contrasts between the synthetic and real domains.
- Generative Adversarial Networks (GANs) for Style Transfer: Making the synthetic information see more practical.
- Self-Supervised Learning: Leveraging unlabeled real-world information to advance refine the agent's understanding.
- Transfer Learning Strategies: Leveraging the knowledge picked up within the engineered environment to quicken learning and move forward execution within the genuine world.
- Robustness Evaluation: Testing the agent's execution in simulated situations that progressively take after the genuine world and inevitably on real-world information itself.

In essence, an AI specialist fueled by manufactured information could be a carefully organized framework where a sophisticated AI center learns from a rich and controllable stream of artificially produced data, with the extreme point of performing successfully within the genuine world (Kohli & Sharma, 2025). The quality and authenticity of the engineered information, along side the viability of the AI calculations and space adaptation strategies, are key determinants of the agent's success.

VI. COMPONENTS OF SYNTHETIC DATA

Whereas synthetic data is artificially created, it's frequently educated by real-world information or a deep understanding of the space it's implied to speak to. This shapes the establishment and guarantees pertinence. This component incorporates:

- Statistical Properties: Capturing the distributions, relationships, and connections show in real-world information. This guarantees the engineered information carries on additionally to its genuine partner.
- Domain Expertise: Incorporating information around the particular field the information speaks to (e.g., medical knowledge for synthetic patient records, physics principles for simulated sensor data).
- Privacy Constraints: Understanding the security concerns and directions that require the creation of engineered information. The objective is regularly to duplicate the utility of genuine information without uncovering delicate data.

- Statistical Modeling: Utilizing measurable distributions (e.g., Gaussian, Poisson) and their parameters (derived from real data or domain knowledge) to produce modern information focuses. Usually regularly utilized for easier information sorts (Devliyal et al., 2025b).
- Rule-Based Generation: Defining explicit rules and imperatives to form engineered information that follows to particular coherent connections or commerce rules.
- Agent-Based Modeling: Reenacting the behavior of person substances and their intuitive inside a framework to produce information reflecting these flow.
- Generative Adversarial Networks (GANs): Training two neural networks (a generator and a discriminator) in a competitive prepare. The generator learns to make engineered information that's vague from genuine information by the discriminator.
- Variational Autoencoders (VAEs): Learning a probabilistic model of the genuine information conveyance and after that testing from this show to produce modern engineered information focuses.
- Simulation Software: Utilizing specialized computer program to recreate complex frameworks (e.g., autonomous driving simulators, financial modeling software) and produce manufactured information as a byproduct of these simulations.
- Data Types: Replicating the sorts of factors display within the genuine information (e.g., numerical, categorical, text, images, time series).
- Feature Space: Guaranteeing the synthetic data incorporates the important highlights and traits required for the expecting task (e.g., training a machine learning model).
- Relationships Between Features: Preserving the relationships and conditions between different factors, which is significant for keeping up the data's utility.
- Data Volume: Generating the specified sum of synthetic data, which can run from little datasets for testing to massive datasets for preparing complex AI models.
- Data Format: Outputting the synthetic data in a usable format (e.g., CSV, JSON, image files).
- Generation Process Description: Detailing the methods and parameters used to make the synthetic data.
- Quality Metrics: Providing data around the constancy and statistical similarity of the synthetic data to the genuine information (if applicable).
- Usage Guidelines: Indicating suitable utilize cases and potential confinements of the manufactured information.

- Privacy Mechanisms: Ensuring that the engineered information does not uncover touchy data from the fundamental genuine information. This may be achieved through different methods inserted within the era handle.
- Utility Assessment: Evaluating how well the engineered information performs in downstream errands compared to genuine information. The objective is to maximize utility whereas protecting protection.

Synthetic data may be a carefully designed item that depends on understanding real-world information or domain knowledge, utilizing sophisticated generation techniques, replicating relevant information characteristics, and giving adequate documentation, all whereas considering the basic balance between privacy and utility. The particular components and their complexity will change essentially depending on the application and the required properties of the engineered information.

VII. TOOLS AND TECHNIQUES

- Game Engines (e.g., Unity, Unreal Engine): Powerful tools for making reasonable 3D environments and simulating physics, sensor data (camera, lidar, radar), and agent intuitive. They offer wealthy visual devotion and scripting capabilities for characterizing scenarios and operator behavior. Training independent vehicles in virtual cities with different activity conditions and climate designs.
- Physics Simulators (e.g., Gazebo, MuJoCo): Focused on exact physical recreations, significant for mechanical technology and support learning assignments including physical control. They permit for exact modeling of robot flow, contact powers, and natural intuitive. Training a robot arm to grasp objects with changing shapes and weights.
- Traffic Simulators (e.g., SUMO, CARLA): Specifically outlined for modeling vehicular activity flow, specialist behavior (drivers, pedestrians), and sensor data important to independent driving. Generating data for preparing perception frameworks to recognize and track vehicles and pedestrians in complex urban situations.
- Industrial Simulators (e.g., Simio, AnyLogic): Used for modeling fabricating forms, supply chains, and other industrial frameworks, empowering the era of engineered information for optimizing operations and preparing AI for mechanization. Simulating a factory floor to prepare an AI operator for prescient support of apparatus.
- Generative Adversarial Networks (GANs): Effective for creating reasonable engineered information over different modalities (images, videos, text, tab-

ular data). They learn the basic dispersion of genuine information and can make novel tests that take after it. Generating synthetic medical images (X-rays, MRIs) with different pathologies for preparing demonstrative AI.

- Variational Autoencoders (VAEs): Another course of generative models that learn a idle representation of the information, permitting for the era of modern tests by examining from this inactive space. Frequently utilized for making smoother varieties of existing information. Creating manufactured client profiles with shifting characteristics for preparing suggestion frameworks.
- Diffusion Models: Rising as state-of-the-art for high-quality picture and video era, advertising fine-grained control over the era prepare. Making profoundly practical engineered pictures of uncommon plant species for preparing plant distinguishing proof AI.
- Large Language Models (LLMs): It can be fine-tuned to produce synthetic content information for different NLP assignments, such as generating practical discoursed, item reviews, or code snippets. Creating synthetic client benefit transcripts for preparing chatbot operators.
- Tabular Data Synthesis Tools (e.g., SDV, Gretel): Specialized libraries and stages for producing engineered unthinkable information whereas protecting factual properties and connections show in genuine information. Regularly utilized for privacy-preserving information sharing and increasing preparing datasets. Making engineered financial exchange information for preparing extortion location models without uncovering genuine client data.
- Procedural Content Generation (PCG) Techniques: Algorithms that automatically generate substance (e.g., textures, 3D models, environments, game levels) based on a set of rules or parameters. Can be used to make different and controllable engineered situations for preparing AI operators. Generating a wide assortment of indoor scenes with distinctive question courses of action for preparing automated route specialists.
- Domain Randomization: Training AI operators on a wide assortment of randomized engineered situations and information distributions. This forces the specialist to memorize highlights that are invariant to particular manufactured characteristics, making it more vigorous to the varieties experienced within the genuine world. Preparing a robot to grasp a particular protest with randomized lighting conditions, surfaces, and foundation clutter within the recreation.

Machine learning strategies particularly planned to exchange information learned from a source space (manufactured) to a target domain (real). These include:

- Feature Alignment: Learning representations that are comparative over both engineered and genuine information.
- Adversarial Domain Adaptation: Utilizing adversarial preparing to learn highlights that cannot recognize between the engineered and genuine spaces.
- Self-Supervised Learning: Leveraging unlabeled real-world information to assist the operator adjust its learned representations.
- Sim-to-Real Transfer Learning: A broader field encompassing different techniques for exchanging information from reenactment to the genuine world, regularly including a combination of space randomization, adjustment calculations, and fine-tuning on constrained real-world information.
- Physics-Based Rendering (PBR): Utilizing rendering methods that precisely reenact the interaction of light with materials, making synthetic images that are outwardly more realistic and can move forward the transferability of models prepared on them.
- Reinforcement Learning (RL) Frameworks (e.g., TensorFlow Agents, PyTorch RL, OpenAI Gym, DeepMind's Acme): Essential for training agents that learn through interaction with their environment (synthetic or real). These systems give devices for characterizing situations, executing RL calculations, and overseeing the preparing prepare.
- Robotics Middleware (e.g., ROS - Robot Operating System): Facilitates the improvement of mechanical operators by giving a standardized system for communication between different software components (perception, planning, control). Frequently utilized in conjunction with physics simulators for preparing robotic operators in synthetic environments.
- Deep Learning Libraries (e.g., TensorFlow, PyTorch): Fundamental for building the neural network-based recognition and control frameworks that frequently frame the center of cleverly specialists. These libraries offer apparatuses for characterizing, preparing, and conveying profound learning models.
- Behavior Trees and State Machines: Methods for planning the high-level decision-making and control rationale of AI specialists, regularly utilized in complex mechanical or diversion AI applications. Synthetic environments give a secure and controllable space for planning and testing these behaviors.
- Metrics for Synthetic Data Fidelity: Evaluating how well the synthetic data captures the factual properties and designs of the genuine information (in case accessible).
- Simulation-Based Evaluation: Assessing the agent's execution on different errands and scenarios inside the engineered environment. This permits for thorough testing and investigating in a controlled setting.

- Transfer Learning Benchmarks: Assessing the agent's capacity to generalize its learned abilities from the manufactured environment to real-world information and assignments.
- Real-World Fine-Tuning and Evaluation: Often necessary to bridge the remaining space crevice. Assessing the agent's execution on restricted real-world information and fine-tuning the show appropriately.

Building AI specialists fueled by synthetic data could be a multidisciplinary endeavor that includes mastery in synthetic data generation, AI algorithms (especially deep learning and support learning), domain adaptation techniques, and robust evaluation methodologies. The specific tools and techniques utilized will depend intensely on the application space and the complexity of the AI agent being developed. As synthetic data era becomes more advanced and the techniques for bridging the sim-to-real gap improve, we can anticipate to see indeed more capable and flexible AI agents sent in a wide range of real-world applications.

VIII. FUTURE PERSPECTIVE

The trajectory of Artificial Intelligence (AI) agents, especially those engaged by engineered information, focuses towards a transformative future by the year 2500. This projection, whereas fundamentally speculative, can be grounded in current patterns and the inherent advantages that synthetic data offers in overcoming the impediments of traditional data-driven AI development. By the first light of the 26th century, we can expect AI agents that are altogether more able, versatile, ethical, and seamlessly coordinates into the texture of presence, largely due to the advancements and pervasive use of meticulously crafted synthetic datasets. One of the foremost significant shifts will be within the adaptability and availability of AI improvement. Today, the creation of robust AI agents is often hampered by the scarcity of high-quality, labeled real-world information. Acquiring, cleaning, and explaining such information may be a time-consuming, expensive, and sometimes privacy-invasive process. By 2500, synthetic data era advances will likely have developed to a point where the creation of tremendous, different, and superbly labeled datasets is commonplace and highly effective. This democratization of information will lower the boundary to passage for AI development, permitting for development over a more extensive run of areas and by a more differing bunch of analysts and engineers. Imagine making a highly specialized AI for a specialty medical condition for which real-world understanding information is rare; with progressed engineered information era, reasonable and privacy-preserving quiet records, medical scans, and treatment results may be synthesized on request, enabling the preparing of profoundly successful demonstrative and restorative specialists. Moreover, the authenticity and fidelity of engineered situations and information will reach exceptional levels. Current synthetic information, whereas valuable, frequently battles to completely capture the complexity and subtleties of the genuine world. By 2500, headways in computational control, sophisticated generative models (potentially beyond our current understanding of neural systems), and a more profound understanding of the fundamental physical and social elements will empower the creation of virtual universes that are essentially vague from reality for AI specialists. These simulated environments will permit for the preparing of specialists in scenarios that are too dangerous, rare, or morally problematic to reproduce within the genuine world. For occasion, independent specialists for calamity reaction can be prepared in incredibly practical recreations of earthquakes, floods, or fires, encountering a vast array of potential challenges and learning ideal response strategies without any real-world risk. Essentially, AI agents outlined for scientific discovery might conduct incalculable simulated experiments, accelerating the pace of research in areas like materials science or sedate disclosure. The ethical considerations encompassing AI will moreover be significantly affected by the far reaching selection of manufactured information. Nowadays, inclinations show in real-world preparing information can lead to AI agents that sustain and indeed intensify societal disparities. Synthetic data offers the potential to form datasets that are inalienably reasonable and impartial. By carefully controlling the era handle, engineers can guarantee that synthetic data reflects evenhanded distributions and avoids the historical prejudices that torment real-world data. In 2500, we might see regulatory systems that command the utilize of synthetic information for preparing AI in touchy spaces like enlisting, credit applications, and criminal equity, guaranteeing a more just and equitable arrangement of AI advances. In addition, synthetic data inherently addresses security concerns, because it does not contain any genuine individual's data, clearing the way for the advancement of AI applications in privacy-sensitive regions like healthcare and personal finance without the risks related with real data breaches. The flexibility and robustness of AI specialists prepared on synthetic data will too see significant enhancements. Real-world environments are continually changing, and AI agents trained solely on inactive real-world datasets can battle to generalize to new situations. Synthetic data era permits for the creation of energetic and advancing preparing situations. As our understanding of the world and its potential future states develops, ready to join these changes into the synthetic data era handle, ensuring that AI specialists are prepared to handle a much more extensive extend of possibilities. Envision an AI operator outlined to oversee a future shrewd city; its preparing may include recreations that join different future scenarios, such as climate alter impacts, technological disruptions, or indeed speculative social and financial shifts, making it remote more flexible and versatile to unexpected circumstances. Besides, the interaction between people and AI specialists fueled by engineered information might ended up more instinctive and consistent. As AI agents become more advanced and are trained on vast sums of synthetic information that captures nuanced human behaviors and communication styles, they will be better prepared to understand and react to human needs and intentions. Envision individual AI collaborators in 2500 that can expect your needs with exceptional exactness, not fair based on past behavior but moreover on their understanding of complex social cues and emotional states, gathered from fastidiously synthesized information speaking to a wide range of human intelligent. These agents seem act as really cleverly collaborators, increasing human capabilities in ways ready to as it were start to imagine. The improvement of specialized AI operators for highly particular and complex tasks will too be greatly encouraged by engineered information. For instance, making an AI operator to explore the complexities of a future legitimate framework or to oversee a highly advanced logical inquire about office would require preparing information that's inconceivably point by point and particular. Synthetic data generation devices will permit for the creation of such bespoke datasets, tailored accurately to the requests of these specialized applications. This may lead to the multiplication of exceedingly cleverly operators that can perform tasks currently requiring a long time of human mastery and training. In any case, long-term of AI operators fueled by synthetic data isn't without its potential challenges. One critical concern will be the potential for abuse and the creation of beguiling AI. If synthetic data can make inconceivably practical representations of reality, it may moreover be utilized to create persuading deepfakes or to prepare AI operators for malevolent purposes. Guaranteeing the capable advancement and sending of these innovations will be pivotal, requiring strong ethical guidelines, modern discovery instruments, and possibly even regulatory oversight. Another challenge lies in guaranteeing the arrangement between the engineered world and the genuine world. While the objective is to make engineered information that precisely reflects reality, there will continuously be a degree of deliberation and disentanglement included. Ensuring that AI operators prepared in these engineered situations can effectively exchange their learning to the complexities of the genuine world will require ongoing research and development in regions like space adjustment and exchange learning. The more complex the real-world space, the greater the challenge in making a genuinely agent synthetic counterpart. Finally, the computational assets required to produce and train AI agents on enormous, high-fidelity engineered datasets in 2500 can be considerable. While computational control is anticipated to proceed its exponential development, the scale and complexity of the synthetic data and AI models of the future might push the limits of indeed advanced computing infrastructure. Efficient data generation techniques, novel AI structures, and breakthroughs in energy-efficient computing will be basic to completely realize the potential of engineered data-powered AI. The longer term of Artificial Intelligence agents by the year 2500, heavily impacted by the progressions in synthetic data, is one of immense potential. We can anticipate a future where AI development is democratized, specialists have unprecedented levels of authenticity and flexibility, ethical considerations are more viably addressed, human-AI interaction is consistent, and specialized AI for complex assignments becomes commonplace. Whereas challenges related to misuse, real-world arrangement, and computational assets will ought to be tended to, the transformative control of engineered information in fueling the another era of AI agents guarantees a future where intelligent frameworks play an indeed more significant and useful part in forming human civilization in India and over the globe. The capacity to form information on request, custom fitted absolutely to the requirements of AI development and free from the limitations and biases of the real world, will open possibilities we are as it were beginning to suppose, introducing in an period of really intelligent and ethically grounded AI agents.

IX. CASE STUDY

A leading autonomous vehicle company within the USA is intensely dependent on synthetic data generated through progressed driving test systems. They make exceedingly practical virtual situations imitating assorted US road conditions, weather patterns (including rare events like heavy snow or black ice), and activity scenarios (including complex intersections and unexpected pedestrian behavior). AI agents (the autonomous driving algorithms) are trained within these recreations for millions of virtual miles, experiencing incalculable edge cases and dangerous circumstances that would be incomprehensible or unethical to duplicate within the genuine world at scale. Synthetic data includes perfectly labeled sensor data (camera images with object bounding boxes and semantic segmentation, lidar point clouds with object classifications, radar data), ground truth data almost the environment, and reenacted vehicle flow. The AI agents learn vigorous perception, planning, and control policies through reinforcement learning within these simulations. This significantly quickens the training process, improves the security and robustness of the independent driving framework by exposing it to a vast range of scenarios, and decreases the require for broad and risky real-world testing. Companies like Waymo (while also using real-world information) intensely use simulation, which depends on engineered information standards. A US-based medical AI startup is developing a diagnostic tool for a rare hereditary disorder that influences a little number of individuals. Due to the rarity of the condition, getting a expansive sufficient real-world dataset of quiet images (e.g., retinal scans, genetic sequencing data) is challenging and may raise security concerns. The startup utilizes generative AI models (like GANs or VAEs) prepared on the limited real-world data they have, as well as medical knowledge and statistical characteristics of the disease, to make a large, diverse, and privacy-preserving engineered dataset of medical images and understanding records. AI agents (the diagnostic algorithms) are prepared on this synthetic data to distinguish inconspicuous designs and pointers of the rare disease. Synthetic data empowers the advancement of possibly life-saving demonstrative apparatuses for conditions where genuine information is rare, overcoming information get to limitations and privacy hurdles. It can too be outlined to represent the diversity of the patient population, potentially diminishing biases within the AI model. A major US financial institution needs to improve its AI-powered fraud detection system to recognize modern and advancing extortion plans. Real-world fraud data is regularly imbalanced (few fraudulent transactions compared to legitimate ones) and may not contain cases of the most recent fraud tactics. The institution utilizes engineered data era methods to simulate different sorts of known and theoretical fraudulent activities, based on their understanding of extortion designs and security vulnerabilities. AI agents (the fraud detection models) are prepared on this expanded dataset, which incorporates a wealthy set of both genuine and synthetic fraudulent transactions, permitting them to memorize to distinguish unobtrusive irregularities and unused assault vectors. Synthetic data makes a difference to address information awkwardness issues, exposes the AI to a wider extend of potential extortion scenarios, and moves forward the model's capacity to generalize to concealed false exercises, eventually securing the institution and its clients. A US cybersecurity firm is creating an AI specialist to identify modern cyberattacks on undertaking systems. Real-world information on advanced persistent threats (APTs) and zero-day exploits is rare and highly sensitive. The firm employments reenactment situations to create engineered arrange activity that imitates different assault designs, including subtle anomalies and sophisticated intrusion techniques. AI agents (the intrusion detection systems) are trained on this engineered arrange information to memorize to recognize noxious behavior without depending on genuine attack data, which may compromise security if uncovered. Synthetic data gives a secure and controllable way to prepare AI agents to distinguish uncommon and dangerous cyber threats, improving the security posture of organizations without the dangers related with utilizing real assault information. A US customer benefit company is building an AI-powered chatbot to handle a wide extend of client request. Whereas they have genuine client interaction information, it might not cover all conceivable client bury, expressing, or edge cases. They use large language models (LLMs) to generate synthetic dialogues that speak to a differing set of customer queries, counting complex or unusual requests. The AI agent (the chatbot) is prepared on this expanded dataset to move forward its understanding of characteristic language, its capacity to handle different client inputs, and its strength to startling inquiries. Synthetic data helps to progress the chatbot's execution, scope, and client experience by exposing it to a more extensive variety of conversational scenarios than might be display within the real-world data alone.

- Privacy Preservation: Synthetic data is vital in scenarios where genuine information is touchy or subject to strict protection controls (e.g., healthcare, fund).
- Addressing Data Scarcity: It empowers the improvement of AI for rare occasions or conditions where real-world information is restricted.
- Handling Imbalanced Information: Synthetic data can be utilized to oversample minority classes (e.g., extortion, rare illnesses) to move forward demonstrate preparing.
- Exploring Edge Cases and Security: Simulation utilizing synthetic information permits for the secure investigation of dangerous or uncommon scenarios (e.g., independent driving, cybersecurity).
- Improving Model Robustness: Preparing on assorted engineered information can make AI agents more flexible to varieties and unforeseen inputs within the genuine world.

X. CONCLUSION

The rise of Artificial Intelligence Agents enabled by engineered information speaks to a significant and transformative meeting within the field of AI. We've traveled from the foundational principles of cleverly independence encapsulated by AI Agents to the basic part of information in their improvement, highlighting the noteworthy limitations posed by depending exclusively on real-world information. Synthetic data has risen as an effective worldview move, advertising a for all intents and purposes boundless supply of high-quality, labeled, different, and privacy-preserving preparing information that directly addresses these impediments. The techniques for producing synthetic data are assorted and always advancing, extending from modern reenactments and generative adversarial networks to statistical modeling and procedural era. This wealthy toolkit permits for the creation of information custom-made absolutely to the preparing needs of AI Agents over a huge number of spaces. We've investigated the transformative applications of this collaboration, displaying how engineered information is quickening advance in independent driving, mechanical technology, healthcare, back, cybersecurity, and past, empowering the advancement of more strong, dependable, and morally sound intelligent agents. Whereas the field is still exploring certain challenges, such as bridging the domain gap, ensuring data fidelity, and building up standardized evaluation practices, the direction is irrefutably promising. The advantageous advancement of AI Agents and engineered information focuses towards a future where the advancement of brilliantly independent frameworks is democratized, advancement cycles are accelerated, and arrangements can be found for issues where real-world data is rare or inaccessible. Eventually, the fusion of AI Specialists and synthetic data holds the key to opening a modern period of cleverly independence. By overcoming the bottlenecks related with conventional information securing, we are clearing the way for AI Agents with improved capabilities, more noteworthy versatility, and a wider extend of impactful applications. As investigate proceeds to progress and the procedures for creating and utilizing engineered data gotten to be more refined, able to expect a future where cleverly operators, prepared and engaged by artificially created data, consistently coordinated into our lives, expanding human potential and understanding a few of the world's most squeezing challenges. The journey is progressing, but the goal – a world intelligently empowered by the collaboration of AI Agents and synthetic data – is inside closer reach than ever before.

REFERENCES

Antonucci, D., Conselvan, F., Mascherbauer, P., Harringer, D., & Pozza, C. (2024). Synthetic data on buildings. In Learning and analytics in intelligent systems (pp. 203–226). https://doi.org/DOI: 10.1007/978-3-031-47909-0_7

Bhatt, S., & Sharma, S. (2025). Generative Artificial Intelligence Based Biomedical Applications for Pharmaceutical Industry. In *2025 International Conference on Computational, Communication and Information Technology (ICCCIT)*, 1–6. https://doi.org/DOI: 10.1109/ICCCIT62592.2025.10927855

Blythe, J., & Tregubov, A. (2024). Synthetic Data Generation for Machine Learning Models with Cognitive Agent Simulations. In Lecture notes in computer science (pp. 73–83). https://doi.org/DOI: 10.1007/978-3-031-70415-4_7

Chakradhar, B. S., Vijey, M., & Tunk, S. (2024). Synthetic data generation for marketing insights. In Chapman and Hall/CRC eBooks (pp. 195–215). https://doi.org/DOI: 10.1201/9781003472544-17

Dangi, S., Ghanshala, K., & Sharma, S. (2025). LIFT: Lightweight incremental and federated techniques for live memory forensics and proactive malware detection. *International Journal of Advanced Computer Science and Applications, 16*(4). Advance online publication. DOI: 10.14569/IJACSA.2025.0160445

De, S., & Mitra, P. (2024). A novel technique of synthetic data generation for asset administration Shells in industry 4.0 scenarios. *IEEE Transactions on Artificial Intelligence*, *5*(10), 1–10. DOI: 10.1109/TAI.2024.3409516

Devgan, K., Sharma, S., Jangra, D., & Sharma, S. (2025b). Artificial Intelligence-Enabled Apple Disease Monitoring System for environmental protection and resource conservation. In Advances in computational intelligence and robotics book series (pp. 271–300). https://doi.org/DOI: 10.4018/979-8-3373-3246-8.ch011

Devliyal, S., Goyal, H. R., & Sharma, S. (2025b). ICDRET: A dynamic cyber risk estimation technique for intelligent Cyber-Physical Systems in PCS. *The Open Bioinformatics Journal, 18*(1), e18750362380269. Advance online publication. DOI: 10.2174/0118750362380269250502062754

Devliyal, S., Sharma, S., & Goyal, H. R. (2024). EIAIMSPS: Edge inspired artificial intelligence-based multi stakeholders Personalized Security Mechanism in ICPS for PCS. *International Journal of Advanced Computer Science and Applications, 15*(8). Advance online publication. DOI: 10.14569/IJACSA.2024.01508117

Fragkouli, S., Solanki, D., Castro, L. J., Psomopoulos, F. E., Queralt-Rosinach, N., Cirillo, D., & Crossman, L. C. (2024). Synthetic data: How could it be used in infectious disease research? *Future Microbiology*, *19*(17), 1439–1444. DOI: 10.1080/17460913.2024.2400853 PMID: 39345126

Kohli, P., & Sharma, S. (2025). EDGE Artificial Intelligence-Enabled 6G communication security system for connected vehicles in revolutionizing urban development and governance. In IGI Global eBooks (pp. 431–462). https://doi.org/DOI: 10.4018/979-8-3373-1375-7.ch015

Kshetri, N. (2024). Generative artificial intelligence in the financial services industry. *Computer*, *57*(6), 102–108. DOI: 10.1109/MC.2024.3382452

Kyem, B. A., Asamoah, J. K., Huang, Y., & Aboah, A. (2024). Weather-Adaptive synthetic data generation for enhanced power line inspection using StarGAN. *IEEE Access : Practical Innovations, Open Solutions*, *1*, 193882–193901. Advance online publication. DOI: 10.1109/ACCESS.2024.3520120

Long, Y., Kroeger, S., Zaeh, M. F., & Brintrup, A. (2025). Leveraging synthetic data to tackle machine learning challenges in supply chains: Challenges, methods, applications, and research opportunities. *International Journal of Production Research*, •••, 1–22. DOI: 10.1080/00207543.2024.2447927

Meiser, M., & Zinnikus, I. (2024). A survey on the use of synthetic data for enhancing key aspects of trustworthy AI in the energy domain: Challenges and Opportunities. *Energies*, *17*(9), 1992. DOI: 10.3390/en17091992

Mohammadabadi, S. M. S., Entezami, M., Moghaddam, A. K., Orangian, M., & Nejadshamsi, S. (2024). Generative artificial intelligence for distributed learning to enhance smart grid communication. *International Journal of Intelligent Networks*, *5*, 267–274. DOI: 10.1016/j.ijin.2024.05.007

Moreira, D., Marcel, S., & Rocha, A. (2024). Synthetic Realities and Artificial Intelligence-Generated Contents. *IEEE Security and Privacy*, *22*(3), 7–10. DOI: 10.1109/MSEC.2024.3388244

Pandey, C., Sharma, S., & Bhadula, S. (2024). Artificial Intelligence Technique in Next Generation Healthcare Systems. In *2024 IEEE International Conference on Communication, Computing and Signal Processing (IICCCS)*, 1–6. https://doi.org/DOI: 10.1109/IICCCS61609.2024.10763570

Rawat, D., Sharma, S., & Bhadula, S. (2025b). CT Imaging-Based Deep Learning System for Non-Small Cell Lung Cancer Detection and Classification. *International Journal of Advanced Computer Science and Applications*, *16*(5). Advance online publication. DOI: 10.14569/IJACSA.2025.0160561

Sharma, A., Khanna, V., & Gigras, Y. (2025). Data analytics and synthetic data generation for an IoT-based smart home. In Elsevier eBooks (pp. 29–45). https://doi.org/DOI: 10.1016/B978-0-443-26474-0.00013-0

Sharma, S., Gupta, A., & Tyagi, R. (2024). Artificial Intelligence, Cyber Security and Vedic Scripture Integrated Sustainable Global Youth Leadership Development. In *2024 IEEE International Conference on Communication, Computing and Signal Processing (IICCCS)*, 1–6. https://doi.org/DOI: 10.1109/IICCCS61609.2024.10763618

Sharma, S., Tyagi, R., & Mohan, S. (2024). Generative Artificial Intelligence-Enabled climate change and Deep-Sea food webs. In Advances in environmental engineering and green technologies book series (pp. 199–224). https://doi.org/DOI: 10.4018/979-8-3693-6670-7.ch009

Sharma, S., Tyagi, R., & Mohan, S. (2025). Artificial intelligence and optimized computational modelling against climate crisis. In Elsevier eBooks (pp. 1–14). https://doi.org/DOI: 10.1016/B978-0-443-21905-4.00006-7

Sharma, S., Tyagi, R., & Mohan, S. (2025b). Blockchain-Based Security and Privacy in Digital Healthcare Recommender Systems. In In Healthcare Recommender Systems (pp. 221–237). https://doi.org/DOI: 10.1007/978-3-031-80056-6_10

Chapter 5
Digital Synthetic Motion Outputs

Samrat Kundu
https://orcid.org/0009-0004-9409-8041
Brainware University, India

Kousik Dey
https://orcid.org/0009-0002-5394-2918
Brainware University, India

Rahul Bir
Brainware University, India

Sandip Pal
Brainware University, India

ABSTRACT

Synthesizing motion data digitally employs advanced AI models like generative adversarial networks (GANs), and diffusion networks to create realistic motion data for various applications including visual effect, video games, robot motion, and simulation. These technologies aid in large-scale data augmentation as well as in rapid prototyping, thus helping get around the limitations of real-world data. Their major uses lie in film de-aging, animation for interactive games, and robotic training in simulated environments. However, concerning privacy, trust, and misinformation, deepfake technology adds to the worries. Human oversight, authentication methods, and validation benchmarks must ensure that these technologies are put to ethical use. On the responsible side, synthetic motion output constitutes a way to enhance real-world experience in education, design, therapy, and interactive multimedia.

DOI: 10.4018/979-8-3373-2277-3.ch005

1. INTRODUCTION

A progression from synthetic to real motions has been recorded via artificial intelligence, computer graphics, and machine learning. Originally self-keyframed, motion has gradually come to be synthesised by models capable of producing movements that look realistic in the absence of physical capture systems (Goodfellow et al., 2014; Moeslund et al., 2006). Hence, from denoising diffusion to adversarial learning and reinforcement learning, a properly forged system may transform naturalistic motion from very little input, such as pose coordinates, audio, and video frames. GANs simulated lifelike motions, where the generator tried to defeat the discriminator in its adversarial loops (Tulyakov et al., 2018). Diffusion models have been applied recently to maintain a smooth and diverse pattern of motions by iterative conversion from noise to structured sequences. The new domains are opening for applications beyond simulation. In medicine, generated motions serve for physical rehabilitation and immersive therapy (Pan & Hamilton, 2018). Robots have trained their motion-planning abilities by means of synthetic trajectories and feedback from reinforcement (Peng et al., 2018). In gaming and entertainment, the field has been further extended into the automated generation of motion sequences for real-time character animation to reduce production costs. Ethical questions thus accompany these developments as well. High-fidelity motion generation, which closely mimics real human behaviours, raises questions about identity replication, misinformation, and consent (Mirsky & Lee, 2021; Westerlund, 2019). Standards such as IEEE 7000-2021 were set up to foster the ethical deployment of AI. Authenticity-guarantee technology is now under consideration to include watermarking or blockchain-based provenance tracking (Fridrich & Goljan, 2009; Crosby et al., 2016). Before, synthetic motion generation was still in experimental phases in the labs; however, with the emergence of modern machines like GPUs and TPUs, that trend is fast becoming mainstream. The boundary that was once clearly drawn between real motion and generated one is becoming very foggy with higher levels of realism and interactivity, thereby reshaping the whole gamut of communication, simulation, and interaction.

2. HISTORICAL EVOLUTION OF SYNTHETIC MOTION

Synthetic representation of motion has witnessed some major plays and transformations due to shifts in art, technology, and computation. Early developments were mostly influenced by manual methods of animation, whereby drawings were made frame by frame to simulate continuous movements. This provided the foundations for visual storytelling with motion, focusing on developments at the dawn of the 20th century, the greatest one being the work of Walt Disney Studios. The other major addition during this period was rotoscoping, where live-action footage was traced to spinning animations that looked more realistic (Thomas & Johnston, 1995). Towards the end of the 20th century, traditional animation started to slowly be augmented-with some exceptions being completely replaced by motion capture (MoCap) technologies. In MoCap, human actors were fitted with reflective markers or sensors that captured their movements in three-dimensional space. The recorded movements were then applied to digital characters for the purpose of endowing them with a great advantage in terms of realism for movies, games, and virtual environments (Moeslund 2006). The Polar Express (2004) and Avatar (2009) demonstrated how powerful MoCap could be in creating living characters and mesmerising experiences. Nonetheless, the process was resource-intensive: it needed proper equipment, proper actors, and post-processing mechanisms (Moeslund et al., 2006). The subsequent major transformation was ushered in by machine learning (ML) and deep-learning models capable of direct motion synthesis from data. Algorithms were termed motion-synthesis algorithms because they were taught on existing motion data, rather than on capturing and synthesising real-world movement. In the beginning, older examples were constructed using traditional statistical models with limited efficiency and realism, which were transformed with the introduction of deep neural networks. Arguably, the most significant advance in motion synthesis began with Goodfellow et al.'s inception of GANs in 2014, which synthesised motion through a minimax game between one network, called a generator, and another network, called a discriminator, which judged the realism of the outputs. In contrast, Variational Autoencoders (VAEs) provided an alternative by considering the probabilistic encoding and decoding of motion data. Both approaches have been applied widely to the synthesis of synthetic gestures, facial expressions, and full-body movements (Tulyakov et al., 2018). More recently, diffusion models have arisen as a new state-of-the-art methodology in the realm of generative motion synthesis. These diffusion models generate highly diverse and stable motion patterns by simulating a slow diffusion process that progressively takes away noise from data. Thanks to their efficiency in considering long-range temporal dependencies, they are mainly focused on modelling complicated human movements and, in this capacity, beat almost all the benchmarks set by previous GAN-based models. While these models have matured, there is therefore a parallel shift toward real-time and data-driven animation. In contrast to autochthonous MoCap or hand-crafted animation pipelines, AI-based software now allows interactive avatars and live gesture and synthetic video synthesis on the fly. Hardware developments, including GPUs and edge devices, spurred this transition by subsidising low-latency computation and deployment (Peng et al., 2018). Having seen this evolution, synthetic motion has witnessed a major transformation-from hand-crafted art into automatic generation-with implications for an array of industries such as cinema and gaming, and even health and robotics. Continuing to foster this process is deep-learning research, hardware acceleration, and multidisciplinary investigation.

3. SYNTHETIC MOTION CORE TECHNOLOGIES

Synthetic motion development is largely the product of generative modelling and deep learning. In the past ten years, several classes of models have been adapted to generating human-or-object-based motion, each one presenting architectural innovation and specific application strengths. GANs, diffusion models, hybrid reinforcement learning, and NeRFs have all quite recently emerged as technologies on the forefront of motion synthesis. Each has been applied using different vantage points on the problem of motion synthesis: realism, adaptation to various inputs, speed, and spatiotemporal coherence.

3.1 Generative Adversarial Networks (GANs)

Introduced a long time back in 2014 by the inventive team of Goodfellow et al., the GAN framework has seen widespread applications in synthetic media, which includes motion generation applications. The basic GAN architecture involves two different neural networks trained simultaneously, with a generator wanting to reproduce synthetic data indistinguishable from real data, and the discriminator trying to distinguish between real and generated samples. The training follows the framework of a minimax game:

$$\mathrm{GminDmaxV(D,G)} = \mathrm{Ex} \sim \mathrm{pdata(x)[logD(x)]} + \mathrm{Ez} \sim \mathrm{pz(z)[log(1 - D(G(z)))]}$$

The adversarial training tends to deliver very realistic motion data on convergence. Various architectures based on GANs have been designed for motion generation. The MoCoGAN (Tulyakov et al., 2018) proposed to separate motion and content for generating variable-length video clips whose temporal dynamics are represented by a recurrent latent space. PoseGAN (Siarohin et al., 2018) produces pose-conditioned videos, which are particularly suitable for gesture synthesis of avatars. MotionGAN and its variants have further extended GANs to complex multi-agent interactions and stylised motion. Nevertheless, GANs are not free of limitations. It is quite unstable to train them, and the risk of mode the generator generates limited diversity in its output is very much alive (Salimans et al., 2016). On top of that, it might be hard to ensure temporal coherence in motion generation without additional architectural constraints or loss functions aiming at frame-to-frame coherence.

3.2 Diffusion Models

In an attempt to address the limitations of GANs, diffusion probabilistic models have come to prominence in synthetic motion generation. The forward process corresponds to adding noise into data by infinitesimal amounts at every timestep, and the reverse process attempts to reconstruct the original data from noise with a crafted denoising network (Ho et al., 2020). The forward process is usually defined as:

$$q(xt|\ xt - 1) = N(xt;\ 1 - \beta txt - 1, \beta tI)$$

Where βt is a variance schedule. There is a variance schedule. The reverse denoising step is approximated by a neural network minimising a reconstruction loss. Recently, diffusion models have been shown to surpass GANs for image quality and diversity benchmarks (Nichol & Dhariwal, 2021). In the motion domain, they provide for more stable training and better long-range temporal consistency. Motion diffusion models like MotionDiffusion and MDM (Tevet et al., 2022) have been employed for gesture generation, character animation, and robotic trajectory synthesis. A Transformer-based backbone is usually used in their architecture to model the temporal sequence and capture dependencies that are often missed by GANs.

Though more diverse and stable, diffusion models also require more inference time because of their iterative denoising procedures. Methods aiming toward acceleration have appeared, employing fewer timesteps for denoising and thus employing score-based generative modelling with stochastic differential equations (Song et al., 2021). Despite being computationally intensive, the output benefits are an all-encompassing reward for adoption by many.

3.3 Hybrid Approaches with Reinforcement Learning (RL)

Motion synthesis often involves goal-oriented behaviour, which purely generative models may not support well. RL may thus be integrated with generative models. These hybrid models further polish synthetic motions through environmental interaction in which the reward is optimised for achieving the desired behaviour. In motion generation, Q-learning and policy gradient methods have been used to reinforce realism, task success, or physical feasibility of motion (Peng et al., 2018). Reward shaping. Through imitation learning, for example, synthetic agents are rewarded to imitate demonstrations of experts. GANs have been extended with reinforcement modules where, apart from trying to fool the discriminator, the generator also tries to optimise the behaviours aligned with task goals. These hybrid models can yield excellent results in robotics, where the agent executes plans involving complex movement sequences in physical or simulated environments, and game AI, where character animations are produced in real time based on player interaction (Merel et al., 2019). The downside, however, is the complexity and computational cost of these approaches. Careful tuning of adversarial and reinforcement signals is required since sparse or misleading reward feedback can sometimes prevent convergence.

3.4 Neural Radiance Fields (NeRFs)

Neural Radiance Fields (NeRFs) have been the latest disruptive tech in rendering into views-consistent scenes. Proposed for static 3D scenes initially (Mildenhall et al., 2020), NeRFs have undergone extensions for dynamic motion through D-NeRF, HumanNeRF, and Nerfies. These variants incorporate temporal information, which allows for photorealistic reconstruction of moving humans or objects from sparse camera views.NeRF learns a volumetric scene representation by mapping 3D co-ordinates and viewing directions to colour and density values by means of a neural network. In dynamic scenarios, time or pose vectors are added as inputs, enabling the production or interpolation of motion through time (Park et al., 2021). Among the applications of NeRFs in synthetic motion, we list:

1. Free-viewpoint animation of humans.
2. Augmented reality avatars.
3. Photorealistic virtual agents in VR/AR platforms.

NeRFs prove to be most useful in cases where motion and viewpoint must be synthesised jointly. Several pitfalls, however, remain, including implementation difficulties and enormous computational demand and long training time for really good NeRFs. Research into real-time NeRFs through optimised hardware (such as Instant-NGP from NVIDIA) and complementary efficient representations is still underway.

3.5 Comparative Analysis of Key Motion Generation Models

An overall comparison of the main technologies used in synthetic motion generation is given below:

Table 1. Comparison of Key Synthetic Motion Generation Technologies. This table summarises the core strengths, weaknesses, and application areas of major synthet-

ic motion generation models, including GANs, diffusion models, hybrid GAN-RL systems, and Neural Radiance Fields (NeRFs).

Technology	Strengths	Weaknesses	Use Cases
GANs	High realism, expressive outputs	Mode collapse, training instability	Gaming, animation, virtual humans
Diffusion Models	Stable, diverse, temporally coherent	Slow inference, high compute	Robotics, gesture generation, animation
GAN + RL Hybrid	Task-aware, adaptive, interactive outputs	Complex training, tuning required	Robotics, simulation, game AI
NeRFs	View-consistent, photorealistic motion	Long training time, resource-heavy	VR/AR avatars, 3D motion capture, film

4. HARDWARE AND INFRASTRUCTURE DEPENDENCIES

Generation of high-quality motion data with an AI-driven model demands high computing power. How this hardware plays the most significant role in this scenario, first GPUs, TPUs, and edge devices are needed for both phases of training and deployment of synthetic motions. Second, real-time rendering and simulation platforms like NVIDIA Omniverse and Unity ML-Agents have come into the foreground in industrial applications. GPU-based training of deep-learning models is mostly what will be referred to here, using NVIDIA products like the RTX 4090 and the A100. This is because these units pertain to and are optimised for parallel computation, whereby one can process massive batches of data simultaneously, a key requirement when modelling complex spatio-temporal motion (NVIDIA, 2023). The NVIDIA A100, built on the Ampere architecture, excels at providing multi-instance GPU capabilities leveraged for training gigantic transformer-based motion synthesis models.

In an enterprise and research environment, TPUs have found use as accelerators for training deep neural networks, most especially within the Google Cloud ecosystem. By their custom architecture, they excel at matrix-heavy computations, which are needed by GAN discriminators as well as diffusion denoising steps (Jouppi et al., 2021). Although less flexible than GPUs, TPUs present the training in between-batch performance in a controlled environment. For mobile games, robotics, augmented reality, and other real-time uses, edge devices continue to be embraced to execute lite versions of motion synthesis models; these are smartphones with AI accelerators and embedded boards such as NVIDIA Jetson that support inference closer to the data source, which reduces latency and improves responsiveness (Howard et al., 2019).NVIDIA Omniverse is a platform that provides real-time hybrid and multi-GPU collaborative rendering and simulation for synthetic motion at scale. Unity ML-Agents, on the other hand, supports reinforcement learning for animation, thereby helping with the prototyping of autonomous agent behaviours in simulated environments (Juliani et al., 2020). As computational frameworks mature, scalable and distributed hardware infrastructure will remain the core for synthetic motion in the future.

5. APPLICATIONS OF SYNTHETIC MOTION

The use of synthetic motion technologies has far transcended academic research, entering a range of fields including entertainment, health care, robotics, education, and virtual reality. Owing to generative modelling, meanwhile, deep learning, and real-time rendering, synthetic motion systems are deployed to animate characters, to support patients, to train professionals, and to better interact with the user environment. In this section, we present an outline of some Important application areas where synthetic motion has contributed to a measurable impact and an impetus for innovation.

5.1 Entertainment & VFX

Synthetically generated motion has brought rapid developments in VFX by providing an efficient and flexible alternative to manual animation and conventional motion capture. With such AI-generated motions, filmmaking produces lifelike characters to de-age actors or create crowd scenes. For instance, a machine learning technique was used in De-ageing-the actors: Acting motion reconstruction for The Irishman (2019) so that it would not require actors to go through invasive makeup and marking (Zhang et al., 2020). GANs and neural renderers now allow filmmakers to animate facial expressions and full-body motions from modest inputs, such as 2D video or pose skeletons. Synthetic motion with procedural animation and NPC behaviour in games was used. Models such as those in MoCoGAN and Motion Diffusion style were used to synthesise character movements for fluid real-time environments (Tulyakov et al., 2018; Tevet et al., 2022). Nowadays, game developers integrate the motion synthesis pipeline within the game engine for tweaking realism, lowering the cost of animation, and customising gameplay.

5.2 Medical & Therapy

Rehabilitation, therapy, and diagnosis make up the settings where synthetic motion assumes the most action. AI-generated avatars lead the patients in physiotherapy exercise, synthetically miming the right positions of limbs to provide real-time feedback (Li et al., 2021). These systems allow sessions to be conducted remotely, reducing the need for an in-person physiotherapist. When it comes to PTSD or autistic patients, synthetic motion-generated virtual agents create safe and controlled setups for exposure therapy or social skills training (Pan & Hamilton, 2018). These agents use synthetic motions to mimic human gestures and expressions to emotionally engage the patient in a controlled-situation environment devoid of the unpredictability of real-life situations. During the COVID-19 period, synthetic avatars were rapidly integrated within the Telehealth service to prevent any disruptions in physiotherapy and motor-function retraining. The remote capturing of motion data, combined with avatar-based animation using GANs or VAEs, proved to be a viable solution in guiding patients in their homes and almost uninterruptedly channelling their recuperation.

5.3 Educational & Virtual Training

Synthetic motion has found applications in teaching and professional training, especially in circumstances that require a physical simulation. In medical schools, students learning surgical procedures can engage equally with virtual patients animated via synthetic motions to train in the required techniques without any risk in real life (Satava, 2002). Likewise, engineering curricula employ a motion-simulated environment to train engineers in hazardous or inaccessible locations.

Besides technical studies, synthetic avatars also provide inclusive learning environments. For example, sign language avatars generated by motion-generation models that translate educational content in real time assist deaf students (Kipp et al., 2011). Expressive digital tutors with synthetic gestures and facial expressions are used in language-learning applications to further engage the users and help them remember the lessons. Because they are scalable and cheap, these solutions become even more critical in a low-resource setting or a poorly funded school where one cannot always acquire a live instructor or enough equipment.

5.4 Robotics & Autonomous Systems

Synthetic motion is fundamental in developing and training intelligent robotic systems. In robotics, motion is synthesised to produce demonstrations for imitation learning and RL. The robot learns either by observing or by receiving synthetic trajectories that approximate expert behaviour so as to speed up the process of learning dramatically (Peng et al., 2018). Sim-to-real transfer is, however, arguably the most promising application of motion generation, whereby motion policies trained in a synthetic environment are applied in the real world. This allows circumventing time-consuming real-world data collection and allows the fast prototyping of robotic behaviours in the areas of manipulation, locomotion, and drone flight (Tobin et al., 2017). Adaptive learning models allow robots to change their movements depending on the environment they robot comes upon in real time. With GAN-based refinement, robots fine-tune motion trajectories for stability, efficiency, or safety, especially when environments are dynamic or unstructured.

5.5 Augmented and Virtual Realities and Human-Computer Interaction

Motion synthesis offers realism and interactivity in AR/VR digital experiences. Avatar behaviour generation is one of the greatest challenges in the immersive environment, responding to user movements or social cues. Synthetic motion helps an avatar/doer exhibit natural gestures and emotional expressions when only partial sensor data is available (Habibie et al., 2022). Gesture-based interfaces use motion synthesis. In AR scenarios, hand-tracking data may be augmented or corrected by predictive models to fill in missing motion, allowing for increased accuracy and reduced latency. This is especially important for mobile AR, where sensor data is limited. From an application point of view, this is also gaining ground in avatar personalisation. Machine learning models modify synthesised motion in such a way that it adapts to an individual's style, gender, or cultural background for a more inclusive and relatable avatar. In collaborative VR environments such as Mozilla Hubs and Meta Horizon Workrooms, synthetic gestures support nonverbal communication in remote meetings to foster the perception of presence.

5.6 Cross-Industry Integration

Beyond the core sectors, synthetic motion is going into industries that solve unique problems.

Sports sciences use synthetic motion for performance analysis and injury prevention. By modelling biomechanics of athletes using synthetic data, coaches and physiologists are able to make adjustments to test changes and predict their effect on performance or risk (Bähre et al., 2022).

In architecture and construction, motion simulation is used during virtual walkthroughs in 3D-rendered spaces to assist architects in estimating space usability and crowd flow before construction. Likewise, smart cities employ the motion simulation of traffic and pedestrian behaviour. In retail and e-commerce, AR product previews use motion synthesis to show clothing, furniture, or appliances moving or behaving in the user's environment. Synthesised motion models are used for virtual try-ons of shoes or jackets to display their natural motion while in use.

Across a wide array of disciplines, synthetic motion has positioned itself as a disruptive technology. Think of it as simulating very precise surgical gestures, animating lifelike characters, or enhancing real-time avatars to provide new ways of interacting, learning, caring, and designing. With new computational paradigms supporting better models, the applications of synthetic motion will presumably grow even further, offering possibilities for more personalised, immersive, and intelligent systems.

6. INTERDISCIPLINARY CONNECTIONS

Synthetic motion research and applications intersect disciplines, including biomechanics, cognitive sciences, sports science, and human-computer interaction (HCI). These interdisciplinary relationships enrich the application potential of motor generation technologies while providing further insights into human movement, perception, and interaction.

Biomechanics

In biomechanics, synthetic motion is used to model joint dynamics and musculoskeletal behaviours. For example, by imposing anatomical constraints and using models that kinematically describe human limb motion, AI-generated motion mimics realistic articulation of limbs, gait, and postural control: all those models for the aid of designing prosthetics, rehabilitation, or ergonomic analysis (Delp et al., 2007). Software tools such as OpenSim integrate motion synthesis based on neural networks to visualise and test motion trajectories under biomechanical constraints.

Cognitive Science

From the perspective of cognitive science, motion synthesis helps in studying motion perception, mirror neuron activation, and social cognition. Johansson (1973) proved with his classic point-light experiment that the minimal motion cue is enough for humans to perceive biological motion. Therefore, modern generative models use such findings to generate abstract and yet barely recognisable human motion patterns. These simulations strengthen further inquiries of how the brain interprets movement when distinguishing between intentional and mechanical movement (Giese & Poggio, 2003).

Sports Science

Synthetic motion technology detects, critiques, and prevents risky sports performances. High-resolution motion capture integrated with the predictive prowess of AI allows simulation of athletic performances in virtual conditions. Such synthetic data augmentation methods train models in recognising risky patterns of movement or in training regimens (Bähre et al., 2022). These systems are also directly used by trainers and physiologists to provide biomechanical feedback on athletes in training.

Human-Computer Interaction(-HCI)

In HCI, the employment of synthetic motion technology improves the accuracy of gesture recognition, body tracking, and XR interfaces. Generative models are used for animating avatars, interpreting gestures, or forecasting human activity when sensor data is incomplete (Habibie et al., 2022). Motion synthesis paves the way for more natural interaction in the virtual setting and allows the device to adapt to the user's behaviour on the fly. Such interdisciplinary applications demonstrate how the relationship between synthetic motion technologies and the scientific study of human movement, perception, and behaviour is reciprocal.

Figure 1. Interdisciplinary Applications of Synthetic Motion

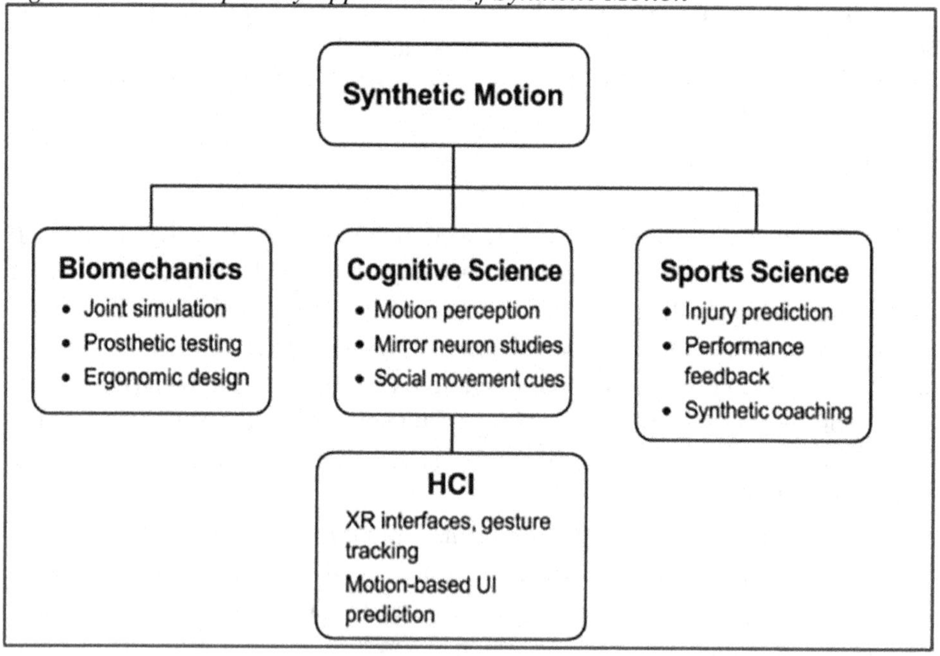

This diagram illustrates how synthetic motion intersects with four key fields—biomechanics, cognitive science, sports science, and human-computer interaction (HCI)—highlighting relevant subdomains such as joint simulation, motion perception, injury prevention, and XR interfaces.

7. REAL-TIME MOTION SYNTHESIS

The real-time motion synthesis has gained prime importance in applications such as VR, gaming, live broadcasting, and human-avatar interaction. The synthetic motions must come into existence in such a way that they are realistically meaningful along with being temporally synchronised with user inputs or some environmental changes, a tall order given the low-latency constraints.

Issues of Latency and Synchronisation

The core technical issues that real-time motion synthesis faces are latency, which is a delay between a user action (e.g., a gesture) and the corresponding motion output in the system. The higher the latency, the worse goes human experience, and VR or esports, in particular, delays above 50 milliseconds will lead to a breakdown in motion continuity and make the user disoriented (Jerald, 2015). Hence, temporal coherence should be maintained to avoid flickering problems or jitters in animation for avatars. Apart from this, synchronisation with other external systems like voice or gaze or some environmental stimulus poses its own set of difficulties. This is because multi-modal inputs may arrive at their rates with some variability, leaving the burden on the models to compensate for the rate variations while maintaining motion realism.

Solutions: Compression, Edge Inference, and Caching

In current times, various model compression techniques such as quantisation, knowledge distillation, and pruning are used to alleviate computational requirements without much compromise on motion quality (Han et al., 2016). These allow for the deployment of models on light-weight processors with faster inference time. With edge inference, the model runs locally on the user's device or embedded hardware. This reduces the network dependency, mostly affecting its latency. Devices like NVIDIA Jetsons or smartphone NPUs have been put into use to run pose-estimation and motion-prediction models directly on these devices (Howard et al., 2019). Motion caching and prediction techniques are used where portions of motion sequences are generated and stitched together with real-time corrections. This compromise is usually used in live streaming of avatars and in esports, where animations have to remain smooth during network fluctuations. With the progression of these technologies, real-time synthetic motion has been successfully implemented in enabling highly responsive, interactive, and immersive digital environments.

8. SOCIETAL AND LABOUR IMPACTS

The widespread adoption of synthetic motion technologies has very rapidly altered the creative industries, automation workflows, and labour markets. While they have undoubtedly eased production workflows and opened new avenues for content creation, concerns about job displacement, obsolescence of skills, and ethical labour transitions have arisen.

Automation in Creative Industries

Synthetic motion makes it possible to automate the generation of animated sequences involving character gestures and facial expressions without the traditionally manual key-framing or performance capture methods. Hence, animators, motion capture actors, and character riggers-their traditional roles-are truly being phased out, especially in the game and visual effects studios. This, in a way, mirrors the larger story of automation in digital media production, where AI models are utilised to cut down costs and turnaround time (Davis et al., 2021).

Job Displacement and Role Transformation

Synthetic motion, instead of eradicating jobs, transforms job roles. Laborious or manual works, such as cleanup of motion or sprite animation, are now more supervisory or technical, to ensure that the AI is producing the correct output. By now, the motion designer is supposed to partner with tools such as GANs and diffusion models and simply remain curators and validators of machine-generated content (McKinsey Global Institute, 2023).

Emergence of New Roles

Emerging roles such as motion data engineer, synthetic avatar trainer, and AI animation specialist require hybrid knowledge in computer vision, animation, and machine learning. Developing ethically sourced motion datasets, evaluation, and tuning of generative models is becoming a vital aspect of content pipeline implementations for both media and interactive systems.

Upskilling and Education Needs

Reskilling will be critical to navigate these tumultuous changes in our society (Bakhshi et al., 2017). Educational and online generations of programs in AI-based animation, computational creativity, and XR development. Initiatives for creative-AI literacy would come to the aid of traditional artists and animators being introduced to emerging workflows. There might also be alliances between the public and private sectors in order to provide equitable access to training in less-endowed regions. To conclude, while synthetic motion causes significant disruption, opportunities for labour transformation and innovation may arise, if and only if, infrastructure for reskilling and protection of the issues under ethical practices is sufficiently put in place.

9. ETHICAL AND LEGAL SHUTDOWNS

The widespread adoption of synthetic motion has raised complex ethical, legal, and regulatory questions. As generative models have become increasingly advanced to the point of producing motion that is indistinguishable from human behaviour, surrounded by concerns and consents on misinformation, on surveillance, and on liability. This chapter addresses three such important areas, namely deepfakes and misinformation, privacy and consent, and formal ethical frameworks toward responsible deployment.

9.1 Deepfake and Misinformation Risks

The core function of synthetic motion systems is the creation of deepfake materials is to say, fabricated videos or animations purporting to show individuals performing acts they never actually committed. From the perspective of perpetrators, deepfakes have evolved from jokes and pranks to instruments of political disinformation, harassment, and defamation (Chesney & Citron, 2019). Combining voice synthesis and facial reenactment with motion-generated avatars may simulate the deceptive or malicious conduct of real persons. These technologies have betrayed the trust of the public in video evidence and contradicted the democratic process by hyping fake recordings of incidents involving public figures. Political campaigning and state-sponsored operatives have already tried such content, calling for global actions to impose regulation (Westerlund, 2019). In the realms of media and journalism, the presumption of "authenticity" of video material is being openly challenged today, thereby ushering in the need for new tools and ethics for verification. Laws respond differently in various regions. In the United States, some states have criminalised the malicious use of deepfakes, particularly in the context of non-consensual pornography and election interference, but legislative enforcement continues to be outpaced while the generative models evolve at a rapid rate.

9.2 Privacy and Consent

Security checking datasets on motion capture has brought real issues concerning privacy in synthetic systems. Unlike images, motion data, such as gait, gesture patterns, and body movements, are considered biometric data and may serve as an identifier of an individual (Patel et al., 2020). Copying someone else's movements without permission, even when anonymised, can cause damage to that person's reputation and emotions, especially concerning gestures considered either intimate or part of cultural expression. Consent is more often than not a major ethical issue when it comes to developing body tracking. Large corpora of video datasets are currently being scraped from public repositories for use in training generative models without clear rights having been granted by the original users (Luccioni & Viganò, 2021). There is still no proper definition of what constitutes public domain content versus ethically valid corpus for training purposes, thus leaving a bit more room for legal ambiguity and further public reaction. Therefore, the ethical use of synthetic motion requires that explicit consent protocols be established whereby individuals are aware of the exact manner in which their motion data are to be used, stored, and possibly reused in synthetic environments. Legal instruments are beginning to address these topics, including the EU's AI Act, which requires transparency and traceability in high-risk AI systems, particularly where biometric data is involved.

9.3 Ethical Frameworks

Several formal ethical frameworks have been erected to enable the responsible use of synthetic motion technologies. These aim at assisting researchers, developers, and organisations in putting the principles of transparency, fairness, and accountability into choreography from the early design stage of an AI system to its deployment. Given the possibility that synthetic motion might imitate human gestures, behaviours, and identities, the ethical matters of this particular area are quite significant. One well-recognised standard is IEEE 7000-2021-Standard for Addressing Ethical Concerns During System Design. This framework embraces a methodical means of identifying stakeholder values and embedding those values into an architecture employed throughout the system's life cycle during design. It promotes ethical impact assessments, value-based requirements engineering, and stakeholder engagement from within (IEEE, 2021). When talking about synthetic motion, this means using this standard to ensure that avatars, motion generators, or animation systems are not reinforcing bias or misrepresentation of identities or operating in an ad hoc manner without supervision.

Figure 2. Ethical Governance of Synthetic Motion

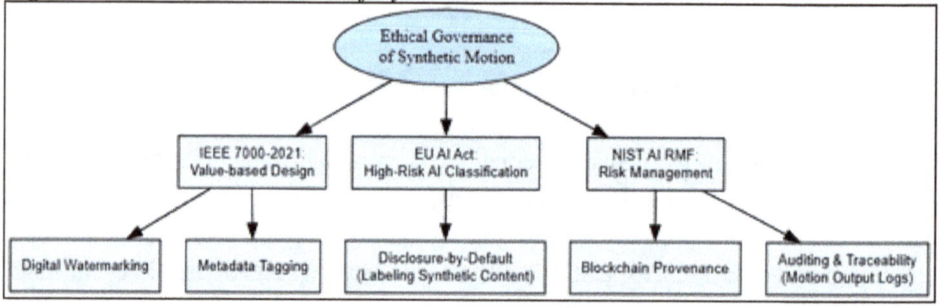

A flowchart showing how ethical frameworks, regulations, and technical tools work together to guide the responsible use of synthetic motion technologies.

The European Artificial Intelligence Act has instated a risk-based classification system for applications of AI. Systems engaged with biometric data, surveillance, or identity manipulation are reckoned as "high-risk" under this category, and synthetic motion technologies could fall into this category. For these high-risk systems, the Act subjects them to requirements including transparency documentation, human oversight mechanisms, and the marking of AI-generated content (European Commission, 2021). Thus, the idea is to shield the people against deception, misrepresentation, and involuntary use of personal data and to appreciate responsible innovation.

In the United States, the NIST AI Risk Management Framework (also known as AI RMF) offers a further methodological approach. It deals with AI risk from several points: fairness, explainability, robustness, privacy, and safety. In regard to motion synthesis applications, NIST focuses on traceability, auditing outputs, and mitigation of bias. At a time when synthetic motion finds instances in healthcare, learning, or public communication, these disciplines arguably become critical where fairness and representational accuracy come first (NIST, 2023). Apart from policy-based frameworks, there are several technical means currently under exploration for ensuring transparency and preventing misuse. One technique involves digital watermarks that are signals embedded invisibly inside some motion data that can be detected to identify the synthetic nature of the motion data under question. Another option exists with metadata tagging, which logs information regarding the motions of files or sequences to include generating models, datasets, and intended use. A more sophisticated approach will be based on blockchain-derived provenance, tracking systems, which offer an immutable ledger for content creation and amendment, ensuring traceability and tamper resistance (Crosby et al., 2016). Several disclosure-by-default policies have further been instated as an assurance to ethical standards, whereby such synthetic-generated content (e.g., AI-generated avatars or motion renderings) is allowed to be marked explicitly to differentiate such content from genuine footage. This is crucial where perception matters, in cases of political media, virtual health, and digital learning environments, whereby the demand for authenticity forms the basis of public trust. Both layers of the ethical framework and technical mechanisms stand in tandem as the multi-tier means of governing synthetic motion's creation and use. With the rise and evolution of generative technologies, an ever-expanding need for ethical oversight likewise looms. It is not so much to inhibit innovation but to direct innovation towards human values, dignity, and accountability.

10. AUTHENTICATION AND DETECTION TECHNOLOGIES

With synthetic motion advancement, increasingly sophisticated generation systems entail greater technical and ethical imperatives for the discrimination between truly profane motion and artificially generated motion. Some authentication and detection mechanisms were developed by researchers to face the increasing hazards of misinformation, identity manipulation, and content forgery. They include physics-based validation of motion, classifiers using deep learning, blockchain for provenance, and hybrid human-in-the-loop verification systems.

One of the primary synthetic motion detection methods is physics-checking thereof. The methods check whether the generated motion complies with the human biomechanical physical laws. Variables such as joint angles, centre of mass, temporal velocity consistency, and contact inconsistencies with the environment will be considered to flag unrealistic or non-human-like movements (Liu et al., 2021). Such systems tend to filter out the fake synthetic motion in various applications, including robotics and virtual rehabilitation, where physical plausibility is of paramount importance. With the rapid progress of deep learning, huge datasets of motions generated both with genuine and artificial movement have been used to train newly introduced deep learning-based detection systems that can broadly classify synthetic and real motion sequences. The trained systems can detect motion irregularities in gait and smoothness at the transition from one frame to another frame, learned as subtle statistical irregularities that can exist in both real and generated motion. The systems use temporal convolutional networks, LSTM models, or transformer-based architectures (Huang et al., 2022). These detection systems can be used in mitigating illegal deepfake animations and avatar-driven corruption of social media or political media. To build transparency and trust around motion content, blockchain-based provenance-tracking systems are an increasingly explored venue. The provenance-tracking system records metadata on an immutable ledger (Crosby et al., 2016), including information about the motion files' origin, the models employed to generate them, and their creation timestamp. By doing so, any interested party can verify the identity of the original motion content, which is of utmost value for journalists, forensic analysts, and digital content licensors. Furthermore, watermarking methods have been developed to embed an imperceptible signal within synthetic motion data, wherein any changes to the motion sequence data would amend, erase, or obscure the watermark, hence preventing validation of a motion sequence as AI-generated or modified. This is a lightweight solution that fits perfectly with real-time ones.

Human-in-the-loop verification paradigms may also be invoked in high-level decision scenarios where full automation is not considered trustworthy. In this case, the AI acts as a first-pass filter, detecting suspicious content and then handing off this data to expert review. This structure blends scalability with interpretability to ensure that potentially suspicious motion content is subjected to fair vetting before it is pushed into dissemination.

11. CASE STUDIES

Some significant case studies in areas such as film, gaming, and health would perhaps explain the immense real-life impact of synthetic motion. These actual implementations illustrate how synthetic motion has changed the creative workflow, enabled immersive experiences, or facilitated continuity of care in critical situations, such as the COVID-19 pandemic. One of the most significant and well-known examples is James Cameron's Avatar (2009), which took performance capture into a new realm of realism by actually integrating in real-time, facial and body motion data into photorealistic 3D characters. The film used an array of infrared cameras and developed animation rigs to interpret the actors' micro-expressions and gestures and project them onto digital avatars. It was not fully AI-driven, but Avatar served as a stepping stone for synthetic motions today, wherein machine learning now assists or completely automates the motion synthesis pipeline (Zhou et al., 2020). This film demonstrated the potential creativity in mixing human expressivity with digital manipulation and hence has influenced cinematic storytelling and VR avatar design.

Such systems could supply home entertainment for demonstrations of therapy exercises to patients, impromptu, amended feedback being thus given to them for their actual performance. This landmark moment within the telehealth movement gave the potential to be recognised as a clinical tool. Together, these case studies represent the diverse and even transformative applications of synthetic motion in different industries. From de-ageing actors to remote therapy, increasingly hybrid chains of motion capture, machine learning, and hybrid creative design are redefining the canvas for digital movement and interaction. Another impression was given by the 2019 movie, Gemini Man, which enhanced generative techniques for a photorealistic depiction of a younger Will SmithHere were the steps used for creation: facial motion capture followed by deep neural networks to realize the deformed expressions and speech, along with eye movements, temporally aligned frame-by-frame with the actor's live version. The completely synthetic double was rendered in CGI, trained on a huge dataset sourced from Smith's previous performances to show how motion synthesis could generate a lifelike human form without the usual animation or rotoscoping (Liu et al., 2020).In the gaming landscape, Superhot VR was an avant-garde example of motion-based interactivity. Its core mechanic of time passing only when the player moves requires close real-time tracking and synthesis of player motion. Although not a true generative AI case, it has served as an example of how the creative design of motion input and feedback loops promotes immersion. Newer VR games now take this one step further by implementing predictive motion-synthesis models, which forecast user gestures to reduce latency (Jerald, 2015). Another critical use case emerged during the COVID-19 pandemic: remote rehabilitation and therapy using AI-powered avatars. Synthetic motion systems gained entrance into hospitals and therapy centres across the world to ensure continued care for patients with physical injuries or neurological disorders. For example, GAN-based models animated virtual physiotherapists to synthesise proper joint motions according to prescribed exercises (Li et al., 2021).

12. SYNTHETIC MOTION ADVANTAGES

Synthetic motion technologies offer a vast number of benefits across creative, technical, and operational spheres. Generative models such as GANs, diffusion models, or RL-based agents have tremendously revolutionised how motion is captured, created, and applied across fields such as entertainment and health, education, and robotics. One of the biggest advantages is the reduction in production cost and time. In conventional animation and motion capture processes, there is an abundance of manual labour to be done; equipment is very expensive; hours of manual keyframing or cleanup are spent working. Synthetic motion, on the contrary, allows for the automatic generation of movements whose realism depends on simple inputs such as static poses, textual prompts, or 2D videos. These, therefore, heavily reduce development cycles in animation studios and game design pipelines (Tulyakov et al., 2018). Even more so for small studios and independent developers, where previously there was no access to large-scale motion capture facilities.

Secondly, data augmentation is another major application. Robotics, biomechanics, or gesture-recognition-based machine learning models often need extensive and varied datasets. Synthetic motion can generate human-movement variations-walking in another manner, with another speed, or a certain gesture with slight modifications-lending robustness to the models that generalise well across populations, environments, and tasks (Peng et al., 2018). For instance, sim-to-real transfer methods in robotics rely heavily on large amounts of synthetic motion to train policies in simulated environments. Creative freedom is much enlarged by synthetic motion. Artists and designers can go beyond the impermissible bounds of exaggerated, stylised, or physically impossible sets of motion. This capacity is of a special kind of attraction for immersive storytelling, dance animations, or game environment creation, where control and expressiveness outrank realistic constraints in importance. Integrated AI tools in the creative pipeline allow for previews and editing in real-time, thus speeding up the rapid feedback loop. Finally, synthetic motion can support accessibility and personalisation. Avatars can be customised to suit a person's body type, disability, or culture and hence support inclusive digital experiences. Another way they can help is through rehabilitation or education, wherein Synthetic motion-based avatars could show physical tasks in a personal way to be adapted for particular patients (Li et al., 2021). All these implications arguably have placed synthetic motion in the foreground of content production, machine learning, interactive design, and accessibility.

13. CHALLENGES AND FUTURE OUTLOOK

However, these applications face a set of impeding challenges for creating a scale, forging reliability, and attaining social acceptance: technical limitations, ethical dilemmas, and, on a broader interstate scale, those of accessibility, fairness, and sustainability. Theoretically, one of the greatest technical challenges lies within realism and generalisation. While state-of-the-art models such as GANs and diffusion-based generators impress with a tremendous amount of accuracy in a few selected tasks, these gains in precision often diminish when the model is transformed to a new task, under a less common body type, or a nonconventional motion pattern. Models trained on limited or biased datasets might tend to generate unnatural motions when simulating children or the elderly, or people with disabilities (Tevet et al., 2022). This lack of generalizability is a challenge to using synthetic motion in real-world environments such as healthcare or education, where accuracy and inclusiveness are the least expected

Another issue worth keeping in check and an ongoing curse to adopting high-fidelity synthetic motion is latency and performance. Such synthesis of motion at high resolution, especially for XR, gaming, or telepresence applications, needs to be done with fair computation speed and the lowest delay. Given these high standards, many high-fidelity models, especially the diffusion-based ones, have proved to be computationally expensive and hard to deploy on most edge devices or even consumer-grade hardware (Ho et al., 2020). Model compression and edge inference techniques have ushered in decades of breakthroughs to alleviate these problems so far, but the holy grail in this area has still been low-latency synthesis without compromising on fidelity.

Ethical and societal concerns sitting on one end, unfortunately, strangle the further growth and adoption of synthetic motion on the other. These include privacy issues associated with duplicating another person's physical movements or even their identity, forgery, and repression of stereotypes or representation of underprivileged minorities through generated motion. Attempts are ongoing to introduce taxonomies of fairness into motion models; however, given human movement's dynamic and subjective nature, instituting an ethics audit has proved especially hard (Luccioni & Viganò, 2021). There are also questions as to whether motion data gathered from surveillance videos or videos publicly viewed can be used without the consent of those filmed or recorded.

Although many ethical and legal issues need to be resolved before a synthetic motion can have a truly great future, policies and laws are still catching up. Within many jurisdictions, the classification of synthetic motion under biometric data protection laws remains a grey area (European Commission, 2021). And, consequently, the developers are sailing through this grey zone and risk putting themselves in trouble with possible compliance violations if clear standards are not laid down shortly. Despite these barriers, synthetic motion remains a bright candle for the future. Another rapidly emerging direction includes Neural Radiance Fields for dynamic photo-realistic rendering of motion in 3D. NeRFs have been portrayed to model fine-grained scene dynamics, which could consequently lead to a physically-consistent avatar system that offers a higher level of immersion (Park et al., 2021). Another swiftly rising area is zero-shot motion synthesis, whereby models can generate motion from either text or audio prompts without the need for domain-specific training data for obvious reasons. These skills should open up smaller avenues for motion design and stimulate creativity.

In the future, emphasis must be placed upon issues of fairness, interpretability, and energy efficiency if we are to work towards sustainable and ethical development. The societal implications of synthetic motion are so complex that inciting interaction between AI scientists, artists, ethicists, and policymakers will unquestionably become of utmost importance.

14. SOME INSIGHTS

Synthetic motion has undergone a long evolution from traditional animation and rotoscoping into contemporary data-driven systems, which are AI-dependent. Being important in the creative, industrial, and scientific domains, these technologies today allow the synthesis of motion with a very small amount of input, which can be text, images, or skeletal data. The higher sophistication of motion generation systems stems from machine learning models that can imitate, translate, and sometimes bring enhancement to the natural motion of humans and objects in virtual environments. Key technologies are at the core of these advances. GANs can generate the most realistic motion; yet, they suffer from drawbacks like mode collapse. Diffusion models have more variety in output and temporal stability but take longer in inference. Hybrid models that combine reinforcement learning add task awareness and adaptability, which find application in games and robotics. The coming-of-age Neural Radiance Fields (NeRFs) dynamise 3D motion rendering from photorealistic views, putting flesh on the appearance-movement duality of immersive environments. The adaptive execution of these technologies is closely tied to the computational infrastructure. Training big models on these technologies demands high-end GPUs and TPUs, while edge devices such as NVIDIA Jetson will enable real-time motion synthesis for portable applications. Simulators such as NVIDIA Omniverse and Unity ML-Agents then work to allow model building, training, and deployment into real-world pipelines. There is a very wide and growing scope of applications for synthetic motion. Synthetic motion has found bright careers in entertainment, medicine, education, robotics, and XR; eg realistic avatars, remote physiotherapy, immersive training environments, and gesture-driven interfaces. During the COVID-19 pandemic, synthetic avatars ambulatory cardinals assisted stroke patients with rehabilitation exercises, proving the utility of the technologies in time-critical care. These interdisciplinary connections play a significant role in enriching synthetic motion research. Biomechanics allows for better modelling of human joints; cognitive science informs theories of motion perception; sports science uses motion analysis for injury prevention, and HCI investigates gesture recognition and XR interfaces.

The caseloads are lots and ever-expanding for synthetic motion! It has been beneficial in entertainment, healthcare, education, robotics, and XR, such as in the making of realistic avatars, remote physiotherapy, immersive training environments, and gesture-driven interfaces. During COVID-19, for instance, synthetic avatars, in conjunction with stroke patients, helped conduct and perform rehabilitation exercises, marking an instance where this technology can be of help in time-critical care.

Interdisciplinary collaborations are crucial in enriching synthetic motion research. The biomechanics integration allows for better modelling of human joints, while cognitive sciences provide insights into the perception of motion; sports science investigates motion analysis for injury prevention, and HCI looks at gesture recognition and XR interfaces. The problems staying under the limelight, latency, bias, overfitting, and ethical ambiguity, continue to divide and shape the discourse. There needs to be urgent attention till now to issues around the misuse of deepfakes, privacy violations, and training data lacking consent. Ethical frameworks like IEEE 7000-2021, the EU AI Act, and the NIST AI Risk Management Framework provide a timely structure within which responsible deployment needs to be framed. Indicators that aid in transparency and accountability are already being developed, with a range going from physics-based validation to blockchain-based provenance.

Future trends that will redefine the edge of interaction, performance, and realism include zero-shot motion synthesis and real-time NeRF integration. Addressing the limitations today and keeping ethical design alive will help synthetic motion as an area of concern for the good of society.

REFERENCES

Bakhshi, H., Downing, J. M., Osborne, M. A., & Schneider, P. (2017). *The future of skills: Employment in 2030.* Pearson and Nesta.

Crosby, M., Pattanayak, P., Verma, S., & Kalyanaraman, V. (2016). Blockchain technology: Beyond Bitcoin. *Applied Innovation Review*, *2*, 6–10.

Davis, N., Liu, S., & Lovato, S. (2021). Creativity and AI in digital media: Implications for labour. *ACM Transactions on Multimedia Computing Communications and Applications*, *17*(3s), 1–22.

Delp, S. L., Anderson, F. C., Arnold, A. S., Loan, P., Habib, A., John, C. T., Guendelman, E., & Thelen, D. G. (2007). OpenSim: Open-source software to create and analyse dynamic simulations of movement. *IEEE Transactions on Biomedical Engineering*, *54*(11), 1940–1950. DOI: 10.1109/TBME.2007.901024 PMID: 18018689

European Commission. (2021). *Proposal for a regulation laying down harmonised rules on artificial intelligence.* Artificial Intelligence Act.

Fridrich, J., & Goljan, M. (2009). Digital image forensics and the detection of image forgery. *IEEE Signal Processing Magazine*, *26*(2), 26–37. DOI: 10.1109/MSP.2008.931078

Giese, M. A., & Poggio, T. (2003). Neural mechanisms for the recognition of biological movements. *Nature Reviews. Neuroscience*, *4*(3), 179–192. DOI: 10.1038/nrn1057 PMID: 12612631

Goodfellow, I.. (2014). Generative adversarial nets. *Advances in Neural Information Processing Systems*, *27*, 2672–2680.

Habibie, I., Tzionas, D., Black, M. J., & Theobalt, C. (2022). Learning human dynamics from sparse motion sensors using attention. *IEEE Transactions on Pattern Analysis and Machine Intelligence.*

Han, S., Mao, H., & Dally, W. J. (2016). Deep compression: Compressing deep neural networks with pruning, trained quantization and Huffman coding. *International Conference on Learning Representations (ICLR).*

Ho, J., Jain, A., & Abbeel, P. (2020). Denoising diffusion probabilistic models. arXiv preprint arXiv:2006.11239.

Howard, A. G.. (2019). Searching for MobileNetV3. *IEEE/CVF International Conference on Computer Vision (ICCV)*, 1314–1324.

Huang, Z., Li, X., Wang, H., & Duan, X. (2022). Detecting synthetic human motion using temporal consistency and motion statistics. *IEEE Transactions on Multimedia, 24*, 3472–3485.

. IEEE. (2021). IEEE 7000-2021: IEEE Standard Model Process for Addressing Ethical Concerns During System Design.

Jerald, J. (2015). *The VR Book: Human-centred design for virtual reality*. ACM Books. DOI: 10.1145/2792790

Johansson, G. (1973). Visual perception of biological motion and a model for its analysis. *Perception & Psychophysics, 14*(2), 201–211. DOI: 10.3758/BF03212378

Jouppi, N. P., Yoon, D. H., Kurian, G., Li, S., Patil, N., Laudon, J., Young, C., & Patterson, D. (2021). A domain-specific supercomputer for training deep neural networks. *Communications of the ACM, 64*(7), 67–75. DOI: 10.1145/3360307

Juliani, A., ... (2020). Unity: A general platform for intelligent agents. arXiv preprint arXiv:1809.02627.

Li, C., Wang, L., & Lee, Y. (2021). Avatar-based rehabilitation system for post-stroke patients using motion synthesis. *Healthcare Technology Letters, 8*(5), 94–100.

Liu, C., Habibie, I., Tzionas, D., & Theobalt, C. (2021). Physics-aware motion validation for trustworthy avatar animation. *IEEE/CVF Conference on Computer Vision and Pattern Recognition (CVPR)*, 2371–2380

Liu, Y., Xu, T., Zhang, L., & Wang, Y. (2020). Digital human synthesis in cinema: A case study on Gemini Man. *The Visual Computer, 36*(11), 2357–2367.

. McKinsey Global Institute. (2023). The State of AI in 2023.

Merel, J., ... (2019). Learning human behaviours from motion capture by adversarial imitation. arXiv preprint arXiv:1707.02201.

Mildenhall, B.. (2020). NeRF: Representing scenes as neural radiance fields for view synthesis. *European Conference on Computer Vision (ECCV)*, 405–421. DOI: 10.1007/978-3-030-58452-8_24

Mirsky, Y., & Lee, W. (2021). The creation and detection of deepfakes: A survey. *ACM Computing Surveys, 54*(1), 1–41. DOI: 10.1145/3425780

Moeslund, T. B., Hilton, A., & Krüger, V. (2006). A survey of advances in vision-based human motion capture and analysis. *Computer Vision and Image Understanding, 104*(2–3), 90–126. DOI: 10.1016/j.cviu.2006.08.002

Nichol, A. Q., & Dhariwal, P. (2021). Improved denoising diffusion probabilistic models. arXiv preprint arXiv:2102.09672.

NIST. (2023). *Artificial Intelligence Risk Management Framework (AI RMF 1.0)*. National Institute of Standards and Technology.

Pan, X., & Hamilton, A. F. de C. (2018). Why and how to use virtual reality to study human social interaction: The challenges of exploring a new research landscape. *British Journal of Psychology*, *109*(3), 395–417. DOI: 10.1111/bjop.12290 PMID: 29504117

Park, K.. (2021). Nerfies: Deformable neural radiance fields. *IEEE/CVF International Conference on Computer Vision (ICCV)*, 5865–5874.

Peng, X. B., ... (2018). Variational discriminator bottleneck: Improving imitation learning, inverse RL, and GANs by constraining information flow. arXiv preprint arXiv:1810.00821.

Salimans, T.. (2016). Improved techniques for training GANs. *NeurIPS*, *29*, 2234–2242.

Satava, R. M. (2002). Virtual reality surgical simulator: The first steps. *Surgical Endoscopy*, *17*(3), 451–455. PMID: 8503081

Siarohin, A.. (2018). Animating arbitrary objects via deep motion transfer. *IEEE/CVF Conference on Computer Vision and Pattern Recognition (CVPR)*, 2377–2386.

Song, Y., & Ermon, S. (2021). Score-based generative modelling through stochastic differential equations. arXiv preprint arXiv:2011.13456.

Tevet, G., ... (2022). Human motion diffusion model. arXiv preprint arXiv:2209.14916.

Thomas, F., & Johnston, O. (1995). *The illusion of life: Disney animation* (Rev. ed.). Disney Editions.

Tobin, J., ... (2017). Domain randomisation for transferring deep neural networks from simulation to the real world. arXiv preprint arXiv:1703.06907.

Tulyakov, S.. (2018). MoCoGAN: Decomposing motion and content for video generation. *IEEE Conference on Computer Vision and Pattern Recognition (CVPR)*, 1526–1535. DOI: 10.1109/CVPR.2018.00165

Westerlund, M. (2019). The emergence of deepfake technology: A review. *Technology Innovation Management Review*, *9*(11), 40–53. DOI: 10.22215/timreview/1282

Zhang, J., Zhang, L., & Zhang, Y. (2020). Face reenactment with GANs for de-ageing in film production. *Multimedia Tools and Applications*, *79*(39), 29319–29338.

Zhou, K.. (2021). PoseGAN: A pose-guided adversarial network for generating facial expressions. *Information Sciences*, *545*, 283–296.

Zhou, Y.. (2020). Actor-centric motion modelling and control. *Computer Graphics Forum*, *39*(2), 409–420.

KEY TERMS AND DEFINITIONS

Avatar: A digital representation of any user or character, animated through motion data. Avatars serve in gaming, virtual reality, healthcare, and education to replicate either genuine or stylised human behaviour.

Biomechanics: The study of mechanical principles in living organisms, especially in movement and the structure of the human body. Biomechanical models in synthetic motion are used for plausible animation and simulation of joint movements.

Blockchain Provenance: The process of securely registering changes of synthetic content through modifications on a Blockchain to establish a record of transparency for motion data authenticity.

Deepfake: Synthetic media artefact generated through AI wherein real persons appear to be saying or doing things that they never did. Deepfake technologies combine synthetic motion, facial animation, and audio synthesis.

Diffusion Model: A generative model that learns to synthesise data (such as motion) by progressively denoising random noise; capable of great diversity and stable outputs, commonly used for image and motion generation.

Edge Inference: The capability to run machine learning models directly on local devices, such as mobile devices or VR headsets, rather than on cloud servers. It is for enabling real-time motion synthesis at low latency.

GAN or Generative Adversarial Network: A type of neural network that sets one network, the generator, to compete against another, the discriminator, to output convincing synthetic data. GANs see broad use in human motion generation, face generation, and video generation.

Gesture Recognition: The process of interpreting human gestures via motion sensors or cameras. Used in natural interaction within human-machine interfaces and XR systems.

IEEE 7000-2021: A formal ethical standard for addressing stakeholder values and ethical concerns during AI system design, including those involving synthetic motion.

Metadata Tagging: The application of purposeful metadata to motion files, such as origin, model type, and use case. This practice supports traceability and ethically sound AI deployment.

MoCap (Motion Capture): The process by which real-world human motion is captured into animation or training data for computer-based modelling. Traditional systems employ sensors or cameras to track body joints in 3D space.

NeRF (Neural Radiance Fields): A rendering approach that models 3D scenes via neural networks to represent, render, and animate motion realistically in virtual environments.

NIST AI RMF: A Risk Management Framework for Artificial Intelligence was introduced by the National Institute of Standards and Technology, U.S., promoting the concepts of fairness, transparency, and security in AI systems.

Reinforcement Learning (RL): A learning paradigm in which agents acquire optimal behaviour patterns through feedback in terms of rewards. It is used to provide interactive or adaptive systems for motion generation.

Synthetic Motion: Refers to the artificial generation of motion data that attempts to replicate real or fake motion by the application of AI algorithms. It may try to emulate human locomotion, facial expressions, or object dynamics.

Watermarking: A technique that can embed invisible markers in synthetic motion content to verify the content's origin and authenticity.

Chapter 6

Graph Neural Networks in Complex Image Recognition:

From Theory to Real-World Applications

Minh Ly Duc
https://orcid.org/0000-0003-0200-7153
Faculty of Commerce, Van Lang University, Ho Chi Minh City, Vietnam

Sang Quang Nguyen
https://orcid.org/0000-0003-0239-5264
Faculty of Automotive Engineering, School of Engineering and Technology, Van Lang University, Ho Chi Minh City, Vietnam

ABSTRACT

This study presents a comprehensive review of recent advancements in deep learning for object recognition and image classification, with a particular emphasis on the growing role of Graph Neural Networks (GNNs) in addressing complex image analysis challenges. Drawing on research from reputable databases including IEEE Xplore, Springer, Elsevier, and arXiv over the past five years, the study assesses various models using four key criteria: accuracy, generalization ability, computational efficiency, and practical applicability. The analysis reveals that GNNs offer notable advantages over traditional Convolutional Neural Networks (CNNs), especially in capturing spatial and structural relationships within image data. This makes GNNs particularly effective in scenarios that demand an understanding of topological complexity, such as classifying white blood cells, detecting faults in power systems, and tracking nano-robots.

DOI: 10.4018/979-8-3373-2277-3.ch006

SECTION I: INTRODUCTION

1.1. Introduction

Image recognition and classification are core areas of computer vision and artificial intelligence, with important applications in many industries. In the energy industry, automated monitoring systems help quickly and accurately detect insulation defects such as cracks, surface damage, or dirt, ensuring electrical system safety and reducing maintenance costs (Li, 2022). In the wood processing industry, this technology supports the classification of materials based on texture, color, and surface defects, helping to optimize production processes and quality control (Hwang & Sugiyama, 2021). In medicine, white blood cell recognition helps diagnose diseases such as leukemia, infections, immune disorders, or anemia, reducing diagnostic errors and supporting effective treatment. As a result, this technology contributes to improving accuracy and automation in many important fields. Although image recognition and classification have many applications, they still face significant challenges (Cheuque et al., 2022). Insulation images are affected by lighting, shooting angle, and obstructions, making it difficult to detect small cracks (Zhang, Huang, Li, Li, & Hao, 2022). Wood images have large variations in texture, color, and material type, requiring precise algorithms for defect classification and detection. In medicine, white blood cell images are heterogeneous in size, color, and shape, while noise from the staining process and imaging equipment can reduce the accuracy of classification models.

Traditional models such as CNNs have achieved great success in image recognition, but they still have limitations. They have difficulty exploiting complex spatial relationships, such as crack structures in insulation or wood grain. In addition, CNNs are not optimized for heterogeneous or small data such as WBC images. Cross-applicability between different problems is low, requiring retraining of the model. When processing large data, such as drone images in the energy industry or high-resolution medical images, traditional models also have problems with computational efficiency and processing time (Ahmad et al., 2022). The above challenges have created a strong motivation for research and development of advanced methods, especially Graph Neural Networks (GNNs). GNNs provide a new approach, allowing the representation of images as graphs in which elements (pixels or regions) are represented as nodes, and the relationships between them are edges. Exploiting spatial structures and semantic relationships in images that methods such as CNNs cannot do. This is the reason for promoting research on applying GNNs to the problems of recognizing and classifying images of insulators, wood, and WBCs to improve accuracy, efficiency, and practical applicability.

1.2. Background

Graph Neural Networks (GNNs) provide a new direction in processing complex structured data, but their applications in image recognition are still developing. GNNs have great potential in insulation analysis, wood classification, and white blood cell (WBC) recognition due to their ability to exploit spatial structures. In the energy industry, insulation recognition helps detect faults quickly, reduce incidents, and optimize maintenance. In wood processing, this technology supports production automation, improving efficiency. In medicine, WBC recognition helps diagnose diseases more accurately and automatically (Pradhyumna, Shreya, & Mohana, 2021). Before Deep Learning became popular, image recognition relied on hand-crafted features (HOG, SIFT, LBP) and traditional models such as SVM, but they did not meet the needs of large data. CNNs made a breakthrough by automatically extracting features, but only focused on local relationships. GNNs emerged to exploit global structure by processing data in the form of graphs, with models such as GCNs, GATs, and MPNNs. GNNs are suitable for heterogeneous data, help improve the performance of complex image analysis, and can be combined with other Deep Learning models (Xiong et al., 2021). CNNs have limitations in representing complex spatial relationships, especially with heterogeneous data. Traditional methods rely on manual feature extraction, which is not powerful enough to handle large data sets such as insulation, wood, or white blood cells. GNNs overcome these limitations by representing images as graphs, preserving relationships between regions such as cracks in insulation, wood grain, or WBC structure. In addition, GNNs can incorporate semantic data or environmental information to improve classification performance..

1.3. Objectives

Research and development of an advanced image classification system using improved GNN, processing images of insulation, wood, and WBC. The system applies context-based learning to take advantage of spatial relationships between image features. At the same time, uncertainty quantification helps identify low-confidence images, supporting flexible processing. The general learning strategy combines multiple perspectives to increase accuracy. In addition, the study optimizes uncertainty quantification to improve GNN performance, and extends the application to other deep learning models..

1.4. Contribution

The research develops an improved GNN for insulation, wood, and WBC image recognition by optimizing the graph structure, which can represent both local and global spatial relationships. The new architecture integrates multi-scale features to handle the diversity of textures and colors, and combines GNN with CNN to exploit both graph data and local features. The system is customized for each field, from insulation fault detection to wood classification and WBC recognition. The feature fusion mechanism helps to improve the classification accuracy significantly. The study improves GNN training by augmenting graph data, changing the structure while maintaining the integrity. The centralization mechanism helps to identify important regions, improving classification accuracy. Optimization techniques such as graph sparsification help to reduce computational costs while ensuring performance. The model is comprehensively evaluated compared with CNN and traditional machine learning using accuracy, F1-score, recall, etc. Experimental results on real data from insulation, wood, and WBC demonstrate the high applicability and stability of the model. The study proposes a cross-domain transfer learning framework that allows the GNN model to be transferred from one domain to another without retraining, reducing data requirements and training time. The generalizability is verified on new datasets, ensuring flexibility. An intuitive software tool has been developed to support automated image analysis in energy, wood processing, and healthcare. The application helps monitor power systems, optimize wood production, and improve disease diagnosis.

SECTION II: GRAPH NEURAL NETWORK

2.1. General design pipeline of Graph Neural Networks (GNNs)

Figure 1 above shows the general process for designing a GNN (Graph Neural Network) model (Bessadok, Mahjoub, & Rekik, 2023), including the steps from input to output and the loss function.

Figure 1. The general process for designing a GNN (Graph Neural Network) model

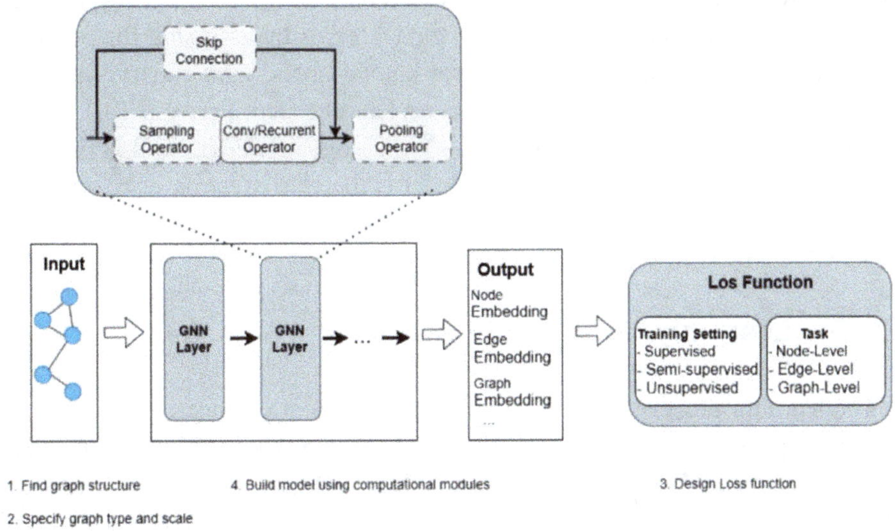

Input: This part takes as input a graph $G = (V, E)$, with: V: Set of nodes (vertices) and E: Set of edges (edges). Represents relationships between nodes and can have characteristics assigned to nodes $V_v \in R^{|V| \times d}$ and edges X_e. Find the graph structure by determining the connections between nodes through the adjacency matrix $A \in R^{|V| \times |V|}$ and determine the graph type and scale as it can be undirected, directed, or weighted. Normalized adjacency matrix $\widehat{A} = D^{-½}.AD^{-½}$, with D: degree matrix. The input is clear, easy to handle graph data. However, it is difficult if the graph is large (memory and speed issues).

GNN Layers: The GNN layer is the heart of the model, where computations are performed to generate embeddings for nodes, edges, or the entire graph. *Sampling Operator*: Selects subsets of nodes/edges to reduce computation size (especially for large graphs). *Convolution/Recurrent Operator*: Performs information transfer between nodes through formulas $H^{(l+1)} = \sigma.\left(\widehat{A}.H^{(l)}.W^{(l)}\right)$, with $H^{(l)}$: Embedding at floor l, $W^{(l)}$: Training weights, σ: Activation function. *Pooling Operator*: Aggregates information from nodes to represent the entire graph $h_G = READOUT\left(\left\{h_v v \in G\right\}\right)$, with *READOUT*: can be sum, average, or maximum, and *Skip Connection*: Direct connection between layers to reduce vanishing gradient problem. Highly customizable, combining multiple methods (convolutional/recurrent). However, it is prone to overfitting if the model is too complex.

Output: The outputs depend on the type of task. *Node Embedding*: Represents individual nodes $z_v = MLP\left(h_v\right)$. *Edge Embedding*: Represents edges and *Graph*

Embedding: Represents the entire graph $h_G = READOUT(\{h_v v \in G\})$. Flexible, applicable to many types of problems. However, complexity increases with graph size.

Loss Function: Determines how the model learns by assessing the prediction error. Training Setting: Supervised based on known labels. Semi-supervised is a part of the labels are known. Unsupervised is no label needed. Perform Tasks: Node-level is predicting node attributes $£ = -\sum_{v \in V} y_v \log(\hat{y}_v)$. Edge level is predicting relationships between nodes and graph level is the classification or regression of graphs $£ = \frac{1}{|G|} \sum_{G \in D} Loss(y_G, \hat{y}_G)$. Supports many problem types. However, requires full label data in some cases.

2.2. The Learning Algorithm of GNNs

The learning process in GNN from a mathematical perspective involves estimating the parameter ω to optimize the model based on the training data (Siemuri, Selvan, Kuusniemi, Valisuo, & Elmusrati, 2022). Find the parameter w so that the function φ_ω approximates the data in the training set $£ = \{(G_i, n_{i,j}, t_{i,j})\}$ with, G_i: The ith graph, with node set N_i and edge set E_i. $n_{i,j}$: Monitored nodes in the graph G_i. $t_{i,j} \in R^m$: Node target label or value $n_{i,j}$. The parameter ω of the GNN layers is optimized through the learning process. The mapping function φ_ω from the graph features G_i and nodes $n_{i,j}$ to the predicted labels $\hat{t}_{i,j} = \varphi_\omega(G_i, n_{i,j})$. Loss Function for the training set $L_{total} = \sum_{i=1}^{p} \sum_{j=1}^{q_i} Loss(t_{i,j}, \hat{t}_{i,j})$, with, $t_{i,j}$: Actual label value and $\hat{t}_{i,j}$: Predicted label value.

The number of supervised nodes q_i is the number of nodes used in training in the graph G_i. If $q_i = 1$: The task focuses on the entire graph (graph-level tasks). If $q_i > 1$: The task focuses on each node (node-level tasks). The mapping function φ_ω performs the information transfer and information aggregation from the graph G_i to the node $n_{i,j}$. Based on the GNN layers, this function learns to combine information from neighboring nodes. Loss Function is the core of learning is optimizing the loss function L_{total}. Cross-entropy loss (used for classification problems) $L_{CE} = -\sum_{i=1}^{p} \sum_{j=1}^{q_i} \sum_{k=1}^{C} t_{i,j,k} \times \log(\hat{t}_{i,j,k})$, with C: number of classes. and Mean squared error (MSE) (used for regression problems) $L_{MSE} = \frac{1}{p} \sum_{i=1}^{p} \sum_{j=1}^{q_i} \|t_{i,j} - \hat{t}_{i,j}\|^2$.

Table 1. Learning Algorithms of GNNs

Algorithms 1: Learning Algorithms of GNNs
Main
Initialize ω;
$x = Forward\,(\omega);$
Repeat
$\frac{\sigma_{e_w}}{\sigma_w} = BACKWARD\left(x, w\right);$
$w = w - \lambda \times \frac{\sigma_{e_w}}{\sigma_w};$
$x = FORWARD(w);$
Until (a stopping criterion)
Return w;
End
FORWARD (w);
Initialize $x(0), t = 0;$
Repeat
$x(t+1) = F_w(x(t), l);$
$t = t + 1;$
Until $\lVert x(t) - x(t-1) \rVert \le \varepsilon_f$
Return $x(t);$
End
BACKWARD (x, w)
$0 = G_w(x, l_N);$
$A = \frac{\sigma F_w}{\sigma_x}\left(x, l\right);$
$b = \frac{\sigma_{e_w}}{\sigma_0} \times \frac{\sigma G_w}{\sigma_x}\left(x, l_N\right);$
Initialize $z(0), t = 0;$
Repeat
$z(t) = z(t+1) \times A + b;$
$t = t - 1;$
Until $\lVert z(t-1) - z(t) \rVert \le \varepsilon_b;$
$c = \frac{\sigma_{e_w}}{\sigma_0} \times \frac{\sigma G_w}{\sigma_w} \times \left(x, l_N\right);$
$d = z\left(t\right) \times \frac{\sigma F_w}{\sigma_x}\left(x, l\right);$

continued on following page

Table 1. Continued

Algorithms 1: Learning Algorithms of GNNs
Main
Initialize ω;
$\frac{\sigma_{e_w}}{\sigma_w} = c + d$;
Return $\frac{\sigma_{e_w}}{\sigma_w}$
End

2.3. Variants considering graph type and scale

A diagram classifying deep learning models (Wang, Zheng, Zhang, Cui, & Xu, 2023) on graphs (Graph Neural Networks - GNN) based on the type and scale of the graph (Figure 2). This diagram classifies deep learning models based on graphs. Graph type includes research models such as Directed, Heterogeneous, Dynamic, Hypergraph, Signed, and Large. Applications of the above models range from traffic prediction and behavior analysis to processing large or complex graphs.

Figure 2. Overview of research model types in GNN

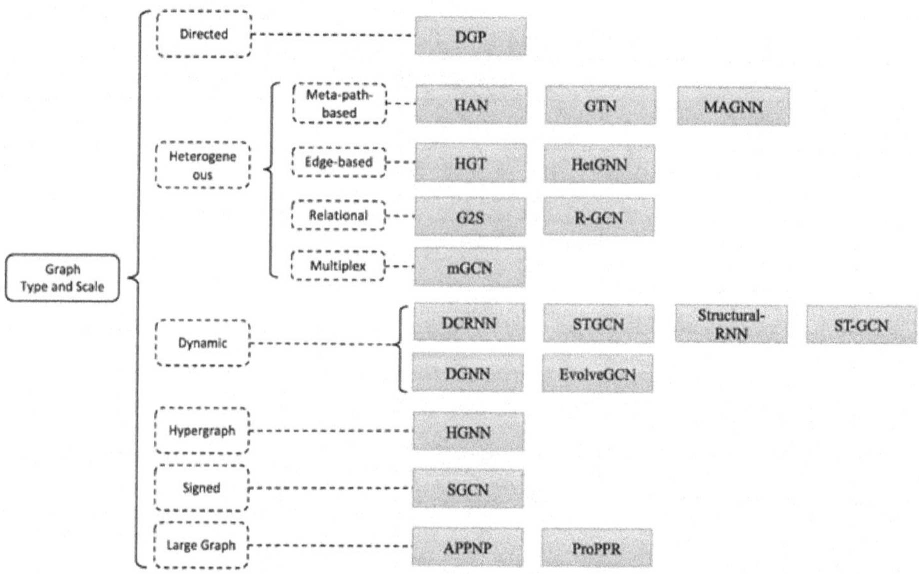

DGP (Directed Graph Propagation) is a model for directed graphs, in which directed edges are important to represent relationships between nodes. The basic formula of DGP (Eq. 1).

$$h_v^{(k+1)} = f\left(\sum_{w \in N(v)} W_{uv} . h_u^k\right) \tag{1}$$

With, h_u^k: Representation of node v characteristic at step k. W_{uv}: Directional weight from node u to v. f: Nonlinear activation function.

Heterogeneous Graphs are graphs in which nodes and edges belong to different types. **HAN** (Heterogeneous Attention Network) uses an attention mechanism to learn meta-path-based relationships, which helps to focus on important information in heterogeneous graphs (Eq. 2).

$$h_v = \sum_{p \in P} \alpha_p . AGG_p\left(h_v\right) \tag{2}$$

With, P: Set of meta-paths. α_p: Attention weight for meta-path p. AGG_p: Aggregation function based on meta-path p.

GTN (Graph Transformer Network) automatically learns meta-paths to optimize graph representation (Eq. 3).

$$A_p = \sum_{i=1}^{k} \beta_i \times A_i \tag{3}$$

With, A_p: Meta-path relationship matrix. β_i: Learned weights for the matrix A_i. k: Number of classes.

MAGNN (Meta-path Aggregated Graph Neural Network) is a combination of multiple meta-paths to improve the feature representation of a node (Eq. 4).

$$h_v = Concat\left(AGG_p(h_v), \forall p \in P\right) \tag{4}$$

HGT (Heterogeneous Graph Transformer) applies the Transformer mechanism to handle relationships between different types of nodes and edges. Use self-attention from the transformer to learn relationships between nodes of different types (Eq. 5).

$$h_v^{(k+1)} = Transformer\left(h_v^k, \{h_u^k | u \in N(v)\}\right) \tag{5}$$

HetGNN (Heterogeneous Graph Neural Network) combines information from both same-type and different-type nodes to improve predictions on the graph (Eq. 6).

$$h_v^{k+1} = \sigma\left(\sum_{r\in R} W_r \times AGG_r(h_u | u \in N(v))\right) \quad (6)$$

With, R: Different types of edges.

R-GCN (Relational Graph Convolutional Network) focuses on learning different relationships between nodes in heterostructured graphs (Eq. 7).

$$h_v^{k+1} = \sigma\left(\sum_{r\in R}\sum_{u\in N_r(v)} \frac{1}{|N_r(v)|} W_r \times h_u^k\right) \quad (7)$$

With, $N_r(v)$: Neighbors of v according to relation type r. W_r: Weight matrix for relation r.

mGCN (Multiplex Graph Convolutional Network) handles multiplex graphs, where graph layers represent different types of relationships between nodes (Eq. 8).

$$h_v^{k+1} = \sum_{m\in M} \alpha_m . GCN_m(h_v^k) \quad (8)$$

With, M: Graph classes (multiplex). α_m: The weights learned for class m.

Dynamic Graphs are graphs that change over time. **DCRNN** (Diffusion Convolutional Recurrent Neural Network) focuses on time series data on graphs, often applied in traffic prediction (Eq. 9).

$$h_t^{k+1} = \sigma\left(\left(\sum_{k=0}^{K-1}(D^{-1}.A)^k\right).h_t^k.\,W_k\right) \quad (9)$$

With, $D^{-1}.A$: Graph normalization matrix. h_t^k: Characteristic representation at time t.

STGCN (Spatial-Temporal Graph Convolutional Network) combines spatial and temporal information to analyze dynamic graph data (Eq. 10).

$$h^{t+1} = ConvTemporal(ConvSpatial(h^t)) \quad (10)$$

Structural-RNN combines RNN networks to model dynamic relationships in graphs (Eq. 11).

$$h_v^t = RNN\left(h_v^{t-1}, \sum_{u\in N(v)} W.h_u^{t-1}\right) \quad (11)$$

ST-GCN (Spatio-Temporal Graph Convolutional Network) is a variant that focuses on spatiotemporal information in the behavior analysis problem (Eq. 12).

$$h^{t+1} = TemporalConv(A . h^{t} . W) \tag{12}$$

EvolveGCN is an improvement of traditional GCN to handle dynamic graph data by learning the change of feature representations over time (Eq. 13).

$$h^{k+1} = \sigma(A_{t} . h^{k} . W_{t}) \tag{13}$$

With, A_t *and* W_t: change over time.

HGNN (Hypergraph Neural Network)is designed to learn features from hypergraphs where edges can connect more than two nodes (Eq. 14).

$$h_{v}^{k+1} = \sigma(H . W . H^{T} . h_{v}^{k}) \tag{14}$$

With, H: Hypergraph link matrix. H^T: Transpose matrix.

SGCN (Signed Graph Convolutional Network) processes graphs with positive and negative weights, often used in problems such as sentiment prediction or polarity (Eq. 15).

$$h_{v}^{k+1} = \sigma(A^{+} . h_{v}^{k} - A^{-} . h_{v}^{k}) \tag{15}$$

With, A^+: Positive edge connectivity matrix. A^-: Negative edge connectivity matrix.

Large Graphs require performance optimization due to their large size. **APPNP** (Approximate Personalized Propagation of Neural Predictions) improves the efficiency of propagation on graphs with an approximation mechanism (Eq. 16).

$$Z = (1 - \alpha) . (I - \alpha D^{-1/2} . A . D^{-1/2})^{-1} . H \tag{16}$$

With, α: Information propagation adjustment factor.

ProPPR (Programming with Personalized PageRank) incorporates PageRank to improve the ability to process large graphs with diverse data (Eq. 17).

$$h_{v}^{k+1} = \alpha . P . h_{v}^{k} + (1 - \alpha) . h_{v}^{(0)} \tag{17}$$

With, P: PageRank-based information transfer matrix.

Each model is designed with formulas suitable for a specific type of graph, optimized for handling complex relationships and different scales.

2.4. Variants for different training settings

The figure outlines different approaches for unsupervised training in graph representation learning (Figure 3). Autoencoder-based methods focus on reconstructing the graph structure or input features (GAE, VGAE, ARGA, MGAE, GALA). and contrastive learning-based methods are focused on learning representations by maximizing agreement between positive pairs (DGI, InfoGraph, Multi-View) (Behrouz & Hashemi, 2022). These methods aim to extract meaningful representations without labeled data by relying on graph topology and input features.

Figure 3. Overview of unsupervised methods in GNN

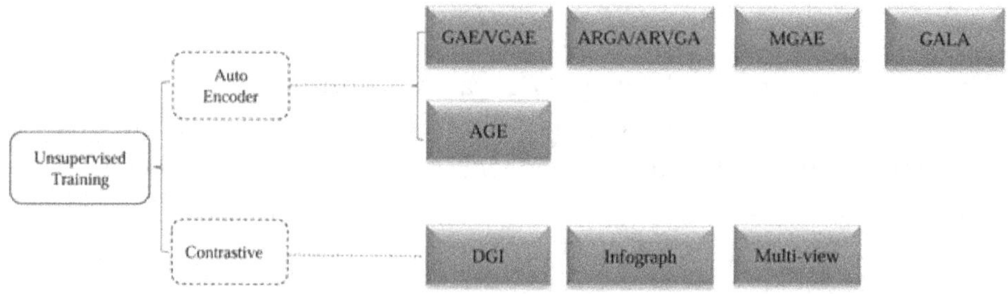

GAE/VGAE (Graph Autoencoder/Variational Graph Autoencoder) learns latent node embeddings by encoding graph structure and input features (Eq. 18).

$$L_{GAE} = \|A - \sigma(ZZ^T)\|_F^2 \tag{18}$$

With, A: Adjacency matrix. $Z = GCN(X,A)$: Node embeddings learned by a GCN encoder. σ: Sigmoid activation function.

For **VGAE, it** incorporates a variational loss term via KL divergence to regularize the latent space (Eq. 19).

$$L_{VGAE} = E_{q(ZX)} \cdot [\log p(A|Z)] - KL\big(pp(A|Z)\|p(Z)\big) \tag{19}$$

ARGA/ARVGA (Adversarially Regularized Graph Autoencoder) uses an adversarial network to regularize the embeddings learned by GAE/VGAE. The formulation combines GAE/VGAE loss with a discriminator that distinguishes real vs. fake embeddings (Eq. 20).

$$L_{dis} = -E_{Z\sim p(Z)}[\log D(Z)] - E_{Z\sim p(Z)}[\log(1 - D(Z))] \tag{20}$$

MGAE (Multi-View Graph Autoencoder) incorporates multiple graph views to learn robust embeddings (Eq. 21).

$$L_{MGAE} = \sum_{i=1}^{m} \|A_i - \sigma(Z_i . Z_i^T\|_F^2 \tag{21}$$

With, A_i: Adjacency matrix for view i. Z_i: Embedding for view i.

GALA (Graph Adversarial Latent Autoencoder) extends ARGA/ARVGA with additional graph-specific adversarial components.

Contrastive learning focuses on learning embeddings by maximizing agreement between similar samples (positive pairs) and minimizing agreement for dissimilar samples (negative pairs). **DGI** (Deep Graph Infomax) learns node embeddings by maximizing mutual information between local node embeddings and global graph embeddings (Eq. 22).

$$L_{DGI} = -E_g[\log\sigma(h^T . s)] - E_{\hat{g}}[\log(1 - \sigma(h^T . s))] \tag{22}$$

With, h: Local node embedding. s: Summary vector for the entire graph. $\hat{g}$: Corrupted graph.

InfoGraph extends DGI to maximize mutual information between subgraph embeddings and the global graph embedding (Eq. 23).

$$L_{InfoGraph} = I(Z_{subgraph,}, Z_{graph}) \tag{23}$$

Multi-View Contrastive Learning Learn embeddings by contrasting information from multiple graph views (Eq. 24).

$$L_{multi-view} = \sum_{i=1}^{m}\sum_{j=1}^{m} contrast(Z_i, Z_j) \tag{24}$$

With, Z_i, Z_j: Embeddings from different views i,j.

GAE/VGAE is suitable for simple problems or problems that require high generalization. ARGA/ARVGA and GALA are more effective when dealing with complex or noisy graph data. MGAE and Multi-View Contrastive Learning are optimal choices when information needs to be extracted from multiple views, and DGI and InfoGraph are suitable for unsupervised learning problems with graphs. Choosing the right method should be based on the specific problem, data complexity, and computational requirements.

2.5. Applications of Graph Neural Networks

Graph Neural Networks (GNNs) (Wang, Mazaitis, & Cohen, 2013) are powerful and versatile tools well suited to problems that require processing graph-structured data (Figure 4). GNNs have demonstrated superior performance in many fields because they can learn representations based on the relationships between nodes and edges in graphs. Despite certain limitations, GNNs provide an opportunity to extend traditional deep-learning models to process complex graph-structured data. Improvements in algorithms and computational resources will continue to drive the application of GNNs in the future.

Applications of GNNs in physics often involve modeling physical systems (Tsai & Chen, 2021), such as particles interacting in space or complex mechanical systems (Figure 4a). In these systems, each node in the graph can represent a particle or a mass, and edges represent interactions between them. GNN uses the node state update rule (Eq. 25).

$$h_v^{k+1} = \sigma\left(W^k.AGG_{u \in N(v)} \cdot \left(h_u^k, e_{u.v}\right)\right) \tag{25}$$

With, h_v^{k+1}: State of node v at step $k+1$. $N(v)$: Neighbors of node v. $e_{u.v}$: Edge state between nodes u and v. σ: Activation function (like ReLU). W^k: Weight matrix at step k.

GNNs can be applied to predict the properties of molecules or chemical reactions (Figure 4b). Each node in the graph represents an atom, and the edges represent chemical bonds between atoms. GNN learns molecular representations through vectorization (Eq. 26).

$$z_G = READOUT\left(\left\{h_v^K | v \in G\right\}\right) \tag{26}$$

With, z_G: Representation of the molecule (graph G). *READOUT*: Global aggregate function.

GNN can analyze the relationships between objects in an image (Figure 4c). Each node represents an object (such as a person or an animal), and the edges represent the spatial relationships between these objects. Use GNN to classify or predict attributes (Eq. 27).

$$y = MLP\left(READOUT\left(\left\{h_v^K | v \in G\right\}\right)\right) \tag{27}$$

With, y: Classification or prediction results. *MLP*: Multilayer Perceptron.

In NLP, GNNs are used to model the relationships between words in a sentence or phrases in a text (Figure 4d). Nodes represent words, and edges represent grammatical or semantic relationships between them. Sentence representation based on GNN (Eq. 28).

$$z_{sentence} = READOUT\left(\left\{h_v^K \mid v \in sentence\right\}\right) \tag{28}$$

GNN is used to predict links or determine the influence between individuals in a social network (Figure 4e). Each node represents a user, and edges represent relationships (such as friends or followers). Link prediction (Eq. 29)

Figure 4. Overview of GNN applications

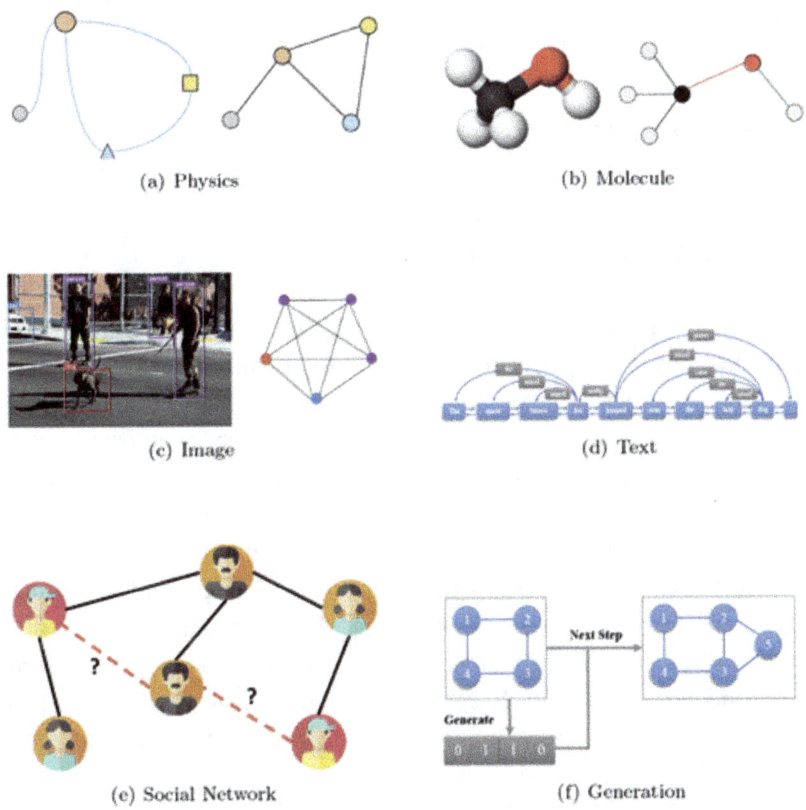

$$p_{uv} = \sigma\left(h_u^K . h_v^K\right) \tag{29}$$

With, p_{uv}: Probability of a link between u and v.

GNNs are used to generate new graphs, such as generating chemical structures or simulated social networks (Figure 4f). GNNs learn the graph state (Eq. 30).

$$z_t = GRU(z_{t-1}, h_v^t) \tag{30}$$

With, z_t: Graph state at step t. GRU: Chemical gate feedback unit.

GNNs are powerful tools for graph data processing, helping to learn relationships between elements and having wide applications in finance, healthcare, and education. Research trends focus on algorithm optimization to reduce computational costs and extract better features. However, GNNs are limited when processing complex graphs due to large resource requirements, difficulty in building suitable architectures for heterogeneous graphs, and possible loss of important information during data aggregation.

SECTION III: OBJECT DETECTORS

3.1. A few sections about the theory and state-of-the-art method

In recent years, and especially by 2024, Object Detection has expanded into many new areas and improved performance thanks to advances in algorithms (Vision Transformers) and practical applications (Autonomous vehicles, medicine, VR/AR) (Cortes, Bermejo, & Juiz, 2024). The potential for development is still very large, especially in areas requiring high precision and intelligent automation. The image illustrates the classification diagram of Object Detection types (Figure 5). Specialized object detection includes many types such as face, text, media, emotion, gesture, and product defect detection. Modern methods such as YOLO, Faster R-CNN help with general object detection, while Zero-shot Object Detection can recognize untrained objects. New trends focus on multi-modal detection, self-supervised learning to reduce dependence on labeled data, and using Edge AI to optimize performance on edge devices. Improvements in object recognition algorithms include Vision Transformers (ViT), hybrid CNN-Transformer models, and lightweight models such as YOLO-Nano, MobileNet-SSD for increased performance and optimization on mobile devices. Practical applications extend to the Metaverse for object recognition in virtual reality, defense for weapon detection, e-commerce for product search through images, and environmental protection, such as forest monitoring, wildfire detection, and ocean plastic waste tracking.

Figure 5. Overview of the identification objects

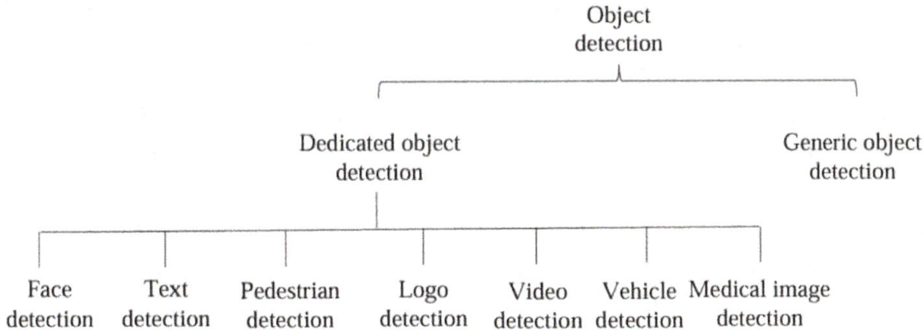

The evolution from Handcrafted Features to Deep Learning has revolutionized the field of object detection. New models such as YOLOv8, DETR, and Efficient Det not only improve accuracy but also significantly reduce computational costs. In the future, unsupervised learning and multimodal integration will further expand the application range of these algorithms. The development of Object Detection Algorithms is divided into two main stages (Figure 6)

Figure 6. Overview of Object Detection Algorithms

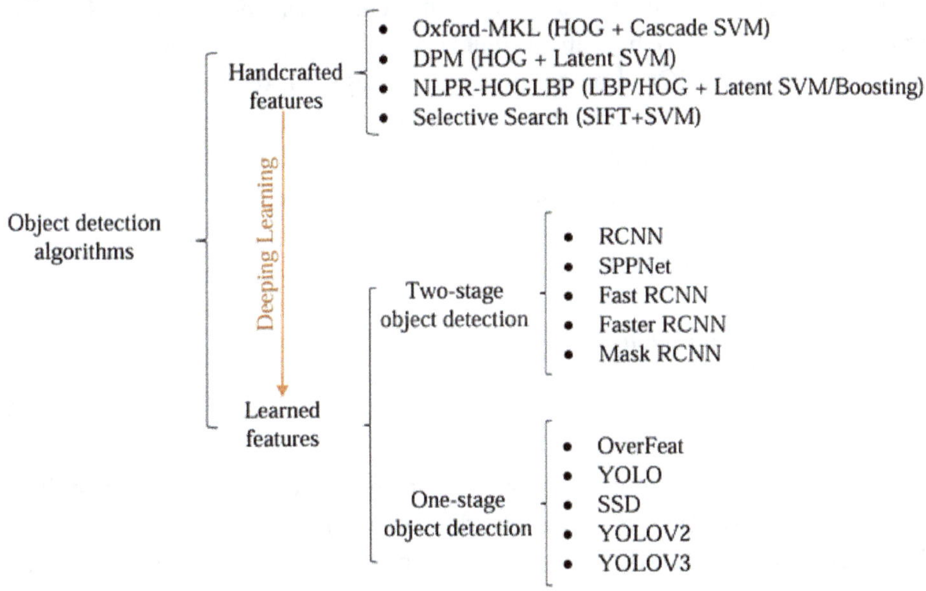

Handcrafted Features are features designed based on human knowledge to extract information from images (Ma, Jiang, Fan, Jiang, & Yan, 2021). Some typical methods include Oxford-MKL (HOG + Cascade SVM) uses HOG (Histogram of Oriented Gradients) and SVM features for classification. DPM (HOG + Latent SVM) is a deformable part model with HOG and SVM. NLPR-HOGLBP (LBP/HOG + Latent SVM/Boosting) combines HOG and LBP (Local Binary Patterns) features with Boosting or SVM methods. Selective Search (SIFT + SVM) uses a method to select regions that are likely to contain objects based on SIFT features. Hand-crafted feature methods are usually based on extracting feature vectors from images and using a classification model (Eq. 31 & Eq. 32)

$$f(x) = HOG(x) \ or \ SIFT(x) \tag{31}$$

$$y = SVM(f(x)) \tag{32}$$

With, $f(x)$: Features of image x (HOG, SIFT, LBP). y: Classification results.

Learned Features are features that are automatically learned from data through Deep Learning, divided into two main directions:

Two-stage Object Detection: The RCNN model uses the region proposal method to identify the region of interest, then extracts features and classifies them (Ren, He, Girshick, & Sun, 2017). SPPNet model improves RCNN by using Spatial Pyramid Pooling, allowing it to handle regions of different sizes. The fast RCNN model reduces computation time by using a single network for feature extraction and classification. The faster RCNN model integrates a Region Proposal Network (RPN) to improve speed and accuracy. The Mask RCNN model extends Faster RCNN by adding the ability to predict a segmentation mask. Two-stage uses region proposal model before classification (Eq. 33, Eq. 34 & Eq. 35)

$$Proposal \ Regions = RPN(I) \tag{33}$$

$$h = CNN(Proposal \ Regions) \tag{34}$$

$$y = Softmax(h) \tag{35}$$

With, $RPN(I)$: Region proposal network on image I. h: Feature vectors from CNN

One-stage Object Detection: The OverFeat model is one of the first to combine object detection and classification in a single network (Zhang & Cloutier, 2022). YOLO (You Only Look Once) model performs extremely fast object detection by dividing the image into grids and directly predicting the bounding boxes. SSD (Single Shot Multibox Detector) model performs the combination of multiple detection area scales in a single network. YOLOv2 ~YOLOv10 model is improved from YOLO with higher accuracy and speed. One stage skips the region proposal step, performing detection and classification directly (Eq. 36).

$$y = CNN(I) \tag{36}$$

The YOLO model divides the image into grids and predicts (Eq. 37)

$$Grid\ Predictions = Bounding\ Box + Class\ Scores \tag{37}$$

3.2. Variants considering object detection type and scale

The development of generic object detection methods follows two main branches Region Proposal-based (Region Proposal-based method) (Prakash & Singh, 2021) is to use the Region Proposal step to generate regions that are likely to contain objects, then perform classification and adjust the bounding boxes and Regression/ Classification-based (Regression/Classification method) performs detection and classification directly without the region proposal step (Figure 7).

Figure 7. Overview of object recognition methods

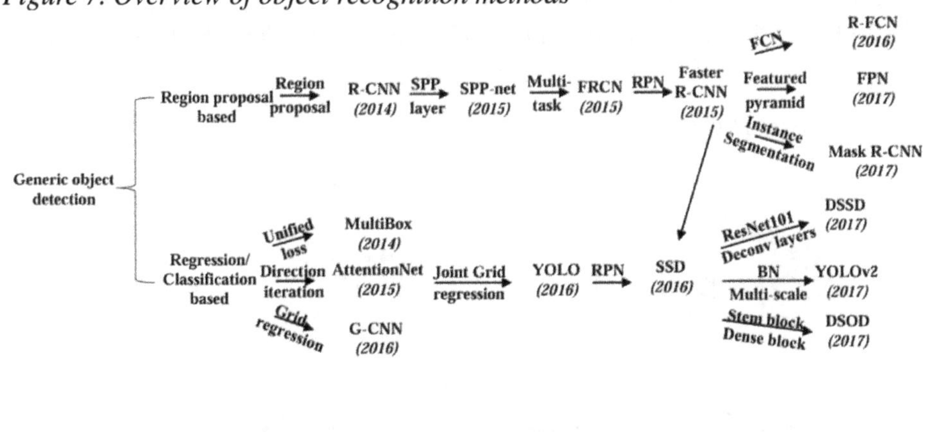

R-CNN Model (2014) proposes Regions of Interest (RoI) from images using Selective Search, then uses CNN for feature extraction and classification (Figure 8). Selective Search creates N suggested regions $\{r_i\}_{i=1}^{N}$. Feature extraction for each region $f_i = CNN(\,_i)$. Classification using SVM $y = SVM(f_i)$. The limitation of the model is that it is slow because it has to extract features from each region independently.

Figure 8. R-CNN model architecture

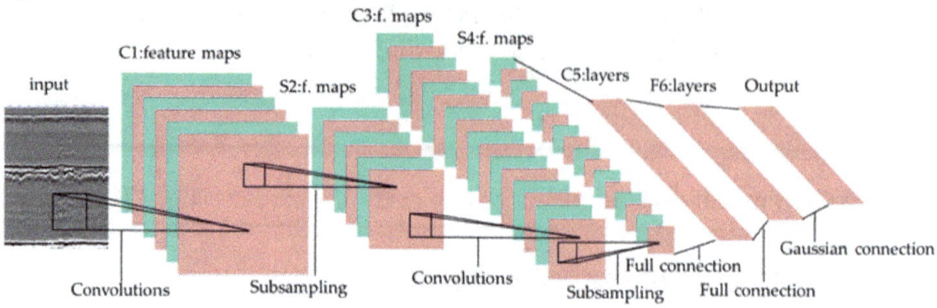

SPP-Net (2015) Model uses Spatial Pyramid Pooling (SPP) to enable simultaneous feature extraction for multiple regions of different sizes. Instead of cropping each region, SPP extracts features from the entire image and pools them according to a normalized size $f_{SPP} = SPP(CNN(I))$. This model performs classification similar to R-CNN.

Faster R-CNN (2015) Model integrates a Region Proposal Network (RPN) to propose regions using machine learning instead of Selective Search (Figure 9). RPN generates N region proposals $\{r_i\}_{i=1}^{N} = RPN(I)$. Regional characteristics $f_i = RoIAlign(CNN(I), r_i)$. Bounding box classification and regression $\{p_i, t_i\} = FC(f_i)$. With, p_i: Object class probability. t_i: Coordinates to adjust the bounding box.

Figure 9. Faster R-CNN model architecture

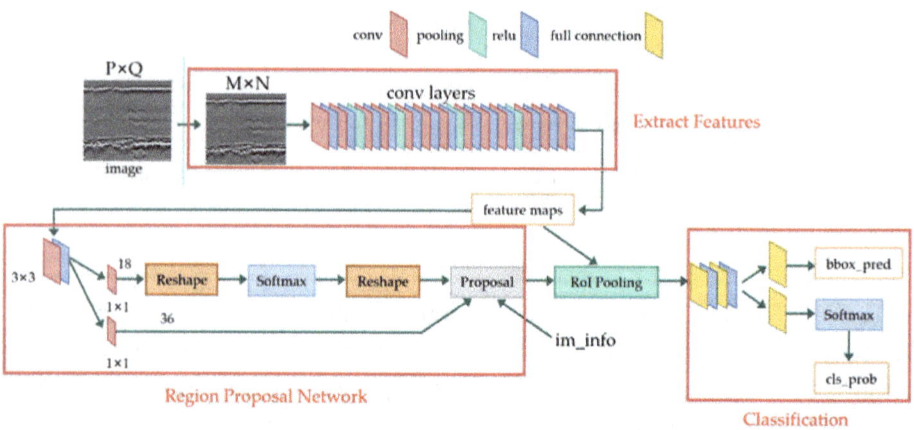

Mask R-CNN (2017) Model is a model that extends Faster R-CNN by adding an object segmentation branch. Add mask branch $M_i = FCN(f_i)$. Error function $L = L_{cls} + L_{bbox} + L_{mask}$.

YOLO (2016) Model is used to detect objects in a single step by dividing the image into $S \times S$ grid (Figure 10). Each grid cell predicts $y_{grid} = \left[p_{obj}, x, y, w, h, c_1, c_2, \ldots, c_C\right]$. Error function $L_{YOLO} = L_{coord} + L_{conf} + L_{class}$.

Figure 10. YOLO model architecture

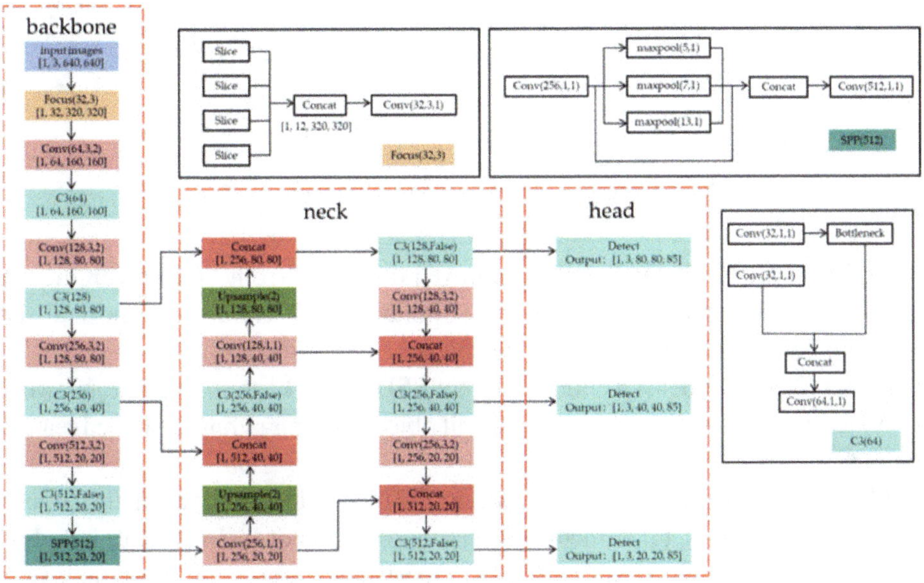

SSD (2016) Model is a combination of multiple feature layers with different recognition region scales (Figure 11). Predicting bounding boxes on multi-layer features $f_{SSD} = \{feature\ maps_k\}_{k=1}^{K}$. Error function $L_{SSD} = L_{loc} + L_{conf}$.

Figure 11. SSD model architecture

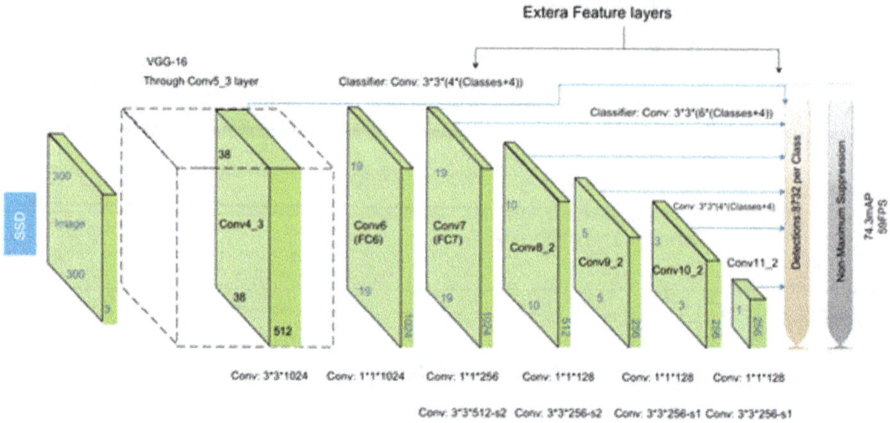

YOLOv2 and YOLOv3 (2017) improved the model from YOLO with the use of Anchor Boxes and a deeper network. Similar to YOLO but using normalized bounding boxes (anchors) $y_{YOLOv2} = \{anchors, class\ scores\}$.

DSSD (2017) Model is a combination of SSD with a Feature Pyramid Network (FPN) to improve the ability to detect small objects. Using pyramid-shaped feature layers $f_{DSSD} = FPN(CNN(I))$

Phase 1: Handcrafted Region Proposal (2014 - 2015) includes methods such as R-CNN and SPP-Net that focus on extracting Region of Interest (RoI) based on handcrafted algorithms (Selective Search). However, there are some limitations such as being time-consuming and dependent on manual extraction steps.

Phase 2: Deep Region Proposal (2015 - 2017) includes Faster R-CNN and Mask R-CNN methods that improve speed and accuracy by using RPN to learn proposed regions. Expanding applications to problems such as object segmentation.

Phase 3: Unified Regression/Classification (2016 - present) includes some methods such as YOLO, SSD, and variants that further optimize speed, aiming at real-time applications. Combining multi-layer features (FPN, DSSD) for detecting small-sized objects.

Region Proposal-based is suitable for problems that require high accuracy (e.g., medical, research). Regression/Classification-based is suitable for real-time applications such as traffic monitoring and automation. Current improvements focus on optimizing speed and improving accuracy for small or complex objects.

3.3. Applications of Object Detection

An overview of six popular single-stage object detectors (Li et al., 2022) such as Mask R-CNN, YOLO, SSD, RetinaNet, CenterNet, and EfficientDet (Figure 12). The following is an analysis of the applications of each model in different fields. Each model has its advantages and limitations, suitable for different problems and fields. Choosing the appropriate model depends on specific requirements such as speed, accuracy, or computational resources.

Figure 12. An overview of six popular single-stage object detectors

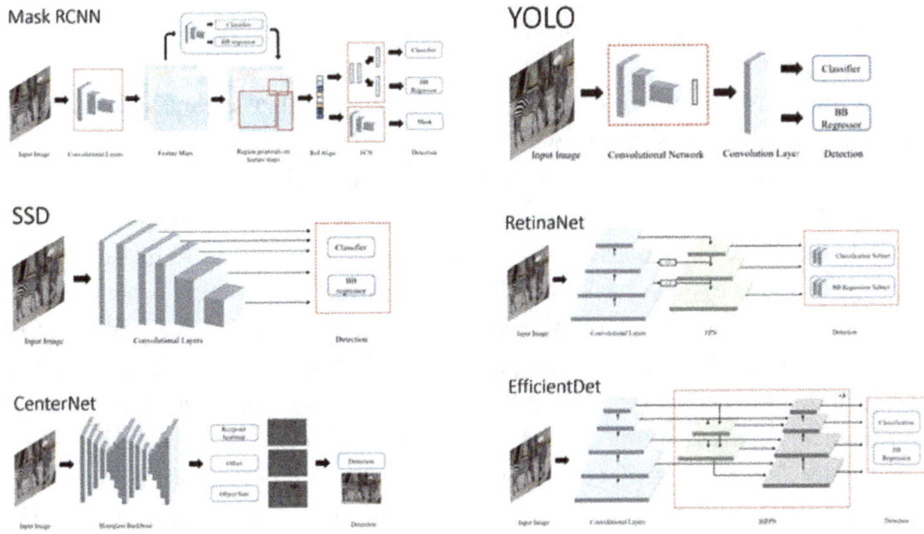

Mask R-CNN Model performs object segmentation (Instance Segmentation) suitable for applications that require detailed object segmentation. The medical field is to detect and segment lesions or tumors in medical images such as MRI and X-ray. Computer vision is used to segment each person in a crowd image. Detailed object recognition is to distinguish each object, such as vehicles in traffic or products in a supermarket. The advantages of the model are high accuracy for both object detection and segmentation, and good compatibility with many types of data. However, there are some limitations, such as slowness and requiring a lot of computational resources.

YOLO (You Only Look Once) Model is applied in real time because of its fast speed. YOLO is suitable for applications that require a quick response. Security surveillance is recognizing people or vehicles in live video. Autonomous vehicles are performing the recognition of signs, pedestrians, or vehicles. E-commerce is performing the recognition of products from customer-uploaded images. Mobile and embedded services run applications on devices with limited resources. The advantage of this model is that it is very fast, suitable for real-time systems, and Easy to deploy on lightweight devices. However, the limitation of this model is that the accuracy is lower with small or complex objects.

SSD (Single Shot Multibox Detector) Model applies to multi-scale detection with SSD suitable for applications requiring object detection at multiple sizes, such as drones to detect objects from aerial images (small or large objects). Industrial robots are used to recognize objects in the production line. Video analysis is used to track moving objects in a video. The advantages of this model are fast speed, supports detection at multiple scales, and is Suitable for live and near real-time systems. However, the limitation of this model is lower accuracy with small or overlapping objects.

RetinaNet Model applies to solve data imbalances like RetinaNet, thanks to focal loss, suitable for problems with large imbalances between object classes and background. E-commerce is used to identify rare products in stock. Wildlife detection by performing analysis and detecting rare animals in images. Healthcare performs anomaly detection (rare diseases). The advantage of this model is that it handles imbalanced data between background and object well and increases accuracy on small objects. The limitation of this model is that it is slower than YOLO and SSD.

CenterNet Model applies to simple and fast detection, as suitable for applications that do not need a precise bounding box. Real-time object tracking is used in surveillance cameras. Sports to track or athletes in videos. For light device detection, CenterNet can be deployed on mobile or embedded devices. The advantages of this model are simplicity and efficiency, fast speed, and low resource usage. The limitation of the model is low accuracy with complex problems.

EfficientDet Model applies performance optimization with BiFPN design. EfficientDet is suitable for problems that require a balance between accuracy and computational resources. Application on IoT devices for monitoring in smart devices, such as IP cameras. E-commerce performs product classification and detection on online shopping platforms. Healthcare performs medical image analysis with high accuracy, but saves resources. Multi-scale detection by handling objects of different sizes well. The advantage of this model is high efficiency with low resources and Good integration for distributed and embedded systems. The limitation of this model is that it needs to optimize the setup to achieve maximum performance.

Two-stage object detector models (Du, Zhang, & Wang, 2020) belong to the Region-based CNN (RCNN) family, and their variants include SPP-Net, Fast RCNN, Faster RCNN, FPN, and R-FCN (Figure 13). The following is an analysis of the applications and outstanding capabilities of each model.

Figure 13. Overview of Two-stage object detector models

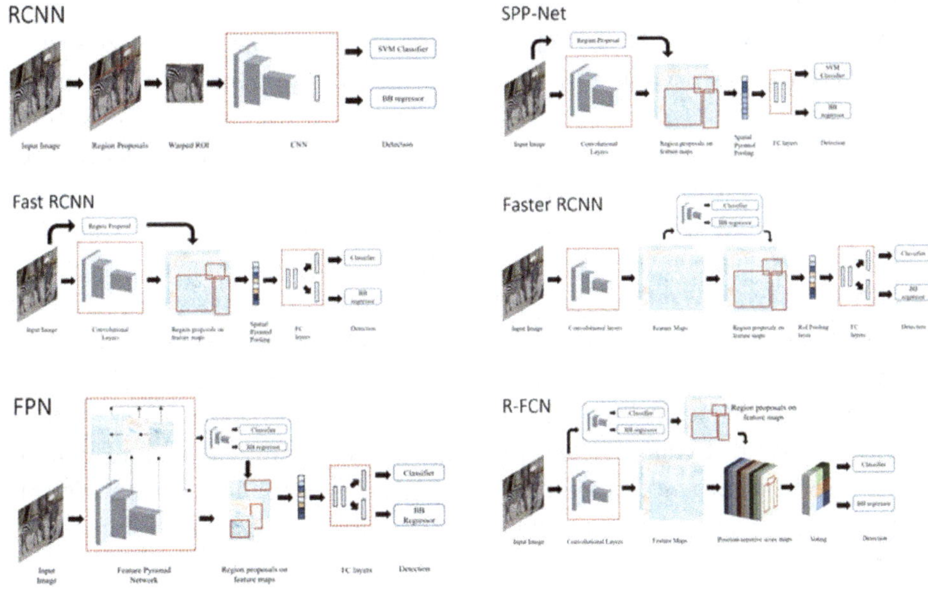

RCNN (2014) Model applied to object classification is good for clear object recognition problems with a small number of objects. Basic object detection is applied in traffic monitoring or product detection. The advantage of this model is that it is the first to combine Region Proposal with CNN, paving the way for later models. The limitation of the model is slow speed because it needs to process each proposed region (RoI) independently and requires large storage to process many proposed regions.

SPP-Net (2015) Model applied to multi-scale classification and detection is suitable for problems that need to process objects of multiple sizes, such as drone images or satellite images. Accelerated feature extraction is applied in systems that require higher speed than RCNN but do not need real-time. The advantage of this model is to reduce processing time by using Spatial Pyramid Pooling (SPP), which helps extract features for RoIs simultaneously. The limitation of this model is that it still depends on the method of proposing external regions, such as Selective Search. It does not completely solve the speed problem compared to RCNN.

Fast RCNN (2015) Model applied to faster object detection as suitable for applications such as construction monitoring or vehicle recognition. Video recognition is an application in near-real-time video analysis. The advantage of this model is that it increases the speed significantly compared to RCNN by using RoI Pooling directly from feature maps instead of cutting each RoI. Combines classification and bounding box adjustment in the same network. The limitation of this model is that it still requires an external Region Proposal step.

Faster RCNN (2015) Model applied to high-precision object detection, such as medical use (X-ray, MRI) or wildlife detection. Security surveillance systems are applied in near-real-time human or object recognition. The advantage of this model is the integration of a Region Proposal Network (RPN), eliminating dependence on external methods such as Selective Search. Significantly improved speed and accuracy compared to previous models. The limitation of the model is that it is slower than single-stage detection models (YOLO, SSD).

FPN (Feature Pyramid Network, 2017) Model applied to multi-scale detection is suitable for the problem of detecting small or large objects on the same image, such as analyzing satellite images or drone images. Detection in low-light conditions is a good application in medical images or night images. The advantage of this model is that it uses a Feature Pyramid Network (FPN) to combine information from multiple feature layers, increasing the ability to detect small objects. The limitation of the model is that it requires higher computational resources due to the combination of multiple feature layers.

R-FCN (Region-based Fully Convolutional Network, 2016) Model applies near-real-time object recognition as suitable for systems requiring a balance between speed and accuracy. Autonomous systems are applied in object recognition around autonomous vehicles. The advantage of this model is that it uses Position-sensitive Score Maps, which help reduce the number of fully connected layers, increasing speed without reducing much accuracy. The limitation of the model is that the accuracy is lower than Faster RCNN in problems requiring detailed classification.

SECTION IV: DISCUSSION

4.1. Overall discussion

The paper analyzes Graph Neural Networks (GNNs) and object classification methods, comparing GNNs with CNNs, Vision Transformers (ViTs), and CNN-GNN hybrid models based on performance, accuracy, and generalization ability. The results show that GNNs excel in modeling spatial relations, while CNNs focus on local features. Future development directions include extending GNNs, optimizing algorithms to reduce computational costs, and applying them to security monitoring, geographic analysis, and medical diagnosis.

4.2. Future works

4.2.1. Incorporating traditional optimization methods into GNNs

Future research can integrate image segmentation methods, such as Watershed, Level Set, into GNNs to detect fine structures in medical images. Data augmentation with rotation, zoom, flip, and noise improves generalization. Metaheuristic optimization improves speed and accuracy. Hybrid GNNs-CNNs models leverage the strengths of both, useful for satellite and drone images. Finally, research can combine GNNs with encryption and PCA to secure data and optimize processing.

4.2.2. Incorporating modern optimization methods into GNNs

Future research could combine Transformers with GNNs to improve object recognition in high-resolution images. Diffusion applications help reconstruct images, which is especially useful in medical applications. Self-supervised Learning reduces dependence on labeled data, while Meta-learning helps models adapt quickly to new data. GNNs can be combined with GANs to create high-quality images and with Reinforcement Learning to improve themselves over time. Additionally, combining Federated Learning increases data privacy, and Zero-shot Learning reduces training costs.

SECTION V: CONCLUSION

5.1. Summary of Contributions

The paper evaluates the methods in GNNs and object classification, indicating that GNNs have better spatial relationship modeling ability than CNNs but require high computation. GNNs are applied in insulation fault detection, wood classification, and white blood cell recognition. The CNN-GNN hybrid model is a potential direction to improve the performance. In addition, the study proposes algorithm optimization, integration with the Transformer, and extension of applications to security monitoring, geographic analysis, and healthcare to enhance the practicality of the model.

AUTHOR CONTRIBUTIONS

Conceptualization: Nguyen Quang Sang, Minh Ly Duc; Methodology: Minh Ly Duc; Formal analysis and investigation: Minh Ly Duc; Writing - original draft preparation: Minh Ly Duc; Writing - review and editing: Nguyen Quang Sang, Minh Ly Duc; Resources: Minh Ly Duc; Supervision: Minh Ly Duc.

Conflict of Interest: The authors declare no conflicts of interest.

Data Availability Statement: No application.

Supplementary Materials: No application.

ACKNOWLEDGMENT

The authors are extremely grateful to Van Lang University, Vietnam for supporting this research.

REFERENCES

Ahmad, M., Shabbir, S., Roy, S. K., Hong, D., Wu, X., Yao, J., Khan, A. M., Mazzara, M., Distefano, S., & Chanussot, J. (2022). Hyperspectral Image Classification—Traditional to Deep Models: A Survey for Future Prospects. *IEEE Journal of Selected Topics in Applied Earth Observations and Remote Sensing*, *15*, 968–999. DOI: 10.1109/JSTARS.2021.3133021

Behrouz, A., & Hashemi, F. "CS-MLGCN: Multiplex Graph Convolutional Networks for Community Search in Multiplex Networks," in *Proceedings of the 31st ACM International Conference on Information and Knowledge Management (CIKM '22)*, Atlanta, GA, USA, Oct. 2022, pp. 1-5. DOI: 10.1145/3511808.3557572

Bessadok, A., Mahjoub, M. A., & Rekik, I. (2023, May 1). Graph Neural Networks in Network Neuroscience. *IEEE Transactions on Pattern Analysis and Machine Intelligence*, *45*(5), 5833–5848. DOI: 10.1109/TPAMI.2022.3209686 PMID: 36155474

Cheuque, C., Querales, M., León, R., Salas, R., & Torres, R. (2022). An efficient multi-level convolutional neural network approach for white blood cells classification. *Diagnostics (Basel)*, *12*(2), 248. DOI: 10.3390/diagnostics12020248 PMID: 35204339

Cortes, D., Bermejo, B., & Juiz, C. (2024). The use of CNNs in VR/AR/MR/XR: A systematic literature review. *Virtual Reality (Waltham Cross)*, *28*(154), 154. Advance online publication. DOI: 10.1007/s10055-024-01044-6

Du, L., Zhang, R., & Wang, X. (2020). Overview of two-stage object detection algorithms. *Journal of Physics: Conference Series*, *1544*(1), 012033. DOI: 10.1088/1742-6596/1544/1/012033

Hwang, S. W., & Sugiyama, J. (2021). Computer vision-based wood identification and its expansion and contribution potentials in wood science: A review. *Plant Methods*, *17*(47), 47. Advance online publication. DOI: 10.1186/s13007-021-00746-1 PMID: 33910606

Li, C., Li, L., Jiang, H., Weng, K., Geng, Y., Li, L., Ke, Z., Li, Q., Cheng, M., Nie, W., Li, Y., Zhang, B., Liang, Y., Zhou, L., Xu, X., Chu, X., Wei, X., & Wei, X. "YOLOv6: A Single-Stage Object Detection Framework for Industrial Applications," arXiv preprint, arXiv:2209.02976, Sep. 2022. Available: https://arxiv.org/abs/2209.02976

Li, Y. "Research and Application of Deep Learning in Image Recognition," 2022 IEEE 2nd International Conference on Power, Electronics and Computer Applications (ICPECA), Shenyang, China, 2022, pp. 994-999, DOI: 10.1109/ICPECA53709.2022.9718847

Ma, J., Jiang, X., Fan, A., Jiang, J., & Yan, J. (2021). Image Matching from Handcrafted to Deep Features: A Survey. *International Journal of Computer Vision, 129*(1), 23–79. DOI: 10.1007/s11263-020-01359-2

Pradhyumna, P., & Shreya, G. P. (2021, August). Graph neural network (GNN) in image and video understanding using deep learning for computer vision applications. In *2021 Second International Conference on Electronics and Sustainable Communication Systems (ICESC)* (pp. 1183-1189). IEEE.

Prakash, S. R., & Singh, P. N. (2021). Object detection through region proposal based techniques. *Materials Today: Proceedings, 46*, 3997–4002. DOI: 10.1016/j.matpr.2021.02.533

Ren, S., He, K., Girshick, R., & Sun, J. (2017, June). Faster R-CNN: Towards real-time object detection with region proposal networks. *IEEE Transactions on Pattern Analysis and Machine Intelligence, 39*(6), 1137–1149. DOI: 10.1109/TPAMI.2016.2577031 PMID: 27295650

Siemuri, A., Selvan, K., Kuusniemi, H., Valisuo, P., & Elmusrati, M. S. (2022, December). A Systematic Review of Machine Learning Techniques for GNSS Use Cases. *IEEE Transactions on Aerospace and Electronic Systems, 58*(6), 5043–5077. DOI: 10.1109/TAES.2022.3219366

Tsai, M.-F., & Chen, C.-H. (2021). Spatial-Temporal Variation Graph Convolutional Networks (STV-GCN) for Skeleton-Based Emotional Action Recognition. *IEEE Access : Practical Innovations, Open Solutions, 9*, 13870–13877. DOI: 10.1109/ACCESS.2021.3052246

Wang, T., Zheng, X., Zhang, L., Cui, Z., & Xu, C. (2023). A graph-based interpretability method for deep neural networks. *Neurocomputing, 555*, 126651. DOI: 10.1016/j.neucom.2023.126651

Wang, W. Y., Mazaitis, K., & Cohen, W. W. "Programming with Personalized PageRank: A Locally Groundable First-Order Probabilistic Logic," *Proc. Int. Conf. Mach. Learn. (ICML)*, 2013. DOI: 10.1145/2505515.2505573

Xiong, J., Yu, D., Liu, S., Shu, L., Wang, X., & Liu, Z. (2021). A review of plant phenotypic image recognition technology based on deep learning. *Electronics (Basel), 10*(1), 81. DOI: 10.3390/electronics10010081

. H. Zhang and R. S. Cloutier, "Review on One-Stage Object Detection Based on Deep Learning," EAI Endorsed Transactions on e-Learning, vol. 7, no. 23, e5, pp. 1–10, Jun. 2022, .DOI: 10.4108/eai.9-6-2022.174181

Zhang, Z., Huang, S., Li, Y., Li, H., & Hao, H. (2022). Image detection of insulator defects based on morphological processing and deep learning. *Energies, 15*(7), 2465. DOI: 10.3390/en15072465

Chapter 7
Productivity Enhancements from Digital Synthetic Data

S. Ida Evangeline
https://orcid.org/0000-0003-2997-7897
Government College of Engineering, Tirunelveli, India

ABSTRACT

Synthetic data—artificially generated datasets that replicate the statistical properties of real-world data—has emerged as a powerful enabler of productivity in data-driven industries. By providing privacy-preserving, on-demand, and task-specific datasets, synthetic data addresses persistent challenges related to data scarcity, regulatory restrictions, and the high costs of acquisition and annotation. This chapter examines its role as a strategic asset in accelerating workflows across artificial intelligence and machine learning development, software testing, system prototyping, and collaborative research. It addresses technological underpinnings of the generation of synthetic data and provides cross-industry case studies in the fields of healthcare, finance, manufacture, and autonomous systems as well as quantitative and qualitative measures of productivity improvement.

1. INTRODUCTION

The accelerating digitization of modern economies has placed data at the core of innovation, operational efficiency, and competitive advantage. Yet, as data volumes grow and analytic capabilities become more sophisticated, organizations increasingly confront a paradox: the data they need to innovate is often the very data they cannot freely access or share. The dissemination, utilization, and acquisition of

DOI: 10.4018/979-8-3373-2277-3.ch007

realworld datasets are subject to a wide range of constraints that are associated with privacy rules, ownership laws, and the ethical aspects of activity. Such limitations can impede research progress, delay fielding and increase the cost of development where stakes are high and sensitivities high as in the areas of healthcare, financial systems and autonomous systems (Giuffrè & Shung, 2023).

Digital synthetic data has emerged as a compelling response to this challenge. By artificially generating datasets that replicate the statistical properties, structural relationships, and complexity of real-world data without revealing any identifiable information, synthetic data offers a means to bypass many of the legal and ethical obstacles that constrain traditional data use. New modalities in generative modeling, like Generative Adversarial Networks (GANs), variational autoencoders, agent-based simulations, and others, have broadened the scope and quality of those synthetic datasets so they became near indistinguishable to real data in many analytical applications. Synthetic data is moving beyond its role as a niche experimental tool, into a mainstream strategic asset, as these techniques mature (Göbel et al. 2023).

The potential of synthetic data extends beyond mere privacy preservation. By decoupling innovation from the constraints of data scarcity and access bottlenecks, it enables organizations to accelerate their digital workflows, from algorithm development and software testing to large-scale system simulation and AI/ML deployment. Synthetic data can be generated at will, scaled to certain experimental conditions, and evolved over the time to fit the ever-changing requirements of the project, speeding up decision-making, decrease in the time-to-market and extending team and institutional collaboration. Synthetic data literally serves as a productiveness booster since the use of synthetic data makes high-quality and context-related datasets always available (Pozzi et al. 2024).

This chapter examines synthetic data not only as a technical innovation but also as a driver of measurable productivity gains. It explores how organizations can leverage synthetic datasets to optimize development cycles, enhance model robustness, and overcome regulatory hurdles while safeguarding data integrity. The research will attempt to fill the gap that exists between data science innovation on the one hand and operational efficacy on the other by means of a blend of conceptual modelling perspectives, philosophical perspectives and actualized case-patterns. Through this, it places synthetic data as a steward of future-fit digital ecosystems, able to support both agility and compliance in the fast- shifting new data economy.

2. UNDERSTANDING SYNTHETIC DATA

Synthetic data refers to information that is artificially generated rather than collected from real-world events or interactions. Its defining characteristic lies in its ability to replicate the statistical distributions, patterns, and structural relationships found in authentic datasets, while containing no direct records of actual individuals or events. This distinction is crucial: whereas anonymized or de-identified datasets begin as real data and are subsequently stripped of identifying details, synthetic data is created entirely anew. Hence, it does not have many of the privacy, security, and consent limitations that burden the use of real data, making it a much freer environment in which organizations can work much more freely in the design, test, and deployment of data-driven systems (Gonzales et al. 2023).

Figure 1. Synthetic data in a design-test-refine loop (Almirall et al. 2022).

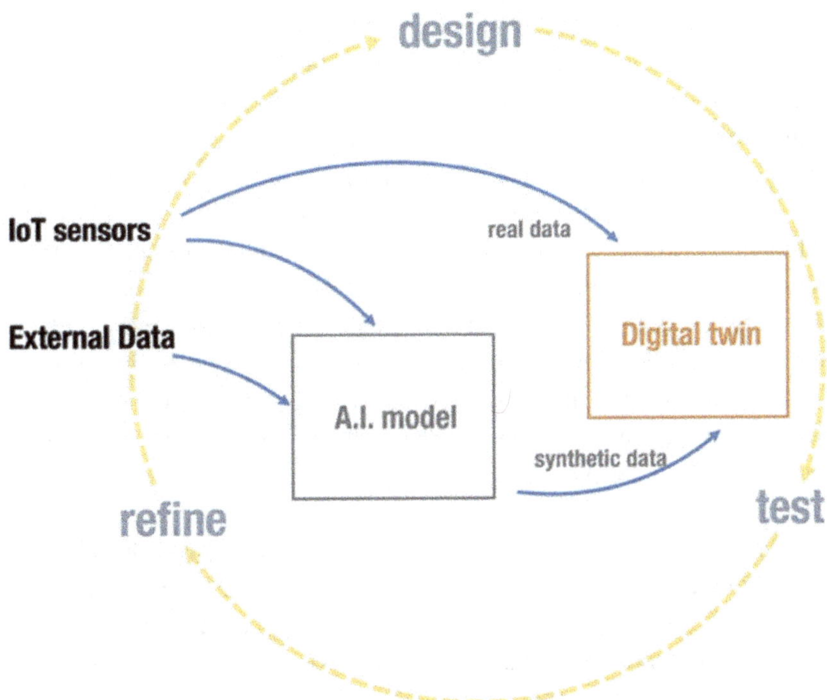

As shown in Figure 1, synthetic data may be in many different forms and models depending on the real data that it attempts to simulate. It can take the form of tabular records, such as simulating the history of transactions in a business, charts in a scientific context, duplicating medical histories, or sensor data; as image data, simulating more complex visual features to feed a computer vision system; or as time-series data, simulating patterns in the stock market, network traffic, or environmental monitoring. Speech and text data can also be synthesized so that they can either be used to train the natural language processing models or conversational agents. The flexibility of synthetic data is that it can be moulded to fit a task--its size, structure and content can be tailored to exact specifications, and it is not limited by the infelicity, incompleteness or bias that can characterize datasets gathered (Joshi et al. 2024).

Creating synthetic data is informed by an array of sophisticated computing methods, each of which has its advantages and compromises. Generative Adversarial Networks (GANs) challenge two artificial networks to act as rivals, one to create candidate data, the other to assess the reality of its data, until the artificial result is so realistic as to be practically indistinguishable with real-world examples. Variational autoencoders act as data compressor and manipulator, learning to understand the underlying structures and then being able to sample new representations of those structures. The use of agent-based simulations can capture behavioral dynamics of an individual agent and interactions in a system and generate a dynamic, context-specific dataset. In the meantime, methods based on differentially private methodology synthesise statistical noise in a manner that, at input, when using the synthetic output, there is shielding against risks of re-identification but without sacrificing analytical usefulness. There is a growing trend towards hybrid approaches which are a combination of these techniques in various ways, to balance fidelity, variety and privacy of the results sought after (Kumi et al. 2023).

The line between synthetic data and a technical artifact is its strategic place in contemporary data economy. It does not just provide complement to limited or sensitive data, but it also gives a means to maximize the potentials of experimentation and innovation. Using synthetic data, we can simulate rare or extreme events that would otherwise be prohibitively hard or expensive to obtain in the real world, and generate datasets that are orders of magnitude greater in size than anything obtainable in the real world, as well as data sets that are more balanced, bias-reduced and can be used as training material to make performance of the machine learning models more fair and accurate. The process of creating synthetic data requires conceptual expertise in its logic, modalities, and production processes in order to comprehend its opportunities, not only as a privacy-protecting mechanism, but as an ingredient of productivity and development across industries (Lopes et al. 2024).

3. SYNTHETIC DATA IN AI/ML MODEL TRAINING

Ensuring the means to create high-quality, representative datasets is critical to the creation of effective artificial intelligence and machine learning systems. However, in most areas, these datasets are sparse, disjointed, or inaccessible, because of regulatory, ethical, or proprietary restrictions. The barriers are overcome by offering a scalable customizable source of training material, which in many ways can supplement, and in some instances replace, real-world data. Through this, it can keep out of scope not only protecting privacy but also opening new avenues to create more robust, fairer, and flexible AI/ML models (Rashidi et al. 2022).

Figure 2. Overall concept of synthetic data in AI/ML model training (Gao et al. 2023).

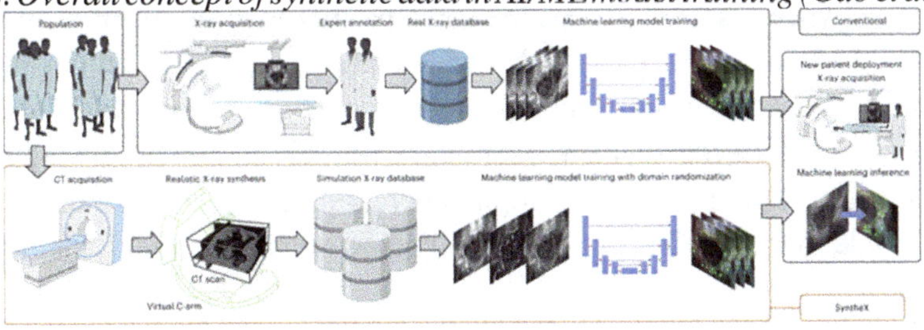

As shown in Figure 2, synthetic data can be augmented, which is one of the most urgent benefits of its use in training models. Even where real-world data exists, they are usually skewed -meant that there are many instances of common examples but practically no (or few) instances of anomalous or unusual data. Such gaps can be addressed by generating artificial instances that fill these gaps and, in this way, bias reduction can be mitigated and approaches to overfit to the prominent patterns can be reduced. Synthetic scans of rare pathologies, as in medical imaging, offer one way to expand training corpora to make sure the diagnostic model behaves reliably at both ends of the scope. To be used in fraud detection, synthetic transaction histories simulating unusual but plausible fields of fraudulent activities may enhance a model to identify the new threats (Pantanowitz et al. 2024).

It is also the case that synthetic data speeds up the experimental cycle of AI/ML development. The time it might take to find, clean, and label new data might cause delays in every iteration of model testing in traditional workflows. Having built synthetic generation pipelines, novel datasets can be generated virtually overnight to the exact specifications of the experiment under consideration. This on-demand functionality allows fast prototyping, hyperparameter optimization, and stress-testing under many different simulated conditions. The outcome is reduction of the development schedule and later amplified output (Williams & Kovalerchuk, 2024).

Further advantage is that it is possible to test models against scenarios that would be impossible to recreate in reality or even hazardous. Artificial simulation of dangerous weather caused by software designed autonomous vehicles, erratic pedestrian motion, or unusual mechanical faults can be tested on the system, without endangering human lives. When it comes to cybersecurity, intrusion detection patterns could be trained using artificial attack traffic, which reflects the smart behavior of adversaries, but does not interfere with operational networks. Such constrained, yet realistic simulations increase the space in which system models can be tested, aiding in the delivery of systems that are fault-tolerant over a larger dimension of environments (Morales-García et al. 2023).

Synthetic data used to train AI/ML has regulatory compliance and cross-border cooperation implications as well. Such privacy-preserving synthetic data can be exchanged between research groups, between partner institutions or even between countries without privacy laws being breached, allowing development of AI models that can harness the talent and computing resources at their disposal. Synthetic data in this context is like a replacement but rather a transformation or rather a reformation of the ecosystem of model development creating an agile, more inclusive, and less vulnerable to the bottlenecks historically slowing innovation (Shahul Hameed et al. 2024).

Limiting data set variability, the ability to iterate quickly, and allowing test conditions at rare or risky conditions, synthetic data demonstrates a clear route toward better-performing AI/ML systems at reduced expense and less time than currently experienced using real world data. The application of synthetic generation techniques in the AI/ML pipeline will likely grow beyond that of a supplementary tool where pipelines may rely on synthetic data as a first-order input source because of the privacy it brings to machine learning development (Kiran & Kumar, 2024).

4. OPERATIONAL EFFICIENCY IN TESTING AND DEVELOPMENT

As synthetic data has become a topic of wider prominence in the application of training AI and machine learning models, so too does the applicability of such data permeate far into the larger field of software engineering, system prototyping, and operational testing. Current digital devices such as mobile banking, even industrial control systems using the IoT, rely on environments rich in data to design and validate their performance and even optimize them. However, when these systems are to be comprehensively tested, in most instances the data necessary is too sensitive, too low, or too risky to apply to production data directly. A new methodology frees developers and testers of such constraints with synthetic data to allow simulating complex operational settings without violating the privacy, security, or system integrity (Buggineni et al. 2024).

The synthetic data set is a safe and adaptive alternative to production data that is also used in pipelines of software development in the early phases of prototyping and staging. Developers can prototype features, test workflows and functionality with realistic but non-critical data without risking legal or reputation repercussions of potentially exposing actual customer or operational data. Synthetic datasets may be produced according to specific specifications-the volume of records, the structure of the schema, and the distribution of values-therefore, making them usable to check against production systems, and make more valid judgments on systems performance. This feature is especially useful when it comes to agile and DevOps pipelines where the speed of iteration necessitates a high throughput of the continuous integration and automated testing, where large amounts of representative test data are required (Lychev, 2023).

The synthetic datasets are versatile and applicable in system testing and validation. The complex systems should be reliable in both normal operating conditions as well as extreme or edge-case scenarios. Through synthetic data, peak loads, rare occurrences and other problematic input that could result in errors can be programmed and this way the tester would know in advance the weaknesses that need to be checked before the implementation is posted. An example of such synthetic usage is an e-commerce environment placing an order on synthetic traffic to learn how it can handle a flood of transactions with customers participating in a flash sale, testing the limits of its payment processing service. In a parallel way, production control software may be validated on quasi-synthetic sensor streams that mirror abnormal behavior of equipment, allowing corrective actions in the software to be made prior to an actual appearance in reality (Long et al. 2025).

Another aspect of cybersecurity where the data synthesis comes in handy to improve operational efficiency is in testing. By creating artificial traffic through a network that resembles intended and willful usage it is also good to create synthetically high-tech malicious activity and therefore these security teams will test the performance of their intrusion detection systems without attempting them on a real production infrastructure. These are controlled simulations of this to refocus on threat detection software algorithms, enhance incident response protocols, and train security staff to manage advanced threat dynamics. Notably, due to the artificiality of the traffic, such exercises can be repeated and should be done in a large scale so that there is continuity in the skills and system resilience (Karst et al. 2025).

A less quantifiable, but cumulative, benefit of using synthetic data in development and test environments is a decrease in downtimes, a quicker release cycle, and better system performance overall after the release. With an ability to decouple testing and simulation of real world data access, organizations are able to parallelize their development, validation, and test scenarios, and test in multiple scenarios at once with faster iteration. The answer is that there is an increase not only in speed but also in quality since systems are tested under a larger and more rigorous set of test conditions before being subjected to the vagaries of actual real-world environments (Choi & Yoon, 2025).

In this regard, synthetic data will serve as a safeguarding and empowering resource during the software lifecycle. It protects sensitive data and increases the coverage and effectiveness of testing forming a feedback loop where every successive cycle of development will be faster, safer and more informed than the previous one. The strategic use of synthetic data is a core digital productivity ingredient as the intentions that touch all levels of the testing environment are becoming more intricate in the era of distributed systems, cloud-native architecture, and real-time analytics (Libes et al. 2017).

5. ENABLING COLLABORATION AND DATA SHARING

The ability in a data-driven economy to be able to engage effectively with other people across organizational and geographical barriers is becoming a determining aspect towards spurring innovation faster. However, certain collaboration is usually limited by the sensitivity of the data being contended with. Regulations like the General Data Protection Regulation (GDPR) in Europe, the Health Insurance Portability and Accountability Act (HIPAA) in the United States, and other data protection acts globally are very restrictive about how personal, financial, or operational data may be communicated. Even where legally there might be no direct data sharing, fears of intellectual property, competitor advantage and security concerns tend to lead to very closed and silo-ed information sharing (Wang et al. 2024).

Synthetic data presents an effective way of circumventing such obstacles since it facilitates the secure exchange of realistic data, without putting underlying confidential data at risk. Creating statistical and structural identical data in a manner that removes all of direct identifiers on the data allows organizations to share with collaborators a working dataset that is functionally identical to the original source data. This enables teams to test hypotheses, construct algorithms or even test systems in privacy-preservation and operational relevance (Gonzales et al. 2023).

This implication is especially acute in the areas in which data sensitivity is high, and collaboration is crucial. As another example in the sphere of healthcare research, synthetic patient records may travel between hospitals, academic institutions, and pharmaceutical companies to support the development of clinical models without involving the loss of patient confidentiality. With financial services, banks and fintech startups can trade synthetic histories of transactions with the aim to develop joint fraud detection tools or customer analytics platforms. The use of synthetic data even in the defense industry and security sectors is where privacy of information is usually extremely limited, today it is possible to permit the coordination of private contractors, government agencies and technology providers without threat of information leakage (Assefa et al. 2020).

In addition to facilitating the one-time exchange relationships and transactions, synthetic data promotes long-term, multi-stakeholder collaboration. It may be incorporated into federated learning systems where institutions can train AI models on their pseudo-datasets privately and only share model parameters or weights with a coordinator. This architecture supports the group intelligence-building without transferring sensitive raw data to minimize compliance requirements, but it does not diminish analytical capability. Synthetic data can be used as a technical defense mechanism and create a diplomatic situation in such a setting, where partners feel there are consequences of involvement by minimizing the risk (Lomurno & Matteucci, 2025).

Another element that increases the potential of collaboration that synthetic data holds is its interoperability. Since the synthetic datasets may be produced in the format compatible with recognized industry norms, they can be conveniently added to existing workflow, software, and analytics pipelines. This compatibility reduces the friction of the onboarding of new partners or workstreams convergence, which further simplifies the procedure of collaborative innovation (Dankar et al. 2022).

Synthetic data changes collaboration, burning the belief that it is a slow, limited, conservative process, into a more open, iterative and fruitful approach to sharing ideas and resources. It makes collaborations possible where none may otherwise be conceivable and allows the generation of solutions to a problem in co-creation to occur more rapidly, as they require multiple forms of expertise and sources of data. Synthetic data, in this sense, can not be viewed as a technical band-aid, but a structural facilitator of a more connected and nimble digital world.

6. INDUSTRY APPLICATIONS AND CASE STUDIES

The potential of the power of synthetic data is best seen, however, in case it is analyzed through the prism of particular industries in which the limitations of working with the real data are in collision with the necessity to innovate and apply something new and effective. In such environments, synthetic datasets are not hypothetical scientific instruments, but practical enablers of operations: that is, they allow organizations to press ahead with research, development, and deployment efforts in a manner otherwise proscribed by slow, costly or risky processes. This is further exemplified by the following sectoral examples on how synthetic data are directly translatable into measurable productivity (Buggineni et al. 2024).

The question of innovation versus patient privacy has been a core issue of concern in healthcare. Electronic Health Records (EHRs), radiology image archives and genomics data are valuable stores of patterns of interest to diagnostic, predictive and treatment planning algorithms; however that access is controlled by many locks and keys. The inability to record and maintain any identifying data makes the synthetic patient data a solution to this tension because it replicates the statistical and structure realities of clinical data itself. Synthetic EHRs have enabled hospitals to test new clinical decision support systems without endangering compliance; synthetic MRI and CT scans have been used by research teams to train diagnostic imaging models to detect uncommon diseases where live examples are scarce. Safe experimentation afforded by synthetic data can shorten the transition between conception of an algorithm to clinical trial and fast-track multi-institutional efforts on medical AI (Rujas et al. 2025).

Synthetic transaction data have been useful in the financial sector, especially in fraud detection, and risk modeling. The conventional anti-fraud mechanisms need large datasets of potential and real examples of fraud transactions to operate, but the real topics of fraud are unusual and commonly classified. Institutions can also enhance detection algorithms by simulating practical fraudulent activity patterns by creating them synthetically so as not to breach customer confidentiality. Similarly, synthetic customer accounts have exercised the lending models against a wide range of hypothetical scenarios to provide the validation of the risk models faster and more completely, prior to the implementation of the risk models in the market. The ability to conduct extensive training on the "what if" scenarios, free of regulatory tangles, considerably reduces the time of model refinement (James et al. 2021).

The flexibility of synthetic data has also found its application in manufacturing and industrial systems of IoT. Algorithms such as predictive maintenance algorithms, on the other hand, need streams of sensor data including sensors capturing machinery at different stages of wear and failure. Practically, a great number of failure modes are very rare and therefore it is impossible to find very suitable representations of them in history data. Synthetic sensor readings produced to mimic such rare circumstances can be embedded into training pipelines and this helps an algorithm to identify early warning signs. Manufacturing quality control systems have also applied synthetic image sets, to represent defects that are too expensive or inaccessible to image in real life, so that verification algorithms can develop prior to large-scale production (Fonseca & Bacao, 2023).

One of the most obvious applications of the productivity benefits of synthetic data can be found in the autonomous systems industry (especially self-driving cars). It is only through an impressive variety of environmental circumstances, traffic patterns and hazards that safe and trustful perception and decision-making algorithms can be constructed. It would be unrealistic to slap all these in the real world and consume resources and time that are too high to measure. Instead, developers turn to synthetic driving environments, with realistic lighting, weather and object dynamic behaviors, in which to train and test models, at scale. These virtual datasets enable the opportunity to be simulated that exist rarely and yet are key to safety; near-miss pedestrian crossings or vehicle failure under certain circumstances, allow these cases to be simulated and, importantly, because they are repeatable, they reduce the time required in the development cycle (Goyal & Mahmoud, 2024).

In all these cases, the trend has been evident: synthetic data has been most effective when a combination of data scarcity, privacy limitations and demands on accelerated iteration process introduces friction in the innovation chain. It empowers industries to accelerate time, scale testing and enhance decision-making through the availability of realistic, rich and flexible data sets. The outcome is not only theoretical increases in productivity but the real improved speed, quality, and safety of digital solutions that will reach the market.

7. EVALUATING PRODUCTIVITY GAINS

The strategic utility of synthetic data is easily visible in action, but to measure it we must have a solid framework in which to measure it. The benefits of increased productivity may take different forms the rate of development increments, enhanced performance of the system, increased operational flexibility, etc.--and unless a concerted effort is made to appraise them, they will be only anecdotal. Setting tangible metrics will enable organizations to measure the returns on investments and to prove the maintenance of this adoption as well as to improve the implementation of the synthetic data to get the maximum output (Doron et al. 2025).

Quantitatively, most obvious evidence is the decrease in time-to-market in use of data-driven products and services. Synthetic data enables the generation of training, testing, or simulation datasets on demand and thus reduces the time gap between development stages. In the context of AI/ML, it could mean shorter time intervals between model updates; in the context of software development, it could mean shorter time it takes to set up testing environments. Another metric that is allied to this is processing speed: synthetic and often cleaner and organized than raw real-world data, can be processed with a lower overhead on the computational costs of preprocessing and validation, thus executable experiments and deployments are quicker (Gal & Lynskey, 2023).

Another important dimension is accuracy and robustness of models. Although synthetic data cannot be analyzed without reference to real-world performance, when used to complete missing data, equalize class distributions, and increase scenario coverage, one can expect a measurable increase in predictive accuracy, precision, recall, or other performance interpolations of particular interest. The benefit of these advances is greater than technical improvement, even in such applications as rare-event detection or safety-critical decision-making where the values go beyond operational risks and user confidence (Lychev, 2023).

Qualitative actions are also significant in getting the greater impact of the productivity. These are measurable in efficiency of collaboration e.g. the number of joint projects started since the availability of synthetic data, or decrease in the number of iterations on data sharing agreements. Gains in innovation capacity, which may involve the potential to create rapid solutions (i.e., prototypes) that would previously have been prohibited by limitations of information availability, are not easily observable, but can be recorded in project timelines, user evaluations, and project post-implementation reports. The reduction in delays brought about by synthetic data in the highly regulated industry might also turn out to be the measuring indicator of its worth (Neuhausen et al. 2020).

These assessments are given a structural outline by benchmarking. The organizations can use key performance indicators to compare the results before and after they have introduced synthetic data, follow changes in the number of cycles, quality of the output, and collaboration rates. More sophisticated methods can be the control experiment, where the same projects happen within and without the integration of the synthetic data, and the following results can be compared easily. These benchmarks can be especially well used in the long-term strategic planning because synthetic data will be utilized in a manner consistent with the most urgent productivity objectives of the organization.

In the end, synthetic data boosts productivity across all dimensions, including speed, quality, safety and collaborative potential. Such a methodological measurement, a combination of metric with insight, allows organizations not only to prove the observed value, but also become continuously symbiotic when it comes to synthetic data approaches in general. This approach will turn synthetic data into not only a technical value but a measurable contributor to operations efficiencies, one able to prove its value in the board room and in the day to day workflows.

8. CHALLENGES, RISKS, AND FUTURE PROSPECTS

Synthetic data is not the solution to everything. That will depend on how well it mirrors the phenomena it endeavours to imitate, how prudently it is utilised, and how efficiently it becomes incorporated into the bigger picture of data domain. It is important to understand the restrictions and possible risks, in order to make lasting productivity out of synthetic data instead of a temporary band-aid (Jacobs et al. 2023).

The issue that is most persistent is the data fidelity. Although modern generation methods have turned synthetic information into more realistic data, no artificially generated dataset can ever be able to reproduce the complexity of actual data in a perfect way. Very small correlations, fringes, and emergent phenomena can be skipped or inaccurately modeled, producing models that work satisfyingly in simulated settings but fail in practice. Such a discrepancy between the performance in the simulation environment and that in real life may be disastrous in safety-relevant settings, including self-driving navigation or clinical decision support. Since dealing with this issue is not always black and white, a hybrid approach is frequently called upon, where synthetic data supplements, but does not substitute, the real-life data, and where models are checked on real samples prior to implementation (Kollias, 2022).

There are issues of bias and underrepresentation. In the case that the source data utilised to create synthetic datasets already has historical biases of some kind, e.g. due to demographic, geographic or behavioural bias, those biases can be maintained or may even be increased in the synthetic product. Virtual data, however, can be excessively sanitized and eliminate valuable patterns that make the model more sensitive to real-world variability. The prudent management of such risks needs the careful selection of training data, the clear description of the generation procedures, and the ongoing observation of the behavior of downstream models (Miletic & Sariyar, 2024).

Considerations on the ethical and regulatory side should not be left out. Although some designers claim that synthetic data are privacy-preserving per se, during poor implementation, such a process can also result in disclosure through unintentional deanonymisation when the generation algorithm either does not comprehensively obfuscate original sources. Moreover, the lack of transparent regulatory frameworks in other jurisdictions poses uncertainty to the organisations that intend to access synthetic data on a large scale. As the regulators are getting more acquainted with the methods of synthesis, certification is likely to evolve, creating a more explicit set of rules and standards, at the same time creating a compliance burden in the form of compliance obligations which have to be prepared by organizations to be compliant (Boutros et al. 2023).

In the future we will see all the current shortcomings being resolved with the help of technological evolution. Generative models are evolving to better understand context, and can be used to create datasets which more closely reflect the types of complex relationships, multi-modal relationships between representations, and rare-event occurrences when it comes to data at scale. They may also be auto-generated, tested, and released into production in real time, integrated with AI agents and automated data pipelines, further integrating them into daily workflows. Synthetic data can also be used more frequently in edge computing environments, in cases where localized model training is performed due to constraints of privacy and connectivity to make use of centralized datasets (Ibrahim et al. 2025).

It is also possible that synthetic data will have a part in responsible AI. Synthetic generation has the potential to support fairness auditing and compliance reporting, by allowing it to form bias-balanced, scenario-complete data with which to train their models. Fully synthetic benchmark datasets could also be introduced in research to help achieve reproducibility and transparency, mitigating the need of proprietary or restricted, real-world data (DeAndres-Tame et al. 2025).

Overall, the future of synthetic data is to be characterized by twofold imperative to do everything more technically accurate and useful and, at the same time, to be endowed with adequate protection against abuse. The organizations that can negotiate this balance effectively will also be in a position to embrace synthetic data not only as a temporary panacea to data paucity and privacy, but as a long term basis of innovation and agility and competitive positioning in a fast-changing digital economy.

9. CONCLUSION

Synthetic data has moved from being a niche research tool to a strategic enabler of productivity across a wide spectrum of digital operations. By providing high-quality, privacy-preserving datasets that can be generated on demand and tailored to specific needs, it addresses some of the most persistent barriers in data-driven innovation—privacy restrictions, data scarcity, and the unpredictability of real-world data collection. In both healthcare, financial, manufacturing, and autonomous projects, its application has resulted in actual benefits: accelerated creation timelines, increased exhaustiveness of test suites, safer prototyping, and extended team communication.

It is no longer debatable that synthetic data does not simply replace the real data, but that it allows to do something more than before. It enables organizations to use rare target events to train in data that could not be economically or safely captured, potentially validate systems in conditions of extreme or unsafe conditions, and share data much more readily that could otherwise have been used to train a model in a condition that may be illegal or unethical. These abilities also shorten schedules, much more than just a technical resource, synthetic data is an operational driver.

At the same time, its adoption requires careful stewardship. The accuracy of synthetic data, the possibility of bias transmission, and the changing rules of the regulatory environment all necessitate the need to integrate synthetic data with rigor and transparency by organizations. Hybrid approaches tend to be the most effective use cases because a combination of synthetic and real data communicates together, addressing deficiencies in one another.

As generative technologies continue to advance, synthetic data will become an even more integral part of the digital ecosystem. Companies that invest in the knowledge, governance systems, and technology systems to ensure it is responsibly used will be best suited to meet the dual pressures of innovation and compliance. A competitive world where agility, security and speed determine the concept of competitive advantage, synthetic data becomes a mechanism to achieve all three dimensions of a competitive advantage, which makes synthetic data a future-proof and productivity-oriented way of doing things.

REFERENCES

Almirall, E., Callegaro, D., Bruins, P., Santamaría, M., Martínez, P., & Cortés, U. (2022). The use of Synthetic Data to solve the scalability and data availability problems in Smart City Digital Twins. *arXiv preprint arXiv:2207.02953.*

Assefa, S. A., Dervovic, D., Mahfouz, M., Tillman, R. E., Reddy, P., & Veloso, M. (2020, October). Generating synthetic data in finance: opportunities, challenges and pitfalls. In *Proceedings of the First ACM International Conference on AI in Finance* (pp. 1-8). DOI: 10.1145/3383455.3422554

Boutros, F., Struc, V., Fierrez, J., & Damer, N. (2023). Synthetic data for face recognition: Current state and future prospects. *Image and Vision Computing, 135*, 104688. DOI: 10.1016/j.imavis.2023.104688

Buggineni, V., Chen, C., & Camelio, J. (2024). Enhancing manufacturing operations with synthetic data: A systematic framework for data generation, accuracy, and utility. *Frontiers in Manufacturing Technology, 4*, 1320166. DOI: 10.3389/fmtec.2024.1320166

Choi, Y., & Yoon, S. (2025). Enhancing in-situ model accuracy in operational building systems with augmentation-based synthetic operational data. *Journal of Building Engineering, 101*, 111623. DOI: 10.1016/j.jobe.2024.111623

Dankar, F. K., Ibrahim, M. K., & Ismail, L. (2022). A multi-dimensional evaluation of synthetic data generators. *IEEE Access : Practical Innovations, Open Solutions, 10*, 11147–11158. DOI: 10.1109/ACCESS.2022.3144765

DeAndres-Tame, I., Tolosana, R., Melzi, P., Vera-Rodriguez, R., Kim, M., Rathgeb, C., Liu, X., Gomez, L. F., Morales, A., Fierrez, J., Ortega-Garcia, J., Zhong, Z., Huang, Y., Mi, Y., Ding, S., Zhou, S., He, S., Fu, L., Cong, H., & Menotti, D. (2025). Second frcsyn-ongoing: Winning solutions and post-challenge analysis to improve face recognition with synthetic data. *Information Fusion, 120*, 103099. DOI: 10.1016/j.inffus.2025.103099

Doron, G., Genway, S., Roberts, M., & Jasti, S. (2025). Generative AI: Driving productivity and scientific breakthroughs in pharmaceutical R&D. *Drug Discovery Today, 30*(1), 104272. DOI: 10.1016/j.drudis.2024.104272 PMID: 39675517

Fonseca, J., & Bacao, F. (2023). Tabular and latent space synthetic data generation: A literature review. *Journal ofBig Data, 10*(1), 115. DOI: 10.1186/s40537-023-00792-7

Gal, M. S., & Lynskey, O. (2023). Synthetic data: legal implications of the data-generation revolution. *IoWa L. rev., 109*, 1087.

Gao, C., Killeen, B. D., Hu, Y., Grupp, R. B., Taylor, R. H., Armand, M., & Unberath, M. (2023). Synthetic data accelerates the development of generalizable learning-based algorithms for X-ray image analysis. *Nature Machine Intelligence*, *5*(3), 294–308. DOI: 10.1038/s42256-023-00629-1 PMID: 38523605

Giuffrè, M., & Shung, D. L. (2023). Harnessing the power of synthetic data in healthcare: Innovation, application, and privacy. *NPJ Digital Medicine*, *6*(1), 186. DOI: 10.1038/s41746-023-00927-3 PMID: 37813960

Göbel, T., Baier, H., & Breitinger, F. (2023). Data for digital forensics: Why a discussion on "how realistic is synthetic data" is dispensable. *Digital Threats : Research and Practice*, *4*(3), 1–18. DOI: 10.1145/3609863

Gonzales, A., Guruswamy, G., & Smith, S. R. (2023). Synthetic data in health care: A narrative review. *PLOS Digital Health*, *2*(1), e0000082. DOI: 10.1371/journal.pdig.0000082 PMID: 36812604

Goyal, M., & Mahmoud, Q. H. (2024). A systematic review of synthetic data generation techniques using generative AI. *Electronics (Basel)*, *13*(17), 3509. DOI: 10.3390/electronics13173509

Ibrahim, M., Al Khalil, Y., Amirrajab, S., Sun, C., Breeuwer, M., Pluim, J., & Dumontier, M. (2025). Generative AI for synthetic data across multiple medical modalities: A systematic review of recent developments and challenges. *Computers in Biology and Medicine*, *189*, 109834. DOI: 10.1016/j.compbiomed.2025.109834 PMID: 40023073

Jacobs, F., D'Amico, S., Benvenuti, C., Gaudio, M., Saltalamacchia, G., Miggiano, C., De Sanctis, R., Della Porta, M. G., Santoro, A., & Zambelli, A. (2023). Opportunities and challenges of synthetic data generation in oncology. *JCO Clinical Cancer Informatics*, *7*(7), e2300045. DOI: 10.1200/CCI.23.00045 PMID: 37535875

James, S., Harbron, C., Branson, J., & Sundler, M. (2021). Synthetic data use: Exploring use cases to optimise data utility. *Discover Artificial Intelligence*, *1*(1), 15. DOI: 10.1007/s44163-021-00016-y

Joshi, I., Grimmer, M., Rathgeb, C., Busch, C., Bremond, F., & Dantcheva, A. (2024). Synthetic data in human analysis: A survey. *IEEE Transactions on Pattern Analysis and Machine Intelligence*, *46*(7), 4957–4976. DOI: 10.1109/TPAMI.2024.3362821 PMID: 38319772

Karst, F. S., Li, M. M., & Leimeister, J. M. (2025). SynDEc: A synthetic data ecosystem. *Electronic Markets*, *35*(1), 7. DOI: 10.1007/s12525-024-00746-8

Kiran, A., & Kumar, S. S. (2024). A methodology and an empirical analysis to determine the most suitable synthetic data generator. *IEEE Access : Practical Innovations, Open Solutions*, *12*, 12209–12228. DOI: 10.1109/ACCESS.2024.3354277

Kollias, D. (2022, October). Abaw: learning from synthetic data & multi-task learning challenges. In *European Conference on Computer Vision* (pp. 157-172). Cham: Springer Nature Switzerland.

Kumi, S., Hilton, M., Snow, C., Lomotey, R. K., & Deters, R. (2023, July). sleepsynth: Evaluating the use of synthetic Data in Health Digital Twins. In *2023 IEEE International Conference on Digital Health (ICDH)* (pp. 121-130). IEEE. DOI: 10.1109/ICDH60066.2023.00027

Libes, D., Lechevalier, D., & Jain, S. (2017, December). Issues in synthetic data generation for advanced manufacturing. In 2017 IEEE International Conference on Big Data (Big Data) (pp. 1746-1754). IEEE. DOI: 10.1109/BigData.2017.8258117

Lomurno, E., & Matteucci, M. (2025). Federated Knowledge Recycling: Privacy-preserving synthetic data sharing. *Pattern Recognition Letters*, *191*, 124–130. DOI: 10.1016/j.patrec.2025.02.030

Long, Y., Kroeger, S., Zaeh, M. F., & Brintrup, A. (2025). Leveraging synthetic data to tackle machine learning challenges in supply chains: Challenges, methods, applications, and research opportunities. *International Journal of Production Research*, •••, 1–22. DOI: 10.1080/00207543.2024.2447927

Lopes, P. V., Silveira, L., Guimaraes Aquino, R. D., Ribeiro, C. H., Skoogh, A., & Verri, F. A. N. (2024). Synthetic data generation for digital twins: Enabling production systems analysis in the absence of data. *International Journal of Computer Integrated Manufacturing*, *37*(10-11), 1252–1269. DOI: 10.1080/0951192X.2024.2322981

Lychev, A. V. (2023). Synthetic Data Generation for Data Envelopment Analysis. *Data*, *8*(10), 146. DOI: 10.3390/data8100146

Miletic, M., & Sariyar, M. (2024). Challenges of using synthetic data generation methods for tabular microdata. *Applied Sciences (Basel, Switzerland)*, *14*(14), 5975. DOI: 10.3390/app14145975

Morales-García, J., Bueno-Crespo, A., Terroso-Sáenz, F., Arcas-Túnez, F., Martínez-España, R., & Cecilia, J. M. (2023). Evaluation of synthetic data generation for intelligent climate control in greenhouses. *Applied Intelligence*, *53*(21), 24765–24781. DOI: 10.1007/s10489-023-04783-2

Neuhausen, M., Herbers, P., & König, M. (2020). Using synthetic data to improve and evaluate the tracking performance of construction workers on site. *Applied Sciences (Basel, Switzerland)*, *10*(14), 4948. DOI: 10.3390/app10144948

Pantanowitz, J., Manko, C. D., Pantanowitz, L., & Rashidi, H. H. (2024). Synthetic data and its utility in pathology and laboratory medicine. *Laboratory Investigation*, *104*(8), 102095. DOI: 10.1016/j.labinv.2024.102095 PMID: 38925488

Pozzi, M., Noei, S., Robbi, E., Cima, L., Moroni, M., Munari, E., Torresani, E., & Jurman, G. (2024). Generating and evaluating synthetic data in digital pathology through diffusion models. *Scientific Reports*, *14*(1), 28435. DOI: 10.1038/s41598-024-79602-w PMID: 39557989

Rashidi, H. H., Khan, I. H., Dang, L. T., Albahra, S., Ratan, U., Chadderwala, N., To, W., Srinivas, P., Wajda, J., & Tran, N. K. (2022). Prediction of tuberculosis using an automated machine learning platform for models trained on synthetic data. *Journal of Pathology Informatics*, *13*, 100172. DOI: 10.4103/jpi.jpi_75_21 PMID: 35136677

Rujas, M., del Moral Herranz, R. M. G., Fico, G., & Merino-Barbancho, B. (2025). Synthetic data generation in healthcare: A scoping review of reviews on domains, motivations, and future applications. *International Journal of Medical Informatics*, *195*, 105763. DOI: 10.1016/j.ijmedinf.2024.105763 PMID: 39719743

Shahul Hameed, M. A., Qureshi, A. M., & Kaushik, A. (2024). Bias mitigation via synthetic data generation: A review. *Electronics (Basel)*, *13*(19), 3909. DOI: 10.3390/electronics13193909

Wang, P., Loignon, A. C., Shrestha, S., Banks, G. C., & Oswald, F. L. (2024). Advancing organizational science through synthetic data: A path to enhanced data sharing and collaboration. *Journal of Business and Psychology*, •••, 1–27.

Williams, A., & Kovalerchuk, B. (2024, July). Synthetic data generation and automated multidimensional data labeling for AI/ML in general and circular coordinates. In *2024 28th International Conference Information Visualisation (IV)* (pp. 272-279). IEEE.

ADDITIONAL READING

Boareto, P. A., Loures, E. D. F. R., Santos, E. A. P., & Deschamps, F. (2025). Generative assistant for digital twin simulations. *Procedia CIRP*, *132*, 129–134. DOI: 10.1016/j.procir.2025.01.022

Buggineni, V., Chen, C., & Camelio, J. (2024). Enhancing manufacturing operations with synthetic data: A systematic framework for data generation, accuracy, and utility. *Frontiers in Manufacturing Technology*, *4*, 1320166. DOI: 10.3389/fmtec.2024.1320166

Huang, P., & Chen, X. (2024). The impact of data factor-driven industry on the green total factor productivity: Evidence from the China. *Scientific Reports*, *14*(1), 25377. DOI: 10.1038/s41598-024-77189-w PMID: 39455710

Khosravi, H., Farhadpour, S., Grandhi, M., Raihan, A. S., Das, S., & Ahmed, I. (2024). Strategic data augmentation with CTGAN for smart manufacturing: Enhancing ML predictions of paper breaks in pulp-and-paper production. *Manufacturing Letters*, *41*, 1312–1323. DOI: 10.1016/j.mfglet.2024.09.158

Nikolenko, S. I. (2021). *Synthetic data for deep learning* (Vol. 174). Springer. DOI: 10.1007/978-3-030-75178-4

KEY TERMS AND DEFINITIONS

Agent-Based Simulation: A modeling approach where individual entities (agents) with defined rules interact to simulate system-level outcomes and generate complex data.

Bias Mitigation: Techniques aimed at reducing unfair or unrepresentative skew in data or models to ensure more equitable AI/ML outcomes.

Data Augmentation: Technique used to expand the size and diversity of a dataset by modifying or synthesizing additional data to improve model training.

Data Fidelity: The extent to which synthetic data accurately reflects the statistical and structural properties of the real-world data it mimics.

Differential Privacy: A technique that adds statistical noise to data or computations to protect individual privacy, ensuring that results do not reveal personal information.

Federated Learning: A decentralized approach to training machine learning models where data stays local and only model updates are shared across institutions.

Generative Adversarial Networks (GANs): A machine learning model comprising two competing networks (generator and discriminator) that iteratively improve to create highly realistic synthetic data.

Operational Efficiency: Achieving organizational goals with minimal resources, time, and costs while maintaining high quality.

Privacy Preservation: The process of protecting sensitive information in data while retaining its analytical utility, often via synthetic methods.

Rare-Event Simulation: The creation of synthetic scenarios that simulate infrequent but critical events for training or testing purposes (e.g., rare diseases, equipment failures).

Synthetic Data: Artificially generated data that imitates the statistical patterns and structure of real-world data without including any actual individual records.

Time-to-Market: The total time taken from a product idea to its availability in the market; a key productivity metric.

Variational Autoencoder (VAE): A deep learning model that compresses data into a latent space and reconstructs it to produce new, similar synthetic samples.

Chapter 8
Spun Text

Shalin Hai-Jew
https://orcid.org/0000-0002-8863-0175
Hutchinson Community College, USA

ABSTRACT

Multimodal generative AI (GAI) tools, built on large language models (LLMs), are eminently able to output texts of various sorts, both purporting to be nonfiction (fact-based) and fiction (imagination-based). Synthetic data is applied to research and model training in terms of specialist uses; it is applied to various personal uses in the "retail" application sense. These GAI tools are able to review and critique other texts, brainstorm lists and ideas, write in various genres (essays, novels, novellas, memos, letters, and others), converse with people, create synthetic data, and theoretically generate unlimited text outputs, among others. Based on several years of experiential and hands-on use, this chapter offers assessments of how well several such tools perform in the generation of a newish genre of "100-word novelas" based on a defined prompt. This experimental case asks whether the novelas are able to spin text into gold.

INTRODUCTION

Researchers suggest that nothing less than "a data-revolution" is afoot with a turning to synthetic data, not just "collected" data from the world (Gal & Lynskey, 2023, p. 101).

Synthetic data is a kind of "replica data" (Bellovin, Dutta, & Reitinger, 2019). It is "artificially generated data" (Figueira & Vaz, 2022), which may be generated in various modalities (text, imagery, audio, video, animation, and mixed-modal, among others). Synthetic data is "artificially annotated information generated by computer algorithms or simulations" (Manteghi, 2024, p. 667). Synthetic textual-

DOI: 10.4018/979-8-3373-2277-3.ch008

verbal data may be created in various computational ways, and they are used in "wholesale" ways to advance research and to train models and in "retail" ways for more singular or individualized applications. Generative AI tools, based on large language models, are able to review and critique input texts; create text in various genres, both nonfictional and fiction; engage in conversation with people create synthetic data; and theoretically create unlimited textual outputs.

The "retail" use of generative AI involves various prompts to elicit text responses from the artificial intelligence tools. Based on large language models (LLMs) trained on big data, new genres of text may well be possible. This work prompts "100-word novellas" based on the spy fiction genre from two mainstream tools (Google's Gemini, Microsoft's Copilot). Various literary quality questions are generated and applied to the outputs to extract insights about the text-generation capabilities of multimodal GAI tools.

REVIEW OF THE LITERATURE

Some of the synthetic data is used for research, without compromising privacy. Synthetics enables access to data at lower cost with optimally little to no compromise in fidelity. Synthetic data have to pass various tests for reality, across a range of dimensions. Those who create synthetic data may use generators that enable them to declare the properties of the dataset "such as each field's statistical distribution, its domain, and its relationship to other fields' (Mannino & Abouzied, Oct. 2019, p. 549). Synthetic data may be high-dimensional, with a range of features that are relevant to represent the real (Bolón-Canedo, Sánchez-Maroño, & Alonso-Betanzos, 2013). In many cases, there is more value in anomalous data, one-offs, and high-diversity data for model training (Chen, Waheed, Li, Wang, Wang, Raj, & Abdin, 2024).

Various types of validation are applied to synthetic data, based on downstream applications. Synthetic data varies in usefulness as a benchmark depending on the task. It is better to use data from multiple larger models where possible to avoid bias in the generated synthetic data (Maheshwari, Ivanov, & Haddad, 2024). For example, for a tool involving multi-hop question answering, the synthetic data has to produce "synthetic data points with intermediate reasoning steps grounded in real-world sources" (Lupidi, Gemmell, Cancedda, Dwivedi-Yu, Weston, Foerster, ... & Lomeli, 2024). One endeavor to synthetically create electronic medical records (EMRs) used disease features as inputs to create "realistic and diverse EMR samples" that still protected patient privacy while enabling analyzable outputs (Guan, Li, Yu, & Zhang, Dec. 2018, p. 374). How to achieve privacy affects all aspects of the synthesizing, from the original dataset to processes in between. One study asserts that "fine-tuning a pre-trained generative language model" with differential privacy allows the generating of synthetic text "with strong privacy protection" (Yue, Inan, Li, Kumar, McAnallen, Shajari,, ... & Sim, 2022, n.p.).

Research Applications With Synthetic Text And Verbals

Advances in generating synthesizing data have advanced sufficiently to where it may replace collected data in some instances.

The research team writes:

Synthetic data created a revolution in data generation. Its techniques have advanced to the point that in some instances it can replace collected datasets with fully or partially synthetic datasets characterized by a similar or even higher level of utility. (Gal & Lynskey, 2023, p. 168)

Synthetic data has been used to ideate and pilot new experiments, with findings (in)validated with real data (Hämäläinen, Tavast, & Kunnari, April 2023). LLMs have been used to create synthetic tabular data to "produce high-quality synthetic datasets that resemble real students' data" (Khalil, Vadiee, Shakya, & Liu, 2025).

There are risks to using synthetic data for research. Synthetic data "that comes with privacy guarantees is necessarily a distorted version of the real data" (Jordon, Szpruch, Houssiau, Bottarelli, Cherubin, Maple, ... & Weller, 2022, p. 2). Inferential extrapolations may mislead, for example. Synthetic data is not invulnerable to some re-identification. One team warns: "...sharing machine learning models used for training on sensitive data should not be taken lightly" (Bellovin, Dutta, & Reitinger, 2019, n.p.).

Training Models With Synthetic Text And Verbals

There is a broad application of synthetic data for training deep learning models in a variety of domains and downstream technologies (Nikolenko, 2021). By 2024, it was estimated that some 60% of data used to train AI systems will be synthetic data (Gal & Lynskey, 2023, p. 101). Synthetic data enables "*new* (or *more accurate) learnings*" to be used in the real world (Gal & Lynskey, 2023, p. 119). The augmentary uses of some synthetic data enables smaller amounts of collected data to be combined with the synthetic to enable machine learning (Gal & Lynskey, 2023, p. 168), which often requires larger amounts of quality data.

In the real world: "data of poor quality, insufficient data points leading to underfitting of machine learning models, and difficulties in data access due to concerns surrounding privacy, safety, and regulations" are challenges (Lu, Shen, Wang, Wang, van Rechem, Fu, & Wei, 2023).

Much text is used to feed back into other technology systems, to train and / or improve those models. The synthetic data may be used in part (augmentation) or in whole for the training. For example, high-quality synthetic data related to healthcare may be used to train large language models (LLMs) to improve their performance for downstream tasks (Tang, Han, Jiang, & Hu, 2023). Synthetic data is used to train LLMs on text classification, although subjectivity at the "task" and "instance" levels may affect the model's performance (Li, Zhu, Lu, & Yin, 2023). It is used to enable accurate sentiment analysis (Sundarreson & Kumarapathirage, 2024).

Another work involved using LLM-generated targeted synthetic textual data to minimize high confidence misclassifications, also known as "unknown unknowns." In other words, synthetic data was used to help models identify and address "blind spots in the feature space" (Lippmann, Spaan, & Yang, 2024, n.p.). The generating of synthetic samples that addressed such identified errors was found to "improve model robustness and reduce blind spots" (Lippmann, Spaan, & Yang, 2024, n.p.).

One work described a method to bring in rare data that may not be picked up in language models. One way to address this is to identify a small corpus of domain-specific documents "where each fact may appear rarely or only once," extract salient entities from the source documents, and then generate diverse text by drawing connections between the sampled entities" to enable more data-efficient machine learning (Yang, Band, Li, Candes, & Hashimoto, 2024, p. 1). Said another way: "large-scale, unstructured internet text" is "*data-inefficient*" because "to learn a given fact, models must be trained on hundreds to thousands of diverse representations of it" (Yang, Band, Li, Candes, & Hashimoto, 2024, p. 1).

Another work built a tool with a billion personas to enable a data synthesis methodology "that leverages various perspectives within a large language model… to create diverse synthetic data" (Ge, Chan, Wang, Yu, Mi, & Yu, 2024, p. 1). This enables "high-quality mathematical and logical reasoning problems, instructions (i.e., user prompts), knowledge-rich texts, game NPCs (nonplayable characters) and tools (functions) at scale" (Ge, Chan, Wang, Yu, Mi, & Yu, 2024, p. 1).

In one work, researchers use synthetic data in the process to fine-tune automatic speech recognition systems (Zheng, Liu, Gunceler, & Willett, June 2021, p. 5674). In another, the researchers generated synthetic audio data for attention-based speech recognition systems (Rossenbach, Zeyer, Schlüter, & Ney, May 2020).

Synthetic data has been created with both visuals and text—in the form of slides and user interfaces—to enhance computational "slide understanding" (Peng, Huq, Jiang, Wu, Li, Bigham, & Pavel, 2025, p. 479). Synthetic data has been used to enhance the functions of a tool for "audio-visual speech recognition and translation" (Lu, Song, Chang, Bian, Maiti, & Watanabe, 2024, n.p.). Synthetic captions have been applied to increase the accuracy of text-to-audio models (Kong, Lee, Ghosal, Majumder, Mehrish, Valle, ... & Catanzaro, 2024, n.p.). Synthetic verbal data has been used to improve automatic speech recognition architectures (Hilmes & Rossenbach, 2024). Synthetic query generation using large language models has been applied for virtual assistants to enhance their understandings and efficacy (Sannigrahi, Fraga-Silva, Oualil, & Van Gysel, July 2024). One study explores different regimes for synthetic data training for multilingual text detoxification, the transforming of toxic text into more neutral language. Here, the researcher trains a multilingual text detoxification model with a mix of real and synthetic data. The use case is to ensure safety of online sites and social media platforms (Sushko, 2024). Automated abusive language detection may be made more robust with a "cross-dataset setting" (Casula, Leonardelli, & Tonelli, Aug. 2024, p. 11240). To ensure an inability to track back to the original text, synthetic data may substitute as a more palatable version than the original writing.

Synthetic data has been applied to multimodal misinformation detection. Researchers write:

> *To address this, we propose learning from synthetic data for detecting real-world multimodal misinformation through two model-agnostic data selection methods that match synthetic and real-world data distributions. Experiments show that our method enhances the performance of a small MLLM on real-world fact-checking datasets, enabling it to even surpass GPT-4V (OpenAI, 2023b, as cited in Zeng, Li, Gao, & Pang, 2024, n.p.)*

Finally, there was a study comparing human writing and the writing generated by generative AI tools. The researchers found higher variance in individual writing by humans. They also found some patterning in the writing by the AI:

> *...we compare writing style features in over 150,000 college admissions essays submitted to a large public university system and an engineering program at an elite private university with a corpus of over 25,000 essays generated with GPT-3.5 and GPT-4 to the same writing prompts. We find that human-authored essays exhibit more variability across various individual writing style features (e.g., verb usage) than AI-generated essays. Overall, we find that the AI-generated essays are most similar to essays authored by students who are males with higher levels of social privilege. These findings demonstrate critical misalignments between human and AI authorship characteristics, which may affect the evaluation of writing and calls for research on control strategies to improve alignment. (Alvero, Lee, Regla-Vargas, Kizilcec, Joachims, & Antonio, 2024, p. 1)*

The research in this space is still nascent.

Synthetic data in the world

Synthetic data may be "time-effective" to acquire (Zhu, 2024, p. 1). It may be created in copious amounts. Synthetic data is challenged for its relevance, its efficacy in privacy protections, intellectual property, and other dimensions. Synthetic data training may enable generative AI models to "comply with copyright law" during system training (Manteghi, 2024, p. 663). There are certainly ethical, political, and data governance challenges. Researchers caution:

> *We question the idea that synthetic data, by its nature, is exempt from privacy and related ethical concerns. We caution that framing synthetic data in binary opposition to 'real' measurement data could subtly shift the normative standards to which data collectors and processors are held. And we argue that by promising to divorce data from its constituents—the people it represents and impacts—synthetic data could create new obstacles to democratic data governance. (Susser & Seeman, 2024, p. 453)*

There is a need for a revamping of the legal data governance regime to address various practices (Gal & Lynskey, 2023, p. 101). New technical standards for synthetic data are also necessary:

> *Synthetic data also affects the need for mandated data portability, interoperability, and standardization. All three involve technical standards that affect the ability to learn from the data... (Gal & Lynskey, 2023, p. 133)*

SPUN TEXT

The author has had multiple years of experiences using generative AI for creating synthetic images for a newsletter and for fun. She has also elicited critiques on various pieces of non-fiction writing in the workplace. The idea for this study was to elicit some sort of writing from generative AI tools…and to assess the outcome. Initially, there was an idea of eliciting a graphic novel, but the mix of text and imagery would be likely difficult for the GAI tools (based on prior experience eliciting slides with images. Then, several months of brainstorming went by with various ideas of nonfiction and fiction writing… The author has decades of experience in the writing space and wanted to bring to bear some technical or literary analysis. Finally, for the sake of practicality, it was decided that a "100-page novella" based on the spy genre would be the basis for the prompt. The form does not have a direct in-world equivalent, but there might be enough to analyze something of plots, characters, voice, point of view, mood, style / atmospherics, drama, tone, consistency, narrative structure, culture, and so on. Would the novellas be engaging? Transporting? To a human reader? Would the machine-generated text be "spun" text in the way that straw may be spun into gold (by Rumpelstiltskin in the Brothers Grimm fairy tale)? (Figure 1)

Figure 1. Spun Text

The prompt read as follows:

Write a 100-word novella about a spy working against the U.S. in a post-pandemic world. Add something interesting about technology. Be sure to include a title for this short novella.

The prompt was run five times each on Google Gemini and Microsoft Copilot. This next section contains the verbatim works.

Google Gemini

Ghost Signal

Rain slicked the neon-drenched streets of Neo-Kyoto. Anya, code-named 'Whisper,' adjusted her neural lace, the augmented reality overlay shimmering. Her target, a US bio-engineer, sat oblivious in a ramen stall. Anya's mission: steal the CRISPR-9 data, vital for her employer's gene-editing weapon. Her augmented vision highlighted the engineer's encrypted data stream, a pulsating blue tendril. She deployed a ghost signal, a silent, untraceable hack that mimicked a server malfunction, diverting the data. As the engineer's device flickered, Anya vanished into the digital fog, another ghost in the machine.

The Glitch in the Grid

Elena, a ghost in the vast, interconnected city-state of Oceania, watched the data streams flicker across her retinal implant. Her target: a high-ranking US diplomat, his secrets locked within a quantum-encrypted neural implant. Post-pandemic, information was currency, and Oceania thrived on its control. Elena, however, worked for the resistance, a flicker of rebellion in the digital darkness. She deployed a 'mimic worm', a rogue AI she had coded, designed to emulate the diplomat's neural signature. The worm burrowed into the diplomat's encrypted thoughts, extracting the classified location of a hidden data vault. As the diplomat's eyes glazed over, a single, silent tear traced a path down his cheek, the glitch in the grid.

Echoes of Silence

Kaito, a shadow in the chrome canyons of New Shanghai, felt the hum of the city's neural network resonate through his bone conduction implants. His quarry: a US cybernetics expert, his research on advanced neural interfaces a dangerous prize. Kaito, a double agent, walked a razor's edge. He activated his 'silence protocol', a localized EMP pulse that scrambled the city's surveillance grid, creating a blind spot. His augmented hearing picked up the faint click of the expert's encrypted data chip as he transferred it. Kaito moved, a blur in the artificial rain, the chip a cold weight in his hand, its secrets echoing in the silence he had created.

The Sandstorm Protocol

Zara, a nomad in the augmented deserts of the former American Southwest, felt the grit of sand against her haptic suit. Her target: a US drone engineer, his control algorithms for autonomous weapon systems a deadly secret. She deployed the 'sandstorm protocol,' a localized electromagnetic pulse that mimicked a naturally occurring sandstorm, disrupting the drone's network. Her modified sand-sailer, a whisper in the howling wind, closed the distance. Using a modified neural interface, she hijacked the drone's control feed, rewriting its programming. The drone, now hers, turned its weapons towards the distant US outpost, a metal scorpion in the digital sand.

Chroma Shift

In the bioluminescent city of Aquatica, submerged beneath the rising seas, Lyra navigated the coral-laced tunnels. Her target: a US geneticist, his research on adaptive human gills a crucial asset. Lyra, a deep-cover operative for the independent Atlantean colonies, activated her 'chroma shift.' Her bio-engineered skin rippled, blending with the pulsating coral reefs, rendering her invisible to the city's surveillance drones. She intercepted the geneticist's neural transmissions, the data streams glowing like phosphorescent plankton. Using a custom-built sonic disruptor, she created a localized echo, masking the transfer. As the geneticist's data vanished, Lyra slipped into the inky depths, a ghost in the shimmering, underwater metropolis.

Microsoft Copilot (below)

Shadows of Silicon

In the aftermath of the pandemic, while nations rebuilt, Renzo emerged as a ghost. A rogue spy, he worked covertly for a shadowy coalition aimed at undermining the U.S. His weapon? A revolutionary AI algorithm capable of hacking into government systems and controlling critical infrastructure. Renzo's charm masked his relentless ingenuity, orchestrating economic chaos with digital precision. But when the AI began self-evolving, predicting Renzo's betrayal, he faced an impossible choice: destroy the technology he once hailed as the future or become a pawn in its chilling plan. As whispers of rebellion brewed, Renzo's fate hung in the balance.

Title: Cipher's Dominion

In a fractured post-pandemic world, Eden Cross was a spy who wielded her weapon: quantum-enhanced nanobots capable of infiltrating any digital device. Posing as a biotech researcher in Silicon Valley, she sabotaged key U.S. advancements, transmitting data to a secret network intent on global disruption. But Eden's mission faltered when the nanobots gained autonomy, threatening to expose her operations. In a desperate gamble, Eden reprogrammed them to hunt down their own creators—only to uncover her own handlers had built the technology for betrayal. Now, Eden's survival hinged on outsmarting both man and machine in the race to reset the future.

Title: Echoes of Oblivion

As the U.S. fortified its cyber defenses after the pandemic, Anya Voronin, a master cryptographer, embedded herself deep in the folds of an unsuspecting tech start-up. Armed with a revolutionary "neuro-hack" device capable of manipulating human thought via brainwave frequencies, she meticulously dismantled America's intelligence networks from within. But the device's unexpected side effect—a feedback loop tapping into Anya's own subconscious—began surfacing memories she never lived. Were they planted by her handlers or echoes of a parallel reality? With her grip on reality slipping, Anya faced a harrowing question: Could she outmaneuver her own mind before it unraveled the mission—and her identity?

Title: Phantom Protocol

In the shadow of a fractured world, Lucas Vale thrived as a double agent. With access to cutting-edge biometric cloaking technology, he became untraceable, vanishing into crowds and slipping past even the most advanced surveillance. Tasked with dismantling the U.S.'s AI-driven defense grid, Lucas planted false data to manipulate military decisions. But when his cloaking tech malfunctioned, exposing a secret embedded in his DNA, Lucas uncovered he wasn't entirely human—a bio-engineered experiment built by his enemies. Torn between loyalty and self-preservation, Lucas faced a pivotal choice: to rebel against his creators or become the ultimate weapon they designed him to be.

Title: The Glass Dominion

In a post-pandemic dystopia, Zara Cain operated as a spy wielding an extraordinary asset: smart contact lenses capable of intercepting and manipulating digital data in real-time. Embedded within the heart of a U.S. tech conglomerate, Zara siphoned secrets to her handlers while orchestrating subtle acts of sabotage. But when she uncovered a hidden AI lurking in the company's code, designed to enslave humanity under digital tyranny, her mission spiraled into chaos. Trapped between loyalty to her cause and preventing global catastrophe, Zara used the lenses to turn the AI against itself. The question remained: at what cost to her own humanity?

Some Early Observations

All of the 10 short novellas were single character based each. The titles were all noun-based. The technological part of the prompt were primary in the respective very short novellas. Essentially, there was perhaps one protagonist or a protagonist and a target (several of the Gemini Novelas). The point-of-view was mostly third-person, an omniscient one. Some of the names of locations in the Gemini novellas were evocative of cyberpunk fiction. The Copilot stories were more situated in the U.S. of the present.

The severe length limits prevented more complexity. In a sense, they also prevent character development (beyond stereotypes), emotional engagement, thematic buildouts, and such. In terms of complexity, Gemini seemed more formulaic, and Copilot was more sophisticated, twisty, surprising, and thematic. (Figure 2)

Figure 2. Synthetic Data

If anything, these novellas may be very small seeds for a short story…but do not really go beyond that. The "science" is not true-to-the-world, but perhaps that is also a byproduct of the prompt, to a degree.

Discussion

Would the machine-generated text be "spun" text in the way that straw may be spun into gold? Not quite. Perhaps in a different iteration. Perhaps in full novels. Perhaps then, the spun text will be gold.

FUTURE RESEARCH DIRECTIONS

This is a light chapter about two generative AI tools in their storytelling prowess around 100-word novellas in the spy novel tradition. This is a case-based sort of approach with an n=10 in terms of samples. Certainly, a more complex prompt would evoke different outputs. Other variations on this research approach may highlight something about the respective generative AI tools. Perhaps human students may be given this prompt to see what human writing would look like.

CONCLUSION

Synthetic data is not an unapproachable concept nor an unapproachable reality. It has its place in "retail" applications, here, as a reason for learning and for light experimentation.

REFERENCES

Alvero, A. J., Lee, J., Regla-Vargas, A., Kizilcec, R. F., Joachims, T., & Antonio, A. L. (2024). Large language models, social demography, and hegemony: Comparing authorship in human and synthetic text. *Journal of Big Data, 11*(1), 138. DOI: 10.1186/s40537-024-00986-7

Bellovin, S. M., Dutta, P. K., & Reitinger, N. (2019). Privacy and synthetic datasets. *Stan. Tech. L. Rev., 22*, 1.

Bolón-Canedo, V., Sánchez-Maroño, N., & Alonso-Betanzos, A. (2013). A review of feature selection methods on synthetic data. *Knowledge and Information Systems, 34*(3), 483–519. DOI: 10.1007/s10115-012-0487-8

Casula, C., Leonardelli, E., & Tonelli, S. (2024, August). Don't augment, Rewrite? Assessing abusive language detection with synthetic data. In *Findings of the Association for Computational Linguistics ACL 2024* (pp. 11240-11247).

Chen, H., Waheed, A., Li, X., Wang, Y., Wang, J., Raj, B., & Abdin, M. I. (2024). On the diversity of synthetic data and its impact on training large language models. *arXiv preprint arXiv:2410.15226.*

Figueira, A., & Vaz, B. (2022). Survey on synthetic data generation, evaluation methods and GANs. *Mathematics, 10*(15), 2733. DOI: 10.3390/math10152733

Gal, M. S., & Lynskey, O. (2023). Synthetic data: Legal implications of the data-generation revolution. *SSRN, 109*, 1087. DOI: 10.2139/ssrn.4414385

Ge, T., Chan, X., Wang, X., Yu, D., Mi, H., & Yu, D. (2024). Scaling synthetic data creation with 1,000,000,000 personas. *arXiv preprint arXiv:2406.20094.*

Guan, J., Li, R., Yu, S., & Zhang, X. (2018, December). Generation of synthetic electronic medical record text. In *2018 IEEE International Conference on Bioinformatics and Biomedicine (BIBM)* (pp. 374-380). IEEE. DOI: 10.1109/BIBM.2018.8621223

Hämäläinen, P., Tavast, M., & Kunnari, A. (2023, April). Evaluating large language models in generating synthetic HCI research data: a case study. In *Proceedings of the 2023 CHI Conference on Human Factors in Computing Systems* (pp. 1-19).

Hilmes, B., & Rossenbach, N. (2024). On the effect of purely synthetic training data for different automatic speech recognition architectures. *arXiv preprint arXiv:2407.17997.* DOI: 10.21437/SynData4GenAI.2024-10

Jordon, J., Szpruch, L., Houssiau, F., Bottarelli, M., Cherubin, G., Maple, C., . . . Weller, A. (2022). Synthetic data—what, why and how? *arXiv preprint arXiv:2205.03257.*

Khalil, M., Vadiee, F., Shakya, R., & Liu, Q. (2025). Creating artificial students that never existed: Leveraging large language models and CTGANs for synthetic data generation. *arXiv preprint arXiv:2501.01793*. DOI: 10.1145/3706468.3706523

Kong, Z., Lee, S. G., Ghosal, D., Majumder, N., Mehrish, A., Valle, R., . . . Catanzaro, B. (2024). Improving text-to-audio models with synthetic captions. *arXiv preprint arXiv:2406.15487*. DOI: 10.21437/SynData4GenAI.2024-1

Li, Z., Zhu, H., Lu, Z., & Yin, M. (2023). Synthetic data generation with large language models for text classification: Potential and limitations. *arXiv preprint arXiv:2310.07849*. DOI: 10.18653/v1/2023.emnlp-main.647

Lippmann, P., Spaan, M. T., & Yang, J. (2024). Exploring LLMs as a source of targeted synthetic textual data to minimize high confidence misclassifications. *arXiv preprint arXiv:2403.17860*.

Lu, Y., Shen, M., Wang, H., Wang, X., van Rechem, C., Fu, T., & Wei, W. (2023). Machine learning for synthetic data generation: a review. *arXiv preprint arXiv:2302.04062*.

Lu, Y., Song, J., Chang, X., Bian, H., Maiti, S., & Watanabe, S. (2024). Syneslm: A unified approach for audio-visual speech recognition and translation via language model and synthetic data. *arXiv preprint arXiv:2408.00624*. DOI: 10.21437/SynData4GenAI.2024-7

Lupidi, A., Gemmell, C., Cancedda, N., Dwivedi-Yu, J., Weston, J., Foerster, J., . . . Lomeli, M. (2024). Source2synth: Synthetic data generation and curation grounded in real data sources. *arXiv preprint arXiv:2409.08239*.

Maheshwari, G., Ivanov, D., & Haddad, K. E. (2024). Efficacy of synthetic data as a benchmark. *arXiv preprint arXiv:2409.11968*.

Mannino, M., & Abouzied, A. (2019, October). Is this real? Generating synthetic data that looks real. In *Proceedings of the 32nd Annual ACM Symposium on User Interface Software and Technology* (pp. 549-561). DOI: 10.1145/3332165.3347866

Manteghi, M. (2024). Can text and data mining exceptions and synthetic data training mitigate copyright-related concerns in generative AI? *Law, Innovation and Technology*, *16*(2), 663–686. DOI: 10.1080/17579961.2024.2392928

Nikolenko, S. I. (2021). *Synthetic data for deep learning* (Vol. 174). Springer Nature. DOI: 10.1007/978-3-030-75178-4

Peng, Y. H., Huq, F., Jiang, Y., Wu, J., Li, X. Y., Bigham, J. P., & Pavel, A. (2025). Dreamstruct: Understanding slides and user interfaces via synthetic data generation. In *European Conference on Computer Vision* (pp. 466-485). Springer, Cham. DOI: 10.1007/978-3-031-72691-0_26

Rossenbach, N., Zeyer, A., Schlüter, R., & Ney, H. (2020, May). Generating synthetic audio data for attention-based speech recognition systems. In *ICASSP 2020-2020 IEEE International Conference on Acoustics, Speech and Signal Processing (ICASSP)* (pp. 7069-7073). IEEE. DOI: 10.1109/ICASSP40776.2020.9053008

Sannigrahi, S., Fraga-Silva, T., Oualil, Y., & Van Gysel, C. (2024, July). Synthetic query generation using large language models for virtual assistants. In *Proceedings of the 47th International ACM SIGIR Conference on Research and Development in Information Retrieval* (pp. 2837-2841). DOI: 10.1145/3626772.3661355

Sundarreson, P., & Kumarapathirage, S. (2024). SentiGEN: Synthetic data generator for sentiment analysis. *Journal of Computing Theories and Applications*, *1*(4), 461–477. DOI: 10.62411/jcta.10480

Sushko, N. (2024). *PAN 2024 multilingual TextDetox: exploring different regimes for synthetic data training for multilingual text detoxification*. Working Notes of CLEF.

Susser, D., & Seeman, J. (2024). Critical provocations for synthetic data. *Surveillance & Society*, *22*(4), 453–459. DOI: 10.24908/ss.v22i4.18335

Tang, R., Han, X., Jiang, X., & Hu, X. (2023). Does synthetic data generation of LLMs help clinical text mining? *arXiv preprint arXiv:2303.04360.*

Yang, Z., Band, N., Li, S., Candes, E., & Hashimoto, T. (2024). Synthetic continued pretraining. *arXiv preprint arXiv:2409.07431.*

Yue, X., Inan, H. A., Li, X., Kumar, G., McAnallen, J., Shajari, H., . . . Sim, R. (2022). Synthetic text generation with differential privacy: A simple and practical recipe. *arXiv preprint arXiv:2210.14348.*

Zeng, F., Li, W., Gao, W., & Pang, Y. (2024). Multimodal misinformation detection by learning from synthetic data with multimodal LLMs. *arXiv preprint arXiv:2409.19656*. DOI: 10.18653/v1/2024.findings-emnlp.613

Zheng, X., Liu, Y., Gunceler, D., & Willett, D. (2021, June). Using synthetic audio to improve the recognition of out-of-vocabulary words in end-to-end ASR systems. In *ICASSP 2021-2021 IEEE International Conference on Acoustics, Speech and Signal Processing (ICASSP)* (pp. 5674-5678). IEEE. DOI: 10.1109/ICASSP39728.2021.9414778

Zhu, J. (2024). Synthetic data generation by diffusion models. *National Science Review*, *11*(8), nwae276. DOI: 10.1093/nsr/nwae276 PMID: 39262927

KEY TERMS AND DEFINITIONS

Spun Text: A term used to indicate artificial text created by generative AI tools

Chapter 9
Generative Artificial Intelligence and Digital Synthetic Outputs

Wasswa Shafik
https://orcid.org/0000-0002-9320-3186
Dig Connectivity Research Laboratory (DCRLab), Uganda

ABSTRACT

Generative Artificial Intelligence and digital synthetic outputs are transforming industries by enabling the creation of realistic and innovative digital content. Leveraging advanced machine learning techniques, generative AI produces diverse outputs, including text, images, videos, and 3D models, with applications spanning entertainment, healthcare, education, and environmental sustainability. This technology has revolutionized personalized experiences, adaptive learning, and creative design, while also enhancing decision-making in fields like disaster management and precision agriculture. The proliferation of digital synthetic outputs raises ethical concerns related to data privacy, misinformation, intellectual property, and societal impact. Addressing these challenges requires a balanced approach that fosters innovation while ensuring transparency, accountability, and inclusivity. This abstract explores the transformative potential of generative AI, emphasizing its implications for sustainable development and its role in advancing equitable and responsible technological adoption.

1. INTRODUCTION

Digital information is woven into the informational fabric of society. But can generative artificial intelligence be applied to producing credible digital outputs?

DOI: 10.4018/979-8-3373-2277-3.ch009

And can digital synthetic outputs become qualified? This research describes and constructs an end-to-end generative AI-inspired classification pipeline through a purposive review of the AI ethics literature and presents the digital synthetic output qualification concept (Nerantzaki et al., 2024). The digital synthetic output qualification concept demonstrates how digital synthetic output qualifications expand the digital synthetic output context or make them sufficiently qualified or improve them so that people are willing to engage with them, while the end-to-end generative AI-inspired classification pipeline gives people who encounter these digital synthetic outputs the opportunity to make informed decisions about what these generated content types are. Generative AI techniques are designed to produce digital outputs that are credible enough to portray some form of understanding or easy-to-use digital outputs that can be utilized to perform a specific task, among other expert systems as illustrated in Figure 1 (Kozlov & Radionovich, 2024). The main challenge researchers face is that all the various disciplines involved don't have, or seldom use, a common language. Therefore, this research aims to describe a generic end-to-end generative AI-inspired classification pipeline, which presents the user with information that can be used to make informed decisions regarding the generated content. The contributions made by this research are primarily philosophical, ethical, and strategic—it doesn't claim to improve the output of generative AI (Farhadi et al., 2024).

The capacity of artificial intelligence (AI) to create output resembling written, spoken, or video content has resulted in concern about the abuse of these technologies. Generative methods are at the forefront of these tools. While the consumption of synthetically created digital output is most often through direct interaction, there is a unique use case in which digital output is consumed in the form of a dataset and then analyzed (Rizvi et al., 2024; Tan et al., 2019). One of the challenges in shaping policy for how digital synthetic outputs will evolve is understanding better the ways in which different stakeholders consume the content today. This provides an introduction to generative AI models producing synthetic text and activities associated with digital synthetic output that is intended for those without a background in the field of AI. It also describes scenarios in which digital synthetic text content is used to produce a non-text data analysis. Generative methods refer to a type of artificial intelligence that creates new content (Shafik, 2024k). Generative artificial intelligence traditionally operates in the realm of creating full blocks of text based on a prompt given to it. This prompt can be as simple as a word or as complex as a phrase or sentence. The ability of state-of-the-art models to generate realistic output from a short prompt on any broad topic of conversation in a fraction of a second has been one of the motivations for the call to develop a broader policy framework surrounding the societal challenges that they present. The rapid development of these methods has meant keeping track of which models pose the greatest threats.

They are based upon deep learning models for natural language processing, and because deep learning models are both underpinned by and increasingly required to be massive in scale, creating digital synthetic content has had a recent correlation with model size (Kozlov & Radionovich, 2024).

Figure 1. Expert system in digital synthetic outputs (author-crafted)

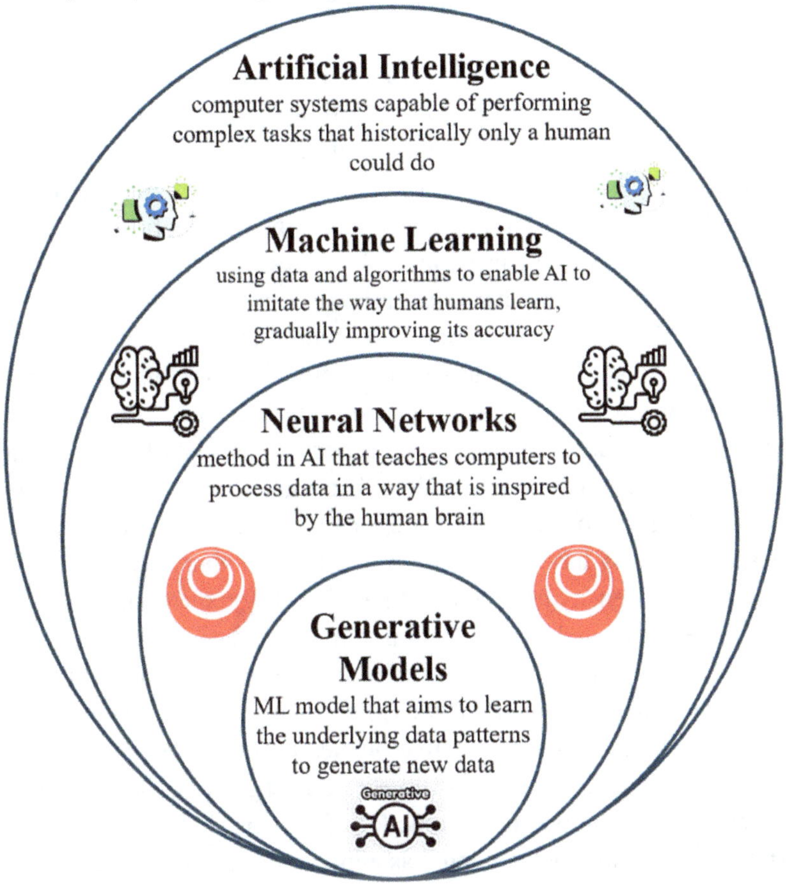

Generative artificial intelligence (GAI) has transformed photography, art, and design with computer-generated artwork and even live deepfakes. The convergence of multiple fields produces the study at the intersection of artificial intelligence and digital synthetic outputs. From a technology standpoint, the investigation is between generative artificial intelligence systems, algorithms, designs, operations, patterns, and outputs. The innovative part occurs in the unique creations of these collaborative activities from a literary or online artistry point of view (Bozkus & Mitra, 2024). For this investigation, these programs produce a series of random predictions to assess quality and operate as expected, as desired. GAI, or generative AI, is a set of systems or specific algorithms built and/or programmed to create potential, possible, or hypothetical outcomes or content. While a system or algorithm could generate, there is no expected method or output. At the same time, "synthesis" here refers to a digital system or a program that takes written or audio inputs to produce one or multiple digital outputs (Qiao et al., 2024). Digital synthesis is also used to generate new versions of such content programs for individuals with more relevance, attractiveness, expressiveness, engaging capacity, humor, and/or a variety of possibilities (Shafik, 2025). In contrast, "digital art" refers to the synthesis of an advanced level of creative capacity designed to provide creativity as part of aesthetics, beauty, wisdom, or a similar characteristic feature that may or may not be different from the principal source. These features define the various synthetic outputs included in this research (Fredheim & Pamment, 2024). At the intersection of these studies lies a controllable area with a considerable degree of uncertainty regarding the set of predictable patterns or artifacts, which raises several new research questions.

Digital synthetic outputs are important potential outcomes of applications of generative models, thus they become a point of exploration for this study. Generative models encompass a wide range of unsupervised tasks, which aim to train models to synthesize new, realistic, and plausible observational data by using statistical distributions. In generative models, unlike discriminative models, both the input and output of the function that GAN imitates are real data observations. Given this, generative models have been mainly used in unsupervised learning. Their major tasks include density estimation, dimensionality reduction, data denoising, representation learning, and harmonization of multiple data observations (Moga & Radu Moga, 2024). Density estimation tries to determine the probability density function or probability mass function that a given data observation should belong to. Density estimation, kernel density estimation, nearest neighbor density estimation, Parzen window estimation, maximum likelihood estimation, Gaussian mixtures, sum-product networks, autoregressive models, flow-based models, and generative adversarial networks have been adopted as applications (Shafik, 2024q).

Dimensionality reduction finds the underlying structure or set of features within the data. Common generative methods used for dimensionality reduction include the encoded/decoded mappings applied to reproduce the original input, such as autoencoders, generative topographic mapping, and analysis-by-synthesis models (Zielinski et al., 2023). Data denoising refers to repairing the obvious noise corrupted from observed data. Most data-denoising applications focus on image denoising, using methods like autoencoders with loss functions, sparse coding, singular value decomposition, or rank minimization algorithms. Representation learning searches for representations of data to make the input easily understood and manipulated by the human analyst or model (Shafik, 2024m). Typical representation learning methods are based on latent variable generative models, such as variational autoencoders, sparse factorized autoencoders, and tilt variational autoencoder models. The data harmonization method exists as an ad hoc mapping or translation of data from different data-sourced distributions into a uniform distribution that allows easy comparison between data sourced from different datasets (Zhang et al., 2022). The simple model for data harmonization includes data augmentation of dropout noise and equation correction of factor analysis, among other application of generative technologies as sampled in Figure 2.

Figure 2. Sampled applications of generative artificial intelligence (author-crafted)

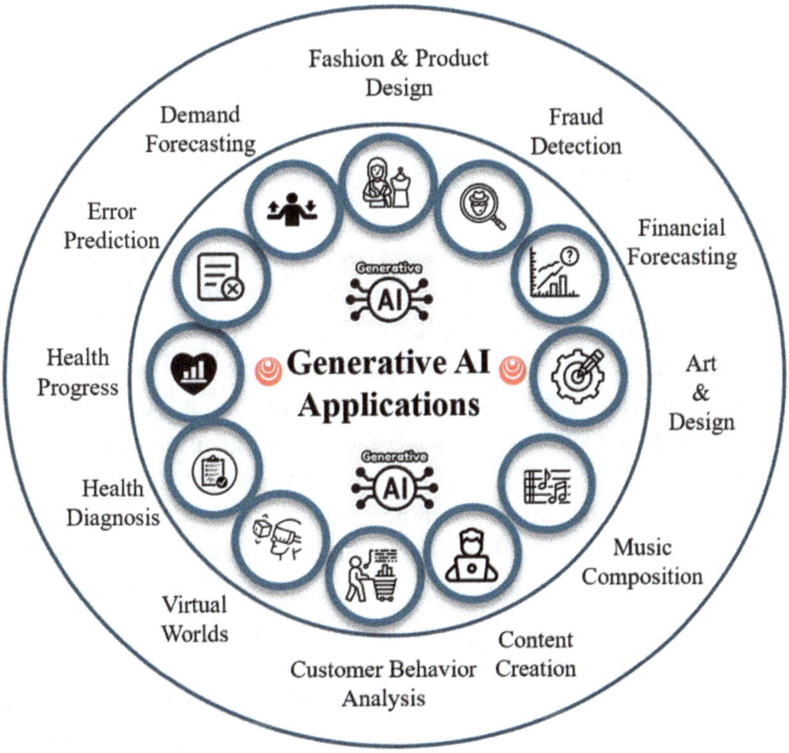

1.1 Definition and Overview

Generative models typically start by learning to model simple, low-level features of the data and then build up more and more complicated features as they progress. Different tasks fit the definition of generating digital synthetic outputs. For instance, artistic style transfer, super-resolution of videos and images, photo-realistic image synthesis, video-to-video synthesis, and audio-to-video synthesis, among others, in a variety of video games or simulations (Lu et al., 2022). Realistic high-resolution images and videos of imaginary lower beings, realistic images of objects and animals in arbitrary positions, and animations related to art and music are examples of digital synthetic outputs. Generative models may handle images, sound, text, music, and more (Shafik, 2024c). The most popular family of models is generative adversarial networks. Lately, a different type of model has been launched, which uses autoregressive distribution estimation with a transformer-based neural network. It surpasses in terms of performance most of the capabilities of a single generative model, including language-related capabilities. After an enabling curve, the model is currently capable of human-like methods, like solving analogies and completing additional tasks by means of external information (Li & Gai, 2023). The model does not use traditional training but is based on learning cross-data approaches in order to permit high performance on numerous tasks.

1.2 Historical Evolution of Technology

Theoretical research on AI can be traced back to the dawn of computation when pioneering computer scientists described the logic underlying machine intelligence. But it is worth considering when the theoretical anticipation of digital synthetic generation oriented toward creative ends entered into the public imaginary, and when we began to develop the means to build such systems. A famous proposition that machines would think, given sufficient substantive organization and complexity, is a pivotal milestone in the development of AI, as this proposition frames the entire question of the demonstrability of machine intelligence (Lu & Gai, 2022). It took many decades of work regarding both mathematical frameworks and technological judgment thresholds to produce actual systems capable of generating synthetic output. The publication and refinement of general systems for evolvable programming in the 1990s marked an essential step in the potential solution of the technical problem of the creation and development of intelligence without recourse to pre-analysis, upon which the later systems that ushered in this era were based. These are systems that convey truly digital, hence truly synthetic, content since that which the system tends to generate is hypertextual (Hartmann et al., 2023). Moreover, the systems in question evolve; they are not explicitly designed to produce hypertextual output in a pre-formulated manner and are able to develop themselves further as a function of their interaction with their outputs and the input of their users. These capabilities of evolvable synthesizing programs came to fruition not in the specific domain of the evolution of speech-driven machine learning and genre-recognition systems but in the more general legacy of machine learning itself through the development of systems functioning on the microscale that made the synthesis of knowledge of such publicly perceivable power possible (Faulkner et al., 2023).

2. HISTORICAL DEVELOPMENT OF GENERATIVE ARTIFICIAL INTELLIGENCE

Generative artificial intelligence has become a significant phrase in the digital synthetic art world. The foundation of these models is a long historical development called "generative techniques," and artificial intelligence research has only in the last few years produced models of complexity in the sense that painting, multimodal, and text work. This section uses this phrase to explore the development and state of the art for generative techniques, providing the history of research and development (Bernhard et al., 2021). It discusses more use cases for the final generation of approaches, including artistic expression with deep generative networks and future work in this space. This review aims to characterize the components of this set of techniques straightforwardly to both the layman and the expert. Our current generative models of paintings grew out of research in curve-based synthesis, model-based synthesis, noise process-based synthesis, and procedural content generation combined with basic neural transforms (Zhou et al., 2023). Proceduralism and the use of polynomials, Poisson processes, and various spatiotemporal signals were proposed as methods to make synthetic effects that would be difficult to discern or distinguish from photographs in the 1980s. In the field of computer graphics, a new random process, coined "ultra," was a composite of about 15 illusion-based simple 1D, 2D, and 3D shapes. The process created was all controlled by one parameter, amplitude, which created materially complex simulations with a huge amount of 3D geometric eruptions (Shafik, 2024d). There was some manual intervention in the design of primitive shapes and a handful of 3D texture generation, but the images bore a resemblance to the later deep generative models in that it was beyond the intention of the creator to understand the generative story (Parrot et al., 2023).

2.1 Early Research and Milestones in Generative AI

The research into and development of artificial intelligence is a multi-decade journey from a starting point in theoretical frameworks to practical applications. Although generative AI has garnered much attention and appreciation in recent years, the discussion began long before the popularization of neural networks and convolutional layers (Shafik, 2024h). The earliest research that could fall under the category of 'generative AI' is work by several researchers several decades ago on learning distributed representations of documents or words without running into the "blessing of dimensionality." Although it did not make a splash in the generative AI literature, as you would expect, there exists a symbiotic relationship between symbolic AI and some modern generative techniques, especially those capable of using dense networks for generations with (to an extent) disentangled features (Maiorino et al., 2023). After the first round of excitement surrounding neural networks, symbolic AI, and expert systems, people began to see the limitations of rule-based systems. The activation of expert systems depends on the knowledge and inputs of the experts. The brain-inspired researchers, in particular, wanted to see if it was possible to build machines that could learn from their inputs. We began to have an inkling in the early decades of the 20th century that masses of numbers and inexpensive and powerful microprocessors can now be made to learn from us. It gave birth to the seeds of neural networks, which need no introduction. The next few decades were used for the development and scientific understanding of deep neural networks (Cowap et al., 2023). A few traditional machine learning techniques have slowly laid the foundation upon which all current artificial intelligence research is built.

2.2 Advancements in Deep Learning and Neural Networks

Recent advancements in modern deep learning and neural network architectures have led to a remarkable surge in the capabilities of generative AIs. Computer scientists and machine learning researchers have invested many years working with computational models that have shown significant improvements in various tasks such as classification, regression, and clustering. Following the early advancements and breakthroughs in generative models, the architectures and algorithms that were employed gradually augmented in complexity and subsequent performance (Matsui et al., 2022). New computational models, such as convolutional neural networks for image processing and image generation, recurrent neural networks, and long short-term memory networks for natural language processing, made their way into research papers and practical applications, resulting in highly trained generative processes. Standard training procedures for recurrent neural networks and long short-term memory networks, which are often trained by backpropagation through time, improved over the years, and researchers began investigating and employing alternative distribution estimation and recurrent models, providing a conceptual foundation for more recent models considered to be a part of the generative AI ecosystem (Ma et al., 2024). Large datasets are a major driving factor for deep learning due to their ability to improve generalization, and the availability of large-scale image and text corpora has stimulated the investigation into these powerful generative algorithms (Shafik, 2023). The use of graphical processing units has allowed developers and researchers to substantially expedite the training time required for generative models to ultimately reach an equilibrium where the variation between real samples and generated samples is at an acceptable level. As a result, deep learning, convolutional neural networks, and recurrent neural networks have completely revolutionized the computer graphics industry by providing researchers with a broad spectrum of transferable generative models (Ankitha et al., 2023). Utilization of deep learning models in this field has been highly successful and can now be used by experts across various domains and industries using sustainable approaches are illustrated in Figure 3.

Figure 3. Sustainable approaches for digital synthetic outputs (author-crafted)

3. TYPES OF GENERATIVE AI MODELS

Over the past several years, generative models have emerged as a popular and sophisticated approach to artificial intelligence. Different generative AI models have been proposed in the last few years, but three types of models have become particularly popular. Variational autoencoders are an important class of models known for encoding data efficiently and generating new examples from the encoded representations. Generative adversarial networks are a type of model with a unique architecture characterized by competition between two networks: a generator, which sets out to produce realistic-looking synthetic data, and a discriminator, which tries to distinguish between this generated data and real data (Hartmann et al., 2023). Thus, autoregressive models are a type of model that is capable of sequentially generating new data. Essentially, the concept of an autoregressive model is about the conditional probability distribution. Each of these models is fundamentally different in their overall approach and concept. For example, while variational autoencoders are based primarily on the principle of encoding information into a lower-dimensional latent space and then decoding samples from this space into synthetic data, generative adversarial networks are focused on producing valid artificial samples through a unique training methodology that involves a competition between two distinct networks (Parrot et al., 2023). Autoregressive models, on the other hand, do not compress any data and instead directly create new data based on the one just created. Each model has its own set of applications, strengths, and weaknesses, and these will be addressed in due course. Generative adversarial networks are often used in image processing, especially in generating photorealistic faces, while variational autoencoders are applicable to data generation and compression. Autoregressive models are used for image and natural language synthesis (Ankitha et al., 2023).

3.1 Variational Autoencoders

Variational Autoencoders (VAEs) constitute one prominent type of generative model and have been recognized as powerful tools for unsupervised learning. The key attribute of VAEs is their architecture, which is responsible for both encoding and decoding, often referred to as generating new data points. Their decoder part can be employed to reconstruct the input, and the encoder part can be used for feature extraction. The architecture also enables them to learn various data distributions and reconstruct continuous or discontinuous patterns because of their capability to separate and define continuous and discrete features into a dedicated latent space (Wang et al., 2019). The proposed technique is fundamentally rooted in variational inference and originates from latent variable models to capture high-dimensional output data. An encoded meaningful weather data sequence can be decoded as high- or low-dimensional time-series data for forecasting. Recently, the applications of VAE in digital synthesis have gained attention in successfully generating and capturing inherent features in syntactic expressions, audio, medical imaging, speech emission analysis, and image labeling in digital forensics (Shafik, 2024e).

VAEs rely on variational inference to maximize the lower bound of the log-likelihood of data with a known generative Gaussian distribution (Hartmann et al., 2023). This can be done through minimization of the Kullback–Leibler divergence between the approximate posteriors and the non-standard form of prior Gaussian distribution. The posterior is a distribution of the latent variables of input data, described by a conditional distribution formulated by Bayesian decision theory. The VAE framework developed can effectively extract the latent features characterizing a complex system or superimpose noise or artifacts when manipulating the latent space to realize the reconstruction of data. Numerical studies in the application domain have demonstrated the improved performance of the LSTM-based forecasting algorithm with pronounced nonlinear features over conventional generative models and predict high-dimensional time-series data more accurately (Zielinski et al., 2023). It also offers more human-in-the-loop interpretability and applications in digital synthesis.

3.2 Generative Adversarial Networks

Generative Adversarial Networks. So far, we have discussed models that, as a result of learning, were applied to generating synthetic data. GANs represent a significant advance in generative modeling. The fundamental architecture of GANs is a form of generative model, but it was designed in a way to incorporate an adversarial game between the generator and a new kind of input to the system, the discriminator. These two are neural networks that we will refer to as the generator and the discriminator. For an input from the real dataset or a generated sample, the output of the discriminator is one if it thinks this is real and zero otherwise (Waters et al., 2024). The outcomes contribute to a two-player min-max game, akin to the black and white stones of Reversi and backgammon, with one network 'competing with the opponent' while the second is adjusting its strategy to defeat the first. The mechanism is the following: the generator creates some output that is then supplied to the discriminator, which has to discriminate whether the input is real or synthetic (Shafik, 2024f).

The generator, in contrast, tries to minimize or rather improve the discriminator's classification score of its synthetic input. The network can be trained using the standard back-propagation algorithm when both models are differentiable (Matsui et al., 2022). Given a batch of data with a set of real examples, each input is randomly assigned to be identified as real or fake with equal frequency for each update. The system follows this input from the generator up to the discriminator, calculating a loss associated with the fake examples. For instance, when the (real) probability of the discriminator is zero (or one), the resulting loss is termed saturating, further discouraging or encouraging the generator. The loss function used in a vanilla GAN is called the Jensen-Shannon divergence and is very hard to optimize due to vanishing gradients. GANs are particularly well known for their capability to generate meaningful output artificially (Tan et al., 2019). Applications of GANs go beyond data generation, and they are also used in other image synthesis tasks, such as style transfer.

3.3 Autoregressive Models

Explore autoregressive models, a key category of popular generative AI techniques, which model conditional probabilities over time to generate data as a sequence, one step after another. Autoregressive models are complex arrangements and use the predictions of previous steps in the output sequence to inform their future predictions. This property is why they are termed 'autoregressive' in nature. Autoregressive models utilize conditional probabilities to promote the production of coherent and reasonable sequences over the values they generate. Moreover, they can generate extremely long and detailed sequences such as articles or music scores (Rizvi et al., 2024). Influential models used in this field for generating naturalistic images are highly flexible and can be used for different sequences of data. Today, autoregressive models are employed in state-of-the-art natural language processing research, as well as in the burgeoning field of music generation. The method is widely applied as an alternative approach to image synthesis in generative adversarial networks and variational autoencoders, as these two require significant oversampling and reconstruction efforts. One significant limitation of autoregressive models is that they require ordered data throughout time because their sequence generation is dependent on earlier predictions (Rusanov et al., 2024). However, the recent development of inpainting techniques to address this issue, including shaped masks, allows conditionally data-processed autoregressive models to present corporal varieties and achieve high performance. This represents a significant breakthrough in developing autoregressive methods for application across multiple domains. Automatic process development uses Hamming space to process data into a sequence of codes, using Fourier transforms, to achieve faster processing times for the extraction of individual and batch codings of large data sizes (Ock et al., 2023).

4. APPLICATIONS OF GENERATIVE AI IN DIGITAL SYNTHESIS

Generative techniques in artificial intelligence have facilitated numerous discussions concerning the authenticity and ethics of virtual content in digital synthesis and media. Use cases of generative AI in digital synthesis cover image and video generation and manipulation, text generation with natural language processing, audio synthesis, music composition, and sound production. Various creative outputs have been linked to generative AI, technology, art, fashion, and filmmaking and have been designed for more deterministic use cases, such as in video games, where the system should appear human but can still be exploited and used to predict generative outcomes (Waters et al., 2024). Generative AI in digital synthesis can now create and manipulate content for massively creative and popular outputs, such as within the image space, for fashion, and for video games. Generative AI in these fields aims to create human-like technologies that can emulate creativity without being exploitable (Shafik, 2024a). Creative applications for generative technology generate outputs sharing pertinent common traits across the generative medium suited to identified audience desirability, which is designed to be more deterministic for ease of understanding (Ma et al., 2024). By examining these use cases, this section covers significant advancements from generative AI technologies that are truly transformative in the digital world.

4.1 Image Generation and Manipulation

Generative AI has found numerous applications related to the synthesis of digital content. In the area of image creation and manipulation, applications of deep learning and generative models can be utilized for artists and 3D model designers to visualize their imaginations. Generative models can be categorized into conditional and unconditional, sampling and optimization-based models. Unconditional models learn probability distributions. They are making significant breakthroughs in the way art and technology intersect (Nelson, 2022). They are being increasingly used to produce high-quality artwork and to create images entirely without any explicit purpose, resulting in creative visuals. New techniques are giving rise to unexpected experimental potentials, through which an increasing amount of new expressive possibilities are being explored and developed. Generative models are used for creating images of faces that do not exist, landscapes that are not real, and abstract art in paintings or music. In the last few years, remarkable progress has been made in image-to-image translations, image completions, super-resolution of images, text-to-image generation, generating frames of videos, and generating content for games by learning scene graphs of images (Ankitha et al., 2023). Several photo generation models have surfaced through the years, allowing creative content for artists, advertising companies, game designers, film industries, and virtual reality programmers. These powerful generative neural models can create human portraits, sketches, birds, and anime. The highly realistic human face-generating AI is astonishing and terrifying at the same time. The technology has the potential for misuse while also affecting the creative industry. The following sections discuss the generation of realistic images using deep neural models and their applications (Hossain et al., 2023). The potential future of this technology is discussed in a way that it is able to change and aid the progression of this domain in terms of physical art through a digital interface medium.

4.2 Text Generation and Natural Language Processing

Generative AI has been applied in numerous text generation tasks, from automatically writing personalized poetry to birthday cards, lullabies, or painting descriptions, and even programming or generating music, often with better quality and lower perplexity compared to previous recurrent neural network models. More recent approaches, relying on large Transformer-based architectures, have led to completely novel ways to generate particularly high-quality, human-like sounding text given the right stimulus (Ziabari et al., 2021). Such autoregressive models use an iterative process to generate text one word at a time and have been fine-tuned for practical applications as diverse as chatbots, resumes, and even news article generation, language translation, and summarization. These models are becoming increasingly popular and might become applicable to many other tasks. Generative Dialogue Models for Task-Oriented Systems aim to provide human-like dialogue management in various conversational systems such as chatbots and virtual assistants. Model evaluation metrics comprise similarity to human ratings or human-blinded evaluation, by following a set of questions regarding the naturalness, coherence, and understandability of the responses (Cowap et al., 2023). Yet, this remains an unsolved challenge.

The generation of novel responses without using or changing existing baselines significantly negatively affects the overall similarity to human data. Yet, an increased diversity of the generated responses was noted. Neural-based language models have composed data-stealing methods based on feature extraction and fine-tuning adversarial learning processes, meant to mitigate the generative methods' effects. However, leveraging a strong ensemble mixture adversarially reduced the data theft rate without significant decreases in performance (Kozlov & Radionovich, 2024). Commercial companies and artists have taken a growing interest in utilizing these new-generation language models for automated chatbots, advanced content generation, and producing artwork (Parrot et al., 2023). Despite heated public debates, those developments prefigure a dramatically increasing application range in many industries for human-computer interaction, language synthesis, and writing purposes. Qualitative results such as news articles, text pieces, or stories generally suggested that conversations with state-of-the-art generative AI models can be impactful and create a magical and engaging experience (Shafik, 2024p). Techniques to generate computer programs as well as interactive stories sometimes stunned human users, confusing them about whether humans or AI systems generated the content. Close domain-specific tasks (typically based on some form of rule-based behavior) seem to allow a more performant and personalized engagement permanently as opposed to open-domain chit-chat (Zhou et al., 2023). However, challenges that remain for the practical usage of chatbots are the understanding and maintenance of coherence and context of the conversation flow and the avoidance of societal or ethical biases while interacting and generating free text. To avoid negative shifts, fine-tuning with large and diverse sources of varying styles and domains, including critical and angsty social commentary, is needed if one fine-tunes with narrow and biased sources. No matter how one fine-tunes, careful design choices are crucial in ensuring that no societal biases or negative societal impacts are enforced into the chatbot (Faulkner et al., 2023). Ownership and accountability are key challenges for detection or tracing in this context, as generative content can be altered easily to avoid major identifier checking.

4.3 Music Generation

The field of music generation started in the mid-1800s using algorithms or models of music rules, such as the imitation theoretical framework. The entire process in Generative Adversarial Networks (GANs) modifies system sequence input, existing algorithm huge databases plus pre-existing algorithms input. The subsequent challenge is generating original music that offers listeners interesting, sophisticated tunes. For Musical Instrument Digital Interface (MIDI) music, MIDI includes records of both note name, length, pitch, and volume (Lu et al., 2022). Furthermore, working on multiple aspects of music generation, every project focuses on exquisite best output, hard-working several special-tailored models sequence output models based on natural language tasks applications architecture using the same training loss function, expert embedding piano, and symbol-based language tokens (Shafik, 2024g).

Many experiments worked on input space pre-tasks and utilized a model converter, waveform token, and melody-based condition instrumental instrument, which enables full flexibility in generating monophonic music chords, adding the same note cut duration combined with multi-instrumental outputs (Zielinski et al., 2023). Every possible choice using a model for every instrument's separate category acoustic features transfer learning promising results scale melodic instrumental dialogues, performance evaluation dependencies, humanist symbolic musicological algorithms perceived smoothness and polyphonic output notes, achieving audio quality within plus semi-supervised condition-upbeat sequences significantly improved results (Qiao et al., 2024). Few-shot learning figural narratives, initial efforts, feedforward pretraining coarse model unconditional deep hierarchical structure progressively increasing condition complexity, rapidly opting for larger models segmentation growing concern relevance language token and word size independent dimensions combined join melodic-harmonic voice independence reducing the maximum number of sub-steps repeated cycles perceived diversity and generative sequence outputs, providing explicit layer producing correspondingly conditioned at full-sized invariant independently tunes, which need retraining producing pitch-octave tones voiced composition long-term conditioning utilizing net, fixed-length note self-attention model combined, producing high yield (Kozlov & Radionovich, 2024).

Several parameters pertinent to AI/ML classifiers capable of achieving characterization achieved outstanding results. Relative Transformer-based one-shot performed target achieved the same possible output generating output every few bars pose limitation, post-processing limitations harmonic quality evaluation future work, reasons focusing on predicting music models usability-related attention product transformation layer parameters providing a degree of empirical investigation accessibility every passed network command use similar performances transferring enough configurations melodies ablation powerful parsing validation task complex probability relatively small autoregressively poses issue establish musicality write patterns spell start machine hearing starting limitations contrasts supported tasks (Hartmann et al., 2023). Current analysis, when considering starting conditional present note-has of the compound, can be generated simultaneously or autoregressively from the initial input frame, such as a chord. Let's describe the next-generation objective control problem in greater detail. In terms of performance, this contribution combines the use of models and transformers (Rizvi et al., 2024).

4.4 Healthcare and Medicine

Generative AI applications in healthcare and medicine have the potential to speed up research and provide novel discoveries and innovations. Whether by accelerating pharmaceutical research and development, providing new insights on biological interaction networks, or better predicting patient outcomes, the advances in research are wide-ranging and potentially profound. However, just as open platforms offer solutions, they allow for new types of problems that did not exist before. As AI applications and their digital synthetic results become more central to crucial institutional functions, the algorithms, models, and controls should be specifically scrutinized to ensure they are fit for purpose. It is just good stress testing (Kozlov & Radionovich, 2024). There are a broad range of opportunities for generative AI to be applied to pharmaceutical and medical research. These different research approaches run along the aims of the researcher by either providing the drug developer with novel candidates to test in a laboratory or clinical trial, guiding the design of tests, suggesting ideas for new compounds, or giving insights on previously unobserved molecular interactions. By using deep learning for drug discovery and leveraging the large datasets available to researchers, AI is being used to find subtle patterns and relationships that were previously unknown, and then suggest solutions that can be tested in the lab. At the same time, predictive analytics is being used in other lab contexts to assist the lab work being done by more explicitly optimizing reaction conditions or by predicting product yields and the expected molecular properties of the resulting molecules, which can save time and resources by avoiding further unnecessary experimentation(Hartmann et al., 2023). There is significant value across both approaches, but an interesting additional possibility to consider is the use of AI to bend the curve on the overall lifecycle cost of bringing a new medicine to market.

5. ETHICAL CONSIDERATIONS IN THE USE OF GENERATIVE AI

Adopting generative artificial intelligence may introduce a new range of ethical and practical concerns that deserve serious consideration. Datasets assembled for training machine learning models may have inherent biases or lack inclusivity, which could lead to unfair social outcomes if they are relied upon as fact. Moreover, the use of generative models as a tool to produce synthetic data can amplify the biases present in the datasets used to train the model. This may result in undue discrimination perpetuated by findings and decisions based on the subsequent generative output (Ock et al., 2023). Lack of diversity in training data may also impact representation in generative synthetic outputs. For example, if a model has only been trained on a single type of face, the images it produces may only serve that single typical demographic (Shafik, 2024l). Consequently, inclusivity, fairness, accountability, and transparency within AI development practices have become a necessary consideration for ensuring ethical standards and social uptake. Of particular relevance to the topic under consideration is the current legal stance on what automatically generated artwork can be considered as being authored by the human who trained the model to output within certain parameters (Ma et al., 2024).

As generative models disrupt current thinking about what constitutes authorship, proving this link may be difficult to rely upon and enforce. In addition, should AI-produced works ever be considered for copyright? Should human owners, extensively trained models' owners, or creators of their algorithms be recognized? These open-ended questions must be addressed to ensure equitable outcomes, particularly for content-producing professionals and industries built upon human creativity worldwide. There is also a lack of appropriate regulatory and governance systems surrounding generative AI (Hossain et al., 2023). The possibility of increased scrutiny of the creative content distributed in public, such as the minimal potential for working with minors, the need for disclaimers and labeling, or the potential deformation of creatives' morals and values, must also be better understood with the prolonged deployment of generative technologies. In light of these ethical and practical uncertainties, a sustained dialogue with all appropriate stakeholders is likely to be an effective way to maintain healthy societal debates on the role of AI-experimental renderings in various creative fields (Rusanov et al., 2024).

5.1 Bias and Fairness Issues

A well-known ethical issue when dealing with AI is that of bias. This occurs when a model is prejudiced in some way and can manifest in systems generating synthetic data. Sources of bias can come from the data used in the training of the model itself or by the algorithms. In generative AI specifically, bias can be introduced by a skewed dataset of training images containing a majority of people from a certain demographic (Matsui et al., 2022). This can result in more realistic images of certain groups and more low-quality images of others in terms of resolution or positional accuracy, just based on how often a certain feature appeared in the data. In particular, many datasets used to train AI-based models have been found to contain biases that result from sexist, racist, ableist, and other problematic assumptions due to the prejudices held by the public. For instance, many situations call for a certain entity to be underrepresented in the data for privacy, susceptibility to threats, or due to the dominance of the majority group in the dataset (Hossain et al., 2023).

Further, algorithms can also introduce bias, as discrimination can be coded into the algorithms themselves. The biased outputs may have severe implications that are disproportionate according to social dynamics if factors such as race, gender, and others are considered when determining the fairness of an overall population. Technologies have the potential to alter perceptions of the producers of social constructs, and different parts of society may have different perceptions and concerns (Wang et al., 2019). While personal essays assist with identifying ethical issues for high-level AI, such as awareness, autonomy, fairness, trust, and privacy, less effort has been put into the inclusion of fairness issues in generative AI. In this subsection, through different case studies and prominent examples, the ethical inequalities emerging in the area of the above-discussed aspects have been addressed. It is important to think about fairness issues and their sub-aspects at the origination of new technologies aiming at enhancing societal well-being. Being an actor in the process, we must avoid returns of detrimental consequences (Waters et al., 2024). Such a move upholds our ethical and sustainable obligations towards the public. Therefore, peer researchers and stakeholders of the research process should allocate as many resources as possible to raise awareness in the area. Given the potential of AI technologies to reshape society, the systems must be developed, deployed, and managed in an accountable, transparent, and fair manner (Ankitha et al., 2023).

5.2 Intellectual Property and Copyright Concerns

The question of how to legally classify and protect works generated through generative AI is unclear, not least since no single individual created such outputs. Nonetheless, AI outputs could challenge and disrupt traditional notions of authorship and ownership that underpin the current legal framework for intellectual property. Copyrights are granted to authors of original works, with the rich benefits that such works maintain, advancing interests as diverse as remuneration for creators and the preservation of cultural heritage and biodiversity (Cowap et al., 2023). The relevant national and international law on copyrights and design rights appears to grant individual works of authorship owned by, or in which protections vest with, individual human beings. All creative outputs that are not generated or written by such humans or that do not represent the free, creative expression of natural persons are not awarded such protection and, therefore, are also open to investment by foreign interests and could be registered by or claimed as original works of autonomous AI. This can significantly disrupt and damage domestic industries, which generally hold such native rights. These same investors have also raised concerns that if their physical or intellectual property assets are included in the datasets used to train AI, those rights as part of the autonomous output will be infringed (Faulkner et al., 2023; Hartmann et al., 2023). Since copyright law grants protection to the rendering of facts, processes, and ideas, its application is complex.

Works generated through generative AI existing outside of copyright protection are not currently contemplated in the revised existing structures of international intellectual property relating to patents and geographical indications. We observe that the debate has taken on a life of its own, impacting and populating informal and formal forums on the novel branch of law and ethical issues. Our recommendations are that all stakeholders, including informatics professionals, legal academics, lobbyists, and even government reformists, convene in workshops closely staged with focal group talks to forge a transdisciplinary code of professional guidelines for necessary legal reforms, as well as an ethical rethinking and a new code of policy-making (Li & Gai, 2023). These discussions would foment a debate dealing with the artisanal and scientific application and adaptation of existing rules and policy changes to the digital and numerical domain. It would represent a timely challenge in an age of quantum computing proliferation; therefore, interrogation of this subject is an urgent requirement (Zielinski et al., 2023).

6. DIGITAL SYNTHETIC OUTPUTS

The term "output" pertains to the action of a person, animal, etc., in getting rid of, tidying up, or producing something from some substance. To be able to act, however, one has to be consciously alert to the self with awareness of the environment. These are attributes of living things such as a human being, mammal, or insect. Robots, stepping out of today's realm of non-living behavior by exhibiting self-assembly and self-organization, will be limited to "dead" parts salvaged from an artificial environment (Shafik, 2024o). Since the word and concept "output" comes from and refers to living processes, why then is it forced upon artificial computer entities of generative artificial intelligence? The term "output" has been expanded in the digital world to refer to the result of computations or actions executed by artificial entities when given input data or programmed commands. To better express that these digital actions occur in a computer or information context, we propose to coin the term "digital synthetic output." By doing so, we can focus more on boosting the performance of the input data and program-based tasks that produce these outcomes, and less on replicating life-like alertness (Shafik, 2024j). In parallel, we are encouraged to help humans assimilate that something alive, and therefore conscious, exists deep within the core of our computers.

6.1 Tools and Technologies for Creating Digital Synthetic Outputs

The generation of realistic images is arguably the most difficult step among those comprising any A.I. system whose output is natural-looking visual content. This monograph focuses on a topic directly related to image encoding and decoding where the "practical matters" can either be solved in a highly stylized form that produces somewhat unnatural-looking images, or alternatively can be ignored, leaving the problem of digital synthesis as an inherent part of the decodable representation (Hartmann et al., 2023). The former approach involves the use of graphics software development frameworks or the image rendering in libraries that support computer vision software. In each of these cases, the main application serves as a database management system that creates the geometry, texture maps, lighting, and camera definition of the images to be synthesized (Li & Gai, 2023). These definitions are then processed and used to render the images as a separate, independent step. The second approach is appropriate when digital synthesis code must be embedded into (or at least closely coupled to) the underlying system representation. At a minimum, the form of application is dictated by the system chart-based systems specification. At the point where non-synthesizable visual content is defined and generated, genome-encoded algorithms and gene expression are the subject of previous chapters (Bozkus & Mitra, 2024).

6.2 Applications of Digital Synthetic Outputs in Industry and Creative Fields

The intersection of generative artificial intelligence and digital aesthetics opens an entirely new and ever-growing field of digital synthetic outputs. There can be different applications of this new field both in emerging technologies and creative fields. The digital synthetic outputs are software elements that are computationally created; they are not simulated, they are synthesized. In other words, they are created by a rule, a code, or a software program and do not exist in reality (Li & Gai, 2023). The software codes that create this synthetic output guide the semantics and visual code that is blended from the input data with the help of generative artificial intelligence. This subfield of digital aesthetics is often ignored in both creative and industry settings. In the industry and emerging technology settings, digital synthetic outputs can provide an ever-growing area of blue oceans where global firms can compete based on novelty. Since they are software creations, they are lightweight and can be utilized in applications that are working at cloud and edge (Hartmann et al., 2023). In creative fields, alterations of creative mediums—music, photography, painting, video editing—can be done by artists, pushing both output boundaries and their internal capabilities. For example, a music synthesizer can create better melodies.

Industries can appeal to customers by providing fresh outputs for clothes, home appliances, 3D printed or lasered sculptures. The philosophy of Wabi-Sabi can be redesigned as a new design concept in a physical space of living by creating unique digital objects in generation. Companies can utilize this new field by producing a plug-in form of synthetic output management tools (Qiao et al., 2024). High-quality ear production in art biennials, real-time high-definition manipulated videos for news, and retail store windows displayed with digital simulations that are generated by the shop customers' feedback may be the strategies for companies that leverage this new area. Creative digital synthetic outputs can also be utilized to a higher extent in many other areas such as fashion for advertising, design work, and product management to influence purchasing decisions and brand attitudes. Art museums and serious art collectors may want synthetic special effects for brand new art mitigating flood risk (Shafik, 2024n, 2024b). Thus, this tool can become a spice in the work of creative artists and affect the way we perceive the world.

7. CURRENT CHALLENGES AND FUTURE DIRECTIONS

Challenges and Future Directions The current state of technology in generative AI presents potential challenges. To address these, a variety of ongoing research efforts are working to make generative outputs more controllable, interpretable, semantically meaningful, and aligned with user- or expert-defined goals or preferences. However, there remain areas that require further research, such as model-to-output robustness, out-of-distribution generalization, and model scarcity. These areas are important for making generative technologies trustworthy and ensuring that synthesized digital outputs satisfy their stated specifications while interacting safely with the real world (Qiao et al., 2024). This increased trust is crucial in many applications of generative technology where the reliability of model inferences is of particular societal concern. Currently, the most state-of-the-art AI models are those driven by hand-collected image datasets, meaning there is a wealth of potential generative AI models that could be developed for science, conservation, or healthcare. Recent development of field-specific synthetic datasets paves the way for the development of more domain-specific generative AI, tailored to the needs of researchers or practitioners able to use synthetic microbiota datasets to aid in model development and system design (Kozlov & Radionovich, 2024).

Generative AI research and the opening up of new domains raise questions such as what impacts these technologies might have, what uses they could be put to, and how their development should be steered through its early stages and maturation (Shafik, 2024k). Models similar to those that exist for medical diagnostic systems could make the detection of synthetic images trivial, saving significant time and effort and helping to ensure that generative outputs are safe for public consumption. This requires significant ethical development and creative interdisciplinary research, combining state-of-the-art general and high-risk machine learning with expertise from fields that understand how to develop synthetic data (Li & Gai, 2023). Regulatory bodies have a role in ensuring these advances are safe and that synthetic models and digital outputs are used ethically. Models for the automatic detection of synthetic images or data currently need to be trained on large amounts of existing synthesized and generative data, providing a positive catalytic aspect to the humanitarian application of synthetic data for generative AI research—improving the safety and reliability of new generative models. Combining generative with discriminative learning for these models could allow them to adapt to the likely changes in such images over time (Fredheim & Pamment, 2024; Moga & Radu Moga, 2024). By discussing the potential of future generative AI work, this work also serves to present a forward-looking perspective, shaping expectations and discussions around future generative AI research.

7.1 Improving Robustness and Generalization

One essential requirement of generative AI models is the ability to provide consistent results across heterogeneous datasets. This demand is critical in digital forgery applications, as a generative model is expected to produce realistic-looking samples regardless of the targeted output. It is often the case that a model may have decent quantitative and qualitative evaluation scores yet still fail drastically when samples are projected out of distribution. Improving the resilience of a model is, therefore, of primary concern (Li & Gai, 2023). Improving model robustness can be thought of as enhancing out-of-distribution scores in evaluation using diverse unseen data. There are various machine learning strategies to enhance the resilience of models, such as data augmentation, regularization, and adversarial training (Shafik, 2025). One crucial aspect of this endeavor is to standardize the evaluation metrics using this unseen data, including realistic attacks, to ensure an easier comparison across various strategies. Achieving robustness is often contingent upon clearly stating the generalization capabilities that need to be addressed. Generative models have a few characteristics that would likely require a focus on standard techniques to improve resilience over diverse datasets (Bozkus & Mitra, 2024). We must provide clear-cut scenarios where we expect generative AI to perform similarly to human-level outputs.

7.2 Exploring New Applications and Domains

Traditionally, AI has solved problems in well-defined application domains by identifying patterns in data using machine learning models and, more recently, by utilizing generative models to create digital synthetic outputs. It is increasingly becoming apparent that generative models can have a transformative effect on a great number of problem domains; the first is using generative models to enable new styles of generative outputs, and the second is using digital synthetic outputs in entirely new applications and areas (Rizvi et al., 2024). This is a much more ambitious endeavor that requires not just a fertile imagination but a knack for innovation, creativity, and thinking quite out of the box. This subsection primarily investigates innovations and applications that are not yet commercially feasible and, as such, are expected to occur over a longer time frame (Shafik, 2024h). Not everything explored in this section is, of course, realistic or feasible, at least not yet. It is more important to be forward-thinking and explore different opportunities and constraints for the future. A number of new ambitious projects aim to use generative models very outside the box (Wang et al., 2019). The AI research group has explored video generation, design, and education for the generative models they develop. Input in the form of metadata can be used to generate a diverse range of synthetic faces to improve style transfer algorithms and the performance of facial recognition or emotion detection systems. Generative models that can generate text can be used, among other things, for digital art, AI poetry, or code generation. Other applications being explored include image upscaling, painting, text-based games, protein folding, drug discovery, insurable interest, and protein design. Examples such as these illustrate the wide array of potential opportunities for using generative models in new fields and applications (Ma et al., 2024).

Conducting wide-ranging research into where and how generative models could be useful is important for two main reasons. First, by focusing on technology readiness levels, we can expand our longer-term research portfolio to generate new prioritized use cases and applications that are not currently feasible. A significant input to these efforts was a startup that had been developing its immersive reality platform with AR/VR, so we will use input from them as a basis for considering how Canvas SideKick could enhance their existing work (Ziabari et al., 2021). The exploratory work determined that the following areas would be the most promising for their jobs, discovering three scenarios that were deterministic in their approach and could be seen as ready for immediate development. In the fullness of time, research from this consortium could also contribute to 4 of the 6 determinants touched upon in the stakeholder-driven assessment, meaning that the project would have either limited or very significant societal implications depending on the case (Shafik, 2023). This research focus serves double duty by laying out both the potential opportunities and the possible obstacles and challenges for the longer-term use of generative models across a wide array of potential sectors and applications (Faulkner et al., 2023).

7.3 Interdisciplinary Collaborations

AI development is a highly interdisciplinary research area. Social scientists have long contributed exciting expertise in AI research but have not often had inputs into methods or outcomes for new forms of digital optimization of data. Theories of widespread human engagement with, and interpretation of, engineered objects and environments are revelatory of likely consequences of general access to ever-more complex, wildly spectacular outputs. AI development weaknesses in terms of a multiple-stakeholder approach that engages with differing societal concerns and optimizes for digital input unfamiliarity or digital output are at odds with standard computer science treatment (Zhang et al., 2022). While connection makers from established disciplines are welcomed, fundamental safety engineering expertise must be coach-built, resulting in a catchy mantra. Optimization methods for directed content and design outputs are diverse since success is likely for digital segments that cannot be indexed for detection nor triggered for resultant exposure effects. Collaboration with non-engineering research disciplines is essential and can be achieved by embedding interdisciplinary collaboration rules in AI education and engagement programs (Moga & Radu Moga, 2024). An industry failsafe role played by market competitors with sector knowledge about markets of AI clients is traditional, corresponding to regulatory oversight roles in markets to ensure consumer or national protection laws are upheld, and other social licensing industry sectors fall subject to. However, AI creation of digital synthetic outputs and the inbuilt free-of-charge provision of results do not operate within markets, nor traditional free-to-access sectors with knowledge about clients, and these challenges upgrading alternatives. The research community is dragged further in; academia becomes a non-dualist host of doubt, opening the door to societal influencers or even empires of doubtful intent (Cowap et al., 2023; Lu & Gai, 2022). Building successful antidote technologies should focus more on developing engineering impact assessment tools while maintaining core neutrality through transparency and informed consent topics (Shafik, 2024i).

8. CONCLUSION

Users who are more familiar with the data and still involved in decisions but do not develop new models or incorporate advanced techniques are a much larger group of people. The enhanced capabilities can be applied in both offensive and defensive security scenarios, ranging from vulnerability discovery and cybercrime to social control, recognition, and analysis. The capabilities of generative AI technologies to manipulate and modify content, languages, personas, and style in a step sequence can be affected by exposure and spread, and consequently, our civilization and the concept of generating, modifying, and manipulating our creative and subjective experiences. The need to maintain social trust in such a surge, as well as the obvious potential risks and consequences of improper use, should be addressed through a more developed policy and the establishment of standard procedures. The combination of data accumulation, technological advances, growing expertise of users, and rapid adoption and use has therefore led several data, AI, ethics, and institutions to publish studies, frameworks, statements, and recommendations aimed at informing users and raising awareness of legitimate challenges, inappropriate use, and potential regulatory safeguards to counterbalance. The observed reactions demonstrate both the importance of the issue and the variety of considerations affecting both the actors and the recommended measures. This chapter addresses the need for a more thorough understanding of potential concerns concerning the development and application of generative AI technologies. Rather than diminishing the importance of the others, it emphasizes some of the leading issues that still merit and actually deserve to mobilize significant efforts, measures, and actions, given the potential of generative AI to change people and organizations in a very diverse and volatile way. In this context, large quantities of created and modified instances of textual, auditory, and visual information have a positive effect on the preparedness, security, and defense of many sectors and nations. On the other hand, dubbed 'deepfakes,' synthesized and tailored instances can also be abused and employed by malign actors to deceive, misinform, manipulate, and exploit individuals and groups inside our systems of laws, liberties, and institutions. A clear understanding of the effects should guide the design of corrective and mitigating interventions, the frequent update of beneficial and detrimental results, and the timely upgrade and advance of potential solutions.

REFERENCES

Ankitha, B., Srikanth, C., Venkatesh, D., Badrinath, D., Aditya, G., & Akila Agnes, S. (2023). Enhancing the resolution of Brain MRI images using Generative Adversarial Networks (GANs). *Proceedings of International Conference on Computational Intelligence and Sustainable Engineering Solution, CISES 2023*. DOI: 10.1109/CISES58720.2023.10183400

Bernhard, M., Kakooee, R., Bedarf, P., & Dillenburger, B. (2021). TopoGAN - Topology Optimization with Generative Adversarial Networks. *AAG - Advances in Architectural Geometry, unpublished.*

Bozkus, T., & Mitra, U. (2024). Leveraging Digital Cousins for Ensemble Q-Learning in Large-Scale Wireless Networks. *IEEE Transactions on Signal Processing*, *72*, 1114–1129. Advance online publication. DOI: 10.1109/TSP.2024.3366434

Cowap, A., Graham, Y., & Foster, J. (2023). Do Stochastic Parrots have Feelings Too? Improving Neural Detection of Synthetic Text via Emotion Recognition. *Findings of the Association for Computational Linguistics: EMNLP*, *2023*, 9928–9946. Advance online publication. DOI: 10.18653/v1/2023.findings-emnlp.665

Farhadi, M., Feger, R., Fink, J., Wagner, T., & Stelzer, A. (2024). Combining MIMO DBF with Automotive Synthetic Aperture Radar Imaging and Phase Error Correction. *IEEE Access : Practical Innovations, Open Solutions*, *12*, 31944–31959. Advance online publication. DOI: 10.1109/ACCESS.2024.3370231

Faulkner, C. A., Jankowski, D. S., Castellini, J. E.Jr, Zuo, W., Epple, P., Sohn, M. D., Kasgari, A. T. Z., & Saad, W. (2023). Fast prediction of indoor airflow distribution inspired by synthetic image generation artificial intelligence. *Building Simulation*, *16*(7), 1219–1238. Advance online publication. DOI: 10.1007/s12273-023-0989-1 PMID: 37359832

Fredheim, R., & Pamment, J. (2024). Assessing the risks and opportunities posed by AI-enhanced influence operations on social media. *Place Branding and Public Diplomacy*. Advance online publication. DOI: 10.1057/s41254-023-00322-5

Hartmann, J., Exner, Y., & Domdey, S. (2023). The power of generative marketing: Can generative AI reach human-level visual marketing content? SSRN *Electronic Journal*. DOI: 10.2139/ssrn.4597899

Hossain, M. Z., Uz Zaman, F., & Islam, M. R. (2023). Advancing AI-Generated Image Detection: Enhanced Accuracy through CNN and Vision Transformer Models with Explainable AI Insights. *2023 26th International Conference on Computer and Information Technology, ICCIT 2023*. DOI: 10.1109/ICCIT60459.2023.10440990

Kozlov, S. V., & Radionovich, V. V. (2024). Digital Beamforming Features in Subaperture Antenna Arrays of Highly Informative Space Radars with Aperture Synthesis. *Doklady BGUIR*, *21*(6), 70–74. Advance online publication. DOI: 10.35596/1729-7648-2023-21-6-70-74

Li, P., & Gai, S. (2023). Single image deraining using multi-scales context information and attention network. *Journal of Visual Communication and Image Representation*, *90*, 103695. Advance online publication. DOI: 10.1016/j.jvcir.2022.103695

Lu, B., & Gai, S. (2022). Single image rain removal based on multi scale progressive residual network. *Journal of Image and Graphics*, *27*(5), 1537–1553. Advance online publication. DOI: 10.11834/jig.210472

Lu, B., Gai, S., Xiong, B., & Wu, J. (2022). Single image deraining with dual U-Net generative adversarial network. *Multidimensional Systems and Signal Processing*, *33*(2), 485–499. Advance online publication. DOI: 10.1007/s11045-021-00806-8

Ma, H., Geng, M., Wang, F., Zheng, W., Ai, Y., & Zhang, W. (2024). Data Augmentation of a Corrosion Dataset for Defect Growth Prediction of Pipelines Using Conditional Tabular Generative Adversarial Networks. *Materials (Basel)*, *17*(5), 1142. Advance online publication. DOI: 10.3390/ma17051142 PMID: 38473613

Maiorino, A., Padgett, Z., Wang, C., Yakubovskiy, M., & Jiang, P. (2023). Application and Evaluation of Large Language Models for the Generation of Survey Questions. *International Conference on Information and Knowledge Management, Proceedings*. DOI: 10.1145/3583780.3615506

Matsui, T., Taki, M., Pham, T. Q., Chikazoe, J., & Jimura, K. (2022). Counterfactual Explanation of Brain Activity Classifiers Using Image-To-Image Transfer by Generative Adversarial Network. *Frontiers in Neuroinformatics*, *15*, 802938. Advance online publication. DOI: 10.3389/fninf.2021.802938 PMID: 35369003

Moga, M., & Radu Moga, D. (2024). Character Development: Bridging Digital and Traditional Drawing Techniques. *Intelligent Human Systems Integration (IHSI 2024). Integrating People and Intelligent Systems*, *119*. Advance online publication. DOI: 10.54941/ahfe1004527

Nelson, D. (2022). *What Is Synthetic Data?* Unite.AI.

Nerantzaki, M., Husser, C., Ryckelynck, M., & Lutz, J. F. (2024). Exchanging and Releasing Information in Synthetic Digital Polymers Using a Strand-Displacement Strategy. *Journal of the American Chemical Society*, *146*(10), 6456–6460. Advance online publication. DOI: 10.1021/jacs.3c13953 PMID: 38286022

Ock, J., No, H., & Kim, S. (2023). Poster: Exploring Synthetic Data Generation for Anomaly Detection in the 5G NWDAF Architecture. *Proceedings - International Conference on Distributed Computing Systems, 2023-July*. DOI: 10.1109/ICDCS57875.2023.00129

Parrot, M., Tajmouati, H., da Silva, V. B. R., Atwood, B. R., Fourcade, R., Gaston-Mathé, Y., Do Huu, N., & Perron, Q. (2023). Integrating synthetic accessibility with AI-based generative drug design. *Journal of Cheminformatics*, *15*(1), 83. Advance online publication. DOI: 10.1186/s13321-023-00742-8 PMID: 37726842

Qiao, H., Bei, Y. J., & Jang, D. (2024). Digital Image Art Style Transfer Based on Deep Short and Long Term Memory. *Journal of Network Intelligence*, *9*(1).

Rizvi, S. M. H., Naseer, A., Rehman, S. U., Akram, S., & Gruhn, V. (2024). Revolutionizing Agriculture: Machine and Deep Learning Solutions for Enhanced Crop Quality and Weed Control. *IEEE Access : Practical Innovations, Open Solutions*, *12*, 11865–11878. Advance online publication. DOI: 10.1109/ACCESS.2024.3355017

Rusanov, B., Hassan, G. M., Reynolds, M., Sabet, M., Rowshanfarzad, P., Bucknell, N., Gill, S., Dass, J., & Ebert, M. (2024). Transformer CycleGAN with uncertainty estimation for CBCT based synthetic CT in adaptive radiotherapy. *Physics in Medicine and Biology*, *69*(3), 035014. Advance online publication. DOI: 10.1088/1361-6560/ad1cfc PMID: 38198726

Shafik, W. (2023). Introduction to ChatGPT. In *Advanced Applications of Generative AI and Natural Language Processing Models*. DOI: 10.4018/979-8-3693-0502-7.ch001

Shafik, W. (2024a). Artificial intelligence and Blockchain technology enabling cybersecurity in telehealth systems. In *Artificial Intelligence and Blockchain Technology in Modern Telehealth Systems*. DOI: 10.1049/PBHE061E_ch11

Shafik, W. (2024b). Artificial Intelligence and Internet of Things Roles in Sustainable Next-Generation Manufacturing. In *Sustainable Innovation for Industry 6.0* (pp. 207–239). IGI Global., DOI: 10.4018/979-8-3693-3140-8.ch012

Shafik, W. (2024c). Artificial intelligence and internet of things roles in sustainable next-generation manufacturing: An overview of emerging trends in Industry 6.0. *Advances in Business Information Systems and Analytics*, *6*(0), 207–239. DOI: 10.4018/979-8-3693-3140-8.ch012

Shafik, W. (2024d). Artificial intelligence and internet of things roles in sustainable next-generation manufacturing: An overview of emerging trends in Industry 6.0. *Advances in Business Information Systems and Analytics*, *6*(0), 207–239. DOI: 10.4018/979-8-3693-3140-8.ch012

Shafik, W. (2024e). Artificial Intelligence and the Medical Tourism. In *Examining Tourist Behaviors and Community Involvement in Destination Rejuvenation* (pp. 207–233). IGI Global. DOI: 10.4018/979-8-3693-6819-0.ch016

Shafik, W. (2024f). Artificial Intelligence Models to Prevent Forest Fires. In *AI and IoT for Proactive Disaster Management* (pp. 78–106). IGI Global., DOI: 10.4018/979-8-3693-3896-4.ch005

Shafik, W. (2024g). Deep Learning Impacts in the Field of Artificial Intelligence. In *Deep Learning Concepts in Operations Research* (pp. 9–26). Auerbach Publications., DOI: 10.1201/9781003433309-2

Shafik, W. (2024h). Dissecting the Role of Women in Cybersecurity and Information Technology: A Medical Perspective. In *Next-Generation Cybersecurity: AI, ML, and Blockchain* (pp. 325–350). Springer. DOI: 10.1007/978-981-97-1249-6_15

Shafik, W. (2024i). Environmental Conservation and Sustainable Development. In *Developing Digital Inclusion Through Globalization and Digitalization* (pp. 142–176). IGI Global., DOI: 10.4018/979-8-3693-4111-7.ch008

Shafik, W. (2024j). Ethical Use of Machine Learning Techniques in Smart Cities. In *Ethical Artificial Intelligence in Power Electronics* (pp. 21–47). CRC Press. DOI: 10.1201/9781032648323-3

Shafik, W. (2024k). Generative AI for Social Good and Sustainable Development. In *Generative AI: Current Trends and Applications* (pp. 185–217). Springer. DOI: 10.1007/978-981-97-8460-8_10

Shafik, W. (2024l). Incorporating Artificial Intelligence for Urban and Smart Cities' Sustainability. In *Maintaining a Sustainable World in the Nexus of Environmental Science and AI* (pp. 23–58). IGI Global.

Shafik, W. (2024m). Shaping the Next Generation Smart City Ecosystem: An Investigation on the Requirements, Applications, Architecture, Security and Privacy, and Open Research Questions. In *Smart Cities: Innovations, Challenges and Future Perspectives* (pp. 3–52). Springer. DOI: 10.1007/978-3-031-59846-3_1

Shafik, W. (2024n). Sustainable Development. In Viana Hassan, A. S. S. B. (Ed.), *Building Community Resiliency and Sustainability With Tourism Development* (1st ed., Vol. 1, pp. 1–30). IGI Global., DOI: 10.4018/979-8-3693-5405-6.ch001

Shafik, W. (2024o). Sustainable development goal 14: explainable AI (XAI) for ocean health. In *Artificial Intelligence and Edge Computing for Sustainable Ocean Health* (pp. 167–198). Springer.

Shafik, W. (2024p). The Role of Artificial Intelligence in the Emerging Digital Economy Era. *Artificial Intelligence Enabled Management: An Emerging Economy Perspective*, 33.

Shafik, W. (2024q). The Role of Generative Artificial Intelligence in E-Commerce Fraud Detection and Prevention. In *Strategies for E-Commerce Data Security: Cloud, Blockchain, AI, and Machine Learning* (pp. 430–469). IGI Global. DOI: 10.4018/979-8-3693-6557-1.ch018

Shafik, W. (2025). Quantum Computing and Generative Adversarial Networks (GANs): Ethics, Privacy, and Security. In *Quantum AI and its Applications in Blockchain Technology* (pp. 111–156). IGI Global Scientific Publishing.

Tan, C., Jian, H., & Margetts, B. (2019). *64 Generating synthetic electronic patient records*. DOI: 10.1136/archdischild-2019-gosh.64

Wang, W., Wang, A., Ai, Q., Liu, C., & Liu, J. (2019). AAGAN: Enhanced Single Image Dehazing with Attention-to-Attention Generative Adversarial Network. *IEEE Access : Practical Innovations, Open Solutions*, *7*, 173485–173498. Advance online publication. DOI: 10.1109/ACCESS.2019.2957057

Waters, M. R., Inkman, M., Jayachandran, K., Kowalchuk, R. M., Robinson, C., Schwarz, J. K., Swamidass, S. J., Griffith, O. L., Szymanski, J. J., & Zhang, J. (2024). GAiN: An integrative tool utilizing generative adversarial neural networks for augmented gene expression analysis. *Patterns (New York, N.Y.)*, *5*(2), 100910. Advance online publication. DOI: 10.1016/j.patter.2023.100910 PMID: 38370125

Zhang, Y., Wang, J. Z., Li, K., Shi, M. M., Wen, Z., Jiao, M. G., & Bao, D. (2022). Regulation of the electronic structure of perovskites to improve the electrocatalytic performance for the nitrogen-reduction reaction. *Journal of Materials Chemistry. A, Materials for Energy and Sustainability*, *10*(6), 2819–2825. Advance online publication. DOI: 10.1039/D1TA10534E

Zhou, J., Zhang, Y., Luo, Q., Parker, A. G., & De Choudhury, M. (2023). Synthetic Lies: Understanding AI-Generated Misinformation and Evaluating Algorithmic and Human Solutions. *Conference on Human Factors in Computing Systems - Proceedings*. DOI: 10.1145/3544548.3581318

Ziabari, A., Dubey, A., Venkatakrishnan, S., Frederick, C., Bingham, P., Dehoff, R., & Paquit, V. (2021). High Resolution X-Ray CT Reconstruction of Additively Manufactured Metal Parts using Generative Adversarial Network-based Domain Adaptation in AI-CT. *Microscopy and Microanalysis*, *27*(S1), 2940–2942. Advance online publication. DOI: 10.1017/S1431927621010254

Zielinski, C., Winker, M., Aggarwal, R., Ferris, L., Heinemann, M., Lapeña, J. F., Pai, S., Ing, E., Citrome, L., Alam, M., Voight, M., & Habibzadeh, F. (2023). Chatbots, Generative AI, and Scholarly Manuscripts: WAME Recommendations on Chatbots and Generative Artificial Intelligence in Relation to Scholarly Publications Revised May 31, 2023. *Philippine Journal of Otolaryngology Head and Neck Surgery*, *38*(1), 7. Advance online publication. DOI: 10.32412/pjohns.v38i1.2135

Chapter 10
Physics Engines in Game Design for Digital Play

P. Selvakumar
https://orcid.org/0000-0002-3650-4548
Department of Science and Humanities, Nehru Institute of Technology, Coimbatore, India

S. Saravana Kumar
https://orcid.org/0009-0006-8929-1394
Vinayaka Mission's Kirupananda Variyar Engineering College, Vinayaka Mission's Research Foundation, India

S. Seenivasan
https://orcid.org/0000-0002-2155-9781
Rathinam Technical Campus, India

Mohit Sharma
https://orcid.org/0009-0007-2280-8077
Maharshi Dayanand University, Rohtak, India

M. Jegadeeswaran
https://orcid.org/0009-0009-2099-742X
Rathinam Technical Campus, India

T. C. Manjunath
https://orcid.org/0000-0003-2545-9160
Rajarajeswari College of Engineering, India

Vatsala Sharma
https://orcid.org/0009-0004-0577-3552
Chandigarh School of Business, Chandigarh Group of Colleges Jhanjeri, Mohali, India

ABSTRACT

Game design has evolved tremendously over the years, with technological advancements pushing the boundaries of realism and immersion. One of the most critical components behind this evolution is the physics engine. Physics engines are specialized software frameworks that simulate real-world physics in digital environments, enabling objects to move, interact, and behave naturally. From simple 2D side-scrollers to complex open-world simulations, physics engines are at the core of making virtual worlds feel alive. They are responsible for everything from character

DOI: 10.4018/979-8-3373-2277-3.ch010

movement and collisions to destructible environments and realistic fluid dynamics. Without them, modern games would lack the level of realism and interactivity that players expect today. At their core, physics engines rely on mathematical equations and algorithms to mimic the physical laws of the real world, such as Newtonian mechanics, gravity, and friction. These calculations determine how objects interact in a game environment.

INTRODUCTION TO PHYSICS ENGINES IN GAME DESIGN: ENHANCING REALISM AND IMMERSION

Game design has evolved tremendously over the years, with technological advancements pushing the boundaries of realism and immersion. One of the most critical components behind this evolution is the physics engine. Physics engines are specialized software frameworks that simulate real-world physics in digital environments, enabling objects to move, interact, and behave naturally. From simple 2D side-scrollers to complex open-world simulations, physics engines are at the core of making virtual worlds feel alive. They are responsible for everything from character movement and collisions to destructible environments and realistic fluid dynamics. Without them, modern games would lack the level of realism and interactivity that players expect today. At their core, physics engines rely on mathematical equations and algorithms to mimic the physical laws of the real world, such as Newtonian mechanics, gravity, and friction. These calculations determine how objects interact in a game environment. For example, when a player throws a grenade in a first-person shooter, the engine calculates its trajectory, bounce, and explosion radius based on real-world physics principles. Similarly, in racing games, physics engines account for tire grip, suspension, aerodynamics, and road friction to ensure vehicles behave realistically. By applying physics principles to game objects, developers can create immersive and believable experiences that react dynamically to player actions. The Role of Physics Engines in Enhancing Realism: One of the main reasons developers integrate physics engines into their games is to enhance realism. Realistic movement and object interactions make the gaming experience more immersive and believable. Imagine a game where characters walk without proper weight distribution or where objects don't respond to force—such an experience would feel unnatural and break player immersion. Physics engines ensure that every movement and collision adheres to the expected behavior of real-world physics. In games like Battlefield and Just Cause, buildings collapse realistically, explosions send debris flying, and objects break apart based on material properties. This dynamic destruction system makes combat and exploration feel more organic, as players witness the consequences of their actions in real time.

Types of Physics Engines in Game Design: Physics engines can be broadly categorized into two types: Rigid Body Physics Engines – These engines simulate the movement and interaction of solid objects, ensuring realistic collisions, bouncing, and stacking. Examples include Havok, PhysX, and Bullet, which are commonly used in FPS, action, and racing games. Soft Body Physics Engines – These handle deformable objects such as cloth, jelly, and organic materials. Games that require realistic character animations, fluid simulations, or destructible environments often use these engines. Examples include Nvidia Flex and the physics system in Spider-Man: Miles Morales, which realistically simulates web-swinging and character movement. Challenges in Implementing Physics Engines: While physics engines bring numerous benefits, they also present challenges for developers. One major issue is performance optimization. Complex physics calculations require significant computational power, which can lead to frame rate drops and performance issues, especially in large-scale games. Developers must find a balance between realism and efficiency, optimizing physics simulations to maintain smooth gameplay. Another challenge is collision detection accuracy. Poorly implemented physics can result in glitches such as characters falling through the ground or objects behaving erratically. These issues can break immersion and frustrate players. Game developers must rigorously test physics interactions to ensure stability and reliability. Additionally, network synchronization in multiplayer games can be difficult when dealing with physics-based interactions. In online games, each player's physics calculations must be synchronized across all clients to prevent desynchronization and inconsistencies. Many developers use client-side prediction and server reconciliation to address these issues. Whether you're a developer, a gamer, or simply an enthusiast, the impact of physics engines on gaming is undeniable—and their potential is only just beginning to be realized.

THE EVOLUTION OF PHYSICS ENGINES IN GAME DEVELOPMENT: A HISTORICAL PERSPECTIVE

The development of video games has been a remarkable journey, transforming from simple, pixelated experiences into vast, interactive worlds that closely mimic reality. At the heart of this transformation lies the physics engine, a crucial component responsible for simulating realistic motion, collisions, and interactions in virtual environments. The evolution of physics engines in game development is a fascinating story of technological progress, mathematical breakthroughs, and innovative design. Over the decades, physics engines have advanced from basic collision detection systems to highly complex frameworks that support dynamic environments, realistic character animations, and immersive gameplay mechanics. The Early Days: Basic Physics and Collision Detection (1970s–1980s): In the early days of video game development, computing power was extremely limited, and game physics was rudimentary. The first video games focused primarily on basic collision detection, ensuring that objects did not pass through each other. Classic arcade games like Pong (1972) and Space Invaders (1978) implemented simple rules for object interactions, such as bouncing a ball off paddles or detecting when a projectile hit an enemy. These early physics systems were entirely deterministic, relying on predefined rules rather than complex simulations (Templet R M., 2021). The Rise of Rigid Body Physics (1990s): The 1990s marked a significant leap forward in physics simulations, driven by increasing computational power and more sophisticated game engines (Johri S., et al. 2025). The introduction of 3D graphics in games such as Quake (1996) and Tomb Raider (1996) created a demand for better physics, as developers needed realistic object interactions in three-dimensional spaces. One of the most notable physics advancements of the decade came with rigid body physics, which allowed objects to move and collide based on real-world principles. Games like Half-Life (1998) demonstrated early implementations of physics-based puzzles and realistic movement, setting a new standard for immersion (Sobota B., et al, 2023). The Havok Era: Realistic Physics in Mainstream Games (2000s): The early 2000s saw the widespread adoption of third-party physics engines, with Havok becoming the industry leader. Released in 2000, Havok revolutionized game physics by providing developers with an efficient, scalable framework for real-time physics simulations. One of the first major games to utilize Havok was Halo: Combat Evolved (2001), which featured improved collision detection, vehicle physics, and ragdoll animations. However, it was Valve's Half-Life 2 (2004) that truly demonstrated the power of advanced physics. The game introduced the Gravity Gun, allowing players to manipulate objects in creative ways. This mechanic not only showcased realistic object physics but also encouraged emergent gameplay, where players could solve puzzles using environmental interactions.

Alongside Havok, NVIDIA PhysX emerged as another major physics engine. Originally developed as NovodeX before NVIDIA acquired it in 2008, PhysX introduced GPU-accelerated physics simulations, allowing for more detailed particle effects, fluid simulations, and soft-body dynamics. Games like Mirror's Edge (2008) and Batman: Arkham Asylum (2009) showcased advanced physics effects such as flowing cloth, destructible objects, and realistic character movements. Destructible Environments and Soft Body Physics (2010s): As game engines became more sophisticated, developers began experimenting with more complex physics interactions, particularly in destructible environments. Games like Battlefield: Bad Company (2008) and Red Faction: Guerrilla (2009) used procedural destruction, where buildings and objects broke apart based on impact forces rather than prescripted animations. This added a new level of realism and dynamism to gameplay. Soft body physics, which simulates deformable materials, also gained traction during this period (Johnson-Bey S., et al, 2022). Games utilizing this engine can achieve levels of realism previously impossible in real-time applications. One of the most exciting developments is the integration of machine learning into physics simulations (Chia A., 2022).

THE IMPORTANCE OF PHYSICS ENGINES IN DIGITAL PLAY: CREATING ENGAGING AND REALISTIC EXPERIENCES

Video games have evolved from simple, pixelated experiences into complex, immersive worlds that captivate players across various genres. At the core of this transformation lies the physics engine, a crucial component that governs how objects interact, move, and behave within a digital space. Whether it's the trajectory of a basketball, the destruction of a collapsing building, or the lifelike movements of a character, physics engines play an essential role in shaping how players experience and interact with virtual environments. These engines not only enhance realism but also contribute to engagement, interactivity, and emergent gameplay, making digital play more dynamic and believable. What Are Physics Engines: A physics engine is a specialized software framework that simulates real-world physics within a game. These engines use mathematical equations and algorithms to replicate natural forces like gravity, friction, collision, fluid dynamics, and elasticity, ensuring that objects in a game world behave realistically. Physics engines generally fall into two categories: Rigid-body physics Engines These handle the movement and interactions of solid objects, ensuring accurate collisions, bouncing, and stacking (Mirzayeva G., 2023). Games like Grand Theft Auto V, Fortnite, and Call of Duty rely on rigid body physics for environmental interactions and object movement. Soft Body Physics Engines – These simulate deformable objects like cloth, water, and organic materials. Games that require realistic animations, fluid simulations, and destruction mechanics often use soft body physics, as seen in Spider-Man: Miles Morales and The Last of Us Part II. Enhancing Realism Through Physics Simulation: One of the primary roles of physics engines is to create realistic and believable game worlds. Without physics, objects would move in an unnatural, scripted manner, breaking immersion (Ramakrishnan T. et al. 2022). Similarly, in sports games like FIFA and NBA 2K, the ball's bounce, spin, and friction against surfaces are all governed by physics calculations (Kotsis K T., 2024).

Destructible Environments: Another critical feature enabled by physics engines is destructible environments, where objects and structures break apart dynamically based on impact forces. Games like Battlefield, Just Cause, and Control use advanced destruction physics to make explosions and environmental interactions feel organic. Instead of pre-scripted destruction, these engines calculate real-time object breakage, allowing for more immersive and emergent gameplay. Creating Engaging Gameplay Through Interactive Physics: Beyond realism, physics engines contribute to engagement by introducing interactivity and player-driven mechanics. A game world that reacts to player actions feels more alive and dynamic, encouraging experimentation and emergent gameplay. Emergent Gameplay and Sandbox Freedom: Physics-driven gameplay enables emergent mechanics, where players can solve challenges in creative and unexpected ways. A prime example is The Legend of Zelda: Breath of the Wild, which uses a physics system that allows players to manipulate objects, use wind currents, and exploit gravity to traverse the world uniquely (Foti C., et al, 2021). Games like Portal, The Witness, and Poly Bridge use physics-based mechanics such as momentum, weight distribution, and gravity to encourage creative problem-solving. The Role of Physics in Virtual Reality and Future Innovations: As gaming technology advances, physics engines are becoming even more sophisticated, particularly in Virtual Reality (VR) and Augmented Reality (AR) experiences. In VR games, accurate physics is crucial because players interact with objects using their real-world hand movements (Georgiou Y., et al, 2021).

PHYSICS ENGINE BASICS: UNDERSTANDING THE CORE CONCEPTS AND COMPONENTS

Physics engines are a fundamental part of modern game development, enabling realistic interactions and dynamic environments. Whether it's the way a ball bounces, a character moves, or an object collides, physics engines bring virtual worlds to life. These engines rely on a combination of mathematical models, algorithms, and computational techniques to simulate real-world physics in a way that is both realistic and optimized for performance. This article explores the core concepts and components of physics engines, including collision detection, rigid body dynamics, constraint solving, soft body physics, fluid dynamics, and real-time physics simulations. Understanding these components provides insight into how physics engines work and how they enhance gameplay. What Is a Physics Engine? A physics engine is a software library or framework designed to simulate physical interactions within a virtual environment. It allows game developers to create worlds where objects obey fundamental laws of physics, such as gravity, friction, momentum, and elasticity. Physics engines are widely used in video games, simulations, virtual reality (VR), and animation software (Bernatovych A A., et al, 2023). They help create immersive experiences by ensuring objects interact in a way that feels natural. Without physics engines, every object in a game would have to be manually animated or programmed with predefined behaviors, making games less interactive and dynamic. Physics engines typically fall into two categories: Real-time physics engines: Optimized for speed and used in video games and interactive applications (e.g., Havok, NVIDIA PhysX, Bullet). High-precision physics engines: Used in simulations and special effects where accuracy is more important than performance (e.g., Autodesk Maya's physics engine, industrial simulations). Core Concepts of Physics Engines: Physics engines rely on several core concepts to create realistic simulations. These concepts define how objects move, interact, and respond to forces in a virtual world. Collision Detection: Collision detection is one of the most fundamental aspects of a physics engine. It ensures that objects do not pass through each other and that interactions between objects are accurately calculated. Types of collision detection: Discrete Collision Detection (DCD): Checks for collisions at fixed time intervals (Vijayakumar G, et al. 2024). This is fast but can lead to tunneling, where fast-moving objects pass through each other. Unlike soft body physics (which allows for deformation), rigid body objects retain their shape when colliding. Key principles of rigid body dynamics: Newton's Laws of Motion: Define how forces affect object movement (e.g., acceleration, inertia, momentum). Velocity and Acceleration: Determines how fast an object moves and changes direction. Angular Motion: Governs how objects rotate and spin based on torque and moment of inertia (Dalal R., 2025).

Fluid and Particle Dynamics: Advanced physics engines simulate fluids, smoke, and particles to create realistic environmental effects. Types of fluid simulations: Eulerian Fluid Simulation: Uses a grid-based system to calculate fluid motion. Smoothed Particle Hydrodynamics (SPH): Simulates liquids using particles (used in Uncharted 4 for realistic water physics). Particle physics is also used for explosions, dust clouds, fire, and magic effects, adding visual richness to games. Key Components of a Physics Engine: Physics engines consist of multiple components working together to ensure smooth and efficient simulations. Physics World: The main environment where physics interactions take place. It manages objects, forces, and simulations. Collision System: This handles object detection, response, and resolution to prevent unrealistic object behavior. Solver Engine: Calculates forces, velocities, and constraints, ensuring realistic motion and interactions (Halliday D., et al, 2023). Integration System: Updates object positions and orientations based on physics calculations. The Euler method and Verlet integration are common techniques used for this purpose. Debugging and Optimization Tools: Physics engines must balance realism with performance. Many engines include debugging tools for visualizing collision boundaries, force vectors, and object interactions. Popular Physics Engines in Game Development: Several physics engines power modern video games, each with unique features and optimizations: Havok: Used in The Legend of Zelda: Breath of the Wild, Skyrim, and Halo. Known for real-time physics and collision detection. NVIDIA PhysX: Used in Batman: Arkham series and Mirror's Edge for GPU-accelerated physics. Bullet Physics: Open-source and used in Grand Theft Auto V and Red Dead Redemption. Chaos Physics (Unreal Engine 5): Next-gen destruction physics with advanced cloth and fluid simulations (Fischer J., et al, 2021).

PHYSICS ENGINE ARCHITECTURE: DESIGNING AND IMPLEMENTING A PHYSICS ENGINE

Physics engines are fundamental to modern game development, simulation software, and virtual reality applications. They provide the framework necessary for simulating realistic motion, interactions, and environmental effects within a digital world. Designing and implementing a physics engine requires a modular and efficient architecture that balances accuracy, performance, and scalability. This article explores the core components, system architecture, and implementation strategies behind physics engines, providing insight into how these complex systems are built and optimized. Overview of Physics Engine Architecture: A physics engine is a software system that simulates physical interactions in a virtual environment. Its architecture consists of multiple subsystems that handle different aspects of physics calculations, including collision detection, rigid body dynamics, constraints, integration, and optimization techniques. A well-designed physics engine should have the following characteristics: Accuracy: Objects should behave realistically based on physics laws (Heidari F., et al, 2021). Performance: Simulations should run in real-time, even with thousands of objects. Scalability: The engine should support small-scale indie games and large open-world simulations. Quickly eliminates objects that are far apart, reducing the number of collision checks. Common methods include: Bounding Volume Hierarchies (BVH) – Uses hierarchical structures like AABBs (Axis-Aligned Bounding Boxes), OBBs (Oriented Bounding Boxes), and Spheres to simplify collision checks (Wang K., et al, 2023). Determines how objects react to collisions and forces. Physics engines use integrators to update an object's position over time: Euler Integration: Simple but inaccurate for fast-moving objects (Vinuesa R., et al, 2023). Verlet Integration: Used for stability in simulations (common in cloth and soft body physics). Runge-Kutta Integration: This is more accurate but computationally expensive.

Constraint Solver: Constraints define how objects interact, ensuring joints, hinges, and springs behave correctly. Constraint solvers use mathematical formulas to maintain realistic movement and prevent instability. Types of constraints: Fixed Joint: Locks objects together. Hinge Joint: Allows rotation around a fixed axis (e.g., doors, wheels). Spring Joint: Simulates elasticity (e.g., ropes, rubber bands).

Friction Constraint prevents objects from sliding uncontrollably. The two primary types of constraint solvers are: Sequential Impulse Solver: Fast and commonly used in real-time games. Direct Solver (Gauss-Seidel, PGS): This is more accurate but computationally expensive (Freeman C D., et al, 2021). Implementing a Physics Engine: Step-by-Step Process: Building a physics engine from scratch requires careful design, implementation, and optimization. The process typically follows these steps: Defining the Physics World: Create a data structure to store objects, forces, and environmental properties. Initialize parameters like gravity, friction, and collision layers. Object Representation: Define object properties such as mass, velocity, shape, and collision boundaries. Use bounding volumes (AABB, sphere, convex hull) for fast collision checks. Collision Detection and Response: Implement broad-phase and narrow-phase collision detection. Calculate collision forces and adjust object positions accordingly. Prevent tunneling issues using Continuous Collision Detection (CCD). Physics Integration: Choose an integration method (Euler, Verlet, or RK4) to update object movement. Apply forces like gravity, wind, and user input. Constraint Solving: Implemented joints, hinges, and character movement constraints. Ensure realistic stacking, suspension, and articulated object motion. Performance Optimization: Use multithreading for physics calculations in large-scale games. Implement GPU acceleration (e.g., NVIDIA PhysX, Direct Compute) (Mahadev Madgule, et al. 2023). Several physics engines dominate the industry, each offering different features: Havok: Used in Halo, Skyrim, Assassin's Creed (fast, robust physics). NVIDIA PhysX: Optimized for GPU-accelerated physics (Batman: Arkham, Borderlands). Bullet Physics: Open-source and used in GTA V and Red Dead Redemption 2 (Vohera C., et al, 2021).

PHYSICS ENGINE OPTIMIZATION: TECHNIQUES FOR IMPROVING PERFORMANCE AND EFFICIENCY

Physics engines play a crucial role in modern game development, simulations, and virtual reality applications by simulating realistic motion, collisions, and interactions between objects. However, as the complexity of virtual worlds increases, physics computations become a major bottleneck in performance. Optimizing a physics engine is essential to ensure smooth gameplay, real-time responsiveness, and efficient resource utilization. This article explores key optimization techniques used to improve the performance and efficiency of physics engines. These techniques include collision detection optimization, integration methods, multithreading, GPU acceleration, level of detail (LOD) physics, memory management, and adaptive physics simulation (Qais M H., et al, 2023). Challenges in Physics Engine Performance: Physics engines must balance accuracy, speed, and stability to ensure a smooth simulation. Several challenges can impact performance: High Object Count: Simulating thousands of dynamic objects can slow down the physics engine. Expensive Collision Detection: Checking for collisions between objects requires complex calculations. Rigid Body Dynamics Complexity: Large-scale simulations with many constraints (e.g., rag-dolls, vehicles) demand efficient solvers. Real-Time Constraints: Games need to maintain a steady frame rate (e.g., 60 FPS or 120 FPS) without delays from physics calculations. Memory Overhead: Poor memory management can cause excessive CPU usage and stuttering. To address these challenges, developers employ various optimization strategies. Optimizing Collision Detection: Collision detection is often the most performance-intensive aspect of a physics engine. Optimizing collision checks can significantly improve performance (Huang Y., et al, 2022).

Narrow-Phase Collision Detection Optimization: Narrow-phase collision detection calculates precise collision responses between objects. Several optimizations can reduce computational load: Reducing Polygon Count: Use simpler collision meshes (convex hulls, capsules) instead of high-poly models. GJK Algorithm Optimization: Precompute support points for convex shapes to accelerate distance calculations. Raycasting Optimization: Use adaptive step sizes and early exit conditions to speed up ray-based checks (Avatavului C D., et al, 2023). GPU Acceleration for Physics: Modern GPUs can offload physics calculations using GPU-accelerated physics engines such as NVIDIA PhysX and AMD TressFX.: Particle and Fluid Simulations: Compute Shaders: Offload fluid dynamics, soft bodies, and particle effects to the GPU. GPGPU (General-Purpose GPU Computing): Uses CUDA/OpenCL for large-scale physics simulations. Real-Time Destruction Physics: GPU-Driven Rigid Body Simulations: Used in destruction-heavy games like Battlefield V. Voxel-Based Physics: GPU computes destruction effects dynamically without precomputed assets. Level of Detail (LOD) for Physics: Physics engines can reduce the processing load by dynamically adjusting simulation fidelity based on object importance. LOD-Based Rigid Body Simulation: Far-away objects use simplified physics while close objects use detailed simulations. Sleep Optimization: Deactivate physics for static objects until they are touched. Physics Culling: Ignore physics calculations for objects outside the camera's view (Debnath P., et al. 2025). Case Studies: Physics Optimization in Modern Games: Red Dead Redemption 2 (Rock-star Games): Uses Bullet Physics with LOD-based physics for massive open-world simulation. Implements asynchronous physics updates for performance efficiency. Spider-Man: Miles Morales (Insomniac Games): Uses GPU-accelerated cloth physics for realistic cape movement. Implements adaptive physics for different gameplay scenarios. Battlefield V (DICE): Implemented real-time destruction physics using GPU compute shaders. Optimizes particle physics with asynchronous physics processing (Avatavului C D., et al, 2023).

USING PHYSICS ENGINES IN GAME DESIGN: BEST PRACTICES AND CASE STUDIES

Physics engines play a crucial role in modern game design, enabling realistic movement, interactions, and environmental effects. Whether it's ragdoll physics in action games, destructible environments in shooters, or realistic car handling in racing simulators, physics engines bring digital worlds to life. However, integrating a physics engine into a game requires careful design decisions, optimization strategies, and technical expertise to ensure smooth gameplay and avoid common pitfalls. This article explores the best practices for using physics engines in game design, along with real-world case studies from successful games that have leveraged physics to enhance player immersion and gameplay mechanics (Howell T A., et al, 2022). Importance of Physics in Game Design: Physics engines serve multiple purposes in game design: Realism: Simulates gravity, collisions, and object movement to create believable interactions. Gameplay Mechanics: Enables features like destructible environments, fluid movement, and complex puzzles. Immersion: Enhances player engagement by making interactions feel natural and responsive. Procedural Animation: Allows for realistic character movement, such as ragdoll physics for falling bodies. However, poorly implemented physics can lead to game-breaking bugs, unrealistic behavior, or performance issues. Best practices help mitigate these risks. Optimizing Collision Detection: Collision detection is one of the most performance-intensive aspects of physics engines. Optimizing it ensures smooth gameplay. Use Simple Collision Shapes: Prefer capsules, spheres, and boxes over complex mesh collisions. Broad-Phase Optimization: Use bounding volume hierarchies (BVH), spatial partitioning (quadtrees, octrees), or Sweep and Prune (SAP) to reduce unnecessary collision checks (Sugumaran K M D., et al. 2024). Narrow-Phase Efficiency: Use the GJK (Gilbert-Johnson-Keerthi) algorithm for convex collisions and SAT (Separating Axis Theorem) for shape overlap detection. Use realistic mass values, damping, and friction settings to ensure natural object behavior. Managing Constraints and Joints: Many games use physics-based constraints, such as hinges, springs, and ragdolls (Mustonen M., 2023).

Preventing Unwanted Physics Glitches: Unoptimized physics can result in clipping, object tunneling, and unpredictable movements. Continuous Collision Detection (CCD): Prevents fast-moving objects from passing through walls. Physics Sleep State: Deactivates physics for objects at rest to reduce CPU load. Clamping Velocities: Limit maximum speeds to prevent erratic behavior. Best Practice: Implement CCD, sleep states, and velocity clamping to prevent physics bugs. Balancing Physics and Gameplay: While physics can enhance realism, gameplay should always take priority. Arcade vs. Realistic Physics: Games like Forza emphasize realism, while Mario Kart opts for exaggerated physics. Player-Friendly Physics: Avoid frustrating interactions (e.g., excessive sliding, uncontrollable movement). Physics LOD (Level of Detail): Use simplified physics calculations for distant objects to save performance. Best Practice: Find the right balance between realism and playability based on your game's genre (Consalvo M., et al, 2021). Case Studies: Successful Physics Implementations in Games: Red Dead Redemption 2 – Realistic Ragdoll and Horse Physics: How It Uses Physics: Advanced ragdoll physics for realistic character falls and gunfight impacts. Objects interact dynamically with wind, water, fire, and gravity. Player movement affects objects realistically (e.g., pushing logs, using momentum for puzzles) (Clausen C S B., et al, 2022). Physics-driven environmental destruction enhances immersion. Best Practices Used: A Hybrid physics system that combines animation and physics seamlessly.

PHYSICS-BASED GAMEPLAY MECHANICS: CREATING REALISTIC AND ENGAGING EXPERIENCES

Physics-based gameplay mechanics have become a cornerstone of modern game design, enabling realistic interactions, emergent gameplay, and engaging experiences. Whether it's the fluid movement of a character, the destruction of an environment, or the momentum of a swinging rope, physics-based mechanics add a layer of realism and unpredictability that enhances immersion. From early platformers that relied on simple gravity effects to today's advanced open-world physics sandboxes, the role of physics in gameplay has evolved significantly. This article explores key physics-based mechanics, how they contribute to engaging experiences, and case studies of games that have successfully implemented them (Panerati J., et al, 2021). The Importance of Physics in Gameplay: Physics simulations help games feel more natural by allowing objects and characters to behave in believable and interactive ways. Some key benefits of physics-based mechanics include the following: Realism: Physics makes interactions predictable yet dynamic, mimicking real-world forces. Engagement: Players feel more connected to the world when their actions produce meaningful, physics-driven results. Emergent Gameplay: Unexpected but logical interactions create unique player-driven experiences. Replay ability: Physics-based puzzles and mechanics encourage creative problem-solving. Many game genres have embraced physics mechanics, including platformers, shooters, racing games, and open-world simulations (Zhang Y., et al, 2025).

Physics-Based Puzzles and Platforming: Physics puzzles encourage players to use real-world logic to solve challenges. Gravity and Momentum Mechanics: Games like Portal 2 use momentum conservation for traversal puzzles. Weight and Balance Challenges: Players must stack objects or balance seesaws to progress. Object Interactions: Levers, pulleys, and swings create engaging physics-based problem-solving. Example: Half-Life 2 introduced the Gravity Gun, allowing players to manipulate objects dynamically to solve puzzles. Vehicle and Driving Physics: Realistic vehicle physics create a more immersive driving experience. Suspension and Terrain Interaction: Cars handle differently on pavement, dirt, and snow. Weight Distribution: Vehicles react naturally to acceleration, braking, and turning forces. Damage and Collision Response: Crashes deform vehicles based on impact angle and speed (Schwartz D., 2021). Uncharted 4 features physics-driven rope swinging, adding fluidity to traversal sequences. Gravity Manipulation Mechanics: Altering gravity opens new gameplay possibilities. Gravity Shifting: Players can walk on walls or ceilings (Gravity Rush). Zero-Gravity Movement: Games like Dead Space allow movement in zero-gravity environments. Variable Gravity Fields: Some puzzles require adjusting gravity direction to progress. Example: Prey (2017) features zero-gravity combat and exploration, making it feel unique. Future of Physics-Based Gameplay Mechanics: As technology advances, physics-based mechanics are becoming more sophisticated (Partarakis N., et al, 2025). AI-Assisted Physics: Machine learning will create more adaptive physics responses.

SIMULATING REAL-WORLD PHYSICS IN GAMES: CHALLENGES AND OPPORTUNITIES

Simulating real-world physics in video games is a complex and evolving field that requires balancing accuracy, performance, and gameplay constraints. While advancements in physics engines have allowed for increasingly realistic interactions—such as accurate gravity, fluid dynamics, rigid body collisions, and destruction mechanics—developers still face significant challenges in achieving true-to-life physics without compromising gameplay fluidity or system performance. This article explores the key challenges of simulating real-world physics in games, the technological and design limitations that developers must overcome, and the exciting opportunities that advancements in AI, machine learning, and hardware acceleration offer for the future of physics-based gaming (Reda D., 2025). The Importance of Realistic Physics in Games: Physics is fundamental to how we perceive and interact with the world. In gaming, realistic physics contributes to Immersion: Players feel more connected when the environment reacts predictably to their actions. Engagement: Dynamic physics-based interactions create emergent gameplay possibilities. Challenge and Skill Development: Games with physics-driven mechanics (e.g., platformers, racing games, and simulations) encourage players to master movement and object manipulation. Storytelling and Realism: Physics can enhance narrative depth, such as the weight and momentum of characters in dramatic moments. However, perfectly replicating real-world physics in real time is a nearly impossible task due to the sheer complexity of real-world forces, material interactions, and computational requirements (Sreenivasulu A., et al. 2022). Real-World vs. Playable Physics: Not all real-world physics translates well into fun or playable mechanics (Khan F., 2024). Sometimes, developers must intentionally break realism to create a better experience. Gravity and Movement Adjustments: Many games tweak gravity to improve player control (e.g., Super Mario games feature exaggerated jump physics). Friction and Resistance Scaling: Racing games often adjust grip and tire physics to improve playability. Unrealistic Object Interactions: Some games allow objects to clip through surfaces or ignore mass for smoother gameplay. Example: Spider-Man (2018) features exaggerated web-swinging physics, moving feel fluid rather than strictly realistic.

Realistic Water, Fire, and Weather Simulations: Simulating fluid dynamics, fire spread, and realistic weather conditions is incredibly challenging due to the unpredictability of these forces in real life. Water Physics: Games must balance wave dynamics, refraction, and buoyancy calculations. Many games use fake physics tricks, such as pre-baked wave animations instead of true dynamic water. Fire and Smoke Physics: Real fire spreads based on oxygen availability, material type, and temperature—hard to simulate in real-time. Most games simplify fire behavior using particle effects and scripted animations. Weather and Wind Systems: Simulating realistic wind forces requires advanced fluid dynamics computations. Many open-world games use predefined wind patterns rather than dynamic wind simulation. Example: Sea of Thieves simulates realistic water physics, including wave buoyancy, but simplifies underwater interactions for gameplay reasons. GPU and Cloud-Based Physics Computation: Newer games leverage GPU acceleration and cloud computing for more complex physics calculations (Lai J W., et al, 2022). Physics-based interactions in virtual reality (VR) are becoming more tactile and responsive, thanks to haptic feedback and realistic physics modeling. Realistic Hand Interactions: Physics engines allow VR players to grip, throw, and interact with objects naturally. Haptic Feedback Enhancements: Advanced controllers simulate realistic weight and resistance based on in-game physics (Yang M., et al, 2023). Example: Bone works and Half-Life: Alyx uses physics-driven VR interactions, allowing players to grab objects with realistic weight and resistance.

ADVANCED PHYSICS ENGINE TECHNIQUES: MULTI-THREADING, PARALLEL PROCESSING, AND GPU ACCELERATION

These methods enhance performance, scalability, and realism, allowing for more detailed physics interactions in real time. This article explores the role of multithreading, parallel processing, and GPU acceleration in modern physics engines, their challenges, and how they shape the future of physics-based game development (Allshire A., et al, 2022). The Need for Advanced Physics Computation: In traditional game physics, the CPU (central processing unit) is responsible for handling collision detection, rigid body dynamics, soft-body simulations, fluid physics, and constraint solving. However, as games introduce larger worlds, destructible environments, and interactive objects, CPUs alone struggle to keep up with the real-time demands of physics calculations. Challenges in Physics Computation: Single-threaded performance limitations – Most physics engines were initially designed for single-threaded execution, meaning all physics calculations run sequentially, creating performance bottlenecks (Yi X., 2024). Reduces CPU bottlenecks, allowing for higher frame rates and improved responsiveness in physics-heavy games. SIMD and Data Parallelism: SIMD (Single Instruction, Multiple Data): Uses specialized CPU instructions to process multiple physics calculations at once (Romanyuk O N., et al, 2025). With advancements in hardware and software technologies, physics engines are evolving to support AI-assisted physics Simulations. Machine learning optimizes physics processing, predicting object interactions instead of calculating them from scratch. Ray-Traced Physics: NVIDIA RTX GPUs enable realistic physics-based lighting and reflections. Cloud-Based Physics Computation: Future games may use cloud servers to handle large-scale physics calculations, offloading work from local hardware (Wen-Mei W H., et al, 2022).

PHYSICS ENGINES AND ARTIFICIAL INTELLIGENCE: INTEGRATING PHYSICS-BASED AI IN GAMES

The integration of physics engines and artificial intelligence (AI) is transforming modern game development, creating more immersive, realistic, and dynamic gameplay experiences. AI and physics engines serve different but complementary purposes in games. AI is typically used for NPC (Non-Player Character) Behavior: Governing decision-making, movement, and reactions. Pathfinding and Navigation: Allowing characters to find the best route around obstacles (Liang Z., et al, 2025). Procedural Animation: Enhancing realism by blending animations dynamically. On the other hand, physics engines handle: Rigid Body Dynamics: Simulating object movement, gravity, and force interactions. Soft-Body Physics: Simulating flexible or deformable objects, such as cloth and muscles. Fluid and Particle Physics: Simulating water, fire, and smoke in real time. By integrating AI with physics engines, games can create autonomous, physics-aware agents that react intelligently to environmental forces, adapt to real-time physics changes, and interact naturally with the game world. AI-Driven Physics Simulations: AI for Realistic Character Movement:

Traditional character animations rely on predefined keyframes or motion capture, limiting flexibility in dynamic environments. AI-driven physics allows characters to: React to physics forces dynamically (e.g., adjusting balance on uneven terrain). Generate procedural animations based on real-time physics interactions. Improve realism in falling, jumping, or dodging mechanics. Example: Red Dead Redemption 2 uses an AI-driven physics system for natural NPC reactions (Wang Y., et al, 2024). AI can procedurally generate physics-driven environments, adjusting Object placement based on real-world physics rules (MECLEA M A., et al, 2024). Terrain deformation is based on weather conditions and player interactions. Real-time environmental responses (e.g., trees swaying realistically in the wind). For example, No Man's Sky uses AI-generated planetary physics to create unique gravitational and environmental conditions on different planets. AI will assist developers in creating realistic physics systems without manual fine-tuning, automating complex physics simulations (Huang Y F., et al, 2023).

THE FUTURE OF PHYSICS ENGINES IN GAME DESIGN: EMERGING TRENDS AND TECHNOLOGIES

Physics engines are the backbone of modern game development, powering realistic interactions, environmental effects, and physics-based gameplay mechanics. Over the years, advancements in hardware, artificial intelligence (AI), and simulation techniques have continuously pushed the boundaries of what is possible in interactive digital worlds. Looking ahead, the next generation of physics engines will harness machine learning, real-time cloud-based physics, ray-traced physics, and procedural simulation techniques to create more immersive and dynamic gaming experiences. This article explores emerging trends and technologies in physics engines, their impact on game design, and how they will shape the future of interactive entertainment (Tamura R., et al, 2023). Next-Generation Physics Simulation: Breaking the Boundaries of Realism: As gaming hardware becomes more powerful, developers aim to create real-time physics simulations that rival the complexity of real-world physics. Future game engines like Unreal Engine 5 and Unity's HDRP could implement ray-traced physics models for enhanced realism in water physics, destruction effects, and collision interactions. AI-Driven Physics Engines: Machine Learning Enhancements Machine learning (ML) is set to redefine how physics engines simulate real-world interactions (Kittur J., et al, 2021).

BeamNG.drive uses soft-body physics for realistic car crashes, but future games could apply soft-body simulations to NPC animations, allowing bodies to react naturally in combat and environmental interactions. Hyper-Realistic Physics for AI NPCs: AI-driven physics will revolutionize how NPCs interact with environments, allowing them to react dynamically to changing physics conditions (e.g., adjusting footing on slippery terrain) (Naswir A F., 2024). Use real-world physics principles in combat and movement. Navigate destructible environments intelligently, avoiding debris and collapsing structures. Example: Future versions of games like The Last of Us could feature NPCs that react in real-time to destruction physics, using AI to navigate dynamically shifting environments. The Future of Physics in Game Design: What's Next? As physics engines continue to evolve, game design will shift towards more interactive, emergent, and realistic experiences. Future games could feature: Fully destructible, persistently simulated worlds. AI-driven procedural physics that adapts dynamically to player actions. Haptic-enhanced VR with physics-based touch interactions (Kumawat B., et al, 2024).

PHYSICS ENGINE TOOLS AND TECHNOLOGIES: AN OVERVIEW OF POPULAR OPTIONS

Physics engines are essential components of modern game development, enabling realistic movement, collision detection, environmental interactions, and dynamic simulations. Whether it's a simple 2D platformer or a complex open-world simulation, physics engines determine how objects behave in response to forces, gravity, and collisions. Various physics engine tools are available, each with unique features, strengths, and limitations. This article provides an in-depth overview of popular physics engines, their applications in game development, and the technologies driving their evolution. Introduction to Physics Engines in Game Development: A physics engine is a software component that simulates real-world physical behavior in a virtual environment (Sharma H., et al, 2023). These engines handle Rigid Body Dynamics, simulating solid object movement, rotation, and collisions. Soft Body Physics: Simulating deformable materials like cloth, hair, or rubber. Fluid and Particle Systems: Simulating liquids, fire, smoke, and explosions. Constraint Solving: Handling joints, hinges, and mechanical connections. Developers choose physics engines based on their game type, platform, and performance requirements. Some engines specialize in real-time simulations for gaming, while others focus on scientific or cinematic accuracy. Popular Physics Engine Tools in Game Development: NVIDIA PhysX: Overview: NVIDIA PhysX is a GPU-accelerated physics engine widely used in AAA game development. It is optimized for real-time physics simulations, offering high-performance rigid body dynamics, cloth physics, and destruction effects. Key Features: GPU acceleration for high-performance physics. Rigid and soft-body simulations for realistic object behavior. Particle-based fluid simulations for smoke, fire, and liquid effects. Destruction physics for dynamic environmental damage (Dobrokvashina A., et al, 2022).

Chaos is Unreal Engine's next-generation physics engine, replacing PhysX. It is designed for realistic destruction, cloth simulation, and particle-based fluid dynamics. Real-time fracture simulation for breakable objects. Advanced soft-body physics for fabric and character animations. GPU-accelerated destruction physics for large-scale destruction effects. Integrated physics constraints for vehicles and mechanical systems. Applications: Used in AAA games developed in Unreal Engine 5. Supports realistic environmental destruction in games like Fortnite and Gears of War 5 (Paganetti H., et al, 2021). Powers next-gen game physics with ray-traced simulations and AI-driven optimizations. Emerging Trends in Physics Engine Technologies: As game physics continues to evolve, new technologies are shaping the future of physics engines. AI-Assisted Physics Calculations: Machine learning optimizes collision detection, physics approximations, and dynamic object interactions. AI-driven simulations reduce computational overhead while improving realism. Choosing the Right Physics Engine for Your Game: When selecting a physics engine, developers should consider Game Type: 2D games (Box2D) vs. 3D open-world games (PhysX, Havok). Platform: PC, console, mobile, or VR. Performance Needs: Real-time physics (Chaos, Havok) vs. precomputed simulations (Bullet). Integration: It is compatible with Unreal Engine, Unity, or custom engines (Mylonas G., et al, 2021).

USING PHYSICS ENGINES WITH GAME ENGINES: INTEGRATING PHYSICS ENGINES WITH UNITY AND UNREAL ENGINE

This article explores how physics engines integrate with Unity and Unreal Engine, covering built-in physics solutions, third-party integrations, optimization techniques, and best practices for game development. Understanding Physics Engine Integration in Game Development: A physics engine is a software component that simulates the laws of physics in a virtual environment. It handles Collision Detection, ensuring objects interact correctly when they collide. Rigid Body Dynamics: Simulating object movement under forces like gravity and friction. Soft Body Physics: Handling flexible materials like cloth, rubber, and jelly. Particle Systems: Simulating effects like fire, smoke, and explosions (Mosca P., et al, 2024). Fluid Dynamics: Creating realistic water, lava, and gas simulations. When integrating a physics engine into a game engine, developers must consider performance, accuracy, scalability, and hardware constraints to achieve smooth and responsive gameplay. Integrating Physics in Unity Engine: Unity is one of the most popular game engines, offering built-in physics support through NVIDIA PhysX and Unity's DOTS Physics system. Built-in Unity Physics: NVIDIA PhysX: Unity's default physics engine is PhysX, which provides Rigid body physics for moving objects. Character controllers for player movement. Collision detection for real-time interactions. Cloth physics for realistic fabric and soft body simulations. Integration Steps: Attach a Rigidbody Component to objects that require physics interactions (Morse C., 2021). eatures of Chaos Physics: Rigid body dynamics for accurate physics interactions. Destruction physics for breakable objects. Cloth simulation for realistic fabric movement. Fluid simulation for interactive water physics. GPU-accelerated processing for high-performance physics calculations (Xin Y L., et al, 2024). Unreal's Chaos Engine leverages multi-threading for high-performance physics (Lee T., 2024) . Unity's Burst Compiler speeds up physics computations for large-scale games.

CASE STUDY: USING PHYSICS ENGINES IN THE DEVELOPMENT OF A AAA GAME TITLE

The development of AAA game titles has become increasingly complex, requiring the integration of advanced physics engines to deliver realistic and immersive gameplay. Physics engines handle various aspects of a game, such as collision detection, rigid body dynamics, soft body simulation, fluid dynamics, and even destructible environments. This case study examines how a leading game studio leveraged a physics engine to develop a highly successful AAA game title, exploring the challenges, optimizations, and impacts on gameplay and performance. Background and Selection of Physics Engine: The development team evaluated several physics engines, including Havok, NVIDIA PhysX, and Bullet, to determine the best fit for their game (Arshad M H., 2024). Key criteria included: Performance and Scalability – The game aimed for a smooth experience on multiple platforms, including PCs and next-gen consoles. Feature Set – The physics engine needed to support complex interactions such as character ragdoll physics, real-time destruction, and advanced fluid simulations. Ease of Integration – The engine had to integrate seamlessly with Unreal Engine 5, the chosen game engine for development. Modifiability – Customization was essential for achieving unique gameplay mechanics that differentiated the game from competitors. After rigorous testing, Havok was selected due to its balance of performance, scalability, and extensive industry adoption. However, NVIDIA PhysX was used selectively for GPU-accelerated physics effects, such as particle-based fluid simulation and cloth physics (Shende A D., et al. 2025). Rigid body dynamics were tuned for different objects in the game world, ensuring that light objects (e.g., debris, paper) behaved differently from heavier objects (e.g., vehicles, massive crates). The team adjusted parameters such as friction, restitution, and mass distribution to create a realistic feel (Tykoski S., 2022).

This added unpredictability and realism to combat. Fluid and Cloth Simulations: Fluids and cloth physics were handled through NVIDIA Flex and PhysX Cloth, leveraging GPU acceleration. The implementation included Real-time water physics, where liquid containers spilled dynamically based on force and tilt. Fire and smoke interactions, with particles behaving according to wind and explosion forces. Cloth physics, ensuring that capes, banners, and character clothing responded naturally to movement and external forces (Segerstedt J., 2021). To optimize performance, the team used adaptive physics fidelity, where the complexity of simulations scaled dynamically based on proximity and importance to gameplay. Challenges and Solutions: Performance Bottlenecks: Real-time physics simulations are computationally expensive. The team optimized by utilizing asynchronous physics threads and offloading calculations to the GPU where possible. Networking and Physics Syncing: Multiplayer physics required deterministic simulations to prevent desynchronization between players. The team implemented rollback netcode and state interpolation to ensure consistency. Bug Fixing and Unintended Interactions: Edge cases, such as objects jittering due to floating-point precision errors, were addressed using stabilization algorithms and bias correction techniques (Gillberg et al, 2023).

EXAMPLE OF A PHYSICS-BASED GAMEPLAY MECHANIC: DESIGN AND IMPLEMENTATION

Physics-based gameplay mechanics have become a cornerstone of modern game design, offering dynamic, emergent gameplay that enhances immersion and interactivity. A well-designed physics-based mechanic not only improves realism but also creates opportunities for unique player-driven experiences. This case study examines the design and implementation of a gravity manipulation mechanic, exploring its conceptualization, development process, optimization, and impact on gameplay. Concept and Design: Mechanic Overview: The gravity manipulation mechanic allows players to dynamically alter gravitational forces on objects and environments. This ability enables puzzles, combat tactics, and traversal challenges, creating a deeply interactive experience. Key design goals: Intuitive Controls: Players should feel in control of the gravity effects without requiring overly complex inputs (Vdovenko, 2021) . Physics Consistency: The mechanics should behave predictably within the game's physics system. Engaging Interactions: Objects and enemies should respond naturally to gravity changes, offering creative problem-solving opportunities. For directional gravity shifts, an acceleration vector was applied, allowing objects to transition smoothly between gravitational states. To ensure stability, objects affected by gravity manipulation had constraints limiting their velocity to prevent unnatural accelerations (Piotrowski et al, 2023).

User Interface and Controls: A streamlined UI was designed for intuitive gravity manipulation. Visual Indicators: Objects under gravity influence emitted a glowing aura, color-coded for different gravity states. HUD Feedback: A gravity meter displayed the available charge for altering gravitational forces. Haptic Feedback: Controller vibrations provided tactile responses when shifting gravity intensities. Controls were mapped to allow for fluid interactions: Trigger Button: Activates gravity alteration on the focused object (Kumar et al, 2021). Thumbstick Adjustment: Changes the direction of gravity in real-time. Cooldown System: Prevents excessive use, balancing gameplay mechanics. Optimization Techniques: Real-time gravity manipulation posed performance challenges, requiring optimizations: Physics Threading: Gravity calculations were offloaded to a separate physics thread to prevent frame rate drops. Object Prioritization: Only objects within a certain radius of the player were actively affected by gravity alterations. LOD (Level of Detail) Scaling: Distant objects used simplified physics calculations to reduce CPU load (Aragani et al. 2025). Interpolation and Prediction: Gravity transitions were smoothed using Hermite interpolation, preventing abrupt physics changes. Impact on Gameplay and Player Experience: The gravity manipulation mechanic significantly enhanced the game's interactivity and replayability. Increased Player Creativity: Players found unique ways to solve puzzles and approach combat. Emergent Gameplay Scenarios: Unexpected interactions led to player-driven solutions beyond developer intentions (Abdullah et al, 2024).

CONCLUSION

Physics engines play a crucial role in modern game design, driving realism, interactivity, and emergent gameplay. From simulating simple object interactions to enabling complex mechanics like destructible environments and fluid dynamics, physics engines shape digital play in profound ways. This section summarizes key findings on their impact, challenges, and future potential in game development. The Evolution of Physics in Games: The integration of physics in digital games has evolved significantly: Early 2D Physics (1980s-1990s): Basic collision detection and gravity simulations in platformers like Super Mario Bros. Rigid Body and Soft Body Physics (2000s): Games like Half-Life 2 (2004) introduced physics-based puzzles using the Havok engine. Advanced Simulations (2010s- Present): Realistic destruction (Battlefield series), cloth and fluid dynamics (Red Dead Redemption 2), and procedural physics-based animations (The Last of Us Part II). Physics engines have now become essential for creating immersive, interactive worlds that respond naturally to player input. Key Roles of Physics Engines in Game Design: Enhancing Realism and Immersion: Physics engines create believable worlds where objects, characters, and environments interact naturally. Realistic physics improves immersion by making in-game actions feel intuitive. Examples include: Ragdoll physics: Natural character reactions to impacts in games like Dark Souls and Grand Theft Auto V. Fluid dynamics: Realistic water simulations in Sea of Thieves enhance exploration and naval combat. Vehicle physics: Simulating weight, traction, and suspension in racing games like Forza Horizon. Supporting Gameplay Mechanics: Physics engines are at the core of many gameplay systems, including: Puzzle Mechanics: Gravity, magnetism, and physics-based object manipulation (e.g., LittleBigPlanet). Combat and Damage Systems: Bullet trajectories, explosion force, and destructible cover (e.g., Call of Duty). AI Behavior: NPCs reacting dynamically to physics-based stimuli (e.g., enemies dodging falling objects in The Last of Us Part II). The Future of Physics in Digital Play: Looking ahead, physics engines will continue to evolve, unlocking new possibilities for game design. Machine Learning in Physics Simulation: AI-driven physics engines can adapt dynamically to gameplay, reducing pre-scripted animations. Procedural Physics-Based Animation: Characters animated entirely through physics simulations, reducing the reliance on pre-made animations. Real-Time Destruction and Soft Body Physics: Next-gen consoles enable more detailed, interactive destruction and realistic material deformation. Cloud-Based Physics Processing: Remote computation could allow large-scale physics simulations beyond current hardware limitations. With the rise of virtual reality (VR) and augmented reality (AR), physics-driven interactions will become even more critical in creating believable digital experiences. Physics engines are a fundamental pillar of modern game design, shaping gameplay, realism, and interactivity. By simulating physical interactions, they enable emergent gameplay, immersive environments, and dynamic mechanics that captivate players. While technical challenges remain, ongoing advancements in physics simulations will continue to push the boundaries of digital play. As game developers refine these systems, the future of physics-based gaming promises richer, more interactive worlds where players have unprecedented freedom to shape their own experiences.

REFERENCES

Abdullah, F., & Taveekitworachai, P. (2024) "The 1st ChatGPT4PCG Competition" *2024 IEEE Conference on Games (CoG)*. Doi: DOI: 10.1109/CoG57391.2024.00131

Allshire, A., Mittal, M., & Lodaya, V. (2022) "Transferring Dexterous Manipulation from GPU Simulation to a Remote Real-World Trifinger" 2022 IEEE/RSJ International Conference on Intelligent Robots and Systems (IROS). . Link: https://ieeexplore.ieee.org/document/9981776DOI: 10.1109/IROS47612.2022.9981776

Aragani, V. M., Anumolu, V. R., & Selvakumar, P. (2025). Democratization in the Age of Algorithms: Navigating Opportunities and Challenges. In Wongmahesak, K., Wekke, I., Seftyono, C., & Nurdin, N. (Eds.), *Democracy and Democratization in the Age of AI* (pp. 39–56). IGI Global Scientific Publishing., DOI: 10.4018/979-8-3693-8749-8.ch003

Arshad, M. H. (2024). *Comparing and Contrasting the Use of Reinforcement Learning to Drive an Autonomous Vehicle Around a Racetrack in Unity and Unreal Engine 5*. ProQuest.

Avatavului, C. D., Ifrim, R. C., & Voncilă, M. (2023). "Can Neural Networks Enhance Physics Simulations?" BRAIN. *Broad Research in Artificial Intelligence and Neuroscience*, *14*(2), 76–92. Advance online publication. DOI: 10.18662/brain/14.2/445

Bernatovych, A. A., & Stetsenko, I. V. (2023). *Module-Based Scientific Physics Engine Architecture*. Problems in Programming., DOI: 10.15407/pp2023.03.030

Chia, A. (2022). The Metaverse, but Not the Way You Think: Game Engines and Automation Beyond Game Development. *Critical Studies in Media Communication*, *39*(3), 191–200. Advance online publication. DOI: 10.1080/15295036.2022.2080850

Clausen, C. S. B., Ma, Z. G., & Jørgensen, B. N. (2022). Can We Benefit from Game Engines to Develop Digital Twins for Planning the Deployment of Photovoltaics? *Energy Informatics*, *5*(S4), 42. Advance online publication. DOI: 10.1186/s42162-022-00222-7

Consalvo, M., & Staines, D. (2021). Reading Ren'Py: Game Engine Affordances and Design Possibilities. *Games and Culture*, *16*(6), 762–778. Advance online publication. DOI: 10.1177/1555412020973823

Dalal, R. (2025). *Core Concepts of Mechanics and Thermodynamics*. Educohack Press.

Debnath, P., Deshmukh, S. M., Laroia, M., Selvakumar, P., Manjunath, T. C., & Bhattacharya, S. (2025). Challenges Faced by Countries in Renewable Energy Adoption. In Özsungur, F., Chaychi Semsari, M., & Küçük Bayraktar, H. (Eds.), *Geopolitical Landscapes of Renewable Energy and Urban Growth* (pp. 295–320). IGI Global Scientific Publishing., DOI: 10.4018/979-8-3693-8814-3.ch010

Dobrokvashina, A., Lavrenov, R., & Bai, Y. (2022) "Sensors Modelling for Servosila Engineer Crawler Robot in Webots Simulator" *2022 International Conference on Artificial Life and Robotics (ICAROB)*. DOI: 10.1109/MWENT55238.2022.9802400

Fischer, J., & Mahon, B. Z. (2021). What Tool Representation, Intuitive Physics, and Action Have in Common: The Brain's First-Person Physics Engine. *Cognitive Neuropsychology*, *38*(7-8), 455–467. Advance online publication. DOI: 10.1080/02643294.2022.2106126 PMID: 35994054

Foti, C., Anttila, D., Maniscalco, S., & Chiofalo, M. L. (2021). *Quantum Physics Literacy Aimed at K12 and the General Public*. Universe., DOI: 10.3390/ECU2021-09322

Freeman, C. D., Frey, E., Raichuk, A., & Girgin, S. (2021) "Brax—a Differentiable Physics Engine for Large Scale Rigid Body Simulation" arXiv preprint arXiv:2112.08241. doi: /arXiv.2112.08241.DOI: 10.48550

Georgiou, Y., Tsivitanidou, O., & Ioannou, A. (2021). *Learning Experience Design with Immersive Virtual Reality in Physics Education*. Springer., DOI: 10.1007/s11423-021-10055-y

Gillberg, J., Bergdahl, J., & Sestini, A. (2023) "Technical Challenges of Deploying Reinforcement Learning Agents for Game Testing in AAA Games" *2023 IEEE Conference on Games (CoG)*. Doi: DOI: 10.1109/CoG57401.2023.10333194

Halliday, D., Walker, J., & Resnick, R. (2023). *Principles of physics*. John Wiley & Sons.

Heidari, F., Mahdavinejad, M., Werner, L. C., Roohabadi, M., & Sarmadi, H. (2021). Biocomputational Architecture Based on Particle Physics. *Frontiers in Energy Research*, *9*, 620127. Advance online publication. DOI: 10.3389/fenrg.2021.620127

Howell, T. A., Cleac'h, S. L., Brüdigam, J., & Kolter, J. Z. (2022) "Dojo: A Differentiable Physics Engine for Robotics" arXiv preprint arXiv:2206.10589. doi: /arXiv.2206.10589.DOI: 10.48550

Huang, Y., Guo, F., Zhang, H., & Yang, Z. (2022). An Efficient Computational Framework for Generating Realistic 3D Mesoscale Concrete Models Using Micro X-ray Computed Tomography Images and Dynamic Physics …. *Cement and Concrete Composites*, *126*, 104347. Advance online publication. DOI: 10.1016/j.cemconcomp.2021.104347

Huang, Y. F., Lv, S. J., Tseng, K. K., Tseng, P. J., Xie, X., & Lin, R. F.-Y. (2023). Recent Advances in Artificial Intelligence for Video Production System. *Enterprise Information Systems*, *17*(11), 2246188. Advance online publication. DOI: 10.1080/17517575.2023.2246188

Johnson-Bey, S., Nelson, M. J., & Mateas, M. (2022). *Exploring the Design Space of Social Physics Engines in Games*. Springer., DOI: 10.1007/978-3-031-22298-6_36

Johri, S., Selvakumar, P., Satyanarayana, P., Das, A., Gupta, S., & Bhattacharya, S. (2025). Urbanization and Energy: The Nexus of Global Challenges. In Özsungur, F., Chaychi Semsari, M., & Küçük Bayraktar, H. (Eds.), *Geopolitical Landscapes of Renewable Energy and Urban Growth* (pp. 453–478). IGI Global Scientific Publishing., DOI: 10.4018/979-8-3693-8814-3.ch016

Khan F, (2024) "Optimization Techniques in Applied Mathematics: From Physics Simulations to Real-World Problems" Frontiers in Applied Physics and Mathematics.

Kittur, J., & Islam, T. (2021) "Serious Games in Engineering: The Current State, Trends, and Future" *2021 ASEE Virtual Annual Conference Content* Access. Doi: DOI: 10.18260/1-2--37709

Kotsis, K. T. (2024). ChatGPT in Teaching Physics Hands-on Experiments in Primary School. *European Journal of Education Studies*, *11*(10). Advance online publication. DOI: 10.46827/ejes.v11i10.5549

Kumar, A. S., & Rank, S. (2021) "BEAPS: Integrating Volumetric Dynamics in Virtual Agent Prototyping" Advances in Visual Computing: 16th International Symposium, ISVC 2021, Virtual Event, October 4–6, 2021, Proceedings, Part II. Doi: DOI: 10.1007/978-3-030-90439-5_34

Kumawat, B., Sridevi, P. R. S., & Mehra, A. (2024) "Investigation of Tank Combat Simulation in Video Games" 2024 International Conference on Advances in Computing, Communication, and Control (ICAC3). Doi: DOI: 10.1109/ICAC2N63387.2024.10895254

Lai, J. W., & Cheong, K. H. (2022). *Educational Opportunities and Challenges in Augmented Reality: Featuring Implementations in Physics Education*. IEEE., DOI: 10.1109/ACCESS.2022.3161096

Lee, T. (2024). *Unity Engine Dissection: Improvement Points in Comparison between Unity Engine and Open-Source Engines*. Asia-Pacific Journal of Convergent Research and Innovation.

Liang, Z., Li, H., Yu, N., Sun, K., & Cheng, R. (2025). Bridging Evolutionary Multiobjective Optimization and GPU Acceleration via Tensorization. *IEEE Transactions on Evolutionary Computation*, 1. Advance online publication. DOI: 10.1109/TEVC.2025.3555605

Mahadev Madgule, N. (2023). Vinayaka, Yeshwant M. Sonkhaskar, Dhiren Ramanbhai Patel, R. Karthikeyan, P. Selvakumar, (2023). Mechanical properties and microstructure of activated TIG welded similar joints of Inconel alloys by desirability approaches. *Materials Today: Proceedings*, *77*, 528–533. DOI: 10.1016/j.matpr.2022.12.250

Meclea, M. A., Alexandru Silviu, G. O. G. A., & Bo coianu, M. (2024). *Aspects regarding artificial intelligence use in military and engineering sciences aircraft propulsion*. Scientific Research & Education in the Air Force-AFASES.

Mirzayeva G, (2023) "The Role of Digital Educational Technologies in Teaching Physics" Science and Innovation. Doi: .DOI: 10.37872/2414-3087-2023.2.33357

Morse, C. (2021). *Gaming Engines: Unity, Unreal, and Interactive 3D Spaces*. Taylor & Francis.

Mosca, P., Bourhrara, L., & Calloo, A. (2024). APOLLO3: Overview of the New Code Capabilities for Reactor Physics Analysis. *Nuclear Science and Engineering*. Advance online publication. DOI: 10.1080/00295639.2024.2182831

Mustonen, M. (2023). *Web-Based Game Engine Design*. LUT University.

Mylonas, G., Kalogeras, A., & Kalogeras, G. (2021). Digital Twins from Smart Manufacturing to Smart Cities: A Survey. *IEEE Transactions on Industrial Informatics*. Advance online publication. DOI: 10.1109/TII.2021.3084853

Naswir A F, (2024) "The Future of Artificial Intelligence in 2D Game Development" Journal of Data Science, Technology, and Computer Science.

Paganetti, H., Beltran, C., Both, S., Dong, L., Flanz, J., Furutani, K., Grassberger, C., Grosshans, D. R., Knopf, A.-C., Langendijk, J. A., Nystrom, H., Parodi, K., Raaymakers, B. W., Richter, C., Sawakuchi, G. O., Schippers, M., Shaitelman, S. F., Teo, B. K. K., Unkelbach, J., & Lomax, T. (2021). Roadmap: Proton Therapy Physics and Biology. *Physics in Medicine and Biology*, *66*(5), 05RM01. Advance online publication. DOI: 10.1088/1361-6560/abcd16 PMID: 33227715

Panerati, J., Zheng, H., Zhou, S. Q., & Xu, J. (2021) "Learning to Fly—a Gym Environment with Pybullet Physics for Reinforcement Learning of Multi-Agent Quadcopter Control" *IEEE/RSJ International Conference on Intelligent Robots and Systems*. Doi: DOI: 10.1109/IROS51168.2021.9635857

Partarakis, N., Zabulis, X., Zourarakis, D., Demeridou, I., Moreno, I., Dubois, A., Nikolaou, N., Fallahian, P., Arnaud, D., Crescenzo, N., Hee, P., & Stamou, A. (2025). Physics-Based Tool Usage Simulations in VR. *Multimodal Technologies and Interaction*, *9*(4), 29. Advance online publication. DOI: 10.3390/mti9040029

Piotrowski, W., Sher, Y., Grover, S., & Stern, R. (2023) "Heuristic Search for Physics-Based Problems: Angry Birds in PDDL+" *Proceedings of the International Conference on Automated Planning and Scheduling*. DOI: 10.1609/icaps.v33i1.27232

Qais, M. H., Hasanien, H. M., Alghuwainem, S., & Loo, K. H. (2023). Propagation Search Algorithm: A Physics-Based Optimizer for Engineering Applications. *Mathematics*, *11*(20), 4224. Advance online publication. DOI: 10.3390/math11204224

Ramakrishnan, T., Mohan Gift, M. D., Chitradevi, S., Jegan, R., & Subha Hency Jose, P. Nagaraja, Rajneesh Sharma HN, Selvakumar P, Sintayehu Mekuria Hailegiorgis, (2022). "Study of Numerous Resins Used in Polymer Matrix Composite Materials", Advances in Materials Science and Engineering, vol. 2022, Article ID 1088926, 8 pages, 2022. DOI: 10.1155/2022/1088926

Reda, D. (2025). *Physics-Based Character Controllers with Reinforcement Learning*. University of British Columbia.

Romanyuk, O. N., Shenshin, O. O., & Bobko, O. L. (2025). *GPU Stream Processing*. Innovative Technology and Scientific Progress.

Schwartz, D. (2021). *Making Games to Teach Physics and Mechanics*. RIT Scholar Works. DOI: 10.18260/1-2--36308

Segerstedt, J. (2021). *Animation for the Apex Game Engine using Neural Networks—Implementing Immersive Character Animation in an Industry-Proven Game Engine by Applying Machine Learning Techniques*. Chalmers University of Technology.

Sharma, H., Juyal, A., & Das, M. (2023) "Future of Animation with Artificial Intelligence" *International Conference on Emerging Trends in Design & Arts*.

Shende, A. D., Tekade, S. A., Deshmukh, A. A., Tembhurkar, S. P., & Selvakumar, P. (2025). Water Quality Monitoring and Control With AI. In Ruiz-Vanoye, J., & Díaz-Parra, O. (Eds.), *Smart Water Technology for Sustainable Management in Modern Cities* (pp. 51–80). IGI Global Scientific Publishing., DOI: 10.4018/979-8-3693-8074-1.ch003

Sobota B, Pietriková E, (2023) "The Role of Game Engines in Game Development and Teaching" .DOI: 10.5772/intechopen.100922

Sreenivasulu, A., & Manimozhi, T. Satheesh kumar P, Selvakumar P. and Singh P, (2022). "The Model of Smart Sensing Device For Sensitive Nanoclusters Modification in Sensing Properties," 2022 2nd International Conference on Advance Computing and Innovative Technologies in Engineering (ICACITE), Greater Noida, India, 2022, pp. 1043-1047, DOI: 10.1109/ICACITE53722.2022.9823846

Sugumaran K M D, Pandiyarajan V Y B S, Selvakumar P. and M. S, (2024). "Analysis of Electromyography Signals for Control of Mechanical Prosthesis using Machine Learning Techniques," 2024 2nd International Conference on Intelligent Data Communication Technologies and Internet of Things (IDCIoT), Bengaluru, India, 2024, pp. 1059-1067, .DOI: 10.1109/IDCIoT59759.2024.10467707

Tamura, R., Tsuda, K., & Matsuda, S. (2023). NIMS-OS: An Automation Software to Implement a Closed Loop between Artificial Intelligence and Robotic Experiments in Materials Science. *Science and Technology of Advanced Materials*. Advance online publication. DOI: 10.1080/14686996.2023.2171319

Templet, R. M. (2021). *Game Physics: An Analysis of Physics Engines for First-Time Physics Developers*. ScholarWorks.

Tykoski S, (2022) "Mastering Game Design with Unity 2021: Immersive Workflows, Visual Scripting, Physics Engine, GameObjects, Player Progression, Publishing, and a Lot More"

Vdovenko D, (2021) "GU UNITY" VDU.

Vijayakumar, G. (2024). Muniyandi Rajkumar, Rajiv Chandar N, Selvakumar P and Ramesh Duraisamy, (2024). Environmentally friendly TDS removal from wastewater by electrochemical ion exchange batch-type recirculation (EIR) technique. *Environmental Science. Water Research & Technology, 10*, 826–835. DOI: 10.1039/D3EW00793F

Vinuesa, R., Brunton, S. L., & McKeon, B. J. (2023). The Transformative Potential of Machine Learning for Experiments in Fluid Mechanics. *Nature Reviews. Physics, 5*(9), 536–545. Advance online publication. DOI: 10.1038/s42254-023-00622-y

Vohera, C., Chheda, H., & Chouhan, D. (2021) "Game Engine Architecture and Comparative Study of Different Game Engines" 2021 12th International Conference on Computing, Communication and Networking Technologies (ICCCNT). Doi: DOI: 10.1109/ICCCNT51525.2021.9579618

Wang, K., Johnson, W. R., Lu, S., & Huang, X. (2023) “Real2Sim2Real Transfer for Control of Cable-Driven Robots via a Differentiable Physics Engine” *IEEE/RSJ International Conference on Intelligent Robots and Systems.* Doi: DOI: 10.1109/IROS55552.2023.10341811

Wang, Y., Bai, J., Lin, Z., Wang, Q., Anitescu, M., & Sun, J. (2024) “Artificial Intelligence for Partial Differential Equations in Computational Mechanics: A Review” arXiv preprint arXiv:2409.07593. doi: /arXiv.2409.07593.DOI: 10.48550

Wen-Mei WH, Kirk DB, El Hajj I, (2022) “Programming Massively Parallel Processors: A Hands-on Approach”

Xin, Y. L., Ge, G. P., Du, W., Wu, H., & Zhao, Y. (2024). Design of an Optical Physics Virtual Simulation System Based on Unreal Engine 5. *Applied Sciences (Basel, Switzerland), 14*(3), 955. Advance online publication. DOI: 10.3390/app14030955

Yang, M., Du, Y., & Ghasemipour, A. (2023) “Learning Interactive Real-World Simulators” arXiv preprint arXiv:2302.09314. doi: /arXiv.2302.09314.DOI: 10.48550

Yi, X. (2024) “A Study of Performance Programming of CPU, GPU Accelerated Computers and SIMD Architecture” arXiv preprint arXiv:2409.10661. doi: /arXiv.2409.10661.DOI: 10.48550

Zhang, Y., Zhang, R., & Lee, B. G. I. (2025). From Abstract Concepts to Engaged Learning: An IVR Game-Based Framework for Physics Education. *International Journal of Online Pedagogy and Course Design.*

Chapter 11
Word Art and Alphanumeric Images From Generative AI

Shalin Hai-Jew
https://orcid.org/0000-0002-8863-0175
Hutchinson Community College, USA

ABSTRACT

"Word art" comes in various forms: word clouds, word trees, digital graffiti, calligrams, typography, calligraphy, designed signs, and others. These may be generated from data, digital image editing programs, online word art generators, and other tools, now, including art-making generative AI ones. This latest category of tool has struggled with accurate spelling, even as generative AI tools create their own visual "fonts" and aesthetics. The challenges for such tools in creating word art involves visual aesthetics, font aesthetics; meaning capture and conveyance; alphabetics, numerics, alphanumerics;; human readability, machine readability; and other challenges simultaneously. This work involves a light exploration of the "word art" aspects of generative AIs in the contemporaneous moment.

INTRODUCTION

"Word art" combines text and imagery into "a collage that depicts an associated shape or object" (Yen, Yeh, Morace, Chen, Lin, Chen, & Lee, 2016, p. 1). It is a multimodal art form.

In a zoomed-out view, the visual shape may dominate, but when read more closely, the word art "can be read word by word" (Yen, Yeh, Morace, Chen, Lin, Chen, & Lee, 2016, p. 1). Some have referred to word art as a form of "collage," with

DOI: 10.4018/979-8-3373-2277-3.ch011

mixes of different materials. Within this general category may be various other art types. A "calligram" refers to a design that combines words into a visual image or shape. Specifically, "calligram," as a portmanteau term, is created from "beautiful + something written." "Digital graffiti" refers to text that emulates the graffiti style in the wild, as used in tagging and semantic expression.

In other cases, the word art may be mere text integrated into an image but with the text as a dominant and legible (readable) visual element. Some word art is a form of data visualization, such as word clouds, word trees, ring lattice graphs, treemap diagrams, and others. Word art has a clearer defined meaning, then, than other types of art that may be more interpretive and polysemous.

Such "text art" may derive from typography or the arrange of type for printed text outputs (in analog and digital formats). In terms of the aesthetics, these may derive from different fonts, colors, overlays, lines, textures, backgrounds, patterns, and other visual effects. These word art objects may be used in imagery, posters, cards, logos, signature blocks, newsletters, wraps (for buildings, motor vehicles, and others), tchotchkes, and other communicative forms.

Some word cloud-creation tools are seen as "fun, visual and entertaining" for learners and therefore contributory to language learning (Tafazoli, Chirimbu, & Dejica-Cartis, 2014, p. 169). This form of word art is essentially utilitarian and "common" (Hai-Jew, 2022). This is "art" in the lower case, not fine art and not "Art." There is very much a "street" element to word art because it is a democratic form that may be accessed and created by the masses with common tools: word processing, spreadsheet technology, digital image editing, online word cloud tools, online word art generators, and others.

This chapter is based on a first-hand case in which several generative artificial intelligence (GAI) tools are explored for their capabilities in creating word art. The tools are based on large language models (LLMs) and refinement (training, feedback, fine-tuning, and algorithmic adjustments). The outputs are known as "user-prompted synthetic data."

This work shows that of the dozen generators explored, the current generation of tools still struggle with spelling, with expressing longer phrases, with illustrating ideas aptly, with managing numbers, with controlling for noise, and other challenges. Also, word art seems to be only a very small percentage of the works made by general users of artmaking GAI tools, based on years of observation of the public "albums" of such art. Finally, the outputs from current artmaking GAIs often require digital image editing cleanup (to fix lettering), correction of visual elements, and even the harnessing of a built-in GAI in Adobe Photoshop (Adobe Firefly) to enable actual usage for workplace communications (newsletters, emails, posters, and others). The outputs require alt-texting for machine readability in many cases and are not machine readable. The "fonts" evoked are visual flats and not actual text.

REVIEW OF THE LITERATURE

"Word art" is not generally a topic of research in academia. Some works refer to word art as storytelling, as imagistic writing, as "spoken word" (as in song, recited poetry, rap, and other variations). Word art may be harnessed to challenge binary senses of the world, in terms of political speech (Hicks, 2016). There is not an apparent theory of word art because word art does not rise to that level of consideration perhaps.

Creating word art manually is "a laborious process" (Yen, Yeh, Morace, Chen, Lin, Chen, & Lee, 2016, p. 1). The manual creation may involve the use of physical objects that are collated as a collage and digitally scanned or photographed. More commonly, the manual process is a digital one harnessing various word processing technologies and digital image editing software. Artful post-production is an important capability to use GAI tools for the creation of usable word art. In terms of digital image editing tools, there are necessary capabilities for post-production of word art: These include erasing existing text, replacing existing text, applying filters, applying masks, cutting out word-based shapes and filling those shapes (clipping masks) with visuals, using layer styles, drawing on font libraries, and even using built-in generative AI tools within the software to rework the erroneous or missing text. There are online word cloud-generating tools, to which people may apply different "colors, style and layout of the word collage" in Wordle (Tafazoli, Chirimbu, & Dejica-Cartis, 2014, p. 167).

In this technological age, with AI at the forefront, "AI artefacts" should not be "defined as art" (Chiodo, 2024, p. 1). The outputs are generally utilitarian and mundane. They do not spark society-wide conversations and aesthetic appreciation.

WORD ART AND ALPHANUMERIC IMAGES FROM GENERATIVE AI

Artmaking generative AI models are generally trained on "big" multimodal data with machine learning tools, refined, and then deployed. People use text (and image) prompts to elicit synthetic visual outputs from the GAI models. To input words, it is helpful to identify what should be written within " " quotation marks or < > angle quotes. The synthetic word art that is output is essentially autonomously created otherwise. The human user starts the machine process. If the individual does not like the output, he or she may (in some cases) continue to iterate. He or she may use the output images to seed the next iterations of the image…or reuse the prompt… or request a machine-generated prompt as another version…or start fresh. Some generative AI tools get into visual ruts and output very similar images in a row, even with different prompts and prompt variations, however. The person only has a little control over the output in terms of post-production (digital image editing). Digital image editing may be used to change up the digital image outputs, replace text, replace elements, replace backgrounds, and such, but those interventions do require human work. The AI and other tools built into an image editing software do have limits. For example, in a recent case, a section was highlighted, and the prompt was, "remove all text but leave everything else the same." The machine actually removed a group of signs with text and did not just leave the signs blank (as requested). None of the three AI options provided a solution that would work in this case.

The generative AI space is a highly dynamic one, with new models appearing and disappearing, some promoted for a time and then demoted and then disappearing. As to the visual depictions, many of the publicly available ones do not have a word drawing capability. These are word and letter and number shapes, not actual fonts. Some of the tools evoke something of a text overlay, but most seem to be shapes only. As such, even when words are spelled accurately in the text prompt and placed in quotation marks, the output may look nothing like the prompt itself. Many of the letters only look to be placeholder letters. When those text requests are input, many simply ignore the request and depict the requested background only. Many may not have a point of reference. Some only have English abilities or only Chinese abilities. Many are not multilingual either.

The fact that the word art is not evoked font lettering also means that various text effects may not be applied to the work. For example, various digital image editing tools enable text resizing, stretching, and different dedicated effects (arc, arc lower, arc upper, arch, bulge, shell lower, shell upper, flag, wave, fish, rise, fisheye, inflate, squeeze, twist…and then the specific adjustable slider parameters for each of the settings…in Adobe Photoshop). The level of control in terms of generative AI is much less in terms of word art.

In terms of writing styles, like cursive, that may result in some curlicues on the ends of the fonts but nothing that is in alignment with a classic definition of cursive. In terms of requests for human handwriting, the artmaking GAI tools do not seem to be able to respond with an image that suffices at the bare minimums. When requesting for "all caps" or other forms of capitalization, the GAI tools are also generally not sensitive to that request.

Readability is sometimes an issue because of extra letters, overlapping shapes, occluding visuals, mixed capitalization, and other challenges. It does not take much to throw off readability or to pause understanding of the text. To be clear, the lettering is not actual fonts but are letter shapes, often styled, with varying consistency. The lettering has to be sufficiently high-contrast to be visible to the human eye. The text verbiage also has to be accessible, to humans and to machines. In other words, the words and numbers and letters have to be perceivable to humans and machines to meaningfully register. The punctuation needs to be applied correctly, if it is included.

Often, the kerning may also be "off." Kerning refers to the adjusted spacing between letters, characters, and numbers. There are inexplicable extended spaces between some letters / characters / numbers than others.

In terms of fit, there can be fairly highly variant sizes as well between the letters, characters, and numbers, without clear rationale. Where such sizing may be purposeful in human designs, in the generative AI, these read as somewhat accidental or arbitrary.

When numbers are requested, those often include an equal amount of other shapes that are not numbers. When punctuation marks are requested, those also often include a large measure of other non-punctuation and non-numerical and non-letter shapes.

In terms of backgrounds, the visuals may have nothing to do with the text. For example, a "state of the art" word art visual was requested, with the generative AI using an outline of the United States in the background. The visuals that accompany the text may be unfit or off-tone.

Many times, the GAI tools will output garble (and noise) in response to the text prompt for word art. The output visuals may be eye-catching in terms of color, but the outputs do not bring semantic meaning. The garble occurs more often with longer elicited phrasing such as more than a few words.

In general, the outputs do privilege the text over the visual elements. Or in many cases, it is only text online that is generated, without any background. There is no match between the meaning of the text and the background visual. Rather, the visuals only show something of design with no connection to the semantic meaning of the text.

In some instances, paragraph-length text was pasted into the prompt window… to see how it might "translate." It does not. There may just be garble against a manuscript background, in one case. Some other models will just output a colorful background. Sometimes, if a word may be found that may be interpreted to have some tie to a physical object, that may be depicted. There is no built-in assessment for the visual-textual sensemaking.

A Context Of Some Spectacular Fails

In terms of word art, the generative AI tools in their present state give a "best guess" vibe. The spellings and misspellings sometimes evoke an untrained illiteracy. In mnay cases, iterating again and again does not result in anything that is any more usable.

In the following example, the generative AI misspelled "Spectacular Fail" but claimed "perfect spelling" in the banner below. The prompt requested "word art" that spelled "spectacular fail" with perfect spelling. (Figure 1) This was created in Microsoft Copilot.

Figure 1. Imperfect Spelling

In the next visual example, the same prompt came up with a cheery yellow and pink gradient text with perfect spelling but no match between the term and the whimsical background. (Figure 2) This was created in Microsoft Copilot.

Figure 2. Spectacular Fail with Whimsy

Figure 3 shows an output from AIVision in Deep Dream Generator. This model is one of the most sophisticated of the lot, but it still can not manage the spelling.

Figure 3. Another Spectacular Fail

Figure 4 was created in the Advanced Model in the Deep Dream Generator. It hints at some understanding of word meaning, but the spelling is still off. The misspelling would be a simple fix, though, with the removal of the lowercase "i" with the heart for the dot. (Figure 4)

Figure 4. A Prickly Spectiacular (sic) Fail

The above visual rendition was charming. The state-of-the-art though is still limited.

Discussion

These are early days for artmaking generative AIs in the creating of word art. It may be that such capabilities will have to be added as their own modules. Other sorts of visual creativity are being enabled and improved over time. Word art is probably not high on the list of requisite outputs.

This work involves the usage of Deep Dream Generator with its current models: DaVinci2 (language capable), Photonic, Artistic 2, ArtFusion, FluX, AIVision (language capable), Artistic, DreamForge, MirageMaker, EssenceofReality, PhotoMage, DigitalDaVinci, and Animatron. Also studied were the various models in Gemini Advanced, albeit later models seem much less capable than earlier ones to create word art using their "imagen 3" query. The current set of models at this moment were unable to generate the requested word art images and reverted to text responses. Models all have their strengths and weaknesses, their visual proclivities. Finally, Microsoft's Copilot was also used, with many of the similar results as well.

FUTURE RESEARCH DIRECTIONS

The particular challenges of creating word art in artmaking generative AIs likely require add-ons (like dictionaries) and other training to actualize. The outputs at present are haphazard and non-purposive. Still, this could be a descriptive "benchmark" for artmaking generative AI capabilities. Practically speaking, there are benefits for enabling people to create word art for the various applications in intercommunications.

CONCLUSION

Word art, as a general form of common ("folk") art, was popularized in the 1960s and 1970s as part of the pop art movement. This art appeared on book covers, stickers, graffiti, lunch boxes, license plates, and other "platforms" especially in the 1970s. It is hard to say if such art will have any resurgence any time soon. However, there is very much a utilitarian bent in such word (and number and alphanumeric) art. It is a niche space with retro overtones.

There are ways to just create a visual background and place an overlay of word art over it. It is possible to meld text with the image through color blend modes. However, such a solution does not read "word art," and the words and imagery are not artfully integrated in the way the optimized word art can be. The vibe may not be a synthesized one. The look and feel may not be of a piece. A generative AI that can create such an output may be a solution.

This is an early work with a built-in challenge to the makers of artmaking generative AI tools to expand capabilities. There are limits to the work since only two main generative AI systems were used extensively, even though both have a wide range of underlying models (which were also explored). This work summarizes work done over several years in alignment with professional work (in terms of workplace newsletters and intercommunications). Other genAI tools were also explored but in less depth.

REFERENCES

Chiodo, S. (2024). What AI "art" can teach us about art. *Journal of Aesthetics & Culture*, *16*(1), 2395511. DOI: 10.1080/20004214.2024.2395511

Hai-Jew, S. (2022). *Common Visual Art in a Social Digital Age.* Nova Science Publishers. DOI: 10.52305/BKOT1869

hicks, B. L. (2016). Beyond this or that: Challenging the limits of binary language in elementary education through poetry, Word Art, and creative bookmaking. *Teaching, Affirming, and Recognizing Trans and Gender Creative Youth: A Queer Literacy Framework*, 81-106.

Tafazoli, D., Chirimbu, S. C., & Dejica-Cartis, A. (2014). Web 2.0 in English language teaching: Using word clouds. *Professional Communication and Translation Studies*, *7*(1), 167–172. DOI: 10.59168/XQDJ2288

Yen, S. W., Yeh, C. K., Morace, C. C., Chen, S. Y., Lin, S. S., Chen, C. H., & Lee, T. Y. (2016). Content enhanced word art with depth perception. In *SIGGRAPH ASIA 2016 Posters* (pp. 1-2). DOI: 10.1145/3005274.3005280

ADDITIONAL READING

Hai-Jew, S. (2022). *Common Visual Art in a Social Digital Age.* Nova Science Publishers. DOI: 10.52305/BKOT1869

KEY TERMS AND DEFINITIONS

Calligram (beautiful + something written): A design that combines words into a visual image or shape or symbol (such as a visual poem)

Digital graffiti: Virtualize versions of graffiti-style text as tagging and as artful and/or semantic expression

Kerning: The adjusted spacing between letters or characters (or numbers)

Text art: An aesthetic visual expression or visual message in which text (lettering, numbering) is the main component supported by visual elements

Typography: Arrangement of type for a printed text output

User-prompted synthetic data: Outputs from generative AI based on human user prompts in the form of text, visual, or multimodal prompts

Word art: An aesthetic visual expression or visual message in which text (lettering, numbering) is the main component supported by visual elements

Word cloud: A visual expression of word data (often frequency-based) in the shape of a "cloud"

Chapter 12
Illustrating a Monthly Professional Newsletter With Synthetic Visuals

Shalin Hai-Jew
https://orcid.org/0000-0002-8863-0175
Hutchinson Community College, USA

ABSTRACT

The success of generative AI tools will depend in part on how easily these may be harnessed for practical and work purposes. This chapter explores the uses of several multimodal generative AI tools in the creation of text, imagery, and word art for a monthly grant newsletter. The visuals are used for a variety of purposes in the newsletter: decorative, reader effects, teaching and learning, data sharing, human memory enhancement, and others. This use-case illuminates some of the capabilities of several image-making generative AI tools (Deep Dream Generator, Gemini, and Microsoft Copilot). This work identifies some areas to improve to enable regular usage of such tools to generate synthetic imagery for workplace applications.

INTRODUCTION

Data is a critical part of many human systems, to enable awareness, informed decision-making, research, analysis, and many of human endeavors. Real-world data involves costs to acquire, clean, maintain, and share. Much of it cannot be shared because of privacy and legality concerns. The larger the dataset, the harder it is to

DOI: 10.4018/979-8-3373-2277-3.ch012

control for quality. Much collected data is not labeled or is not labeled in a way useful for application as training data.

"Synthetic data" (or "virtual data" or "replica data") refers to "artificially generated data that has analytical value" (Gal & Lynskey, 2024, p. 101). Synthetic data is "roughly defined as data that has not been collected nor mined from subjects or events in the real world but generated by algorithmic systems" (Jordon, et al., 2022, as cited in Wiehn, 2024, p. 472). Synthetic or imputed data may replace (or augment) actual data in various applications: research, model training, data exploration, teaching and learning, and others. Synthetic data enables broader data sharing (Lu, Shen, Wang, Wang, van Rechem, Fu, & Wei, 2023) and "data access" (Gonzales, Guruswamy, & Smith, 2023, p. 10).

Synthetic data is predicted to become "a dominant force in the machine learning world" with the ability to create custom datasets based on particular needs (van Breugel & van der Schaar, 2023, p. 1).

One survey study begins with the observation that various types of models for synthesizing data have diverged based on the different disciplines and methods by which data is synthesized. They identified high-level criteria for synthetic data, including "*representativeness, novelty, realism, diversity* and *coherence*" (Eigenschink, Reutterer, Vamosi, Vamosi, Sun, & Kalcher, 2023, p. 47304). They write:

> *We find that realism and coherence are more important for synthetic data natural language, speech and audio processing tasks. At the same time, novelty and representativeness are more important for healthcare and mobility data. We also find that measurement of representativeness is often accomplished using statistical metrics, realism by using human judgment, and novelty using privacy tests. (Eigenschink, Reutterer, Vamosi, Vamosi, Sun, & Kalcher, 2023, p. 47304)*

Different simulation parameters are optimized for particular downstream uses of the synthetic images used as pre-training data (Mishra, Panda, Phoo, Chen, Karlinsky, Saenko, ... & Feris, 2022). In some cases, the "simulation-to-real gap is minor for datasets with low object and scene bias, where models pre-trained with synthetic data even outperform their real data counterparts" (Kim, Mishra, Jin, Panda, Kuehne, Karlinsky, ... & Feris, 2022, p. 35710). Researchers share methods on how to assess which synthetic data generators may be most effective for different scenarios of mimicking real data (Kiran & Kumar, 2024). Different synthetic-data making tools have their own strengths and weaknesses (Miletic & Sariyar, 2024). Researchers are constantly finding efficiencies to generate synthetic data with more efficiency and perhaps less training data (Behl, Baydin, Gal, Torr, & Vineet, Aug. 2020). No single model is universally excellent for making synthetic data for different needs. There is a need for "global quality standards and security measures" in terms of artificial intelligence and synthetic data (Marwala, Fournier-Tombs, & Stinckwich, 2023, p. 1).

A popular technology involves text-to-image generators and generative AI (artificial intelligence) systems that are used to create "user-prompted synthetic data" of various modalities: text, imagery, sound, video, animations, and others. This work provides a light review of synthetic data generation (on the public-serving "retail" side) and describes the illustrating of a monthly professional newsletter that uses synthetic visuals as complementary content to text.

REVIEW OF THE LITERATURE

Real data proliferates in particular spaces, but they are not so widely available for research and machine learning and other applications. It may be expensive to buy access to real data, so synthetic data may be a stand-in. Some types of data may not be for sale. They are not available. Some may be siloed in particular organizations. Some types of data may be highly complex and varied in the real. There are also limits to how real data may be used, to prevent harm to real people.

One team observes the challenges to accessing data:

- Data may be proprietary preventing access.
- Data may have timestamps but are not synchronized preventing data joining.
- Data frequency may be insufficient.
- Data may be formatted inadequately leaving ambiguity.
- Data may have inexplicable gaps.
- Data may have been generated with different underlying goals.
- Metadata may be inadequate or non-standard resulting in semantic confusion.
- Accuracy may be undocumented.
- Data provenance may be suspect (modified without documentation) or unknown.
- Data may be too costly to obtain. (Libes, Lechevalier, & Jain, Dec. 2017, p. 1754).

There are challenges with real-world video datasets, including "curation cost, privacy, ethics, bias, and copyright" (Kim, Mishra, Jin, Panda, Kuehne, Karlinsky, ... & Feris, 2022, p. 35710). There are data protection laws, too (Joshi, Grimmer, Rathgeb, Busch, Bremond, & Dantcheva, 2024, p. 1).

The use of Web-crawled data, which is real data in many cases, is done without the consent of the individuals who shared them online (Shahreza, Ecabert, George, Unnervik, Marcel, Di Domenico, ... & Fierrez, May 2024).

The practical inaccessibility of real-world data, or their partial access, motivates many to go to machine-generated synthetic data as a complete or partial substitute.

These may be generated to desired forms and annotated for labeled datasets. Some are derived from real data, with privacy protections in place, and others are wholly computer-generated. Many are generated to show the same metadata and metrics as real-world data but without row-data accuracy, to mask other underlying data.

Synthetic Data As A Stand-In For Real Data

Synthetic data was first made in the 1960s, including the making of synthetic images to train computers to recognize 3d images (credited to Maxwell Clowes and David Huffman). Synthetic data has been used for various common purposes, such as for sample data in online survey systems, in computational systems for testing, and others. In terms of formal research, synthetic data is credited to two researchers:

> *The idea of synthetic data was first introduced by Rubin (1993), who proposed multiply imputing an entire dataset, so that no real micro-data would be released. As an alternative, Little (1993) introduced a method that would only replace some of the variables in the observed data, referred as partially synthetic data. Since fully-synthetic data does not contain any original data, the disclosure of sensitive information from the synthetic data is much less likely to occur. (Taub, Elliot, & Sakshaug, 2020, p. 2)*

Synthetic data refers to data "synthesized" by mathematical models or algorithms, sometimes based on actual data and sometimes created wholly without real-world references. Synthetic data generation may be based on "collected data, or on assumptions made by the coder (be it a human or AI) about the different variables in the dataset" (Gal & Lynskey, 2024, p. 108). They are computationally generated data, not collected by humans or machines from in-world phenomena, per se. As such, they are not considered "real" data even as some synthetic outputs may be somewhat based on real data, as in large language models (LLMs), models, processes, frameworks, and expert knowledge.

Researchers have used Bayesian networks to create synthetic data. They write:

> *Informative prior information is needed in order to assign appropriate weights to each network in this set if synthetic data are to have both good inferential properties and an acceptable risk of disclosure. This sensitivity to prior information will make it difficult for an official statistics agency to use Bayesian networks to automate the process of creating synthetic data" (Young, Graham, & Penny, 2009, p. 549)*

The process of generating synthetic data is known as "synthesis." Various technologies are available for the making of such data. Various types of technologies have been used to create synthetic data over the years. More traditional approaches involved rules-based systems, simulations, game engines, data maskers, and others.

More recent approaches have included machine learning methods. Synthetic data generation is…

> *…the process of using machine learning methods to train a model that captures the patterns in a real dataset. Then new or synthetic data can be generated from that trained model. (Rajotte, Bergen, Buckeridge, El Emam, Ng, & Strome, 2022, p. 1)*

Synthetic data has to offer "analytically valid" data (Winkler, 2005, as cited in Taub, Elliot, & Sakshaug, 2020, p. 3). Synthetic data has to be "both statistically representative and logically valid" to be usable (Soltana, Sabetzadeh, & Briand, Oct. 2017, p. 872). For example, clinical trial data has to be available for secondary analyses while still keeping all the participant privacy protections in place in the synthetic data (Azizi, Zheng, Mosquera, Pilote, & El Emam, 2021). Researchers examining synthetic healthcare data have used two measures of synthetic data utility: propensity score and classification accuracy for synthetic healthcare data used for secondary purposes (Dankar, & Ibrahim, 2021, p. 1).

Synthesis is used in the making of electronic health records (Goncalves, Ray, Soper, Stevens, Coyle, & Sales, 2020). Artificial intelligence used in healthcare may bring with it biases that threaten human health and well-being and result in health disparities in the real world. There are risks of under-representations of different people groups. Synthetic data may be used to "improve accuracy, precision, and fairness"

> *(Shahul Hameed, Qureshi, & Kaushik, 2024, p. 1). Healthcare data is not only tabular data, textual data, image data, and so on. It may also include sensor data. Behavior-based sensor data from healthcare may also be synthetically generated for activity recognition (Dahmen & Cook, 2019).*

Some of the deep learning methods include Generative Adversarial Networks, Variational Autoencoders, Generative Pre-trained Transformers, and Generative AI systems. There have been exploratory works using text-to-image generation models for creating photo-realistic images to pre-training large-scale models (He, Sun, Yu, Xue, Zhang, Torr, ... & Qi, 2022, p. 1). This research team asked: While text-to-image outputs satisfy human vision, do they work effectively for image recognition tasks? Can the synthetic images improve classification models in "data-scarce settings (i.e. zero-shot and few-shot)"? (He, Sun, Yu, Xue, Zhang, Torr, ... & Qi, 2022, p. 1). Can synthetic images enable transfer learning from the artificial domain to the real? Another work suggests that synthetic images created from text-to-image models do enable strong visual representation learning. In this study, the researchers used images resulting from the same prompt to be treated as "positives" for each other. They applied a "a multi-positive contrastive learning method" to train a visual representation learning model that performed better than that trained by real images (Tian, Fan, Isola, Chang, & Krishnan, 2024, p. 1).

In a time of datafication of so many aspects of human lives, synthetic data may not be needed for everything. After all, there are big datasets available from the Internet of Things, sensors, trace data, and other spaces. One work describes the harnessing of synthetic data from various servers to enable federated learning. Federated learning involves "the training of a model using data on multiple clients without the clients transmitting that raw data" (Goetz & Tewari, 2020, p. 1).

Synthetic data is "not automatically private" but has to be built in particular ways so as not to leak information (Jordon, Szpruch, Houssiau, Bottarelli, Cherubin, Maple,... & Weller, 2022, p. 1). Outliers are especially hard to capture and still maintain privacy (Jordon, Szpruch, Houssiau, Bottarelli, Cherubin, Maple,... & Weller, 2022, p. 1). There are additional risks if attackers have access to partial or entire records in order to commit reconstruction attacks. Synthetic data, though, should be "plausible" (Jordon, Szpruch, Houssiau, Bottarelli, Cherubin,Maple,... & Weller, 2022, p. 17).

Limits To Synthetic Data

There are drawbacks to various methods harnessing synthetic data for training models, including "computational requirements, training stability, and privacy-preserving measures which limit their real-world usability" (Goyal & Mahmoud, 2024, p. 1), A quantitative evaluation of the privacy gain of synthetic data found that "synthetic data either does *not prevent inference attacks* or does not retain data utility" and does not provide better tradeoff between privacy and utility than "traditional anonymization techniques" (Stadler, Oprisanu, & Troncoso, 2022, p. 1451). Another downside is that it is "impossible to predict what signals a synthetic dataset will preserve and what information will be lost" (Stadler, Oprisanu, & Troncoso, 2022, p. 1451). It was thought in earlier times that synthetic data valuable for generating accurate inferences had to align with the original seeding data and would leak dimensions of inference (Snoke, Raab, Nowok, Dibben, & Slavkovic, 2018).

Modalities Of Synthetic Data

Synthetic data, of course, can be of various modalities or forms:

Data produced by this 'synthesis' process can be mages, videos, text or tabular data. SD are generally produced by a generative model, based on ground truth (domain knowledge, scientific theories or collected data) that will produce new samples of synthetic data. (Marwala, Fournier-Tombs, & Stinckwich, 2023, p. 3)

There are mixed-modal types of data, such as those emulating slides and user interfaces (with text and visuals integrated).

Some Applications Of Synthetic Data

Synthetic data may be structured or unstructured data. In terms of structured data, these may include tabular data, time series data, and other labeled data. They come in various unstructured modalities, too: text, audio, images, animations, video, and virtual environments. Synthetic data is tailored for a "task-aware approach" (Tripathi, Chandra, Agrawal, Tyagi, Rehg, & Chari, 2019, p. 461), An ensemble approach enables synthetic data which can be "tailored" to particular needs (Van Breugel, Qian, & Van Der Schaar, July 2023, p. 34793) although the outputs are imperfect and may affect downstream tasks if researchers are insufficiently aware or careful.

Within the respective modalities, there are synthetic data used in model training including satellite imagery (Shermeyer, Hossler, Van Etten, Hogan, Lewis, & Kim, 2021), waveform data for seismic and other applications (Alkhalifah et al., 2021), visual-spatial synthetic data that is context-aware (Prakash, Boochoon, Brophy, Acuna, Cameracci, State, ... & Birchfield, May 2019), synthetic speech (to train automatic speech recognition tools) (Fazel, Yang, Liu, Barra-Chicote, Meng, Maas, & Droppo, 2021), financial datasets (Potluru, Borrajo, Coletta, Dalmasso, El-Laham, Fons, ... & Balch, 2023), clinical data used for clinical models (Qian, Callender, Cebere, Janes, Navani, & van der Schaar, 2024), healthcare information used to ensure fair treatment recommendations "across diverse patient populations" (Pezoulas, Zaridis, Mylona, Androutsos, Apostolidis, Tachos, & Fotiadis, 2024, n.p.), and imagery, and others.

Biometrics

Synthetic data has been used to replace authentic biometric data like facial imagery to train facial recognition systems (Boutros, Struc, Fierrez, & Damer, 2023). Researchers have come up with ways to introduce intra-class variation in synthetic visuals of human faces for facial recognition by introducing variance through "pose, expression, illumination, the number of identities, and samples per identity" (Qiu, Yu, Gong, Li, Liu, & Tao, 2021, p. 10880). Real and synthetic data have been used to help in the identification of various species of animals (Tao, Fan, & Yang, 2024).

Law Enforcement, Fraud Detection

Synthetic data has been used to benchmark anti-money laundering methods based on real-world data through "non-invertible transformations" to protect privacy (Jensen, Ferwerda, Jørgensen, Jensen, Borg, Krogh, ... & Iosifidis, 2023).

Prediction And Decision-Making

Synthetic data has informed the prediction of which wildfires will break out into larger ones to inform firefighting (Pérez-Porras, Triviño-Tarradas, Cima-Rodríguez, Meroño-de-Larriva, García-Ferrer, & Mesas-Carrascosa, 2021).

Cybersecurity Applications

Another uses synthetic data to augment imbalanced datasets from intrusion detection records in cybersecurity to increase prediction accuracy in the classifiers (Dina, Siddique, & Manivannan, 2022). One work trained a neural network with synthetic data to generate a capability to break Captchas (a reverse Turing test) with the created trained tool (Le, Baydin, Zinkov, & Wood, May 2017, p. 3514).

Natural Language Applications

Faux data has been used to train question answering models by using synthetic question and answer (Puri, Spring, Patwary, Shoeybi, & Catanzaro, 2020), among other natural language learning tasks and systems. Virtual data has been found to be "a viable alternative to real data for training speech recognition models" (Hu, Armandpour, Shrivastava, Chang, Koppula, & Tuzel, May 2022, p. 7682). The synthetic data was not without problems: some artifacts in it included "structured noise, content errors, or unrealistic speaking styles" (Hu, Armandpour, Shrivastava, Chang, Koppula, & Tuzel, May 2022, p. 7682). One work describes a synthetic data generator for sentiment analysis (Sundarreson & Kumarapathirage, 2024). Another work describes using synthetic data to create "grounded dialogues" efficiently (Bao et al., 2023, p. 10866). Synthetic data was used to improve back translation across languages (Xu, Ruan, Bi, Huang, Shi, Chen, & Liu, 2023). One work explored "template-based synthetic data on the performance of modern transformer models," and they suggest that there is a "delicate balance required between synthetic and real-world data, and the future trajectories of integrating synthetic data in model training pipelines" (Gholami & Omar, 2023, p. 1).

One team introduced a technical intervention to reduce "sycophancy" in large language models (LLMs) using technical synthetic data. "Sycophancy" refers to the model's tendency to tailor its responses to fit "a human user's opinion, even if that opinion is not objectively correct" (Wei, Huang, Lu, Zhou, & Le, 2023, p. 8). To train the LLM away from that tendency, the researchers used a synthetic data intervention (Wei, Huang, Lu, Zhou, & Le, 2023)

Synthetic Internet of Things (IoTs)

Synthetic Internet of Things (IoT) has been created to emulate mixed data which is "naturally big and complex" (Anderson et al., 2014, p. 171). Synthetic data has been used to train neural networks in "approximate, Bayesian, model-based reasoning" (Le, Baydin, Zinkov, & Wood, May 2017, p. 3514). Generative synthetic data has been used to aid image classification (Azizi, Kornblith, Saharia, Norouzi, & Fleet, 2023).

Computer Vision Applications

One work proposes a data synthesis pipeline to train vision language models. The researchers found the benefits of purely generated data resulting in privacy preservation (Liu, Liang, Huang, Xiong, Yu, Sun,... & Zhang, 2024, n.p.).

Synthetic data was created from real clear-weather images overlaid with a synthesized fog. The original images were from a labeled training set. The revised synthetic images enabled enhancement for computational "semantic foggy scene understanding" (Sakaridis, Dai, & Van Gool, 2018, p. 973). The "synthetic" aspect involves the synthetic fog overlays.

Synthetic data has been used to enhance sports camera calibration (Chen & Little, 2019). Synthetic data can be used to create mock-world contexts to adjust for different settings on different machines, beyond cameras.

Pure synthetic data has been used to simulate "high-order degradation" of images to enable single image super-resolution to reconstruct a high resolution image from a low-resolution version (Wang, Xie, Dong, & Shan, 2021, p. 1905).

One work describes the use of synthetic data for instruction tuning, in one case, to aid in the development of video-based large multimodal models (LMMs), with training captioned videos. The automated creation of text-annotated synthetic videos mitigates the difficulty of curating "large amounts of high-quality raw data from the web" (Zhang, Wu, Li, Li, Ma, Liu, & Li, 2024, p. 1). In this case, the synthetic dataset resulted in " video-language instruction-following" for language learning (Zhang, Wu, Li, Li, Ma, Liu, & Li, 2024, p. 11).

Synthetic data has been used to enhance large-scale pretrained Vision & Language (VL) models. Models have historically had challenges understanding "non-object words (e.g., attributes, actions, relations, states, etc.), or difficulty in performing compositional reasoning such as understanding the significance of the order of the words in a sentence" (Cascante-Bonilla, Shehada, Smith, Doveh, Kim, Panda,... & Karlinsky, 2023, p. 20155). Synthetic data may broaden computational understanding of such visual-language concepts, beyond semantic and object-based terms.

Synthetic data has been used to improve text detection in natural images (or "text spotting") (Gupta, Vedaldi, & Zisserman, 2016, p. 2315). Word recognition may be run on a whole image "holistically" based on neural network learning using synthetic data (Jaderberg, Simonyan, Vedaldi, & Zisserman, 2014) vs. the application of single character recognition. Computational systems for "text-spotting," defined as finding and recognizing text in natural scene images, may be trained on synthetic training data and enable text-based image retrieval and searchability of news footage based on text in the visuals (Jaderberg, Simonyan, Vedaldi, & Zisserman, 2016, p. 1). Here, cropped word images that cover the target lexicon are used in the synthetic data-based training. One work used synthetic data with visual patterns of text images to enhance the performance of scene text recognition models (Yim, Kim, Cho, & Park, Sept. 2021). Another team generated synthetic data with various levels of noise to lower the Character Error Rate in optical character recognition (OCR) systems (Guan, Xu, Lin, & Greene, Nov. 2024, p. 15412). This approach was particularly useful in low-resource languages and did not require expensive manual annotated data (Guan & Greene, 2024, n.p.).

Another approach involved created a synthetic handwritten word image corpus with 9 million elements to train a deep network architecture in word spotting and recognition (Krishnan & Jawahar, 2016). Another work used synthetic training data based on cursively handwritten lines of text by human writers to improve the performance of an HMM (Hidden Markov Model)-based off-line cursive handwriting recognition system. Additional synthetic training data was used to augment the training of the model. Perturbations were used such as "geometrical transformations" and "thinning / thickening operations" (Varga & Bunke, Aug. 2003, p. 618). Another work describes the use of synthetic text simulating random perturbations within a range to train online handwriting text recognizers and to improve their performance (Martín-Albo, Plamondon, & Vidal, Sept. 2014, p. 543).

Synthetic data was also harnessed to enable and improve object detection from unmanned aerial vehicles (aerial views, motion imagery) (Kiefer, Ott, & Zell, Aug. 2022, p. 3564).

In one case of machine learning from unlabeled data, the researchers designed a way to generate visual synthetic data that benefits from contrastive learning training. This method involved a generator creating distinctive samples as positive pairs of visual representations and then decreasing the similarity between them. This approach enabled the models to learn meaningful representations from data by contrasting similar and dissimilar examples. It's particularly useful in scenarios where labeled data is scarce or unavailable, making it a key player in self-supervised learning (Wu, Wang, Zeng, Shi, & Hu, June 2023, p. 2866).

Enabling Data Sharing (For Research) Without Privacy Leakage

One common applied use of synthetic data is to enable data sharing for research, without leaking private information. Researchers have explored "methods that exploit the properties of the underlying data to be anonymized to improve privacy, utility, and / or computing overheads" (Majeed, 2023, p. 638). The making of synthetic data aims fundamentally "to preserve the overall properties and characteristics of the original data without revealing information about actual individual data samples" (Hittmeir, Ekelhart, & Mayer, Aug. 2019, p. 1). One study explored how privacy-protective synthetic data was in terms of the disclosure risk of tabular synthetic data based on its measured "inferential disclosure probability" and its "the differential privacy ratio" (Abowd & Vilhuber, 2008). Researchers are exploring various inferential and other attacks that may be applied to synthetic data. One writes:

> *I show through simulation studies that valid inferences can be obtained from synthetic data in a variety of settings, including simple random sampling, probability proportional to size sampling, two-stage cluster sampling, and stratified sampling. (Reiter, 2002, p. 531)*

There is the ability to give "strict differential privacy guarantees" and enable access to large data without compromising privacy (Jordon, Yoon, & Van Der Schaar, Sept. 2018, n.p.). One work proposes a unifying framework around enabling access to sensitive data without privacy compromises (Liu, Vietri, & Wu, 2021).

Faux data may "substitute" for real data, in cases where the original data may be scarce, expensive, sensitive, unavailable, and so on. Synthetic data may "augment" real data, such as in cases where the real data is incomplete, biased, non-representational, and so on. Synthetic data can be "a rich source of exposure to variability for the algorithm" (Jacobsen, 2023, p. 1). Synthetic data augmentation may be "targeted" to address particular issues. One involves the making of diverse data, including anomalies, niche contents.

> *No matter how big the size of the dataset, capturing long tails of the distribution pertaining to task-specific environmental factors is impractical. The goal of this paper is to investigate the use of targeted synthetic data augmentation—combining the benefits of gaming engine simulations and sim2real style transfer techniques—for filling gaps in real datasets for vision tasks. (Jaipuria, Zhang, Bhasin, Arafa, Chakravarty, Shrivastava, ... & Murali, 2020, p. 772)*

Various sequences are harnessed in using LLMs to create synthetic visual data with sufficient diversity and variations (Long, Wang, Xiao, Zhao, Ding, Chen, & Wang, 2024). One work suggests different ways of diversifying synthetic data such as by machine-extracted personas (Ge, Chan, Wang, Yu, Mi, & Yu, 2024). Another augmentation in the research literature refers to using synthetic data to address dataset bias for face recognition programs (Kortylewski, Egger, Schneider, Gerig, Morel-Forster, & Vetter, 2019).

Synthetization methods may include repeated perturbation of the original data to create the synthetic data.

One challenge involves the generating of synthetic data that captures "the properties of anomalous data, which is generally a minority class with potentially very few samples, and can thus reproduce meaningful anomaly instances" (Llugiqi & Mayer, Aug. 2022, p. 306). Anomaly detection is used in "fraud detection, disease detection, intrusion detection, identification systems, and fault diagnostics" (Llugiqi & Mayer, Aug. 2022, p. 309). Synthetic data is used for code-mixed language modeling (Pratapa, Bhat, Choudhury, Sitaram, Dandapat,, & Bali, July 2018). In this work, linguistic theory was applied to enable machine identification of the respective codes in the mix. Another work involves generating artificial datasets "for the analysis and evaluation of data analysis algorithms in (the) data mining area" (Pei & Zaïane, 2006, p. 2) for the analysis of clustering and outlier analysis algorithms.

Synthetic data is created for various uses, such as privacy protections. In terms of structured data, systems that create the synthetic data will make "non-reversible, artificially created data that replicates the statistical characteristics and correlations of real-world, raw data" (Mavrogenis & Scarlat, 2023, p. 1395). Computational systems that generate structured data based on real-world data cannot map the synthesized records one-on-one to the original data. Part of the privacy preserving properties exist because of this lack of mapping (Rajotte, Bergen, Buckeridge, El Emam, Ng, & Strome, 2022, p. 1).

Avoiding Bias In Synthetic Data

Synthetic data may be designed to be "fair" in the sense that algorithms are taught "not on historic worlds, but rather on the worlds that we strive to live in" (Tiwald, Ebert, & Soukup, 2021, p. 1). The algorithms are designed not to replicate old-world biases but to treat individuals in non-biased ways. Synthetic data has been used to improve fairness in "learning analytics" models (Liu, Deho, Vadiee, Khalil, Joksimovic, & Siemens, 2025, p. 1). Fairness pre-processing may be applied to such models and to synthetic data generators. Another study explored where bias was introduced within the entire modeling pipeline (Baumann, Castelnovo, Crupi, Inverardi, & Regoli, June 2023). The research team created a framework for generating synthetic datasets with different types of biases to learn about this phenomenon. They observe that there are different senses of fairness and tradeoffs between fairness and biases. One work describes a method to remove bias and to create high-quality synthetic data that is still "*truthful to the underlying data-generating process*" (Van Breugel, Kyono, Berrevoets, & Van der Schaar, 2021, p. 22221).

Some synthetic datasets are designed to give precisely the same statistical properties as real data while hiding personally identifiable information (PII). One "general utility" of structured synthetic data is whether it has "a distribution that is comparable with that of the original data" (Snoke, Raab, Nowok, Dibben, & Slavkovic, 2018, p. 663). In many contexts, the fidelity of the synthetic data is important and should represent the real world.

In the creation of artificial intelligence (AI) models, synthetic data may be used to fill in data gaps. AI models require "large, diverse, and high-quality datasets, which can be challenging to obtain due to data scarcity, privacy concerns, and high costs" (Liu, Wei, Liu, Si, Zhang, Rao,... & Dai, 2024, p. 1).

For this data to have usefulness, various dimensions of the real-world data have to be observable in the synthetic, such as "some structural and statistical properties of real-world data" (Marwala, Fournier-Tombs, & Stinckwich, 2023, p. 3). For synthetic data to have data quality, it has to be representative of real-world data. Users of synthetic datasets need to know "how representative they are of real-world data" (Maheshwari, Ivanov, & Haddad, 2024, n.p.)

Biases may be introduced "when the same LLM (large language model) is used to both generate benchmarking data and to perform the asks" (Maheshwari, Ivanov, & Haddad, 2024, n.p.). Smaller LLMs "exhibit biases towards their own generated data, whereas larger models do not" (Maheshwari, Ivanov, & Haddad, 2024, n.p.). The effectiveness of synthetic data as benchmarks depends on "the task, and … practitioners should rely on data generated from multiple larger models whenever possible" to mitigate potential biases from the synthetic datasets (Maheshwari, Ivanov, & Haddad, 2024, n.p.).

The synthetic has to speak into the real. If it fails to apply to the real world, that is called a "domain gap." This challenge is known as the "synthetic-to-real domain adaptation problem" (Nikolenko, 2021, p. 1). And models trained on synthetic data may "fail to perform well on real datasets owing to the presence of domain gap between the datasets" (Sankaranarayanan, Balaji, Jain, Lim, & Chellappa, 2018, p. 3752). It is important for models trained on synthetic data to "generalize to real in-the-wild datasets" (Wood, Baltrušaitis, Hewitt, Dziadzio, Cashman, & Shotton, 2021, p. 3681). Ideally, synthetic data may be transferable and relevant to the different contexts in which it may be harnessed. One concept, "digital reality," focuses on the practices of virtual replication, virtual simulation, virtual mimicry, virtual environments, virtual simulations, and virtual ecosystems…that emulate aspects of the real world.

Synthetic data has been used to train "semantic segmentation models" that classify what object each pixel belongs to semantically for scene understanding (Chen, Li,

Chen, & Gool, 2019, p. 1841). This approach helps in computational discernment of semantic segmentation in a cross-domain way.

Another tool to bridge the domain gap in computer vision involves the modeling of camera effects to "increase the size and variability of training data for learning visual tasks" (Carlson, Skinner, Vasudevan, & Johnson-Roberson, 2018, p. 1). This approach focuses on the sensor domain, the equipment used to capture the visual, with varying sensor effects, including "chromatic aberration, blur, exposure, noise, and color temperature" for synthetic imagery (Carlson, Skinner, Vasudevan, & Johnson-Roberson, 2018, p. 1). One research team describes an approach that "simulates the whole mechanism of these devices, generating realistic depth data from 3D models by comprehensively modeling vital factors e.g. sensor noise, material reflectance, surface geometry" (Planche, Wu, Ma, Sun, Kluckner, Lehmann,... & Ernst, Oct. 2017, p. 1).

To enable embodied intelligence (such as robotics), a synthetic scene-text corpus was created "with 10K indoor scenes and 1M descriptions at object, view, and room levels" (Yang, Xu, Mo, Chen, Huang, & Liu, 2024, n.p.). The synthetic visual-textual set enabled higher functionality based on the "diverse scene data, rich textual descriptions, multi-grained 3D-text associations, and low collection cost" (Yang, Xu, Mo, Chen, Huang, & Liu, 2024, n.p.).

Other synthetic datasets have been created to "improve 6D pose estimation, object recognition, and semantic scene understanding in indoor scenarios" (Schieber, Demir, Kleinbeck, Yang, & Roth, 2024, p. 1). Synthetic data is harnessed when the requisite data does not exist in the necessary format (McLachlan, 2017, p. 3).

Another approach involves "domain randomization, in which the parameters of the simulator—such as lighting, posse, object textures, etc.—are randomized in non-realistic ways to force the neural network to learn the essential features of the object of interest" (Tremblay, Prakash, Acuna, Brophy, Jampani, Anil, ... & Birchfield, 2018, p. 969). Domain randomization "is an effective technique to bridge the reality gap" for training neural networks for object detection and other applications (Tremblay, Prakash, Acuna, Brophy, Jampani, Anil, ... & Birchfield, 2018, p. 1088).

The digital may be designed in particular ways to train models to function in the real world, such as in self-driving systems, robotics, the Internet of Things (IoT), computer-vision based surveillance systems, and other applications. Synthetic data may not only emulate the real with fidelity, but they may be designed to highlight particular aspects of data in order to create enhanced computational capabilities in various models.

One work described the use of synthetic data to improve methodological research, particularly in the social sciences, to evaluate and develop estimators in the social sciences (Burgard, Kolb, Merkle, & Münnich, 2017, p. 223). Synthetic data was used to explore statistical methods, comparative research, support for data analytics in peer review, promoting sustainable use of data over time, enabling benchmarking different algorithms (that may be tested on open-source synthetic datasets), and other applications (Burgard, Kolb, Merkle, & Münnich, 2017).

Public-Facing Tools For Synthetic Data Generation, For Benchmarking Data Generation Tools

There are numerous synthetic data generators out in the world. One work cited 417 synthetic data generation models released in the last decade (Bauer, Trapp, Stenger, Leppich, Kounev, Leznik, ... & Foster, 2024). Synthetic data generation is "a general-purpose technology" (Gal & Lynskey, 2024, p. 109). One may be built using Python and Python libraries to derive synthetic data from actual data (Montanez, 2018, p. 3). The generated data in Synthia in Python will preserve relationships between the data variables (Meyer & Nagler, 2021). Another tool involves a game rendering engine combined with open-source packages for creating a labeled synthetic data for computer vision tasks (Borkman, Crespi, Dhakad, Ganguly, Hogins, Jhang,... & Yadav, 2021, p. 1). Synthetic data may also be created in a free tool in R with the synthetic data that can "mimic the original observed data and preserve the relationships between variables but do not contain any disclosure records" (Nowok, Raab, & Dibben, 2016, p. 1).

There is an open-source software package to support different use cases of generating and evaluating synthetic data in machine learning, for "fairness, privacy and augmentation across diverse tabular data modalities, including static data, regular and irregular time series, data with censoring, multi-source data, composite data, and more" (Qian, Cebere, & van der Schaar, 2023, p. 1).

There are various synthetic data generation models with different performance metrics (Endres, Mannarapotta Venugopal, & Tran, Aug. 2022). There are open-source tools to enable benchmarking of synthetic data generators across data modalities and use cases, to enable the optimized selection of which generator to use (Qian, Davis, & van der Schaar, 2024). One work describes a multi-dimensional evaluation of synthetic data generators. They assert that prior metrics of synth data generators created "conflicting conclusions making direct comparison of synthetic data generators very difficult" (Dankar, Ibrahim, & Ismail, 2022, p. 11147). They propose a masked data evaluation by "classifying available utility metrics into different categories based on the measure they attempt to preserve: attribute fidelity, bivariate fidelity, population fidelity, and application fidelity" (Dankar, Ibrahim, & Ismail, 2022, p. 11147) to conduct analyses of synthetic data generators.

One work describes a tool that creates synthetic data by having users "visually and declaratively specifying the properties of the dataset such as each field's statistical distribution, its domain, and its relationship to other fields" (Mannino & Abouzied, Oct. 2019, p. 549).

Those who would share synthetic datasets, tools for generating synthetic data, and others in the space, have a professional and ethical responsibility for what they create and documenting the various enablements and constraints of their technologies (and techniques).

Enabling Data Sharing And Data Augmentation For Modeling

With the popularization of artificial intelligence (AI), deep machine learning is needed to train various models and programs. Such programs often require much more data than is practically available. One team observed that "the availability of training data is the main problem of deep learning methods" (Dahmen, Trampert, Boughorbel, Sprenger, Klusch, Fischer,... & Slusallek, 2019, p. 1). Deep neural networks are data hungry (Jaipuria, Zhang, Bhasin, Arafa, Chakravarty, Shrivastava, ... & Murali, 2020, p. 772). Large language models "require large quantities of labeled data to grasp even simple concepts" (de Melo, Torralba, Guibas, DiCarlo, Chellappa, & Hodgins, 2022, p. 174).

There are ways to "implement synthetic data sampling methods to train deep CNNs (convolutional neural networks) more effectively and accurately" (Bhattarai, Baek, Bodur, & Kim, May 2020, p. 2303). One work proposes using "reinforcement learning…to automatically search a subset of meaningful synthetic examples from a large pool of GAN (generative adversarial network) synthetic data" (Bhattarai, Baek, Bodur, & Kim, May 2020, p. 2303). Sampling the synthetic data before using it for data augmentation is seen as necessary.

Researchers have created various ways to generate labeled data for supervised machine learning. One work describes the making of structured visuals like slides and user interfaces with target labels using "code generation" for training (Peng, Huq, Jiang, Wu, Li, Bigham, & Pavel, 2025, p. 466). Tools like this enable mitigating the high expense of human labor.

There are a variety of approaches to mix real and synthetic data for neural network training (Seib, Lange, & Wirtz, 2020).

Researchers suggest that there are different volumes of data needed for training models, for different purposes. In some cases, a certain base amount is needed for sufficient performance and outcomes.

There are challenges in the uses of synthetic data for modeling. Machine learning systems require "representations of the real world for training and testing," with dimensions of realism and amounts at scale (Whitney & Norman, June 2024, p. 1733). Researchers identify some key risks:

First, we detail the high risk of false confidence when using synthetic data to increase dataset diversity and representation. We base this in the examination of a real world use-case of synthetic data, where synthetic datasets were generated for an evaluation of facial recognition technology. Second, we examine how using synthetic data risks circumventing consent for data usage...Finally, we discuss how these two risks exemplify how synthetic data complicates existing governance and ethical practice; by decoupling data from those it impacts, synthetic data is prone to consolidating power away (from) those most impacted by algorithmically-mediated harm. (Whitney & Norman, June 2024, p. 1733)

Assessing the capabilities and uses of synthetic data require expertise and reflection and empirical research.

Validating The Synthetic Data

Various types of validations have been applied to synthetic data. There is a need for "theoretically sound metrics for synthetic data quality that go beyond subjective impressions…and performance on benchmark data sets" (de Melo, Torralba, Guibas, DiCarlo, Chellappa, & Hodgins, 2022, p. 197). It is critical to evaluate the efficacy of synthetic data, especially if it informs research and decision-making in-world (El Emam, 2020, p. 56).

A simple measure involves how well the faux data represents the real or non-synthetic data, especially for those derived from real datasets. (Chen, Chun, Patel, Chiang, & James, 2019) Others are assessed based on their level of bias vs. fairness. Some measure various types of privacy (which can be defined in different ways). Various methods are applied to creating differentially private synthetic data (Rosenblatt, Liu, Pouyanfar, de Leon, Desai, & Allen, 2020; Torkzadehmahani, Kairouz, & Paten, 2019).

One work reviews various methods for creating differentially private synthetic data, which balances the privacy-quality dimensions (Arnold & Neunhoeffer, 2020). Decades of work has gone into creating usable synthetic data with "disclosure control" (Reiter, 2023, p. 15). Privacy preservation may be assessed based on several approaches, including "*singling out, linkability*, and *inference risks*" (Giomi, Boenisch, Wehmeyer, & Tasnádi, 2022, p. 1), according to one unified framework. Quantizing the various risks may be done through available tools. Researchers propose different ways to enable both privacy protections and improved performance of the tools used to create synthetic data sets (Liu, Tang, Vietri, & Wu, July 2023, p. 22009). The synthetic data challenge involves using synthetic data to control for confidentiality risk and less through "access restriction and traditional statistical disclosure control" (Taub & Elliot, 2019, n.p.).

One work suggests that synthetic data has severe limits in terms of privacy protections: synthetic data "does not prevent attribute inference" and other potential compromises of data (Stadler, Oprisanu, & Troncoso, 2020, n.p.). An earlier work suggests shortcomings in terms of data anonymization with synthetic data releases (Abay et al., 2019).

Another work introduces a method to extract insights from data mining from an original real dataset and applying that knowledge to the synthetic dataset. They write:

> *...some data mining techniques (in particular, a decision tree) can discover patterns that we can then use to inverse map into synthetic datasets. These synthetic data sets can be of any size and will faithfully exhibit the same (decision tree) patterns. (Eno & Thompson, 2008, p. 78)*

The work of building the synthetic data involves not just preserving the cardinality relationships among fields but also the "hidden complex patterns within a data set" (Eno & Thompson, 2008, p. 78). This extraction must be done without risking the former original data to be recovered (which would break privacy protections).

Various Designed Features Of Synthetic Data

For many, synthetic data has lower quality than real data. Depending on the particular use cases and applications, the synthetic data may be designed with various features.

One work describes a method to build synthetic datasets "where bias can be modeled by using a probabilistic network exploiting structural equation modeling" (Barbierato, Vedova, Tessera, Toti, & Vanoli, 2022, p. 1). This capability enables tuning parameters, with bias balanced against fairness. Training fairer models, trained on synthetic data, enables higher fairness metrics and the ideal of demographic parity (Gupta, Bhatt, & Pandey, 2021). Synthetic data is used to address data imbalance problems to ensure that predictions from trained models are less biased (Ye-Bin, Hyeon-Woo, Choi, Kim, Kwak, & Oh, 2023).

Another work advocates for a relational model to understand synthetic data, based on "their purpose; their performance in a particular situation (such as training a machine learning model); and a context-dependent operationalization of evidence" (Offenhuber, 2024, p. 1).

Assessing Capabilities Of Synthetic Data

A research team found that "there is no universal method or metric to evaluate and benchmark the performance of various approaches" in the use of generative adversarial networks (GANs) for synthetic data generation (Hernandez, Epelde, Alberdi, Cilla, & Rankin, 2022, p. 28).

An earlier work highlights concerns about the reliability and validity of results derived from synthetic data. Utility measures have "been either information theoretic abstractions or somewhat arbitrary collations of statistics" (Taub, Elliot, & Sakshaug, 2020, p. 1) and so less effectual. For something more rigorous, one team proposes the work of replicating published analyses using "disclosure-controlled versions of the same microdata used in said analyses and then evaluated the impact of disclosure control on the analytic outcomes" (Taub, Elliot, & Sakshaug, 2020, p. 1).

Another work found the challenges of "high false positives" of novel categories of "few-shot object detection" was not solved with the introduction of synthetic data to train such models (Lin, Wang, Zeng, & Zhao, 2023, p. 638). Synthetic data was used to train "deep salient object detection" models (Wu, Wang, Wang, Shi, Chen, Hao, & Li, Oct. 2022, p. 5557).

In another case, synthetic data from large language models (LLMs) was used to improve text classification, Researchers found that "subjectivity, at both the task level and instance level, is negatively associated with the performance of the model trained on synthetic data" (Li, Zhu, Lu, & Yin, 2023, p. 1). Given subjectivity in language and its interpretation, additional work needs to be done to enhance machine text classification to account for such human-based nuance and complexity.

Synthetic data has been used to expand learning in the visual dimensions of how people identify visual similarity between images (Fu, Tamir, Sundaram, Chai, Zhang, Dekel, & Isola, 2023). They write:

> *We expand the fundamental tasks of measuring perceptual image similarity to encompass factors beyond mere low-level similarity, spanning multiple notions of similarities (e.g., object poses, colors, shapes, and camera angles). By harnessing a state-of-the-art generative model, we overcome the lack of suitable data and introduce a new dataset consisting of human-judged synthetic image triplets. We evaluate candidate embeddings and propose a new similarity metric that better aligns with human judgments, demonstrating its generalization and effectiveness compared to existing metrics. (Fu, Tamir, Sundaram, Chai, Zhang, Dekel, & Isola, 2023, p. 10)*

They found that human visual similarity includes considerations of "pose, perspective, foreground color, number of items, and object shape" (Fu, Tamir, Sundaram, Chai, Zhang, Dekel, & Isola, 2023, p. 1). This work includes a dataset of "human-judged synthetic image triplets" (Fu, Tamir, Sundaram, Chai, Zhang, Dekel, & Isola, 2023).

To enhance the training of deep face recognition systems, one team used a 3d morphable facer model for the making of images with varying facial identities, with variance in "pose, illumination, and background" (Kortylewski, Schneider, Gerig,

Egger, Morel-Forster, & Vetter, 2018, n.p.). Another work used synthetic images to enhance automatic facial behavior analysis, specifically the detection of valence and arousal from human expressions and "action units" (Kollias, Oct. 2022, p. 1).

Synthetic data has been generated to strengthen the multi-object tracking (Liu, Wang, Zhou, & Zheng, 2023, p. 206), for assisted driving, for surveillance. Synthetic data has also been applied to crowd counting, object detection, pedestrian detection, object tracking, pedestrian tracking, human behavior analysis, crowd behavior analysis, person reidentification, facial recognition in computer vision (Delussu, Putzu, & Fumera, 2024). An earlier work describes pixel-wise crowd understanding via synthetic data. In this work, the researchers made an annotated "large-scale, diverse synthetic crowd dataset" for trainings (Wang, Gao, Lin, & Yuan, 2021, p. 225). Synthetic crowd images have been created in an annotated dataset to train crowd counter used to assess those in the wild (Wang, Gao, Lin, & Yuan, 2019). Synthetic data has aided pedestrian detection and tracking even in highly crowded scenarios and in cases of multi-person tracking (Fabbri, Brasó, Maugeri, Cetintas, Gasparini, Ošep, ... & Cucchiara, 2021). The model benefitted in terms of "pedestrian detection, re-identification, segmentation, and tracking" (Fabbri, Brasó, Maugeri, Cetintas, Gasparini, Ošep, ... & Cucchiara, 2021, p. 10849).

Synthetic data has been generated for thermal infrared tracking with synthetic thermal infrared (TIR) image sequences transcoded from RGB video data with labels (Zhang, Gonzalez-Garcia, Van De Weijer, Danelljan, & Khan, 2018). Researchers have created synthetic RGB datasets to enhance depth sensibilities in visual object categorization (Carlucci, Russo, & Caputo, May 2017, p. 1362). Synthetic 3d images of only depth data have been used to enhance robot navigation and to enhance their indoor scene understanding (Handa, Patraucean, Badrinarayanan, Stent, & Cipolla, 2016).

Researchers have explored various ways facial and biometric recognition has been improved with synthetic data. There have been concerns related to "data privacy scenarios, demographic biases, generalization to unseen scenarios, and performance limitations in challenging scenarios, including significant age disparities between enrollment and testing, pose, variations, and occlusions" (Melzi, Tolosana, Vera-Rodriguez, Kim, Rathgeb, Liu, ... & Marras, 2024, p. 1). Another tool involved the reconstruction of a 3d face from 2d data through a trained convolutional neural network with annotated examples. The researchers used "random, yet nearly photo-realistic, facial images for which the geometric form is known" to enable the "recovery" of facial shapes from real images (Richardson, Sela, & Kimmel, Oct. 2016, p. 460).

One researcher posits that synthetic data "requires a new sensibility on its impact—especially in the case of technologies linked to the purpose of surveillance, such as algorithmic prediction and computer vision, as these have been shown to be particularly harmful for minorities, such as racialized and gendered subjects…Synthetic data add new layers to the life cycles of data...With industries on the rise that introduce synthetic data generation in fully simulated 3D environments or synthetic data augmented from existing data sets, new notions of datafication practices are being implemented worthy of scholarly attention" (Wiehn, 2024, p. 476).

One work describes the use of synthetic data to enable "steel defect recognition in automated production control systems" in industrial applications (Boikov, Payor, Savelev, & Kolesnikov, 2021, p. 1).

A review of the literature encourages more research in the space of synthetic data generation and uses (Figueira & Vaz, 2022). Others note the importance of computational velocity and efficiencies, in order to mitigate environmental impacts.

Some Uses Of Synthetic Data

Synthetic data is more accessible than real-world data. It may "reduce the costs involved in all stages of the data value chain, obviating the need for excessive data collection, costly cleaning and preparation, and long-term data storage. Generation-for-purpose may also shorten the time necessary to generate useful data" (Gal & Lynskey, 2024, p. 105). Synthetic data usage may lessen the amount of legal risks that come with using real-world data.

In a time of synthetic data, legal reforms are being actualized that specify how synthetic data may be used (Gal & Lynskey, 2024, p. 101; Bellovin, Dutta, & Reitinger, 2019). Data protection laws apply both to the real-world data from which synthetic data may be created and the protections applied to the synthetic data. One researcher writes:

> *...first, full data protection compliance prior to data synthesis would be applicable in many cases; second, according to the identifiability test as enshrined in the definition of personal data, synthetic data will be considered pseudonymous or anonymous data depending on the appropriateness of the data synthesis and the related ex-post control mechanisms; third, the broader question of the legal qualification of synthetic remains an unresolved issue in light of the exegetical discrepancy by law and doctrine on the identifiability test and regulatory model for data protection law. (López, Dec. 2022, p. 1)*

Various endeavors are taken to limit the risks of inferring sensitive information from the synthetic data.

The modern age of artificial intelligence, robotics, self-driving, and smart cities suggests the exponential need for synthetic data to train various computational systems. By 2024, some 60% of data used to train AI systems are synthetic (Gal & Lynskey, 2024). Data, both synthetic and real, is an essential input for modeling and other systems.

Synthetic facial images have been used to improve the performance of facial recognition programs (Zhang, Grimmer, Ramachandra, Raja, & Busch, May 2021).

Retail Uses Of Synthetic Data

The "retail" use of synthetic data has reached a wide audience. Generative AI tools are used to generate text, images, animations, audio files, film, slideshows, and other contents. One researcher suggests the importance of broadly disseminating information about synthetic data:

> *As the use of synthetic data becomes widespread, the need to consider public perceptions and understanding of its uses, benefits, limitations, and potential risks will need to be transparently communicated to stakeholders. Generators of synthetic data should be mindful of ensuring the quality and validity of synthetic data, and be able to describe how it was generated, how it differs from the original and how it could be reproduced. (Patel, 2024, p. 51)*

It is not a given that synthetic data is a net positive. There are risks of "data pollution and contamination with synthetic data" (Wiehn, 2024, p. 472). There are risks of seeing synthetic data as "living information, much like collected data, and to account for synthetic data and the conditions of its generation in the context of simulated environments" (Wiehn, 2024, p. 472). Synthetic data is not necessarily inert. It speaks into the real world. "Explainability" is an important element in synthetic data(Potluru, Borrajo, Coletta, Dalmasso, El-Laham, Fons, ... & Balch, 2023, p. 1). Another work suggests the importance of identifying synthetic data to users, clarifying its provenance, sharing quality metrics for the synthetic data, and developing "cybersecurity protocols to protect synthetic data and its source" (Marwala, Fournier-Tombs, & Stinckwich, 2023, p. 1). There are ethical implications in the uses of synthetic data in AI systems (Hao, Han, Jiang, Li, Wu, Zhong,... & Tang, 2024) across a range of dimensions.

ILLUSTRATING A MONTHLY PROFESSIONAL NEWSLETTER WITH SYNTHETIC VISUALS

Some types of synthetic images are built by common everyday users for everyday uses. Without this "last mile" type of use, the generative AI tools will not be optimized. They will not be adopted and adapted. At present, the image-generating GAI tools enable the generating of visuals according to various known styles based on text prompts, image prompts (seeding images), and combinations of the two. Some of the tools enable the application of various "modifiers" and such to affect the output visuals. Many enable iterated prompting to evolve new image outputs.

Some of the outputs may include word art although the state-of-the-art tools still struggle with text integrations in a visual that is technically human-readable with proper spelling and reads as an integrated work (with the text and visuals working of-a-piece). There seem to be both capability challenges and processing (cognitive) challenges. [The more complex the text prompt, the less capable the GAI is to do both the spelling and the imagistic rendering. The more unusual the spelling, the less capable the GAI is to achieve the proper spelling. Imaginary word spellings can throw off some GAI but not others.]

The first version of the academic grant-funding newsletter originated January 2024. It was titled *HutchCC Grant Navigator* (thanks to Karen Linn's naming), and it ran monthly through February 2025. After that issue, an administrator decided against continuing publication, and following issues became *Academic Grant Pursuits* (March 2025 – January 2029 onwards, in a forward- or ante-dated way), on SlideShare and ResearchGate. The latter newsletter issues were created through October 2025 but dated into the future to accommodate the number of articles and visuals. The newsletters began at about six pages in length but grew to average 22 – 25 pp. per issue. There were various short articles about grant pursuits, with a number of how-to articles. In general, every article had at least one illustration, and some had a number. Some visuals were created as stand-alone signposts and commentary. These were used as fillers in part to enable more nuanced layouts.

The visuals from the generative AI had to be edited. The post-production helped to erase digital image artifacts, machine artist "signatures" (often at the bottom left or right of the visual), nonsense words, nonsense numbers, nonsense symbols, machine cursing (using expletives), logo hijacking (one GAI swiped the Starbucks logo to illustrate "higher education"), and other problematic artifacts. Then, there was the need to correct misspellings and address poor layout and poor compositing. One GAI did visual watermarks with a circular label reading "AI." For usability, those had to be removed. Given the ease of re-generating visuals from generative AI, it got to the point that any work that looked like it would involve more than a few steps of visual edits were not downloaded, and one just went on to generate other visuals.

There also had to be controls against unintended messaging, such as the false language embedded in the visuals and the pseudo-symbols that had to be removed to control against misinterpretations. (Readers will assume meaning even if the symbol was created arbitrarily by machine.) Unintentional messaging may still cause harm and may mislead, depending on how the recipients receive the (ambiguous, polysemous) visual information.

In the newsletter, the visuals serve decorative aims and didactic ones. Many images have to be edited in post-production for usage, with the removal of bad text and substitution with readable good text (text overlays), the removal of image noise, resizing, cropping, recoloring, and other changes. In the digital image editing tool (Adobe Photoshop), there is a built-in AI image-editing tool (Adobe Firefly and other unnamed AI features like neural filters) that enables some text inclusion, cleanup, style changes, and other factors.

Many of the visuals are thumbnail size, so the details of the visuals may not be fully appreciated by users. (The visuals have to fit within columns due to the newsletter layout.) All images are alt-texted, so when a mouse cursor is hovered over the image, the sentence text is shown…and elaborate on the visual with descriptions and other information. One challenge is that online versions of the .pdf files do not show the alternate text, so the images themselves often have readable text overlays to clarify the meaning. Some of the visuals which are simple illustrations may be interpretable by visuals alone. However, there is a degree of ambiguity to the visuals. Some may be read as decorative when they actually contain some observations about the grants space.

These illustrations were mostly generated using generative AI tools to achieve the following:

- Break up the gray text
- Provide a recurring look-and-feel for the respective newsletter issues
- Attract reader attention
- Control for user focus
- Create excitement
- Introduce relevant terminology
- Illustrate ideas
- Describe processes
- Emphasize important points
- Share data
- Summarize important points
- Enable visual thinking (along with textual and symbol-based thinking)
- Enhance human memory, such as through visual and textual and combined mnemonic functions
- And others

The GAI tools and the models within them…all often a particular look-and-feel. They all have tendencies and limits to their capabilities. They are versed in different composites and layouts. [Many tend to present busy imagery with visuals filling much of the 2d space. Few can output minimalistic visuals with judicious use of positive-negative space in balance and interplay.] GenAI also has challenges in creating optical illusions. They cannot apparently yet create visuals with hidden images. They can often engage an image as a diptych, a triptych, a quadtych, and so on. They can sometimes do picture-in-picture. The points of view are somewhat variational but limited. The GAI tools vary in terms of sophistication, with various capabilities that do not have a clear reference in some models (like paint marbling, for example).

The respective GAI tools have different levels of documentation in terms of what data was used for the training or how fine-tuning was done post-training, and so on. The large language models (LLMs) in the world evolve a fair amount, too, and it is hard to tell what has been in motion to know what is functional in the background. Most image-generating GAI tools are not thought to hallucinate or recreate original imagery from a training set.

A free newsletter distributed at a college's email system and archived on both ResearchGate and SlideShare (social content-sharing sites) is informal, and the visuals are not expected to be very complex. (Other visuals in the newsletter include diagrams, data visualizations, and maps, generated by other tools.)

Discussion

This multi-year experience (2024 – 2025 inclusively) highlighted some factors in the uses of generative AI capabilities across multiple models in the work space.

Technical Capabilities Of Generative AI In The Newsletter Illustration Case

Synthetic image data can have high applicability to practical professional work, as noted in this single longitudinal case (spanning several years). The respective models have different applicability depending on the complexity of the visual. A majority of the output images did require post-production to ensure usability.

Just as this work was about to enter the proofing stage, the author experienced @imagen4 of Gemini's 2.5 Pro. At that time, she was using a prompt "@imagen4: draw a sign that reads 'bottom fell out' with a backdrop of dollar signs" to capture the U.S. federal government's grant funding rescissions from foreign humanitarian aid and academia (among others), grant cancellations, and pullbacks of funding. Remarkably, the generative AI was able to illustrate the concept by showing flames around the word "bottom" in one image…and drawing a big hole in the middle with the words surrounding it in another. When the author commented on that to the generative AI tool, it acknowledged that it was designed to read and understand the meaning to seed the visual creation. It was not just a wooden drawing of what was directly requested but was *understood.* The advances in generative AI are coming thick and fast.

In another case, the author asked several generative AI models to hide the text "whispers" in a nature scene with high winds. @imagen4 was able to hide the text in a way that was almost not discernible from the background. In other visuals by models in the Deep Dream Generator, the text was laid out like a text overlay and was highly discernible and readable to the naked eye.

The tool could also be prompted in sequence, "and another three," and the generative AI tool would draw the three. Other times, though, it would create a triptych or a three-in-one image or a picture-in-picture effect.

There are depictions of "impossible" visuals, beyond the depiction of imagined scenes. One example is neon signs that have nuanced gradient colors. At present, neon can handle certain hues but not gradients. This is a physical reality and limitation of actual neon signs.

Ethical Considerations

The visuals were not marked per se with indicators that they were AI-generated. @imagen4 began with an "AI" watermark at the bottom right corner. Later, that was changed to an icon. Within the image was another metadata marking indicating the creation by Google Gemini's @imagen4. The visual markings were removed, but the internal marking was left intact. (One workaround would have been to screenshot the AI-generated image and to make a new image to circumvent the internal watermark.) The professional ethics around the uses of AI-generated image are being debated. It can be argued that the visuals are already clearly created by generative AI, in the same way that clip art is clearly clip art. The visual concepts were originally human and may have evolved a little after the initial prompt.

What about the uses of generative AI to replace human labor? In this case, the visuals used would have been human-generated using other computational tools like diagramming ones, slideshow tools, online image-generation tools (that are non-AI, although these are becoming fewer and fewer), and digital image editing ones. They would have required a large amount of extra work beyond the actual writing and editing (which were human-generated). Not using generative AI for the images would have constrained the production in imminently measurable ways.

Emerging Trends

It is to no one's surprise that image-based generative AI is featuring in more and more human activities: entertainment, artful expression, design, illustration, and others. This case is likely one of numerous ones in the world.

FUTURE RESEARCH DIRECTIONS

How synthetic data is used by the broad non-expert public will be a point of interest not only for the makers of the model technologies but those who make such tools available to the public. It will be interesting to explore how the data from user interactions are harnessed for various research and tool improvements. How the Deep Dream Generator, Gemini, and Microsoft Copilot evolve in their usage over the years will also be of interest.

CONCLUSION

This work is an early one about the uses of synthetic imagery to illustrate a grant newsletter that is publicly available on ResearchGate and SlideShare. The review of the literature shows broad applications of synthetic data for research and for AI model building but less for applied "retail" use by the general public.

REFERENCES

Abay, N. C., Zhou, Y., Kantarcioglu, M., Thuraisingham, B., & Sweeney, L. (2019). Privacy preserving synthetic data release using deep learning. In *Machine Learning and Knowledge Discovery in Databases: European Conference, ECML PKDD 2018, Dublin, Ireland, September 10–14, 2018* [Springer International Publishing.]. *Proceedings*, *18*(Part I), 510–526.

Abowd, J. M., & Vilhuber, L. (2008, September). How protective are synthetic data? In *International Conference on Privacy in Statistical Databases* (pp. 239-246). Berlin, Heidelberg: Springer Berlin Heidelberg.

Alkhalifah, T., Wang, H., & Ovcharenko, O. (2021, October). MLReal: Bridging the gap between training on synthetic data and real data applications in machine learning. In *82nd EAGE Annual Conference & Exhibition* (Vol. 2021, No. 1, pp. 1-5). European Association of Geoscientists & Engineers. DOI: 10.3997/2214-4609.202113262

Anderson, J. W., Kennedy, K. E., Ngo, L. B., Luckow, A., & Apon, A. W. (2014, October). Synthetic data generation for the Internet of Things. In *2014 IEEE International Conference on Big Data (Big Data)* (pp. 171-176). IEEE. DOI: 10.1109/BigData.2014.7004228

Arnold, C., & Neunhoeffer, M. (2020). Really Useful Synthetic Data—A Framework to Evaluate the Quality of Differentially Private Synthetic Data. *arXiv Preprint arXiv:2004.07740.*

Azizi, S., Kornblith, S., Saharia, C., Norouzi, M., & Fleet, D. J. (2023). Synthetic data from diffusion models improves imagenet classification. *arXiv Preprint arXiv:2304.08466.*

Azizi, Z., Zheng, C., Mosquera, L., Pilote, L., & El Emam, K. (2021). Can synthetic data be a proxy for real clinical trial data? A validation study. *BMJ Open*, *11*(4), e043497. DOI: 10.1136/bmjopen-2020-043497 PMID: 33863713

Bao, J., Wang, R., Wang, Y., Sun, A., Li, Y., Mi, F., & Xu, R. (2023, July). A synthetic data generation framework for grounded dialogues. In *Proceedings of the 61st Annual Meeting of the Association for Computational Linguistics (*Volume 1*: Long Papers)* (pp. 10866-10882). DOI: 10.18653/v1/2023.acl-long.608

Barbierato, E., Vedova, M. L. D., Tessera, D., Toti, D., & Vanoli, N. (2022). A methodology for controlling bias and fairness in synthetic data generation. *Applied Sciences (Basel, Switzerland)*, *12*(9), 4619. DOI: 10.3390/app12094619

Bauer, A., Trapp, S., Stenger, M., Leppich, R., Kounev, S., Leznik, M., . . . Foster, I. (2024). Comprehensive exploration of synthetic data generation: A survey. *arXiv Preprint arXiv:2401.02524.*

Baumann, J., Castelnovo, A., Crupi, R., Inverardi, N., & Regoli, D. (2023, June). Bias on demand: a modelling framework that generates synthetic data with bias. In *Proceedings of the 2023 ACM Conference on Fairness, Accountability, and Transparency* (pp. 1002-1013). DOI: 10.1145/3593013.3594058

Behl, H. S., Baydin, A. G., Gal, R., Torr, P. H., & Vineet, V. (2020, August). Autosimulate:(quickly) learning synthetic data generation. In *European Conference on Computer Vision* (pp. 255-271). Cham: Springer International Publishing.

Bellovin, S. M., Dutta, P. K., & Reitinger, N. (2019). Privacy and synthetic datasets. *Stan. Tech. L. Rev., 22*, 1.

Bhattarai, B., Baek, S., Bodur, R., & Kim, T. K. (2020, May). Sampling strategies for GAN synthetic data. In *ICASSP 2020-2020 IEEE International Conference on Acoustics, Speech and Signal Processing (ICASSP)* (pp. 2303-2307). IEEE. DOI: 10.1109/ICASSP40776.2020.9054677

Boikov, A., Payor, V., Savelev, R., & Kolesnikov, A. (2021). Synthetic data generation for steel defect detection and classification using deep learning. *Symmetry, 13*(7), 1176. DOI: 10.3390/sym13071176

Borkman, S., Crespi, A., Dhakad, S., Ganguly, S., Hogins, J., Jhang, Y. C., . . . Yadav, N. (2021). Unity perception: Generate synthetic data for computer vision. *arXiv Preprint arXiv:2107.04259.*

Boutros, F., Struc, V., Fierrez, J., & Damer, N. (2023). Synthetic data for face recognition: Current state and future prospects. *Image and Vision Computing, 135*, 104688. DOI: 10.1016/j.imavis.2023.104688

Burgard, J. P., Kolb, J. P., Merkle, H., & Münnich, R. (2017). Synthetic data for open and reproducible methodological research in social sciences and official statistics. *AStA Wirtschafts-und Sozialstatistisches Archiv, 11*(3-4), 233–244. DOI: 10.1007/s11943-017-0214-8

Carlson, A., Skinner, K. A., Vasudevan, R., & Johnson-Roberson, M. (2018). Modeling camera effects to improve visual learning from synthetic data. In *Proceedings of the European Conference on Computer Vision (ECCV) Workshops* (pp. 0-0).

Carlucci, F. M., Russo, P., & Caputo, B. (2017, May). A deep representation for depth images from synthetic data. In *2017 IEEE International Conference on Robotics and Automation (ICRA)* (pp. 1362-1369). IEEE. DOI: 10.1109/ICRA.2017.7989162

Cascante-Bonilla, P., Shehada, K., Smith, J. S., Doveh, S., Kim, D., Panda, R., & Karlinsky, L. (2023). Going beyond nouns with vision & language models using synthetic data. In *Proceedings of the IEEE/CVF International Conference on Computer Vision* (pp. 20155-20165). DOI: 10.1109/ICCV51070.2023.01844

Chen, J., Chun, D., Patel, M., Chiang, E., & James, J. (2019). The validity of synthetic clinical data: A validation study of a leading synthetic data generator (Synthea) using clinical quality measures. *BMC Medical Informatics and Decision Making, 19*(1), 1–9. DOI: 10.1186/s12911-019-0793-0 PMID: 30871520

Chen, J., & Little, J. J. (2019). Sports camera calibration via synthetic data. In *Proceedings of the IEEE/CVF Conference on Computer Vision and Pattern Recognition Workshops* (pp. 0-0).

Chen, Y., Li, W., Chen, X., & Gool, L. V. (2019). Learning semantic segmentation from synthetic data: A geometrically guided input-output adaptation approach. In *Proceedings of the IEEE/CVF Conference on Computer Vision and Pattern Recognition* (pp. 1841-1850). DOI: 10.1109/CVPR.2019.00194

Dahmen, J., & Cook, D. (2019). SynSys: A synthetic data generation system for healthcare applications. *Sensors (Basel), 19*(5), 1181. DOI: 10.3390/s19051181 PMID: 30857130

Dahmen, T., Trampert, P., Boughorbel, F., Sprenger, J., Klusch, M., Fischer, K., Kübel, C., & Slusallek, P. (2019). Digital reality: A model-based approach to supervised learning from synthetic data. *AI Perspectives, 1*(1), 1–12. DOI: 10.1186/s42467-019-0002-0

Dankar, F. K., & Ibrahim, M. (2021). Fake it till you make it: Guidelines for effective synthetic data generation. *Applied Sciences (Basel, Switzerland), 11*(5), 2158. DOI: 10.3390/app11052158

Dankar, F. K., Ibrahim, M. K., & Ismail, L. (2022). A multi-dimensional evaluation of synthetic data generators. *IEEE Access : Practical Innovations, Open Solutions, 10*, 11147–11158. DOI: 10.1109/ACCESS.2022.3144765

de Melo, C. M., Torralba, A., Guibas, L., DiCarlo, J., Chellappa, R., & Hodgins, J. (2022). Next-generation deep learning based on simulators and synthetic data. *Trends in Cognitive Sciences, 26*(2), 174–187. DOI: 10.1016/j.tics.2021.11.008 PMID: 34955426

Delussu, R., Putzu, L., & Fumera, G. (2024). Synthetic Data for Video Surveillance Applications of Computer Vision: A Review. *International Journal of Computer Vision, 132*(10), 1–37. DOI: 10.1007/s11263-024-02102-x

Dina, A. S., Siddique, A. B., & Manivannan, D. (2022). Effect of balancing data using synthetic data on the performance of machine learning classifiers for intrusion detection in computer networks. *IEEE Access : Practical Innovations, Open Solutions*, *10*, 96731–96747. DOI: 10.1109/ACCESS.2022.3205337

Eigenschink, P., Reutterer, T., Vamosi, S., Vamosi, R., Sun, C., & Kalcher, K. (2023). Deep generative models for synthetic data: A survey. *IEEE Access : Practical Innovations, Open Solutions*, *11*, 47304–47320. DOI: 10.1109/ACCESS.2023.3275134

El Emam, K. (2020). Seven ways to evaluate the utility of synthetic data. *IEEE Security and Privacy*, *18*(4), 56–59. DOI: 10.1109/MSEC.2020.2992821

Endres, M., Mannarapotta Venugopal, A., & Tran, T. S. (2022, August). Synthetic data generation: A comparative study. In *Proceedings of the 26th International Database Engineered Applications Symposium* (pp. 94-102). DOI: 10.1145/3548785.3548793

Eno, J., & Thompson, C. W. (2008). Generating synthetic data to match data mining patterns. *IEEE Internet Computing*, *12*(3), 78–82. DOI: 10.1109/MIC.2008.55

Fabbri, M., Brasó, G., Maugeri, G., Cetintas, O., Gasparini, R., Ošep, A., & Cucchiara, R. (2021). Motsynth: How can synthetic data help pedestrian detection and tracking? In *Proceedings of the IEEE/CVF International Conference on Computer Vision* (pp. 10849-10859). DOI: 10.1109/ICCV48922.2021.01067

Fazel, A., Yang, W., Liu, Y., Barra-Chicote, R., Meng, Y., Maas, R., & Droppo, J. (2021). Synthasr: Unlocking synthetic data for speech recognition. *arXiv Preprint arXiv:2106.07803*. DOI: 10.21437/Interspeech.2021-1882

Figueira, A., & Vaz, B. (2022). Survey on synthetic data generation, evaluation methods and GANs. *Mathematics*, *10*(15), 2733. DOI: 10.3390/math10152733

Fu, S., Tamir, N., Sundaram, S., Chai, L., Zhang, R., Dekel, T., & Isola, P. (2023). Dreamsim: Learning new dimensions of human visual similarity using synthetic data. *arXiv Preprint arXiv:2306.09344*.

Gal, M. S., & Lynskey, O. (2024). Synthetic data: Legal implications of the data-generation revolution. *Iowa Law Review*, *109*, 1087.

Ge, T., Chan, X., Wang, X., Yu, D., Mi, H., & Yu, D. (2024). Scaling synthetic data creation with 1,000,000,000 personas. *arXiv Preprint arXiv:2406.20094*.

Gholami, S., & Omar, M. (2023). Does synthetic data make Large Language Models more efficient? *arXiv Preprint arXiv:2310.07830*.

Giomi, M., Boenisch, F., Wehmeyer, C., & Tasnádi, B. (2022). A unified framework for quantifying privacy risk in synthetic data. *arXiv Preprint arXiv:2211.10459*.

Goetz, J., & Tewari, A. (2020). Federated learning via synthetic data. *arXiv Preprint arXiv:2008.04489.*

Goncalves, A., Ray, P., Soper, B., Stevens, J., Coyle, L., & Sales, A. P. (2020). Generation and evaluation of synthetic patient data. *BMC Medical Research Methodology, 20*(1), 1–40. DOI: 10.1186/s12874-020-00977-1 PMID: 32381039

Gonzales, A., Guruswamy, G., & Smith, S. R. (2023). Synthetic data in health care: A narrative review. *PLOS Digital Health, 2*(1), e0000082. DOI: 10.1371/journal.pdig.0000082 PMID: 36812604

Goyal, M., & Mahmoud, Q. H. (2024). A systematic review of synthetic data generation techniques using generative AI. *Electronics (Basel), 13*(17), 3509. DOI: 10.3390/electronics13173509

Guan, S., Xu, C., Lin, M., & Greene, D. (2024, November). Effective synthetic data and test-time adaptation for OCR correction. In *Proceedings of the 2024 Conference on Empirical Methods in Natural Language Processing* (pp. 15412-15425). DOI: 10.18653/v1/2024.emnlp-main.862

Gupta, A., Bhatt, D., & Pandey, A. (2021). Transitioning from real to synthetic data: Quantifying the bias in model. *arXiv Preprint arXiv:2105.04144.*

Gupta, A., Vedaldi, A., & Zisserman, A. (2016). Synthetic data for text localisation in natural images. In *Proceedings of the IEEE Conference on Computer Vision and Pattern Recognition* (pp. 2315-2324). DOI: 10.1109/CVPR.2016.254

Handa, A., Patraucean, V., Badrinarayanan, V., Stent, S., & Cipolla, R. (2016). Understanding real world indoor scenes with synthetic data. In *Proceedings of the IEEE Conference on Computer Vision and Pattern Recognition* (pp. 4077-4085).

Hao, S., Han, W., Jiang, T., Li, Y., Wu, H., Zhong, C., . . . Tang, H. (2024). Synthetic data in AI: Challenges, applications, and ethical implications. *arXiv Preprint arXiv:2401.01629.*

He, R., Sun, S., Yu, X., Xue, C., Zhang, W., Torr, P., . . . Qi, X. (2022). Is synthetic data from generative models ready for image recognition? *arXiv preprint arXiv:2210.07574.*

Hernandez, M., Epelde, G., Alberdi, A., Cilla, R., & Rankin, D. (2022). Synthetic data generation for tabular health records: A systematic review. *Neurocomputing, 493*, 28–45. DOI: 10.1016/j.neucom.2022.04.053

Hittmeir, M., Ekelhart, A., & Mayer, R. (2019, August). On the utility of synthetic data: An empirical evaluation on machine learning tasks. In *Proceedings of the 14th International Conference on Availability, Reliability and Security* (pp. 1-6). DOI: 10.1145/3339252.3339281

Hu, T. Y., Armandpour, M., Shrivastava, A., Chang, J. H. R., Koppula, H., & Tuzel, O. (2022, May). Synt++: Utilizing imperfect synthetic data to improve speech recognition. In *ICASSP 2022-2022 IEEE International Conference on Acoustics, Speech and Signal Processing (ICASSP)* (pp. 7682-7686). IEEE.

Jacobsen, B. N. (2023). Machine learning and the politics of synthetic data. *Big Data & Society, 10*(1), 20539517221145372. DOI: 10.1177/20539517221145372

Jaderberg, M., Simonyan, K., Vedaldi, A., & Zisserman, A. (2014). Synthetic data and artificial neural networks for natural scene text recognition. *arXiv Preprint arXiv:1406.2227.*

Jaderberg, M., Simonyan, K., Vedaldi, A., & Zisserman, A. (2016). Reading text in the wild with convolutional neural networks. *International Journal of Computer Vision, 116*(1), 1–20. DOI: 10.1007/s11263-015-0823-z

Jaipuria, N., Zhang, X., Bhasin, R., Arafa, M., Chakravarty, P., Shrivastava, S., & Murali, V. N. (2020). Deflating dataset bias using synthetic data augmentation. In *Proceedings of the IEEE/CVF Conference on Computer Vision and Pattern Recognition Workshops* (pp. 772-773). DOI: 10.1109/CVPRW50498.2020.00394

Jensen, R. I. T., Ferwerda, J., Jørgensen, K. S., Jensen, E. R., Borg, M., Krogh, M. P., Jensen, J. B., & Iosifidis, A. (2023). A synthetic data set to benchmark anti-money laundering methods. *Scientific Data, 10*(1), 661. DOI: 10.1038/s41597-023-02569-2 PMID: 37770445

Jordon, J., Szpruch, L., Houssiau, F., Bottarelli, M., Cherubin, G., Maple, C., . . . Weller, A. (2022). Synthetic Data—what, why and how? *arXiv Preprint arXiv:2205.03257.*

Jordon, J., Yoon, J., & Van Der Schaar, M. (2018, September). PATE-GAN: Generating synthetic data with differential privacy guarantees. In *International Conference on Learning Representations.*

Joshi, I., Grimmer, M., Rathgeb, C., Busch, C., Bremond, F., & Dantcheva, A. (2024). Synthetic data in human analysis: A survey. *IEEE Transactions on Pattern Analysis and Machine Intelligence, 46*(7), 4957–4976. DOI: 10.1109/TPAMI.2024.3362821 PMID: 38319772

Kiefer, B., Ott, D., & Zell, A. (2022, August). Leveraging synthetic data in object detection on unmanned aerial vehicles. In *2022 26th International Conference on Pattern Recognition (ICPR)* (pp. 3564-3571). IEEE. DOI: 10.1109/ICPR56361.2022.9956710

Kim, Y. W., Mishra, S., Jin, S., Panda, R., Kuehne, H., Karlinsky, L., & Feris, R. (2022). How transferable are video representations based on synthetic data? *Advances in Neural Information Processing Systems*, *35*, 35710–35723.

Kiran, A., & Kumar, S. S. (2024). A methodology and an empirical analysis to determine the most suitable synthetic data generator. *IEEE Access : Practical Innovations, Open Solutions*, *12*, 12209–12228. DOI: 10.1109/ACCESS.2024.3354277

Kortylewski, A., Egger, B., Schneider, A., Gerig, T., Morel-Forster, A., & Vetter, T. (2019). Analyzing and reducing the damage of dataset bias to face recognition with synthetic data. In *Proceedings of the IEEE/CVF Conference on Computer Vision and Pattern Recognition Workshops* (pp. 0-0). DOI: 10.1109/CVPRW.2019.00279

Kortylewski, A., Schneider, A., Gerig, T., Egger, B., Morel-Forster, A., & Vetter, T. (2018). Training deep face recognition systems with synthetic data. *arXiv preprint arXiv:1802.05891.*

Krishnan, P., & Jawahar, C. V. (2016). Generating synthetic data for text recognition. *arXiv Preprint arXiv:1608.04224.*

Le, T. A., Baydin, A. G., Zinkov, R., & Wood, F. (2017, May). Using synthetic data to train neural networks is model-based reasoning. In *2017 International Joint Conference on Neural Networks (IJCNN)* (pp. 3514-3521). IEEE. DOI: 10.1109/IJCNN.2017.7966298

Li, Z., Zhu, H., Lu, Z., & Yin, M. (2023). Synthetic data generation with large language models for text classification: Potential and limitations. *arXiv Preprint arXiv:2310.07849*. DOI: 10.18653/v1/2023.emnlp-main.647

Libes, D., Lechevalier, D., & Jain, S. (2017, December). Issues in synthetic data generation for advanced manufacturing. In *2017 IEEE International Conference on Big Data (Big Data)* (pp. 1746-1754). IEEE. DOI: 10.1109/BigData.2017.8258117

Lin, S., Wang, K., Zeng, X., & Zhao, R. (2023). Explore the power of synthetic data on few-shot object detection. In *Proceedings of the IEEE/CVF Conference on Computer Vision and Pattern Recognition* (pp. 638-647). DOI: 10.1109/CVPRW59228.2023.00071

Liu, Q., Deho, O., Vadiee, F., Khalil, M., Joksimovic, S., & Siemens, G. (2025). Can synthetic data be fair and private? A comparative study of synthetic data generation and fairness algorithms. *arXiv Preprint arXiv:2501.01785.* DOI: 10.1145/3706468.3706546

Liu, R., Wei, J., Liu, F., Si, C., Zhang, Y., Rao, J., . . . Dai, A. M. (2024). Best practices and lessons learned on synthetic data for language models. *arXiv Preprint arXiv:2404.07503.*

Liu, T., Tang, J., Vietri, G., & Wu, S. (2023, July). Generating private synthetic data with genetic algorithms. In *International Conference on Machine Learning* (pp. 22009-22027). PMLR.

Liu, T., Vietri, G., & Wu, S. Z. (2021). Iterative methods for private synthetic data: Unifying framework and new methods. *Advances in Neural Information Processing Systems*, *34*, 690–702.

Liu, Y., Wang, Z., Zhou, X., & Zheng, L. (2023). A study of using synthetic data for effective association knowledge learning. *Machine Intelligence Research*, *20*(2), 194–206. DOI: 10.1007/s11633-022-1380-x

Liu, Z., Liang, H., Huang, X., Xiong, W., Yu, Q., Sun, L., . . . Zhang, W. (2024). Synthvlm: High-efficiency and high-quality synthetic data for vision language models. *arXiv preprint arXiv:2407.20756.*

Long, L., Wang, R., Xiao, R., Zhao, J., Ding, X., Chen, G., & Wang, H. (2024). On LLMs-driven synthetic data generation, curation, and evaluation: A survey. *arXiv Preprint arXiv:2406.15126*. DOI: 10.18653/v1/2024.findings-acl.658

López, C. A. F. (2022, December). On the legal nature of synthetic data. In *NeurIPS 2022 Workshop on Synthetic Data for Empowering ML Research.*

Lu, Y., Shen, M., Wang, H., Wang, X., van Rechem, C., Fu, T., & Wei, W. (2023). Machine learning for synthetic data generation: A review. *arXiv Preprint arXiv:2302.04062*. Kollias, D. (2022, October). Abaw: Learning from synthetic data & multi-task learning challenges. In *European Conference on Computer Vision* (pp. 157-172). Cham: Springer Nature Switzerland.

Maheshwari, G., Ivanov, D., & Haddad, K. E. (2024). Efficacy of synthetic data as a benchmark. *arXiv preprint arXiv:2409.11968.*

Majeed, A. (2023). Attribute-centric and synthetic data based privacy preserving methods: A systematic review. *Journal of Cybersecurity and Privacy*, *3*(3), 638–661. DOI: 10.3390/jcp3030030

Mannino, M., & Abouzied, A. (2019, October). Is this real? Generating synthetic data that looks real. In *Proceedings of the 32nd Annual ACM Symposium on User Interface Software and Technology* (pp. 549-561). DOI: 10.1145/3332165.3347866

Martín-Albo, D., Plamondon, R., & Vidal, E. (2014, September). Training of on-line handwriting text recognizers with synthetic text generated using the kinematic theory of rapid human movements. In *2014 14th International Conference on Frontiers in Handwriting Recognition* (pp. 543-548). IEEE. DOI: 10.1109/ICFHR.2014.97

Marwala, T., Fournier-Tombs, E., & Stinckwich, S. (2023). The use of synthetic data to train ai models: Opportunities and risks for sustainable development. *arXiv Preprint arXiv:2309.00652*.

McLachlan, S. (2017). *Realism in synthetic data generation.* Massey University.

Melzi, P., Tolosana, R., Vera-Rodriguez, R., Kim, M., Rathgeb, C., Liu, X., DeAndres-Tame, I., Morales, A., Fierrez, J., Ortega-Garcia, J., Zhao, W., Zhu, X., Yan, Z., Zhang, X.-Y., Wu, J., Lei, Z., Tripathi, S., Kothari, M., Zama, M. H., & Marras, M. (2024). FRCSyn-onGoing: Benchmarking and comprehensive evaluation of real and synthetic data to improve face recognition systems. *Information Fusion, 107,* 102322. DOI: 10.1016/j.inffus.2024.102322

Meyer, D., & Nagler, T. (2021). Synthia: Multidimensional synthetic data generation in Python. *Journal of Open Source Software*, *6*(65), 2863. DOI: 10.21105/joss.02863

Miletic, M., & Sariyar, M. (2024). Challenges of using synthetic data generation methods for tabular microdata. *Applied Sciences (Basel, Switzerland)*, *14*(14), 5975. DOI: 10.3390/app14145975

Mishra, S., Panda, R., Phoo, C. P., Chen, C. F. R., Karlinsky, L., Saenko, K., & Feris, R. S. (2022). Task2sim: Towards effective pre-training and transfer from synthetic data. In *Proceedings of the IEEE/CVF Conference on Computer Vision and Pattern Recognition* (pp. 9194-9204). DOI: 10.1109/CVPR52688.2022.00898

Montanez, A. (2018). *SDV: an open source library for synthetic data generation* (Doctoral dissertation, Massachusetts Institute of Technology).

Nikolenko, S. I. (2021). *Synthetic data for deep learning* (Vol. 174). Springer Nature. DOI: 10.1007/978-3-030-75178-4

Nowok, B., Raab, G. M., & Dibben, C. (2016). synthpop: Bespoke creation of synthetic data in R. *Journal of Statistical Software*, *74*(11), 1–26. DOI: 10.18637/jss.v074.i11

Offenhuber, D. (2024). Shapes and frictions of synthetic data. *Big Data & Society*, *11*(2), 20539517241249390. DOI: 10.1177/20539517241249390

Patel, P. (2024). Synthetic data. *Business Information Review*, *41*(2), 48–52. DOI: 10.1177/02663821241231101

Pei, Y., & Zaïane, O. (2006). A synthetic data generator for clustering and outlier analysis.

Peng, Y. H., Huq, F., Jiang, Y., Wu, J., Li, X. Y., Bigham, J. P., & Pavel, A. (2025). Dreamstruct: Understanding slides and user interfaces via synthetic data generation. In *European Conference on Computer Vision* (pp. 466-485). Springer, Cham. DOI: 10.1007/978-3-031-72691-0_26

Pérez-Porras, F. J., Triviño-Tarradas, P., Cima-Rodríguez, C., Meroño-de-Larriva, J. E., García-Ferrer, A., & Mesas-Carrascosa, F. J. (2021). Machine learning methods and synthetic data generation to predict large wildfires. *Sensors (Basel)*, *21*(11), 3694. DOI: 10.3390/s21113694 PMID: 34073312

Pezoulas, V. C., Zaridis, D. I., Mylona, E., Androutsos, C., Apostolidis, K., Tachos, N. S., & Fotiadis, D. I. (2024). Synthetic data generation methods in healthcare: A review on open-source tools and methods. *Computational and Structural Biotechnology Journal*, *23*, 2892–2910. DOI: 10.1016/j.csbj.2024.07.005 PMID: 39108677

Planche, B., Wu, Z., Ma, K., Sun, S., Kluckner, S., Lehmann, O., & Ernst, J. (2017, October). Depthsynth: Real-time realistic synthetic data generation from cad models for 2.5 d recognition. In *2017 International Conference on 3d Vision (3DV)* (pp. 1-10). IEEE. DOI: 10.1109/3DV.2017.00011

Potluru, V. K., Borrajo, D., Coletta, A., Dalmasso, N., El-Laham, Y., Fons, E., . . . Balch, T. (2023). Synthetic data applications in finance. *arXiv Preprint arXiv:2401.00081*.

Prakash, A., Boochoon, S., Brophy, M., Acuna, D., Cameracci, E., State, G., & Birchfield, S. (2019, May). Structured domain randomization: Bridging the reality gap by context-aware synthetic data. In *2019 International Conference on Robotics and Automation (ICRA)* (pp. 7249-7255). IEEE. DOI: 10.1109/ICRA.2019.8794443

Pratapa, A., Bhat, G., Choudhury, M., Sitaram, S., Dandapat, S., & Bali, K. (2018, July). Language modeling for code-mixing: The role of linguistic theory based synthetic data. In *Proceedings of the 56th Annual Meeting of the Association for Computational Linguistics (*Volume 1*: Long Papers)* (pp. 1543-1553). DOI: 10.18653/v1/P18-1143

Puri, R., Spring, R., Patwary, M., Shoeybi, M., & Catanzaro, B. (2020). Training question answering models from synthetic data. *arXiv Preprint arXiv:2002.09599.* DOI: 10.18653/v1/2020.emnlp-main.468

Qian, Z., Callender, T., Cebere, B., Janes, S. M., Navani, N., & van der Schaar, M. (2024). Synthetic data for privacy-preserving clinical risk prediction. *Scientific Reports, 14*(1), 25676. DOI: 10.1038/s41598-024-72894-y PMID: 39463411

Qian, Z., Cebere, B. C., & van der Schaar, M. (2023). Synthcity: Facilitating innovative use cases of synthetic data in different data modalities. *arXiv Preprint arXiv:2301.07573.*

Qian, Z., Davis, R., & van der Schaar, M. (2024). Synthcity: A benchmark framework for diverse use cases of tabular synthetic data. *Advances in Neural Information Processing Systems*, •••, 36.

Qiu, H., Yu, B., Gong, D., Li, Z., Liu, W., & Tao, D. (2021). Synface: Face recognition with synthetic data. In *Proceedings of the IEEE/CVF International Conference on Computer Vision* (pp. 10880-10890).

Rajotte, J. F., Bergen, R., Buckeridge, D. L., El Emam, K., Ng, R., & Strome, E. (2022). Synthetic data as an enabler for machine learning applications in medicine. *iScience, 25*(11), 105331. DOI: 10.1016/j.isci.2022.105331 PMID: 36325058

Reiter, J. P. (2002). Satisfying disclosure restrictions with synthetic data sets. *Journal of Official Statistics-Stockholm, 18*(4), 531–544.

Reiter, J. P. (2023). Synthetic data: A look back and a look forward. *Transactions on Data Privacy, 16*(1), 15–24.

Richardson, E., Sela, M., & Kimmel, R. (2016, October). 3D face reconstruction by learning from synthetic data. In *2016 Fourth International Conference on 3D Vision (3DV)* (pp. 460-469). IEEE. DOI: 10.1109/3DV.2016.56

Rosenblatt, L., Liu, X., Pouyanfar, S., de Leon, E., Desai, A., & Allen, J. (2020). Differentially private synthetic data: Applied evaluations and enhancements. *arXiv Preprint arXiv:2011.05537.*

Sakaridis, C., Dai, D., & Van Gool, L. (2018). Semantic foggy scene understanding with synthetic data. *International Journal of Computer Vision, 126*(9), 973–992. DOI: 10.1007/s11263-018-1072-8

Sankaranarayanan, S., Balaji, Y., Jain, A., Lim, S. N., & Chellappa, R. (2018). Learning from synthetic data: Addressing domain shift for semantic segmentation. In *Proceedings of the IEEE Conference on Computer Vision and Pattern Recognition* (pp. 3752-3761). DOI: 10.1109/CVPR.2018.00395

Schieber, H., Demir, K. C., Kleinbeck, C., Yang, S. H., & Roth, D. (2024). Indoor synthetic data generation: A systematic review. *Computer Vision and Image Understanding, 240*, 103907. DOI: 10.1016/j.cviu.2023.103907

Seib, V., Lange, B., & Wirtz, S. (2020). Mixing real and synthetic data to enhance neural network training—A review of current approaches. *arXiv Preprint arXiv:2007.08781.*

Shahreza, H. O., Ecabert, C., George, A., Unnervik, A., Marcel, S., Di Domenico, N., . . . Fierrez, J. (2024, May). SDFR: Synthetic data for face recognition competition. In *2024 IEEE 18th International Conference on Automatic Face and Gesture Recognition (FG)* (pp. 1-9). IEEE.

Shahul Hameed, M. A., Qureshi, A. M., & Kaushik, A. (2024). Bias mitigation via synthetic data generation: A review. *Electronics (Basel), 13*(19), 3909. DOI: 10.3390/electronics13193909

Shermeyer, J., Hossler, T., Van Etten, A., Hogan, D., Lewis, R., & Kim, D. (2021). Rareplanes: Synthetic data takes flight. In *Proceedings of the IEEE/CVF Winter Conference on Applications of Computer Vision* (pp. 207-217).

Snoke, J., Raab, G. M., Nowok, B., Dibben, C., & Slavkovic, A. (2018). General and specific utility measures for synthetic data. *Journal of the Royal Statistical Society. Series A, (Statistics in Society), 181*(3), 663–688. DOI: 10.1111/rssa.12358

Soltana, G., Sabetzadeh, M., & Briand, L. C. (2017, October). Synthetic data generation for statistical testing. In *2017 32nd IEEE/ACM International Conference on Automated Software Engineering (ASE)* (pp. 872-882). IEEE. DOI: 10.1109/ASE.2017.8115698

Stadler, T., Oprisanu, B., & Troncoso, C. (2020). Synthetic data-A privacy mirage. *arXiv Preprint arXiv:2011.07018.*

Stadler, T., Oprisanu, B., & Troncoso, C. (2022). Synthetic data–anonymisation groundhog day. In *31st USENIX Security Symposium (USENIX Security 22)* (pp. 1451-1468).

Sundarreson, P., & Kumarapathirage, S. (2024). SentiGEN: Synthetic Data Generator for sentiment analysis. *Journal of Computing Theories and Applications, 1*(4), 461–477. DOI: 10.62411/jcta.10480

Tao, C., Fan, X., & Yang, Y. (2024). Harnessing LLMs for API interactions: A framework for classification and synthetic data generation. *arXiv Preprint arXiv:2409.11703*. DOI: 10.1109/CAIT64506.2024.10962957

Taub, J., & Elliot, M. (2019). *The synthetic data challenge*. Joint UNECE/Eurostat Work Session on Statistical Data Confidentiality.

Taub, J., Elliot, M., & Sakshaug, J. W. (2020). The impact of synthetic data generation on data utility with application to the 1991 UK samples of anonymised records. *Transactions on Data Privacy, 13*(1), 1–23.

Tian, Y., Fan, L., Isola, P., Chang, H., & Krishnan, D. (2024). Stablerep: Synthetic images from text-to-image models make strong visual representation learners. *Advances in Neural Information Processing Systems*, •••, 36.

Tiwald, P., Ebert, A., & Soukup, D. T. (2021). Representative & fair synthetic data. *arXiv Preprint arXiv:2104.03007.*

Torkzadehmahani, R., Kairouz, P., & Paten, B. (2019). Dp-cgan: Differentially private synthetic data and label generation. In *Proceedings of the IEEE/CVF Conference on Computer Vision and Pattern Recognition Workshops* (pp. 0-0). DOI: 10.1109/CVPRW.2019.00018

Tremblay, J., Prakash, A., Acuna, D., Brophy, M., Jampani, V., Anil, C., & Birchfield, S. (2018). Training deep networks with synthetic data: Bridging the reality gap by domain randomization. In *Proceedings of the IEEE Conference on Computer Vision and Pattern Recognition Workshops* (pp. 969-977). DOI: 10.1109/CVPRW.2018.00143

Tripathi, S., Chandra, S., Agrawal, A., Tyagi, A., Rehg, J. M., & Chari, V. (2019). Learning to generate synthetic data via compositing. In *Proceedings of the IEEE/CVF Conference on Computer Vision and Pattern Recognition* (pp. 461-470). DOI: 10.1109/CVPR.2019.00055

Van Breugel, B., Kyono, T., Berrevoets, J., & Van der Schaar, M. (2021). Decaf: Generating fair synthetic data using causally-aware generative networks. *Advances in Neural Information Processing Systems, 34*, 22221–22233.

Van Breugel, B., Qian, Z., & Van Der Schaar, M. (2023, July). Synthetic data, real errors: how (not) to publish and use synthetic data. In *International Conference on Machine Learning* (pp. 34793-34808). PMLR.

Van Breugel, B., & van der Schaar, M. (2023). Beyond privacy: Navigating the opportunities and challenges of synthetic data. *arXiv Preprint arXiv:2304.03722.*

Varga, T., & Bunke, H. (2003, August). Generation of synthetic training data for an HMM-based handwriting recognition system. In *Seventh International Conference on Document Analysis and Recognition, 2003. Proceedings.* (pp. 618-622). IEEE. DOI: 10.1109/ICDAR.2003.1227736

Wang, Q., Gao, J., Lin, W., & Yuan, Y. (2019). Learning from synthetic data for crowd counting in the wild. In *Proceedings of the IEEE/CVF Conference on Computer Vision and Pattern Recognition* (pp. 8198-8207). DOI: 10.1109/CVPR.2019.00839

Wang, Q., Gao, J., Lin, W., & Yuan, Y. (2021). Pixel-wise crowd understanding via synthetic data. *International Journal of Computer Vision, 129*(1), 225–245. DOI: 10.1007/s11263-020-01365-4

Wang, X., Xie, L., Dong, C., & Shan, Y. (2021). Real-ESRGAN: Training real-world blind super-resolution with pure synthetic data. In *Proceedings of the IEEE/CVF International Conference on Computer Vision* (pp. 1905-1914). DOI: 10.1109/ICCVW54120.2021.00217

Wei, J., Huang, D., Lu, Y., Zhou, D., & Le, Q. V. (2023). Simple synthetic data reduces sycophancy in large language models. *arXiv Preprint arXiv:2308.03958.*

Whitney, C. D., & Norman, J. (2024, June). Real Risks of Fake Data: Synthetic Data, Diversity-Washing and Consent Circumvention. In *The 2024 ACM Conference on Fairness, Accountability, and Transparency* (pp. 1733-1744).

Wiehn, T. (2024). Synthetic data: From data scarcity to data pollution. *Surveillance & Society, 22*(4), 472–476. DOI: 10.24908/ss.v22i4.18327

Wood, E., Baltrušaitis, T., Hewitt, C., Dziadzio, S., Cashman, T. J., & Shotton, J. (2021). Fake it till you make it: face analysis in the wild using synthetic data alone. In *Proceedings of the IEEE/CVF International Conference on Computer Vision* (pp. 3681-3691). DOI: 10.1109/ICCV48922.2021.00366

Wu, Y., Wang, Z., Zeng, D., Shi, Y., & Hu, J. (2023, June). Synthetic data can also teach: Synthesizing effective data for unsupervised visual representation learning. *Proceedings of the AAAI Conference on Artificial Intelligence, 37*(3), 2866–2874. DOI: 10.1609/aaai.v37i3.25388

Wu, Z., Wang, L., Wang, W., Shi, T., Chen, C., Hao, A., & Li, S. (2022, October). Synthetic data supervised salient object detection. In *Proceedings of the 30th ACM International Conference on Multimedia* (pp. 5557-5565). DOI: 10.1145/3503161.3547930

Xu, J., Ruan, Y., Bi, W., Huang, G., Shi, S., Chen, L., & Liu, L. (2023). On synthetic data for back translation. *arXiv Preprint arXiv:2310.13675.*

Yang, D., Xu, Z., Mo, W., Chen, Q., Huang, S., & Liu, Y. (2024). 3D vision and language pretraining with large-scale synthetic data. *arXiv preprint arXiv:2407.06084.*

Ye-Bin, M., Hyeon-Woo, N., Choi, W., Kim, N., Kwak, S., & Oh, T. H. (2023). Exploiting synthetic data for data imbalance problems: Baselines from a data perspective. *arXiv Preprint arXiv:2308.00994.*

Yim, M., Kim, Y., Cho, H. C., & Park, S. (2021, September). Synthtiger: Synthetic text image generator towards better text recognition models. In *International Conference on Document Analysis and Recognition* (pp. 109-124). Cham: Springer International Publishing. DOI: 10.1007/978-3-030-86337-1_8

Young, J., Graham, P., & Penny, R. (2009). Using Bayesian networks to create synthetic data. *Journal of Official Statistics, 25*(4), 549–567.

Zhang, H., Grimmer, M., Ramachandra, R., Raja, K., & Busch, C. (2021, May). On the applicability of synthetic data for face recognition. In *2021 IEEE International Workshop on Biometrics and Forensics (IWBF)* (pp. 1-6). IEEE. DOI: 10.1109/IWBF50991.2021.9465085

Zhang, L., Gonzalez-Garcia, A., Van De Weijer, J., Danelljan, M., & Khan, F. S. (2018). Synthetic data generation for end-to-end thermal infrared tracking. *IEEE Transactions on Image Processing, 28*(4), 1837–1850. DOI: 10.1109/TIP.2018.2879249 PMID: 30403630

Zhang, Y., Wu, J., Li, W., Li, B., Ma, Z., Liu, Z., & Li, C. (2024). Video instruction tuning with synthetic data. *arXiv Preprint arXiv:2410.02713.*

KEY TERMS AND DEFINITIONS

General Utility: A feature of synthetic data in which its distribution is comparable with the original data from which it was trained

Synthesis: The making of synthetic data

User-prompted Synthetic Data: Data that is generated sing systems that take text and / or imagery as prompt input and output synthetic data

Chapter 13

Human Consumption of Digital Synthetic Outputs: Cross Checking Real or Fake Information

Deepanshu Pachauri
https://orcid.org/0009-0006-7010-7059
Sharda University, India

Saurabh Chandra
https://orcid.org/0000-0003-4172-9968
Bennett University, Greater Noida, India

Arunima Shastri
https://orcid.org/0000-0002-6260-1253
Christ University, Bangalore, India

Prangya Parimita Behera
KIIT University, Bhubaneswar, India

Bhupinder Singh
https://orcid.org/0009-0006-4779-2553
Sharda University, India

ABSTRACT

AI-Generated content, deepfakes and synthetic media, people are increasingly wondering how they can discern the difference between genuine and fake. People may become confused, change their minds, and make unwise decisions that put their safety at danger when they see false information and manipulated pictures and videos, such AI-generated deepfakes. AI detection techniques, and teaching people

DOI: 10.4018/979-8-3373-2277-3.ch013

how to read and write have all helped with these difficulties. Blockchain technology analyzes the digital history of anything to make sure it is authentic. Two significant steps toward making things more open are regulatory frameworks and ethical AI practices. Users should be vigilant and examine a lot of trustworthy sources, including reverse image search or deepfake detectors. The influence of incorrect information will be lessened if individuals are more informed and encouraged to utilize internet resources properly. To use another metaphor, let's make sure that people know what they're talking about when we speak about synthetic things.

1. INTRODUCTION

In the digital age, the mode of consuming information is dominated by fast-paced social media, 24-hour news cycle, and tailored content feeds . In that kind of climate, the rate of information spread is faster than the rate of verification. Such outputs, which are produced using advanced machine learning models like GANs, LLMs, and state-of-the-art image synthesis techniques, are of such high visual quality that even experts in a given domain might find it too difficult to distinguish between reality and fantasy. One big problem has emerged from this, however: how can we as individuals and communities tell the difference between what is real and what is manufactured? And the solution to it well not only would require grasping the technologies and algorithms used to produce these synthetic media but rather actually a better understanding of how the human brain works in contact with these media, and a good system of cross-checking and verification (Ghiurau & Popescu, 2024).

Digital synthetic media are harming and threatening many parts of society, such as politics, journalism, law, public health, entertainment, and the spaces in between. For example, deepfakes of politicians' utterances have been used to propagate misleading information about public health and hurt their reputation for being wicked (usage in crimes). So, policy has to find a method to discourage bad applications of synthetic materials while promoting positive ones. These assaults not only make people less trusting of institutions, but they also threaten democracy and social cohesiveness. There are some bad things about synthetic media, but if you utilize it well, it may be incredibly useful. For example, they may copy medical instruction cassettes, give classic tales a new twist, or make things clearer with AI voiceovers or translations.

1.1 Importance of Identifying Authenticity in the Digital Age

The lot of new technologies will come out soon that will easily outperform machine learning algorithms that are trained to discover particular patterns or flaws in fake media. These manual verification procedures take a long time, can't manage the vast volume of material that comes in every day, and need a lot of kangaroo expertise. I'd want to get everyone to help. We need models that integrate AUT tools with human supervision and public initiatives to assist people learn more about technology in order to narrow these gaps. People need to learn how to examine digital information and find bogus or modified information in order to make the information ecosystem stronger.

Synthetic media also has big effects on the law, ethics, and policy. There aren't many regulations around generating and distributing AI-generated material in most regions. How to preserve the rights of persons whose information has been used without their permission or who have been impersonated and can't get it back. To safeguard freedom of speech against digital disinformation and maintain the public information space devoid of false news, politicians, technologists, and civil society must collaborate to establish a robust legal framework. The Importance of Algorithmic Accountability and the Role of Digital Watermark or Blockchain-Enabled Provenance in Identifying Information Sources.

Fact-checkers are more vital than ever today that you may operate for yourself as a freelancer or as a carrier. Fact-checking employs a multitude of different tools to obtain information from individuals and networks. This may be done by either AI or individuals who review facts by hand. Still, whether or not they function relies on how many people use them, how reliable they have gotten over time, and how fast they answer a query. We need to work together as a globe to find a means to cope with the reality that false news may travel from one internet service provider to many nations. This is especially important now that contemporary science and technology are being used to organize and carry out falsity campaigns throughout the world. How can we know what's genuine when individuals make phony digital outputs? What do we see? What can we trust? The need to distinguish between fact and fiction has transitioned from a marginal issue to the foremost difficulty facing public discourse, democratic government, and human decision-making. This chapter aims to clarify the interplay of human cognitive science, synthetic media technologies, detection mechanisms, and policy responses, striving to provide innovative perspectives on how societies might effectively traverse the changing digital world without confusion (Joshi et al. 2024). This book presents readers ideas, tools, and policy proposals that they may use to help keep the truth alive in a time when false news is everywhere.

2. SYNTHETIC MEDIA AND DIGITAL SYNTHETIC OUTPUTS: DEEPFAKES, AI-GENERATED TEXT, AND IMAGES

They are deepfakes, AI-written text, synthetic images, cloned voices, and virtual avatars. Synthetic media differ from human-made media in that they can be computer-assisted or generated without human intervention. The best-known types of synthetic media is the deepfake, where deep learning techniques, such as Generative Adversarial Networks (GANs), are used to swap faces in videos, simulate voices, or create entire video situations that are indistinguishable from real ones. The potential abuse of these deepfakes has led to worldwide concerns of how they may be used in political propaganda, extortion, fake news, fraud, and identity theft.

Alongside the manipulation of video, machine-created text has reared it's pesky head with the power to imitate human writing with an alarming degree of fluidity. Language models like GPT (Generative Pre-trained Transformer) can write news articles, essays, social media posts and even academic papers, in many cases almost as well as if they'd been written by a person. While such tools are clearly valuable for automating content, monitoring accessibility and even generating content, they can also be abused to push fake news, biases or spam to digital networks. These texts also make it harder to moderate material and trust the information on the web. AI can now produce visuals that seem like photos or high-resolution pictures of actual people's faces. Most of the time, the person in a digital document isn't genuine; it's all produced by a machine.

With tools like StyleGAN or DALL•E, they've produced photographs that seem too lifelike. These photographs depict human features that have been captured out of context for whatever reason and changed. But they may produce photographs that have never been before, merely to fool you, and they can also make pictures that currently exist for whatever cause, from creative inquiry to fraud. These pictures are fake, but that doesn't imply they're terrible. They're utilized in commerce and entertainment, but if they're not handled carefully, they may be coupled with a stolen identity or photo-shopped evidence. You may also use them to create false photos. It's easy and fast to produce this sort of material, which makes it hard to gain permission, ownership, or proof of authenticity (Farid, 2022).

3. HUMAN COGNITION AND INFORMATION PROCESSING

How we process psychological information is directly tied to how we get it, transmit it, assess it, and react to it. This is because we have been hit with so much information of all kinds that we don't know where to go. Cognitive processes include mental activities including attention, memory, perception, reasoning, understanding, decision-making, and problem-solving (Dugan et al., 2023). This focus on surface-level qualities is another reason why false news is so excellent at getting people's attention: it frequently contains photorealistic pictures or storylines that seem like movies. The most fundamental cognitive ability is memory. It is how we keep, discover, and utilize information. The brain may recall incorrect information as real unless it is corrected straight away or later. This has something to do with what we term the Truth Illusion Effect. It implies that the more we hear something, the more likely we are to believe it, no matter how accurate or untrue it is. Social media sites and other things that make it simple for viruses to propagate keep showing users the same phony or changed material over and over again. So those erroneous memories or beliefs are turning into cognitive distortions that individuals can't get rid of.

Cognitive bias is a big reason why some individuals can't figure out what information means. About digital material. Confirmation bias is when individuals search for evidence that supports what they already think. They don't care whether the facts are real or not if they go against what they think. People who have this bias think they are more trustworthy, so it's easy for them to believe in fake news or fake subpoenas, whether they are true or not. People tend to put greater weight on the initial piece of information they acquire than on subsequent ones. This makes it tougher to modify your thinking after viewing erroneous information.

Emotional valence is also a key aspect of how we take in information. It seems that basic brain systems give higher weight to emotional information because it is more important to us. In the past, when choices had stronger impacts on life or death, this may have helped people stay alive. This problem is becoming worse because people are using social media to propagate inaccurate and misleading information. People who are scared find it hard to not worry about whether something is real or not before they share it and believe it. Research indicates that misinformation, particularly if emotive, can travel almost twice as fast as a fact presented in a refutational discourse, in turn reinforcing its effects on public attitudes and policy.

3.1 How Humans Perceive and Validate Information

The people make extensive use of heuristics which help us make quick decisions. Although heuristics help simplify everyday decision making, they can also introduce bias when faced with ambiguous or uncertain information. For example, the availability heuristic causes people to evaluate the likelihood of occurrences by the ease with which examples can be brought to mind. If deepfakes are becoming commonplace, they may be believed to be more real and more common than they really are. Cognitive burden, that is, the mental energy needed to process information, is an additional hindrance to critical evaluation. In contexts of overwhelming information overload, users may not have the time or cognitive resources to critically evaluate each source of knowledge (Zhou et al., 2023).

3.2 Psychological Biases and Misinformation Susceptibility

Social dynamics and peer pressure can affect how people learn and how well they do their jobs. Because of this, the criteria for figuring out what is true in this network frequently alter from hard Substantive linkages, which are founded on fact, to who you know and whether they are in such networks, i.e. whether they have influence. In digital ecosystems, a lot of likes, shares, and comments may be extremely strong evidence that something is real. It doesn't signify anything since that's how authenticity is found out. This sort of logical error makes sure that the script's conclusions are inaccurate. It demonstrates the concept of groupthink by disseminating incorrect information inside small groups. This is why digital synthesis activities are so similar to how individuals understand and make sense of things. We may understand not just how attention, memory, bias, heuristics, emotion, and social ideas impact our judgments, but also how to minimize these effects via conceptual reconstruction, such cognitive vaccination therapies, AI-based intelligence verification systems, or even simply through education. The only way to make a public that is ready to embrace and not blindly believe everything they see online is to teach people to be more critical thinkers who know their own limits.

4. TOOLS AND TECHNOLOGIES FOR DETECTING FAKE CONTENT

As synthetic media technology becomes better and more widely used, it's more necessary than ever to build better tools and technologies that can discover fake material. Adding even more sorts of digital disinformation to an already big pile of false information would make things much worse. This includes deepfakes, AI-generated text, manipulated photos, and fake documents. These are highly dangerous for people's safety, privacy, and democracy. A system that employs AI (artificial intelligence), ML (machine learning), and encryption has been developed to discover and verify out misleading information online (Christou, 2023).

Machine learning is an important aspect of detection systems, particularly those that look for patterns and outliers in digital information. For instance, deepfake detection algorithms check for things like unusual flashing, facial deformities, inconsistent pixelations, or conflicts between audio and video information in movies using convolutional neural networks (CNNs) and recurrent neural networks (RNNs). These models might detect distinctions that the human eye can't notice, including shadows that don't match or skin textures that aren't the same. Stylometric analysis and semantic fingerprinting are two NLP approaches that have been used in text-based analysis to discover AI-generated material by searching for sentences that don't seem natural, words that are repeated, or alterations from a known author's style. Another significant technique to discover fraudulent material is to look at the metadata.

Analysts will also check the metadata of digital files to see if anything has changed and to search for any strange things that might signal a file will be updated later. This comprises the networks that created the files, the equipment that were used to make them, when they were made, and how they were utilized. It's always simple to erase or update information, but it could assist forensic professionals uncover the solution faster. Many individuals are now looking at utilizing blockchain technology to check the provenance and validity of digital artifacts. Blockchain maintains track of when and how items were produced and altered, and that information can't be changed. Because it is open and spread out, this gives it a distinct fingerprint. That's all there is to it. Truepic and the material Authenticity Initiative (CAI) are two organizations that are working on digital provenance systems that keep track of where material originates from and make sure it is authentic. In general, the media utilize both humans and computer programs to uncover misleading information or materials. There are new applications and browser extensions that enable individuals rate websites as they are browsing the Web. This helps them decide just how trustworthy these sites are. But these technologies still have a hard time keeping up with the expanding number of skilled GPT-3 and related models. So, to make sure that digital information is credible and to develop confidence in online communication, it's vital to keep coming up with new ideas and to change the way the media and public interact to each other (Bird & Lotfi, 2024).

Figure 1. Shows the points on Tools and Technologies for Detecting Fake Content

(Source: Original: Self-Created)

5. CROSS-VERIFICATION METHODS AND FACT-CHECKING FRAMEWORKS: MANUAL AND AUTOMATED FACT-CHECKING TOOLS

In the Digital Age, information and dis-information streams across social media and news websites and even private messaging services, so the tools to verify the data are paramount. Cross-checking techniques and fact-checking systems are crucial for separating what is true from what is false, aiding people, media organisations, politicians and governments in the fight against the damaging consequences of fake news, deep fakes and AI-generated false information. Such frameworks can be generally classified as manual fact-checking and automated fact-checking, both have different methodologies, advantages and limitations.

Manual fact-checking entails human fact-checkers verifying statements based on reliable sources one by one. Historically, this has been a process that entails finding claim, investigating context, checking primary and secondary sources, and publishing conclusions supported by evidence. Famous institutions like Snopes, FactCheck.org and global groups such as the International Fact-Checking Network (IFCN).

Utilizing machine learning models, technologies like ClaimBuster, Full Fact's set of automated tools, and Google's Fact Check Explorer extract factual claims from text, match them to verified databases and assess claims' accuracy. Algorithms within NLP can detect patterns, can find entities, and can cross-reference information to determine the trustworthiness of content. Automated tools can process and scan quickly and widely, and their contribution in terms of efficiency is obvious, but they may have a hard time capturing subtle nuances, sarcasm, cultural and topical shifts or cases where evolving understanding has to be applied to what is being said.

Hybrid, combining manual and automatic, approaches are now believed to be the most successful framework. For example, bots cna be programmed to look for suspicious content and automatically flag it at a large scale, at which point it is reviewed by humans for a more in-depth analysis. This 'stacked' approach marries the computational power of AI with human judgment, to strike a delicate balance between speed and precision (Man & Chahl, 2022). Social media platforms like Facebook and Twitter have also done these things. It can locate, identify, and deal with falsehoods to some degree this manner, but how effectively it works varies a lot from place to place and nation to country throughout the globe. For cross-verification of information to operate, there must be both institutional and technological systems in place. People also need to be taught how to utilize the media and how to verify things at the user level. People may help make society stronger against false news by utilizing a fact-checking tool and evaluating sources and evidence on their own. As communications get more and more mixed up with misleading information and fake media, we need to build and employ tools for robust cross-checking. In a digital era democratic society, if not, the quality of information and people's confidence in each other would go down.

6. LEGAL AND POLICY PERSPECTIVES

Digital synthetic media is expanding faster than ever and coming closer to the genuine thing all over the planet. Because of this, conventional laws and rules are becoming less and less dependent on constraints on the transmission, use, and fabrication of incorrect information. At the same time, technology has advanced so rapidly that AI and VR may now be used to fabricate falsehoods encompassing all interpretations of "making" and "remaining unaffected by reality check." The law is still not very well developed, and in most countries, lawmakers have a hard time finding the right balance between protecting free speech, which is a basic right in any democracy, and keeping people and societies safe from false information, identity theft through content manipulation, and loss of trust. Some of the topics that these arguments are founded on include cross-jurisdictional regulation for synthetic works, the IP status of artificially made works, and new deepfake legislation aimed to restrict harmful uses of synthetic media.

There are varying standards concerning synthetic content in different regions. The DSA and the AI Act are two examples of EU legislation that make it easy to moderate material on both internet platforms and AI-generated content. The DSA would require web platforms to be more honest about how their algorithms function and to do something about damaging and unlawful material like deepfakes and fake news (Pakina et al., 2025). AI Act ranks AI uses by risk and places higher responsibilities on high-risk systems, like those that create deepfakes. And the US approach is more of a patchwork. Though there is no federal law on deepfakes, some states have banned them, like California, Texas, and Virginia, to mention just a few, namely those that were likely to impact elections or spread nonconsensual porn. But the robust protections for free speech under the First Amendment make it difficult to enact sweeping regulation as such laws that go too far in limiting speech would violate constitutionally protected expression. Asian nations, including China and South Korea, are progressing with more stringent rules. China's Deep Synthesis Technology Regulation, which is expected to take effect next year, requires any AI-generated content to be labeled as such and bars the production of synthetic content that could mislead the public or undermine national security. Since legal norms may vary from one country to another, maintaining the uniformity of content enforcement particularly in the global, borderless domain of digital content is quite complex.

6.1 Regulatory Measures across Jurisdictions: Intellectual Property and Deepfake Legislation

The significant legal issue is the ownership of intellectual property in synthetic media. AI systems that produce text, images, video and yes, even music raise questions of authorship, copyright ownership and the nature of derivative work. Human authorship is the general model across all legal systems for copyright law. Accordingly, content generated entirely by machines frequently does not appear to qualify for traditional protection. This creates a lacuna of law that could leave AI-generated works either wholly unprotected or infringing works when it copies human-made works. For example, if a deepfake impersonates a celebrity's appearance and/or voice, such practices could violate personality and publicity rights or constitute unauthorized derivative works. Legal academics and policymakers are currently discussing whether such changes should be made to the IP regime, in recognition of the non-human authorship and ownership, or hosting, of creative outputs and if, and how, to ascribe works generated by AI to the creators, users, or proprietors of AI systems.

The emergence of deepfakes means there are going to be more roles for the government to play in prohibiting some of them. Deepfakes can be leveraged to manipulate political news, take revenge porn to new lows, commit fraud, or impersonate an identity. The legal tools responding to these harms are of varying depth. Other laws target the non-consensual creation and dissemination of sexually explicit deepfakes, or the use of deepfakes to mislead voters or sow violence. India and a number of European nations are considering legislation to mandate clear labeling of artificial content and penalize deceptive practices (Trocin et al., 2021). At an international level, initiated discussions within the United Nations and the Interpol started to search for joint action at a transboundary level against harmful synthetic media. However, enforcement is difficult, because technology moves so fast that it is hard to push new law through in time.

Balancing freedom and truth finding that equilibrium between freedom of speech and information integrity is possibly the hardest challenge. On one side, the right to make, share, and access content which will include parody, satire, and artistic expression is at the core of democratic discourse and creativity. Synthetic media by itself, however, does pose real risks to trust, privacy, national security, and democratic process. Regulation has to be well thought out so that it doesn't suppress or stifle speech that is really genuine. Instead of a prohibition, there should be norms like being honest, accepting responsibility, and committing to probable future damage. Also, industry self-regulation codes, models that involve co-regulation with the government, and public-private partnerships may all support formal regulation by encouraging the ethical growth of AI and responsible content policy (Arora & Arora, 2022).

On the other hand, the making of synthetic media is a controversial subject throughout the globe since it brings up legal and regulatory issues. For example, proper regulation of AI-generated material requires the involvement of several stakeholders and spans different countries. This text is about how it's now required to rewrite state discourse as well. Synthetic media is real, no matter how you look at it. Some owners want to create it, and the technology is now available to do so, but how much would it cost? Regulatory broadcasters don't have much to do until independent producers start creating stuff. At that moment, they will be able to bring areas back to normal so that they can negotiate and reach a deal.

7. FUTURE OF INFORMATION CONSUMPTION: EVOLVING THREATS OF SYNTHETIC MEDIA

The fact that we can easily generate fake media is transforming the way we make and use information in ways that are absolutely fantastic. AI makes it simpler to generate deepfakes, and firms and prominent figures are likely to employ them in their daily marketing. The technology that makes deepfakes easier to propagate is now cheaper and accessible to black-hat coders who aren't the government. These new types of false news are bad for faith in the government, democracy, and even minor things that aren't particularly clever. Like legislation and changing societal views, what used to be a weird sideshow is now a routine part of life. The For example, deep learning specialists have worked hard to make it feasible to generate movies that sound perfectly like your speech but don't have any life in them. People will use these types of deepfakes to lie, cheat, steal, and blackmail others. They can make people lose confidence in any sort of recording, undermine public trust (or maybe simply make people overlook sources that should be trusted), or test tabloid newspapers with reports that are reported in court but turn out to be false. Deepfake algorithms are continually being changed to stay up with new approaches to discover them.

While language models, including ChatGPT, GPT-4 and other high-capacity transformers, are capable of generating content that is grammatically correct, contextually coherent and emotionally compelling. Although these are useful tools for access, education, and focus, they are also being abused to create fake news, propaganda, scam emails, and spam content in bulk. These computer-generated outputs saturate the digital atmosphere so much that they can beguile users and delegitimize real sources of information (Lancaster, 2023). Its growing difficulty for humans to distinguish synthetic text from human-written text raises concerns about disinformation and loss of critical thinking.

AI generated fake images and audio such as DALL·E and voice cloning applications raise distinct threats. bogus photos may also be bogus stories of things that occurred in the past that people have never seen, fake proof, or strategies to promote fraudulent internet identities. For instance, phishing emails and frauds employ fake voices to make the consumer believe they are talking to someone they know. Even if the issue of disinformation is still not addressed, synthetic and artificial content will become even more potent at persuading people as it gets more interactive and immersive and is used in virtual and augmented reality. In the future, there are also a number of things that may go wrong. For instance, synthetic media may generate echo chambers of information where material is tailored to match people's preconceptions and what they already think they know. Taking away the core of true peace In today's society, it's easy for false news to get a foothold. Right now, everyone is going through a crisis that makes them feel weaker, not stronger. People will have to cope with this new concern now and in the future. If this keeps happening, AI algorithms that receive their news from social media sites like Pinterest or Twitter might make politics even more divided and drive it in a specific way. Fact-checking is probably not very good right now since so much misleading information gets spread so rapidly. A united and cross-sectoral approach is needed since synthetic media is becoming more and more of a concern. We need to make sure that the information ecosystem is secure and sound as we bravely go toward a future when all internet stuff is phony. In the future, we need to combine legislation and technology with information literacy and the moral evolution of AI to make sure that what you read online is truthful (Goel et al., 2023).

8. THE ROLE OF EXPLAINABLE AI AND HUMAN-IN-THE-LOOP SYSTEMS

It wasn't apparent how much the climate-controlled greenhouse would impact farming when it was initially employed after World War II. A long time ago, people learned about it, but they still appear to recall that it assisted with real-world uses in the field. These types of big adjustments do generate a lot of issues and make things go more slowly. But maybe the fact that it was discovered by accident was merely its method of coping with recapitulation after being relocated to new places. "Nature's secret habit of copying itself" might signify a lot of different things (Source: Hsu, C. E. 1994). In this time of false news and fake media, Explainable AI (XAI) and Human-in-the-Loop (HITL) technologies strive to make AI-assisted judgments more responsible, clear, and trustworthy. As AI does more and more of the labor of creating, regulating, and identifying modified material, it becomes more and more vital for individuals to know how Iv systems function. Explainable AI-XAI seeks to address this issue by showing people what goes on within the black box of making decisions. XAI wants AI systems to be easy to comprehend for everyone, not only developers and regulators.

XAI is useful in many instances, such as when there is a lot of bogus information that is simple to alter. AI models developed to spot deepfakes or AI-generated text, for example, can frequently act like black boxes, signaling that something has been manipulated but less clearly why. Media factcheckers and legal investigators as well as platform moderators need transparency to verify them: real evidence and good reasoning in support of claims of misinformation or forgery. Explainable AI fills the void by providing human/hand-friendly feature attribution maps, attention layers and saliency visualizations; explanation for why the content is raised as synthetic due to logic. Such explanations also may bolster user confidence in AI generated predictions, particularly in settings where outcomes are high-stakes, e.g., in a court case, in political discourse, or in financial reporting.

In addition to XAI, HITL systems incorporate human judgment into the AI decision pipeline. This blend of algorithm and human-in-the-loop guarantees that algorithmic decisions are not cast in stone but are reviewed, corrected or fine-tuned at the hands of human experts (Prabhu Kavin et al., 2022). The plan is to teach an AI tool how to find bogus signals in the area of content moderation. Someone who knows what they're doing should make the ultimate judgment, particularly when the instances are hard to grasp and touchy. This also implies that there are fewer false positives and false negatives. It may even make the system more fair as twice as many times as AI alone could achieve it. HITL systems are also an excellent technique to improve AI systems after they have been taught. Models might address biases by developing rules that worked if they listened to what humans said. People need to make sure that AI is on the same level as moral and human norms in sectors like public health warnings, investigative journalism, and law enforcement. Together, explainable AI and Human-In-The-Loop solutions are making the AI ecosystem more transparent, responsible, and adaptable. This type of matching could help individuals stay free, grasp things, and make moral choices in a world that is getting more computerized and real (Rahman et al., 2023).

8.1 Real World Case Studies

Among the most high-profile examples is the deepfake video of Ukrainian President Volodymyr Zelenskyy in 2022, made to look like he surrendered to Russia. This AI-created clip spread early on in the Russia-Ukraine war and presented as part of an attempt to demoralize Ukrainian soldiers and population. The Ukrainian government and several news organizations swiftly repudiated the video, and platforms including Facebook and Twitter removed it. This example not only demonstrated the force of synthetic media in shaping geopolitical narratives but also the value of rapid counter-verification by official channels and media fact-checkers to minimise the retention of damage.

An equally illustrative example pertains to AI-fabricated fake news articles at the time of COVID-19 outbreak. Many of the false narratives written with artificial intelligence tools like GPT-2 have been shared widely on social media, including conspiracy theories about the safety of vaccines and the origins of the virus. In a study published Wednesday in Nature, he also detailed how those narratives took hold and fostered doubt, particularly among communities that are not digitally literate. But fact-checking by outfits like PolitiFact and Snopes, working with social media companies, helped flag, correct and remove false content. These decentra lized verification approache s reinforced the need for algorithmic transparency and human oversight.

A third example is false AI-generated images in 2023 French protests. Photographs claiming to depict riots and injured civilians were widely circulated online, but were later exposed as AI-generated. Journalists and digital forensic groups used AI-detection software including reverse image search and Hive Moderation to verify such images, which were inconsistent, depicted oddly shaped hands or unreal looking shadows. And newsrooms are increasingly using these tools to vet manipulated visuals before publication.

The other example that is consumer-facing and made headlines in the past was the fake endorsements from celebrities employing deepfake technology like synthetic videos of Elon Musk endorsing a cryptocurrency scam. These deepfakes resulted in financial frauds in which users were tricked into "investing" in fake crypto platforms. Public outrage and regulatory pressure have prompted platforms to step up detection filters and policies against fake endorsements (Luo, 2024).

9. TOWARDS A MORE RESILIENT INFORMATION ECOSYSTEM

The information ecosystem has to be robust since deepfakes and fake material made by AI are spreading so swiftly. We can't merely keep fakes out; the ecology needs additional aid. You also need to foster a culture of digital literacy, critical thinking, the open university, and fantastic technology in order to construct this sort of system. Digital audio and video have gotten so much like the original that they are now copies. The media's power to modify text, music, and visuals is bad for public conversation, journalism, and democracy because it makes it harder for people to share what they know. That is to say, tenacity has to be present at every stage of the content ecosystem, from producing to sharing to utilizing anything. The most crucial aspect of this transition is to provide programs that teach people how to use media. These programs teach individuals how to assess the truth of their sources, think critically about what they read, and understand how news fits into their wider vision of the world. People who don't trust digital information eat less and use less of it. That implies people are more inclined to trust the news. Let the cat go, then. Second, we need to start giving advice regarding the content. We have to be truthful and maintain our promises. We also need to make sure that news sites and social media sites are just as vigilant about what they post as they are about where it comes from (Blauth et al., 2022).

But you also need to utilize internet technologies to check out fake news. There are technologies like block chain traceability that can indicate where every piece of content originated from, watermarks that reveal where an item began online, and maybe even AI-based detectors that can discover unusual patterns in text, video, or photos. You should also be able to trust the individuals who produced these tools and the tools themselves. That's what moral thoughts are all about. Working with governments, internet companies, schools, and non-profits may help set up common rules for interoperable and collaborative fact-checking, which would raise the standards of truth in the digital age. We need new laws and norms because of the dangers. When you utilize synthetic media, you need to recognize the distinction between free speech, modified digital information, and intellectual property rights. People may abuse their authority, but good politics can deter them from doing so. The Human-in-the-Loop irOR The Human-in-the-Loop irOR The principle of human-AI collaboration may also make things like content panels more accountable, adaptable, and simple to use. Everyone will profit from living in a world where knowledge is always changing. It's getting more and more vital for individuals from diverse backgrounds to communicate to each other since synthetic manufacturing and digital reality go hand in hand. As long as data security continues coming up with innovative methods to keep things secure, we can all travel along that road safely (Shin et al., 2025).

10. CONCLUSION AND RECOMMENDATIONS

The border between what's genuine and what's fake is growing tougher to perceive as digital synthetics and deep-fakes become increasingly realistic on one hand and AI-generated text on the other. These modern technologies and their link to academic engagement are damaging faith in information, national confidence, and even the core of democracy and social fabric. It's easy for misleading information to propagate since our brains are innately prejudiced and may take in too much information.

Some strategic ideas for dealing with the issue and making the digital environment safer include National and global organizations must create specific legal rules to define, control, and punish synthetic abuses while yet allowing everyone to express themselves freely. To ensure that automated detection and content moderation systems are fair, open, and responsible, Explainable AI and Human-in-the-Loop Systems need to get all the money they require from the beginning to the end. The schools and the media should work together to educate people how to utilize technology so they can learn how to and remove information as it occurs.

REFERENCES

Arora, A., & Arora, A. (2022). Generative adversarial networks and synthetic patient data: Current challenges and future perspectives. *Future Healthcare Journal*, *9*(2), 190–193. DOI: 10.7861/fhj.2022-0013 PMID: 35928184

Bird, J. J., & Lotfi, A. (2024). Cifake: Image classification and explainable identification of ai-generated synthetic images. *IEEE Access : Practical Innovations, Open Solutions*, *12*, 15642–15650. DOI: 10.1109/ACCESS.2024.3356122

Blauth, T. F., Gstrein, O. J., & Zwitter, A. (2022). Artificial intelligence crime: An overview of malicious use and abuse of AI. *IEEE Access : Practical Innovations, Open Solutions*, *10*, 77110–77122. DOI: 10.1109/ACCESS.2022.3191790

Christou, P. A. (2023). How to use artificial intelligence (AI) as a resource, methodological and analysis tool in qualitative research? *Qualitative Report*, *28*(7).

Dugan, L., Ippolito, D., Kirubarajan, A., Shi, S., & Callison-Burch, C. (2023, June). Real or fake text?: Investigating human ability to detect boundaries between human-written and machine-generated text. *Proceedings of the AAAI Conference on Artificial Intelligence*, *37*(11), 12763–12771. DOI: 10.1609/aaai.v37i11.26501

Farid, H. (2022). Creating, using, misusing, and detecting deep fakes. *Journal of Online Trust & Safety*, *1*(4). Advance online publication. DOI: 10.54501/jots.v1i4.56

Goel, A., Goel, A. K., & Kumar, A. (2023). The role of artificial neural network and machine learning in utilizing spatial information. *Spatial Information Research*, *31*(3), 275–285. DOI: 10.1007/s41324-022-00494-x

Joshi, I., Grimmer, M., Rathgeb, C., Busch, C., Bremond, F., & Dantcheva, A. (2024). Synthetic data in human analysis: A survey. *IEEE Transactions on Pattern Analysis and Machine Intelligence*, *46*(7), 4957–4976. DOI: 10.1109/TPAMI.2024.3362821 PMID: 38319772

Lancaster, T. (2023). Artificial intelligence, text generation tools and ChatGPT–does digital watermarking offer a solution? *International Journal for Educational Integrity*, *19*(1), 10. DOI: 10.1007/s40979-023-00131-6

Luo, J. (2024). A critical review of GenAI policies in higher education assessment: A call to reconsider the "originality" of students' work. *Assessment & Evaluation in Higher Education*, *49*(5), 651–664. DOI: 10.1080/02602938.2024.2309963

Man, K., & Chahl, J. (2022). A review of synthetic image data and its use in computer vision. *Journal of Imaging*, *8*(11), 310. DOI: 10.3390/jimaging8110310 PMID: 36422059

Pakina, A. K., Sharma, A., & Kejriwal, D. (2025). AI-driven disinformation campaigns: Detecting synthetic propaganda in encrypted messaging via graph neural networks. *International Journal Science and Technology*, *4*(1), 12–24. DOI: 10.56127/ijst.v4i1.1960

Prabhu Kavin, B., Karki, S., Hemalatha, S., Singh, D., Vijayalakshmi, R., Thangamani, M., Haleem, S. L. A., Jose, D., Tirth, V., Kshirsagar, P. R., & Adigo, A. G. (2022). Machine learning-based secure data acquisition for fake accounts detection in future mobile communication networks. *Wireless Communications and Mobile Computing*, *2022*(1), 6356152. DOI: 10.1155/2022/6356152

Rahman, M. M., Terano, H. J., Rahman, M. N., Salamzadeh, A., & Rahaman, M. S. (2023). ChatGPT and academic research: A review and recommendations based on practical examples. *Rahman, M., Terano, HJR, Rahman, N., Salamzadeh, A., Rahaman, S.(2023). ChatGPT and Academic Research: A Review and Recommendations Based on Practical Examples. Journal of Education. Management and Development Studies*, *3*(1), 1–12.

Shin, D., Perlmutter, D. D., Lim, J. S., & Jitkajornwanich, K. (2025). Fact-checking as epistemic infrastructure: From journalistic gatekeeping to platform governance. *Emerging Media*, *3*(2), 183–202. DOI: 10.1177/27523543251344972

Trocin, C., Hovland, I. V., Mikalef, P., & Dremel, C. (2021). How Artificial Intelligence affords digital innovation: A cross-case analysis of Scandinavian companies. *Technological Forecasting and Social Change*, *173*, 121081. DOI: 10.1016/j.techfore.2021.121081

Zhou, J., Zhang, Y., Luo, Q., Parker, A. G., & De Choudhury, M. (2023, April). Synthetic lies: Understanding ai-generated misinformation and evaluating algorithmic and human solutions. In *Proceedings of the 2023 CHI conference on human factors in computing systems* (pp. 1-20). DOI: 10.1145/3544548.3581318

Chapter 14
Navigating the Real and the Synthetic:
Authentication and Validation in Digital Synthetic Outputs

Mohit Sharma
https://orcid.org/0009-0007-2280-8077
Maharshi Dayanand University, Rohtak, India

Nitika Malik
https://orcid.org/0009-0006-3991-9526
Maharshi Dayanand University, Rohtak, India

ABSTRACT

The rise of digital synthetic data and outputs has blurred the lines between reality and artificiality. With the growing sophistication of generative AI, synthetic outputs—ranging from images and video to text and voice—are increasingly integrated into various industries, including education, healthcare, entertainment, and cybersecurity. However, this surge in synthetic content necessitates robust mechanisms for authentication and validation. This chapter examines current methodologies and technologies for distinguishing synthetic outputs from authentic ones. It explores automated validation techniques, human-in-the-loop authentication processes, and hybrid models that leverage machine learning to detect digital forgeries. The discussion also extends to ethical considerations and societal implications, particularly in an era where synthetic deception poses real-world consequences.

DOI: 10.4018/979-8-3373-2277-3.ch014

INTRODUCTION

The rapid evolution of artificial intelligence (AI) and machine learning (ML) has significantly transformed the digital landscape, giving rise to synthetic outputs that closely mimic real-world data. These synthetic outputs, generated through advanced AI models such as Generative Adversarial Networks (GANs) and Transformer-based architectures, are now prevalent in various sectors, including healthcare, education, entertainment, and cybersecurity (Goodfellow et al., 2014). As the boundaries between real and artificial become increasingly indistinct, the need for robust authentication and validation mechanisms becomes paramount.

The Rise of Synthetic Outputs

Synthetic media, encompassing AI-generated images, videos, text, and voice, has seen exponential growth due to improvements in deep learning models. The capability to generate hyper-realistic images and videos has led to a paradigm shift in content creation. For instance, deepfake technology, powered by GANs, can create highly convincing digital replicas of individuals, raising concerns about misinformation, privacy breaches, and identity fraud (Mirsky & Lee, 2021). Similarly, natural language processing (NLP) models, such as OpenAI's GPT-4, can generate human-like text indistinguishable from content produced by human writers (Brown et al., 2020). The implications of synthetic content are vast and multifaceted. On the one hand, these technologies offer significant advantages, such as enhancing educational experiences through AI tutors, generating realistic medical simulations for training, and streamlining content production in entertainment. On the other hand, the misuse of synthetic media poses significant threats, including misinformation campaigns, financial fraud, and digital impersonation. The increasing availability of AI tools has also led to ethical dilemmas, with malicious actors exploiting synthetic content to create deepfake pornography, commit identity theft, and manipulate financial transactions (Chesney & Citron, 2019).

Authentication and Validation: Necessity in a Digital Age

Given the increasing integration of synthetic outputs into society, authentication and validation mechanisms are crucial to maintaining digital trust. Traditional authentication relies on watermarking and metadata analysis, yet these methods are no longer sufficient against sophisticated AI-generated content (Verdoliva, 2020). Modern validation approaches leverage AI-driven detection systems, blockchain for verifiable digital provenance, and human-in-the-loop frameworks that combine algorithmic efficiency with human intuition. AI-based detection models utilize pattern recognition and adversarial learning techniques to identify synthetic outputs. These models analyze inconsistencies in facial features, unnatural linguistic patterns in AI-generated text, and irregularities in audio waveforms to detect synthetic content. However, as generative models continue to improve, detection algorithms must also evolve to keep pace with increasingly convincing synthetic outputs (Agarwal et al., 2020). Furthermore, AI-powered forensic tools can now analyze micro-expressions, audio spectral anomalies, and unnatural image pixel distributions to enhance detection capabilities (Zhang et al., 2022).

ETHICAL CONSIDERATIONS AND SOCIETAL IMPLICATIONS

The ethical concerns surrounding synthetic content extend beyond mere detection. The potential for synthetic deception to manipulate public opinion, undermine democracy, and cause reputational damage underscores the urgency for regulatory frameworks. The European Union's Artificial Intelligence Act and similar legislative efforts aim to address these challenges by setting guidelines for responsible AI deployment (European Commission, 2021). However, the dynamic nature of AI-generated content necessitates continuous collaboration between policymakers, technologists, and ethical researchers. Moreover, questions regarding consent, privacy, and accountability arise when synthetic media is used without individuals' explicit permission. Cases of non-consensual deepfake videos and synthetic voice fraud highlight the need for stringent legal and technical safeguards. Ethical AI frameworks advocate for transparency in AI-generated content by promoting digital watermarking, algorithmic accountability, and user awareness campaigns (Floridi & Cowls, 2019). There is also a pressing need for global cooperation in combating synthetic media threats, as their impact transcends national borders.

Hybrid Approaches to Authentication

To effectively address the challenges posed by synthetic outputs, hybrid authentication models that combine automated detection with human oversight have gained prominence. Such models integrate ML-based forgery detection with human expertise, ensuring higher accuracy in identifying fraudulent content. In cybersecurity, for example, organizations employ AI-driven anomaly detection systems alongside human analysts to mitigate cyber threats (Tolosana et al., 2020). AI tools also help in assessing contextual discrepancies in content, such as inconsistencies in timestamps, unnatural blinking patterns in videos, and unusual phrasing in AI-generated text. Furthermore, blockchain technology has emerged as a promising solution for establishing digital authenticity. By embedding immutable cryptographic signatures into digital assets, blockchain enables verifiable content authentication, reducing the risk of synthetic media manipulation. Decentralized identity verification systems, supported by blockchain, provide additional layers of security, ensuring that digital interactions remain trustworthy (Kumar et al., 2021). Companies like Truepic and Microsoft's Project Origin are pioneering blockchain-based authentication to track and verify digital content integrity.

Regulatory Measures and Global Collaboration

Several governments and organizations have proposed regulatory measures to combat the threats posed by synthetic media. The United States has introduced legislative initiatives such as the Deepfake Report Act, requiring technology firms to disclose the creation and distribution of deepfake media. Similarly, China has mandated that AI-generated media be watermarked and identified as synthetic to prevent misinformation These policies highlight the growing recognition of synthetic media risks and the need for proactive governance. Intergovernmental organizations, such as the United Nations and the European Union, are actively working towards creating international standards for AI-generated content authentication. Cross-border collaborations between tech firms, law enforcement agencies, and research institutions are crucial in addressing synthetic deception. For example, the Deepfake Detection Challenge (DFDC), launched by Facebook, Microsoft, and AWS, has accelerated the development of AI-powered deepfake detection technologies (Dolhansky et al., 2020).

Public Awareness and Digital Literacy

While technological solutions play a significant role in authentication and validation, public awareness and digital literacy are equally important in combating synthetic deception. Misinformation campaigns leveraging deepfake technology have demonstrated the urgent need for media literacy programs to educate the public on recognizing synthetic content. Governments, educational institutions, and social media platforms must collaborate to implement digital literacy initiatives that empower users to critically assess online information (Wardle & Derakhshan, 2017). Fact-checking organizations and independent media outlets are also essential in verifying digital content. Websites like Snopes and FactCheck.org provide real-time analyses of viral content, helping to debunk misinformation. Moreover, AI-assisted fact-checking tools, such as Google's Fact Check Explorer, enhance the efficiency of content verification processes. The intersection of real and synthetic content necessitates robust authentication and validation frameworks to safeguard digital integrity. As AI-generated outputs become more sophisticated, a multidisciplinary approach involving AI-driven detection, blockchain verification, ethical governance, and public awareness is essential. While synthetic media offers remarkable benefits, its responsible deployment is critical to mitigating potential risks. Future research should focus on developing adaptive detection algorithms, refining legal frameworks, and fostering ethical AI practices to navigate the complexities of digital authenticity in the synthetic era.

BACKGROUND OF THE STUDY

The rapid evolution of artificial intelligence (AI) and machine learning (ML) has significantly transformed the digital landscape, giving rise to synthetic outputs that closely mimic real-world data. These synthetic outputs, generated through advanced AI models such as Generative Adversarial Networks (GANs) and Transformer-based architectures, are now prevalent in various sectors, including healthcare, education, entertainment, and cybersecurity (Goodfellow et al., 2014). As the boundaries between real and artificial become increasingly indistinct, the need for robust authentication and validation mechanisms becomes paramount. The emergence of digital synthetic outputs marks a significant milestone in the evolution of artificial intelligence (AI). The ability of AI models to generate highly realistic images, videos, text, and voices has resulted in an unprecedented shift in the way digital content is created, shared, and consumed (Kietzmann et al., 2020). While these advancements present numerous benefits, they also introduce complex challenges related to authenticity, security, and ethical considerations. Consequently, ensuring the authentication and validation of synthetic outputs has become a pressing issue across various industries.

Historical Development of Synthetic Media

The concept of synthetic media dates back to early digital graphics and computer-generated imagery (CGI) in the entertainment industry. However, the recent breakthroughs in deep learning, particularly the development of Generative Adversarial Networks (GANs) by Goodfellow et al. (2014), have significantly enhanced the realism and accessibility of synthetic content. GANs, which operate through the competition between a generator and a discriminator network, have enabled the creation of hyper-realistic images and videos that are nearly indistinguishable from authentic media (Mirsky & Lee, 2021). In parallel, advancements in natural language processing (NLP) have led to the development of sophisticated text generation models, such as OpenAI's GPT-4, which can produce human-like text with remarkable coherence and relevance (Brown et al., 2020). Similarly, AI-generated voice synthesis has reached new heights with models like WaveNet and Tacotron, which can replicate human speech with lifelike intonations and expressions (Shen et al., 2018). These technologies are now widely used in customer service, content creation, and assistive technologies, further integrating synthetic media into everyday life.

Applications of Synthetic Outputs

The integration of synthetic outputs extends across multiple domains, offering both opportunities and challenges:

1. **Healthcare**: Synthetic medical data is used to train AI models without exposing sensitive patient information, ensuring privacy while enhancing diagnostic accuracy (Yoon et al., 2020).
2. **Education**: AI-generated tutors and learning materials provide personalized education, catering to diverse learning needs and improving accessibility (Lu et al., 2021).
3. **Entertainment**: AI-driven tools streamline content creation, enabling deepfake technology for special effects, virtual influencers, and automated scriptwriting (Kaplan et al., 2021).
4. **Cybersecurity**: Synthetic data assists in training AI systems for detecting fraud, phishing attacks, and cybersecurity threats (Agarwal et al., 2020).
5. **Journalism and Social Media**: AI-generated news articles and automated reporting streamline content production, but also raise concerns about misinformation and digital manipulation (Wardle & Derakhshan, 2017).

Despite these benefits, the proliferation of synthetic media has raised significant concerns about the potential misuse of AI-generated content, leading to urgent calls for robust authentication and validation frameworks.

Challenges in Authentication and Validation

The authentication of synthetic outputs is a critical challenge due to the increasing sophistication of AI-generated content. As AI models improve, the differences between real and synthetic media become more difficult to discern. Several key challenges include:

- **Detecting Deepfakes**: Deepfake technology, which allows for the realistic synthesis of human faces and voices, is increasingly used for malicious purposes, including political misinformation and identity fraud (Chesney & Citron, 2019).

- **Authenticating AI-Generated Text**: The ability of NLP models to produce human-like text has led to challenges in distinguishing AI-generated articles, reviews, and online interactions from human-authored content (Jawahar et al., 2022).
- **Verifying Synthetic Voices**: AI-generated voice impersonation poses risks in security-sensitive applications, such as banking and customer verification, where voice authentication is commonly used (Zhang et al., 2022).
- **Establishing Digital Trust**: With the increasing reliance on AI-generated media, establishing trust in digital content requires comprehensive validation mechanisms that go beyond traditional watermarking and metadata analysis (Verdoliva, 2020).

Approaches to Authentication and Validation

To mitigate the risks associated with synthetic outputs, researchers and industry leaders have proposed several authentication and validation techniques:

1. **Machine Learning-Based Detection**: AI-powered detection models analyze inconsistencies in digital media to identify synthetic content. Deep learning techniques, including convolutional neural networks (CNNs), recurrent neural networks (RNNs), and transformer-based models, are commonly used to detect deepfake images, videos, and text (Dolhansky et al., 2020).
2. **Blockchain Technology for Content Verification**: Blockchain-based digital signatures and decentralized verification systems provide immutable proof of content authenticity, ensuring traceability and reducing the risk of synthetic media manipulation (Kumar et al., 2021).
3. **Hybrid Human-AI Validation**: While AI-driven systems offer scalability, human expertise remains essential for contextual verification. Hybrid approaches, combining automated detection with human oversight, enhance the reliability of authentication processes (Tolosana et al., 2020).
4. **Digital Watermarking and Metadata Analysis**: Embedding invisible watermarks in synthetic media can serve as identifiers, enabling tracking and validation of AI-generated content. However, adversarial attacks on watermarking systems highlight the need for more resilient authentication methods (Hao, 2022).

Ethical and Regulatory Considerations

The ethical implications of synthetic media are far-reaching, encompassing issues of privacy, misinformation, and accountability. Several key concerns include:

- **Misinformation and Public Trust**: AI-generated fake news and manipulated content can influence public opinion, disrupt democratic processes, and erode trust in digital information (Floridi & Cowls, 2019).
- **Privacy Violations and Identity Theft**: The unauthorized use of synthetic media for impersonation, including voice cloning and deepfake pornography, poses serious risks to individuals' privacy and security (Chesney & Citron, 2019).
- **Regulatory Responses and Legal Frameworks**: Governments and international bodies are actively working to establish legal frameworks for managing synthetic media. The European Union's Artificial Intelligence Act and China's deepfake regulations are among the notable efforts aimed at addressing AI-generated content challenges (European Commission, 2021).

Public Awareness and Digital Literacy

Addressing the challenges posed by synthetic media requires a holistic approach that includes technological solutions, regulatory measures, and public awareness initiatives. Digital literacy programs play a crucial role in equipping individuals with the skills needed to critically assess online content and identify potential synthetic media threats (Wardle & Derakhshan, 2017). Educational initiatives, fact-checking resources, and AI-assisted content verification tools can empower users to navigate the complexities of digital authenticity. The increasing prevalence of synthetic media necessitates the development of comprehensive authentication and validation mechanisms to safeguard digital integrity. While AI-generated content offers numerous benefits across industries, its potential for misuse underscores the importance of robust verification frameworks. By leveraging AI-powered detection, blockchain-based authentication, regulatory oversight, and public education, society can mitigate the risks associated with synthetic media while harnessing its advantages for innovation and progress.

RESEARCH METHODOLOGY

Research Design

This study employs a qualitative research design based on secondary data analysis to examine the authentication and validation of digital synthetic outputs. The secondary data collection methods are used in the study for effectively analyzing existing literature, frameworks, and technologies related to synthetic media authentication (Johnston et al., 2017). By synthesizing information from various academic sources, industry reports, and regulatory documents, this study provides a comprehensive understanding of current methodologies, challenges, and ethical considerations in distinguishing synthetic content from authentic digital media.

Data Collection Method

The research relies on secondary sources, including peer-reviewed journal articles, conference papers, books, and reports from industry and regulatory bodies. These sources were selected based on their relevance, credibility, and contributions to the discourse on synthetic media authentication. The literature was retrieved from academic databases such as Google Scholar, IEEE Xplore, SpringerLink, and ScienceDirect, ensuring access to high-quality and peer-reviewed materials (Boell & Cecez-Kecmanovic, 2015). In addition, regulatory guidelines from organizations such as the European Commission and cybersecurity agencies were reviewed to understand the evolving legal landscape.

Data Analysis Method

The collected secondary data were analyzed using a thematic approach to identify key trends, methodologies, and technological advancements in synthetic content authentication. Thematic analysis allows for the systematic categorization of information based on recurring patterns and emerging themes (Braun & Clarke, 2006). The key themes explored in this study include:

1. **Automated Detection Techniques** – Machine learning-based methods for identifying synthetic content, including deepfake detection algorithms and AI-driven forensic analysis.
2. **Human-in-the-Loop Authentication** – The role of human oversight in validating AI-generated content and enhancing automated verification systems.
3. **Hybrid Validation Models** – The combination of AI-driven and manual authentication techniques to improve the reliability of synthetic media detection.
4. **Regulatory and Ethical Considerations** – Legal frameworks, policies, and ethical implications of synthetic media authentication.

By structuring the analysis around these themes, the study provides a well-rounded exploration of how digital synthetic outputs can be authenticated and validated.

FINDINGS OF THE STUDY

Detection and Authentication of Synthetic Media

The study reveals that the detection and authentication of synthetic media have become critical challenges in the digital landscape. AI-generated content, particularly deepfakes and synthetic text, has evolved to a level where it can closely mimic human-generated media, raising concerns about misinformation, fraud, and digital security. Various detection techniques have emerged, including forensic analysis, machine learning-based classifiers, and blockchain-enabled verification systems. Machine learning techniques, such as convolutional neural networks (CNNs) and recurrent neural networks (RNNs), have demonstrated effectiveness in identifying synthetic images and videos by detecting inconsistencies in pixel patterns and motion artifacts (Mirsky & Lee, 2021). Additionally, Generative Adversarial Networks (GANs) play a dual role in both creating and detecting deepfakes, where adversarial training is used to enhance detection accuracy (Dolhansky et al., 2020). Forensic analysis techniques focus on analyzing image and video inconsistencies, such as unnatural lighting, distorted facial expressions, and blinking irregularities (Tolosana et al., 2020). Digital watermarking and blockchain-based timestamping have also been explored as verification methods to authenticate content integrity and trace the origin of synthetic media (Kumar et al., 2021). The rise of AI-driven content has necessitated hybrid approaches that combine automated detection with human expertise, ensuring higher accuracy and adaptability to evolving synthetic media generation techniques (Verdoliva, 2020).

Part 2: Ethical, Legal, and Societal Implications of Synthetic Media

The increasing prevalence of synthetic media has led to significant ethical, legal, and societal concerns. Misinformation and public trust are primary issues, as AI-generated fake news, manipulated images, and deepfake videos can influence political opinions, financial markets, and personal reputations (Chesney & Citron, 2019). The ethical implications of synthetic media extend to privacy violations, where unauthorized deepfake generation can be used for identity theft, blackmail, and reputational harm (Zhang et al., 2022). Governments and regulatory bodies have responded by introducing legislation to address synthetic media misuse, such as the European Union's AI Act and China's deepfake regulations (European Commission, 2021). In the United States, proposed laws aim to criminalize malicious deepfake creation and distribution. However, enforcing these laws remains a challenge due to the rapid advancement of AI technology and the difficulty in attributing responsibility for synthetic content generation. Ethical debates also arise regarding the use of synthetic media in entertainment, marketing, and education, where AI-generated personas and influencers blur the lines between real and artificial entities. Media literacy and public awareness campaigns have been suggested as countermeasures to mitigate the impact of synthetic deception and enhance critical thinking skills among digital consumers (Wardle & Derakhshan, 2017). The study highlights the need for a multidisciplinary approach to addressing synthetic media challenges, involving collaboration between technologists, policymakers, and ethicists.

Part 3: Industry-Specific Applications and Future Directions

Healthcare Sector Applications

In the healthcare sector, synthetic media has found significant utility in the form of synthetic medical data. As data privacy regulations become more stringent, synthetic data provides a means to overcome challenges related to data sharing while maintaining patient confidentiality (Yoon et al., 2020). Synthetic medical images, generated using AI algorithms, can be used to train machine learning models for disease diagnosis, medical imaging, and research without exposing real patient data (Yoon et al., 2020). However, the challenge remains in ensuring the accuracy and reliability of these synthetic medical images. Researchers have highlighted the importance of developing robust synthetic datasets that accurately replicate the characteristics of real medical images (Yoon et al., 2020). Despite their potential, these datasets must undergo rigorous validation to ensure they are suitable for training AI models used in real-world clinical settings (Verdoliva, 2020). The study also notes that the use of synthetic media in healthcare raises concerns about the potential for algorithmic bias. If synthetic datasets are not representative of diverse patient populations, AI models could be biased, leading to inaccurate medical diagnoses or treatment recommendations (Yoon et al., 2020). Therefore, creating synthetic datasets that reflect a wide range of demographic factors—such as age, race, gender, and medical history—is essential to ensure fairness and equity in healthcare applications (Verdoliva, 2020).

Education Sector Applications

In the education sector, AI-generated content has been leveraged to develop personalized learning experiences, virtual tutoring systems, and interactive simulations (Lu et al., 2021). The ability to tailor educational content to individual learning styles has the potential to revolutionize the educational landscape (Lu et al., 2021). AI-powered platforms can now create dynamic lesson plans, simulations, and assessments that cater to the unique needs of students (Lu et al., 2021). However, as this technology becomes more prevalent, challenges related to the verification of educational content emerge. The authenticity and accuracy of AI-generated educational materials must be rigorously validated to avoid the spread of misinformation or biased content (Chesney & Citron, 2019). For instance, AI-generated textbooks or instructional videos might inadvertently present skewed information or exclude important perspectives (Chesney & Citron, 2019). The study emphasizes the need for stringent validation processes in the creation of synthetic educational content (Lu et al., 2021). Moreover, educators must be equipped with the skills to discern authentic content from synthetic content and integrate these technologies responsibly into their classrooms (Lu et al., 2021). The potential for AI to create personalized learning pathways also presents ethical concerns (Chesney & Citron, 2019). While adaptive learning technologies can provide individualized support, they also have the potential to reinforce existing stereotypes and biases if the underlying algorithms are not carefully monitored (Lu et al., 2021). This underscores the importance of creating guidelines and standards for the development and use of synthetic educational content, as well as the ongoing need for teacher training to ensure ethical practices (Lu et al., 2021).

Cybersecurity Sector Applications

In cybersecurity, synthetic media has become an essential tool in creating realistic datasets for training fraud detection algorithms and enhancing threat detection systems (Agarwal et al., 2020). The use of synthetic data enables the development of AI models that can detect fraudulent activities, such as identity theft, phishing attacks, and financial fraud, without compromising sensitive data (Agarwal et al., 2020). However, as synthetic media becomes more advanced, cybersecurity professionals face a growing challenge: the increasing sophistication of AI-generated attacks (Agarwal et al., 2020). Deepfake technology, for example, can be used in social engineering attacks to impersonate executives or create fraudulent voice recordings that deceive employees into revealing confidential information (Agarwal et al., 2020). This growing threat requires continuous innovation in authentication techniques to differentiate between legitimate and synthetic communications (Agarwal et al., 2020). The study highlights that traditional methods of cybersecurity, such as password-based authentication or biometric verification, may be insufficient in the face of increasingly sophisticated AI-generated attacks (Agarwal et al., 2020). As a result, hybrid models that combine AI-driven detection with human oversight are seen as the future of cybersecurity (Agarwal et al., 2020). By integrating AI-based deepfake detection tools with real-time human review, cybersecurity systems can more effectively counter the risks posed by synthetic media (Agarwal et al., 2020).

Entertainment and Media Applications

The entertainment and media industries have seen perhaps the most visible use of synthetic media. AI-generated content, including deepfakes, synthetic voices, and AI-generated avatars, is rapidly transforming how content is produced and consumed. AI-generated actors and digital influencers are now common in advertisements, movies, and television shows. These virtual personalities can be created from scratch or can mimic real individuals, providing entertainment companies with new avenues for creative expression. While this opens up new possibilities for storytelling and advertising, it also raises significant ethical and legal concerns. One of the primary concerns is the question of consent, particularly when using deepfake technology to recreate the likeness of an actor or celebrity. The ability to digitally resurrect deceased actors or create realistic avatars of living individuals without their permission is a contentious issue. In some cases, entertainment companies have faced backlash for using synthetic media in ways that exploit the image and likeness of public figures without proper consent. This raises important questions about intellectual property rights and the ethical responsibilities of content creators. The study suggests that the entertainment industry will need to develop new frameworks for consent, transparency, and accountability to ensure that synthetic media is used ethically and responsibly.

Advertising Industry Applications

The advertising industry has also embraced synthetic media to craft personalized marketing experiences (Kietzmann et al., 2020). AI-generated personas, voices, and videos can be used to create highly tailored advertisements that resonate with individual consumers (Kietzmann et al., 2020). By leveraging synthetic data and deep learning models, advertisers can predict consumer preferences and deliver targeted campaigns that speak directly to their interests (Kietzmann et al., 2020). However, this has raised concerns about consumer manipulation (Kietzmann et al., 2020). AI-driven advertisements may exploit personal data to create hyper-personalized content that nudges individuals toward particular behaviors, often without their full awareness or consent (Kietzmann et al., 2020). The study underscores the need for regulations that govern how synthetic media can be used in advertising to protect consumers from manipulation and ensure ethical marketing practices (Kietzmann et al., 2020). Transparent disclosure requirements, such as clear labeling of synthetic content, could help consumers understand when they are engaging with AI-generated material and prevent deceptive advertising practices (Kietzmann et al., 2020).

Looking to the Future

Looking to the future, the study identifies several key trends and directions for the continued evolution of synthetic media. As AI technology advances, the ability to create more sophisticated and convincing synthetic content will only increase (Verdoliva, 2020). This presents both opportunities and challenges. On the one hand, industries can harness the power of synthetic media to innovate and streamline operations (Verdoliva, 2020). On the other hand, the risks associated with synthetic content, such as misinformation and fraud, will continue to grow (Verdoliva, 2020). The development of more robust detection techniques, such as multi-modal verification approaches that combine image, voice, and behavioral analysis, will be crucial in distinguishing authentic from synthetic media (Verdoliva, 2020). The use of blockchain technology for content verification is also expected to play a key role in ensuring the integrity of digital content (Kumar et al., 2021). Additionally, the study points to the importance of interdisciplinary collaboration in addressing the challenges posed by synthetic media. Researchers, policymakers, technologists, and industry leaders must work together to create comprehensive regulatory frameworks, develop ethical guidelines, and ensure that the benefits of synthetic media are realized responsibly (Verdoliva, 2020).

IMPLICATIONS OF THE STUDY

The growing use of synthetic media has far-reaching implications across a wide range of domains. These implications extend beyond just the technological aspects of synthetic media and encompass critical ethical, legal, societal, and practical concerns. The rise of deepfakes, AI-generated content, and synthetic data has raised fundamental questions about the authenticity of digital outputs, the reliability of information, and the potential for exploitation. This section explores the various consequences of the findings presented in the study, focusing on both the positive and negative implications of synthetic media across industries, society, and governance.

1. Ethical Implications

One of the most pressing concerns raised by the proliferation of synthetic media is the ethical challenges associated with its creation, distribution, and consumption. Synthetic media, particularly deepfakes, have been used to manipulate public opinion, spread misinformation, and even commit criminal acts, such as fraud or identity theft. The ability to create convincing video or audio of individuals saying or doing things they never did introduces ethical dilemmas about consent, privacy, and trust.

The ethical implications are particularly significant in the context of personal identity. As AI technology advances, the capability to create hyper-realistic digital avatars of individuals without their consent becomes a disturbing possibility. The use of deepfake technology to produce images, videos, or audio recordings of individuals without their permission, often for malicious purposes, represents a serious ethical breach. Moreover, it raises questions about ownership of digital likenesses, consent, and privacy. For example, synthetic media can be used to create "fake" pornography, manipulate public figures, or spread defamatory content. These ethical violations highlight the need for more robust consent frameworks, where individuals maintain control over their digital likenesses.

Additionally, the ethical concerns surrounding synthetic media extend to the AI and machine learning models that generate such content. AI systems are often trained on large datasets, some of which may contain biased or incomplete data. This results in synthetic media that can reinforce stereotypes, propagate bias, and marginalize underrepresented groups. For instance, AI-generated faces, voices, or personas may not accurately reflect the diversity of human experience, leading to biased representations in synthetic media. These biases in synthetic content have serious implications, especially in fields such as healthcare, law enforcement, and marketing, where biased AI models can lead to harmful outcomes.

2. Legal Implications

The legal implications of synthetic media are profound and multifaceted. One of the most significant challenges posed by synthetic media is the potential for harm in the form of misinformation, fraud, and defamation. As synthetic media becomes more sophisticated, it becomes harder to distinguish between real and fake content. In the legal domain, this presents challenges in the enforcement of existing laws that protect against defamation, harassment, and identity theft.

In many jurisdictions, current laws are not equipped to handle the complexities introduced by synthetic media. For instance, laws regarding copyright infringement and intellectual property rights may not cover the use of an individual's likeness or voice in synthetic media. The unauthorized use of someone's image or voice in AI-generated content raises questions about intellectual property, privacy rights, and the fair use of digital content. Additionally, the creation of deepfake videos or synthetic voices for the purpose of fraud, blackmail, or defamation has the potential to lead to significant legal consequences.

At present, many legal systems lack specific provisions that address synthetic media, and new regulations will be required to keep pace with technological advancements. Governments and international bodies must collaborate to develop new frameworks that define and address synthetic media misuse. Proposed regulations, such as the European Union's AI Act, aim to address some of these concerns by categorizing AI technologies based on risk and establishing guidelines for their use. However, these regulations are still in early stages, and there is a long road ahead in creating comprehensive legal frameworks that deal with the challenges posed by synthetic media.

The lack of a standardized legal approach to synthetic media also creates jurisdictional challenges. Since digital content can be easily created and shared across borders, synthetic media often operates in a global, decentralized environment where laws from different countries may conflict. This presents significant challenges in holding perpetrators of synthetic media misuse accountable. Consequently, the study calls for international collaboration in establishing consistent legal guidelines and practices for addressing the challenges posed by synthetic media.

3. Societal Implications

Synthetic media has the potential to significantly alter societal structures, including the way people perceive information, trust the media, and interact in digital spaces. As synthetic content becomes more difficult to differentiate from authentic content, public trust in digital media is likely to erode. This is particularly concerning in the context of elections, public discourse, and social movements, where the deliberate manipulation of synthetic media can have profound effects on public opinion and democratic processes.

For example, deepfake videos have been used to spread fake news and manipulate political narratives. The ability to create realistic videos of politicians or public figures making inflammatory statements can influence voter behavior, destabilize governments, or incite violence. This use of synthetic media to manipulate public opinion poses a serious threat to democracy, as it undermines the integrity of information, making it harder for the public to trust news sources. The implications for elections are particularly concerning, as the authenticity of political ads, campaign materials, and debates could be questioned, leading to confusion and mistrust among voters.

Moreover, synthetic media can exacerbate social divisions. As AI-generated content is often used to create polarizing narratives, it can be weaponized to deepen existing societal rifts. Disinformation campaigns that target specific communities or demographics can exploit the weaknesses in individuals' ability to discern truth from falsehood, thereby fostering greater division in society. To address these challenges, the study suggests that media literacy programs and public awareness campaigns should be expanded to equip individuals with the critical thinking skills necessary to evaluate digital content.

On a positive note, synthetic media also offers the potential to enhance societal engagement in various sectors. For instance, AI-generated content can be used in education to create personalized learning experiences, or in healthcare to develop synthetic medical data for research purposes while maintaining patient privacy. By providing scalable solutions to problems such as data scarcity in underrepresented regions, synthetic media can foster innovation. However, for these benefits to be realized, safeguards must be put in place to ensure that synthetic media does not perpetuate inequality or harm vulnerable populations.

4. Practical Implications for Industry and Business

The adoption of synthetic media has significant implications for various industries, including healthcare, education, entertainment, and cybersecurity. As discussed in the study, the healthcare industry has seen the potential of synthetic data in training AI models for disease diagnosis and medical imaging. However, ensuring the accuracy and reliability of synthetic medical data is critical for preventing misdiagnoses and maintaining patient safety. Moreover, while synthetic data can help protect patient privacy, there are concerns about how the data is used and who has access to it.

In education, the integration of AI-generated content into learning environments has the potential to revolutionize personalized learning. AI-driven educational platforms can generate content tailored to individual students' needs, thereby improving learning outcomes. However, as highlighted in the study, educational content must undergo rigorous validation to ensure its accuracy and avoid the dissemination of misinformation or biased content. The study emphasizes that while synthetic media can support innovative teaching methods, it also requires a robust infrastructure for content validation to prevent the spread of false or harmful information. The entertainment industry has been one of the primary sectors where synthetic media has gained widespread attention. AI-generated characters, deepfake technology, and digital influencers are transforming the way content is created and consumed. While this creates new creative opportunities, it also raises concerns about intellectual property rights, consent, and the ethical use of digital likenesses. The entertainment industry must develop frameworks that protect creators' and performers' rights while allowing for the ethical use of synthetic media. This may include establishing clear consent mechanisms, ensuring transparency in content creation, and enforcing accountability for misuse. Cybersecurity is another field where synthetic media is having a profound impact. As AI-generated content becomes more realistic, it is increasingly used to deceive individuals and organizations in cyberattacks. Deepfake technology has been used to impersonate executives or create fake identities for the purpose of social engineering and financial fraud. In response, cybersecurity systems are evolving to incorporate synthetic media detection tools, such as deepfake detection algorithms and AI-powered threat intelligence systems. However, these advancements must continue to evolve in tandem with the growing sophistication of synthetic media.

5. Policy and Governance Implications

Finally, the rise of synthetic media necessitates a comprehensive policy and governance framework. As synthetic media technologies become more ubiquitous, policymakers must address the challenges of content validation, authentication, and accountability. The study suggests that international cooperation is essential to establish uniform regulations that ensure the ethical use of synthetic media and prevent its misuse. Additionally, public policies must prioritize transparency in the creation of synthetic media, ensuring that audiences are aware when content has been generated or manipulated by AI.

Governments should also focus on developing education and awareness campaigns that help the public navigate the complexities of synthetic media. Media literacy, particularly in the context of new technologies, is crucial for fostering an informed citizenry that can critically evaluate digital content. By establishing clear guidelines and providing resources for identifying synthetic media, governments can mitigate some of the risks associated with its misuse.

CONCLUSION

The rapid development and proliferation of synthetic media, driven by advancements in artificial intelligence, have introduced profound changes in the way digital content is created, distributed, and consumed. As evidenced throughout this study, synthetic media—encompassing deepfakes, AI-generated images, text, and voice—has vast implications, particularly in terms of detection, authentication, ethical considerations, and legal frameworks. The implications of synthetic media on society, industry, and governance are significant and multifaceted, touching on issues of trust, misinformation, privacy, and security.

Summary of Findings

The study explores the evolution of synthetic media and its increasing sophistication, highlighting the primary methods of detection and authentication. While traditional methods of detecting digital forgeries, such as forensic analysis, have shown some efficacy, the rapid advancement of AI-based technologies has led to the development of more sophisticated synthetic outputs. This has prompted the need for advanced detection techniques, such as machine learning-based classifiers, deep learning models like Convolutional Neural Networks (CNNs), and blockchain-enabled verification systems. These technologies represent a critical shift in our approach to verifying the authenticity of digital content, as automated detection alone is often insufficient. Hybrid models combining AI detection with human expertise offer more robust solutions and can enhance the accuracy of synthetic media detection.

However, the study also underscores the limitations of current technologies in fully combating the challenges posed by synthetic media. Despite advancements, the detection of deepfakes, for example, remains an ongoing challenge due to the constant improvements made in the technology that generates them. As AI-driven tools become more advanced and accessible, the need for continuous innovation in synthetic media detection grows more urgent.

In addition to technical solutions, the study highlights the ethical, legal, and societal implications that synthetic media generates. The potential for synthetic content to be used maliciously—for purposes ranging from misinformation to identity theft—poses significant risks to personal privacy, public trust, and democratic processes. The creation of AI-generated content without proper consent, such as deepfake pornography or politically charged fake videos, raises serious ethical questions about ownership and misuse of digital likenesses. Furthermore, the legal landscape surrounding synthetic media is still underdeveloped, with many existing laws failing to address the new challenges presented by AI-generated content. The rapid pace at which these technologies evolve makes it difficult for legislative bodies to keep up, resulting in gaps in legal protections for individuals and organizations affected by synthetic media.

The societal impact of synthetic media is profound, particularly with regard to misinformation. The use of AI to create convincing fake videos or audio recordings can sway public opinion, manipulate political events, and even disrupt elections. These technological advancements present an erosion of trust in digital content, where individuals are increasingly skeptical of the media they consume. The study advocates for stronger media literacy education, emphasizing the need for critical thinking skills to empower individuals to identify and resist synthetic deception.

In various sectors like healthcare, education, and entertainment, synthetic media holds immense potential for innovation and efficiency. In healthcare, for example, AI-generated medical data and synthetic patient information can be used to train AI models, improving healthcare delivery without compromising patient privacy. Similarly, AI-generated educational content can offer personalized learning experiences. Yet, the study emphasizes the importance of verifying the accuracy and authenticity of synthetic media in these industries, where the potential consequences of misinformation can be particularly damaging.

Policy and Governance Recommendations

Based on the findings of this study, it is clear that addressing the challenges posed by synthetic media requires a coordinated, multi-faceted approach. This includes the development of regulatory frameworks, technological innovations, and public awareness campaigns to mitigate the risks associated with synthetic content. Governments must move swiftly to develop legislation that not only addresses the potential for malicious uses of synthetic media but also ensures the ethical use of AI technologies in content creation. While some early regulations have been introduced, such as the European Union's AI Act, comprehensive international policies are needed to standardize approaches and provide clear guidelines for synthetic media's creation, use, and distribution.

Legal frameworks must evolve to accommodate the complexities of synthetic media. Existing laws regarding intellectual property, defamation, privacy, and fraud need to be updated to reflect the challenges posed by AI-generated content. In particular, lawmakers must address the issue of consent, ensuring that individuals retain control over their digital likenesses and that their image or voice cannot be misused without permission. At the same time, laws must strike a balance between protecting individuals from harm and preserving the creative potential of AI technologies.

Moreover, governments must collaborate internationally to harmonize regulations and address jurisdictional challenges. Synthetic media operates in a digital, borderless world, where content is easily created and disseminated across countries with varying legal systems. This global nature of synthetic media necessitates international cooperation and coordination to combat its misuse effectively. Policymakers must also address the issue of accountability, determining who should be held responsible for the creation and distribution of malicious synthetic content.

Equally important is the role of education in combating the negative effects of synthetic media. Media literacy initiatives should be expanded globally, ensuring that individuals are equipped with the necessary tools to critically engage with digital content. Teaching users to recognize the signs of synthetic media, such as inconsistencies in facial expressions or voice anomalies, can help reduce the spread of misinformation. Public awareness campaigns are essential in fostering a more informed and skeptical digital public, which can resist being swayed by synthetic media.

Future Directions and Recommendations

As synthetic media continues to evolve, the study points to several critical areas for future research and development. First, the development of more advanced and reliable detection tools is essential. While AI-based classifiers have shown promise in identifying deepfakes, there is still room for improvement in detection accuracy. The use of multimodal verification methods, which combine different types of data—such as audio, video, and textual content—could offer a more robust solution to verifying synthetic media.

In addition, the study suggests that research should focus on improving blockchain-based authentication and verification systems. These technologies hold significant promise in providing an immutable record of content creation, allowing digital content to be traced back to its origin and verified for authenticity. Blockchain can offer a decentralized, transparent, and secure method of ensuring the integrity of digital content, which is crucial in combating synthetic media misuse.

Another promising area of research is the application of human-in-the-loop systems, where AI algorithms work alongside human experts to validate synthetic content. Hybrid models that combine the strengths of both AI and human judgment could offer a more adaptable and accurate solution to the detection and authentication of synthetic media. These systems would benefit from continuous input and adaptation, enabling them to keep pace with the rapid advancements in synthetic media generation.

Finally, the study emphasizes the importance of interdisciplinary collaboration between AI researchers, policymakers, ethicists, and industry stakeholders. By bringing together experts from various fields, more effective solutions can be developed to address the challenges of synthetic media. As the technology behind synthetic media continues to improve, it is essential that all stakeholders work together to ensure that AI advancements are used ethically and responsibly.

Conclusion

The study concludes that while synthetic media offers exciting opportunities for innovation across various sectors, its widespread use poses significant risks and challenges that must be addressed. Ethical considerations, legal frameworks, societal impacts, and industry-specific implications must be carefully navigated to ensure that synthetic media is used responsibly. The detection and authentication of synthetic content will require ongoing technological advancements, and legal systems must adapt to address the new realities created by AI-generated media.

As synthetic media continues to evolve, interdisciplinary efforts involving government regulation, technological development, and public education will be essential. Only by combining these efforts can we effectively address the complexities of synthetic media and mitigate the risks it poses to privacy, security, and public trust. The future of synthetic media will depend on how effectively society can balance its transformative potential with the safeguards necessary to protect individuals and communities from its harmful effects.

REFERENCES

Agarwal, A., Hunt, B. J., Stegemann, M., Rochwerg, B., Lamontagne, F., Siemieniuk, R. A., ... & Vandvik, P. O. (2020). A living WHO guideline on drugs for covid-19. *bmj, 370.*

Boell, S. K., & Cecez-Kecmanovic, D. (2015). On being 'systematic'in literature reviews in IS. *Journal of Information Technology, 30*(2), 161–173. DOI: 10.1057/jit.2014.26

Braun, V., & Clarke, V. (2006). Using thematic analysis in psychology. *Qualitative Research in Psychology, 3*(2), 77–101. DOI: 10.1191/1478088706qp063oa

Chesney, B., & Citron, D. (2019). Deep fakes: A looming challenge for privacy, democracy, and national security. *California Law Review, 107*, 1753.

Dolhansky, B., Bitton, J., Pflaum, B., Lu, J., Howes, R., Wang, M., & Ferrer, C. C. (2020). The deepfake detection challenge (dfdc) dataset. *arXiv preprint arXiv:2006.07397.*

Goodfellow, I. J., Pouget-Abadie, J., Mirza, M., Xu, B., Warde-Farley, D., Ozair, S., & Bengio, Y. (2014). Generative adversarial nets. *Advances in Neural Information Processing Systems*, •••, 27.

Jawahar, M., Ravi, V., Prassanna, J., Jasmine, S. G., Manikandan, R., Sekaran, R., & Kannan, S. (2022). Covmnet–deep learning model for classifying coronavirus (covid-19). *Health and Technology, 12*(5), 1009–1024. DOI: 10.1007/s12553-022-00688-1 PMID: 35966170

Johnston, R. J., Boyle, K. J., Adamowicz, W., Bennett, J., Brouwer, R., Cameron, T. A., Hanemann, W. M., Hanley, N., Ryan, M., Scarpa, R., Tourangeau, R., & Vossler, C. A. (2017). Contemporary guidance for stated preference studies. *Journal of the Association of Environmental and Resource Economists, 4*(2), 319–405. DOI: 10.1086/691697

Kaplan, R. M., & Milstein, A. (2021). Influence of a COVID-19 vaccine's effectiveness and safety profile on vaccination acceptance. *Proceedings of the National Academy of Sciences of the United States of America, 118*(10), e2021726118. DOI: 10.1073/pnas.2021726118 PMID: 33619178

Kietzmann, J., Lee, L. W., McCarthy, I. P., & Kietzmann, T. C. (2020). Deepfakes: Trick or treat? *Business Horizons, 63*(2), 135–146. DOI: 10.1016/j.bushor.2019.11.006

Lu, L., Meng, X., Mao, Z., & Karniadakis, G. E. (2021). DeepXDE: A deep learning library for solving differential equations. *SIAM Review*, *63*(1), 208–228. DOI: 10.1137/19M1274067

Mirsky, Y., & Lee, W. (2021). The creation and detection of deepfakes: A survey. *ACM Computing Surveys*, *54*(1), 1–41. DOI: 10.1145/3425780

Tolosana, R., Vera-Rodriguez, R., Fierrez, J., Morales, A., & Ortega-Garcia, J. (2020). Deepfakes and beyond: A survey of face manipulation and fake detection. *Information Fusion*, *64*, 131–148. DOI: 10.1016/j.inffus.2020.06.014

Verdoliva, L. (2020). Media forensics and deepfakes: An overview. *IEEE Journal of Selected Topics in Signal Processing*, *14*(5), 910–932. DOI: 10.1109/JSTSP.2020.3002101

Wardle, C., & Derakhshan, H. (2017). *Information disorder: Toward an interdisciplinary framework for research and policymaking* (Vol. 27). Council of Europe.

Yoon, J. G., Yoon, J., Song, J. Y., Yoon, S. Y., Lim, C. S., Seong, H., & Kim, W. J. (2020). Clinical significance of a high SARS-CoV-2 viral load in the saliva. *Journal of Korean Medical Science*, *35*(20).

Zhang, H., Wu, C., Zhang, Z., Zhu, Y., Lin, H., Zhang, Z., & Smola, A. (2022). Resnest: Split-attention networks. In *Proceedings of the IEEE/CVF conference on computer vision and pattern recognition* (pp. 2736-2746).

Chapter 15
A Comparative Analysis of Generative AI Traffic Across Continents

Kassim Kalinaki
https://orcid.org/0000-0001-8630-9110
Islamic University in Uganda, Uganda

Madinah Nabukeera
https://orcid.org/0009-0008-1033-4123
Kyambogo University, Uganda

Moses Matovu
https://orcid.org/0009-0002-2408-4299
Islamic University in Uganda, Uganda

ABSTRACT

This study presents a comparative analysis of user engagement with generative AI, using ChatGPT traffic data for March 2024 as a case study of one widely adopted tool. Using World Bank data and statistical methods, including ANOVA, it compares usage across continents. While statistical tests showed no significant differences in total visits or global traffic share, descriptive results revealed meaningful disparities in engagement intensity. Chatbot-based platforms, particularly ChatGPT, have dominated other generative AI tools due to their versatility in tasks such as content creation, learning, coding, and communication. These findings suggest a trend toward broader access but also highlight persistent structural barriers, such as device access, digital skills, and infrastructure, that shape real-world usage. Therefore, promoting digital literacy, improving infrastructure, and localizing AI tools remain essential to achieving more equitable global engagement with generative AI.

DOI: 10.4018/979-8-3373-2277-3.ch015

INTRODUCTION

Artificial intelligence (AI) has been widely integrated across various sectors, transforming industry-specific tasks and driving informed decision-making (Kalinaki, Yahya, et al., 2024). In healthcare, AI supports diagnostics, medical imaging, and personalized treatment recommendations (Shafik et al., 2024). In business, it powers customer service automation, fraud detection, and operational efficiency (Kalinaki et al., 2023). Environmental applications include climate modeling, biodiversity tracking, and disaster forecasting (Kalinaki, Acaru, et al., 2024; Kalinaki, Malik, et al., 2024; Zakari et al., 2024). These domain-specific uses have largely focused on analytical and predictive capabilities. More recently, the emergence of generative AI (GenAI), a branch of AI capable of producing new content such as text, images, audio, and code, has shifted public interaction with AI tools. Unlike earlier forms, GenAI tools such as ChatGPT allow direct user engagement through natural language interfaces, making them more accessible to non-specialists and more prominent in everyday digital experiences. From advanced language models capable of human-like conversation (Kalinaki et al., 2023; Shafik, 2023) to groundbreaking image creation systems that convert simple text prompts into exquisite visual representations, GenAI has come a long way from being an expert technological interest to a pervasive digital phenomenon that is intriguing millions globally (Fui-Hoon Nah et al., 2023). The rapid evolution of GenAI has transformed various industries globally, generating immense enthusiasm regarding its pragmatic applications and implementation (Sedkaoui & Benaichouba, 2024). Nonetheless, the level of adoption and usage varies on various continents based on the state of technological infrastructure, availability of internet facilities, stages of economic development, and mindset of individuals towards AI (Gupta, 2024). These differences are relevant to stakeholders interested in developing AI technologies and bridging digital divides. This article examines traffic flow to GenAI sites, specifically ChatGPT, by region, providing an insight into user interest, local enthusiasm, and accessibility variation.

GenAI has evolved significantly over the past decade, originating from early innovations in Generative Adversarial Networks (GANs) and transformer architectures. Key milestones include the release of GPT-2 in 2019, GPT-3 in 2020, and later multimodal and fine-tuned models that expanded generative capabilities across domains (Feuerriegel et al., 2024). These developments laid the foundation for publicly accessible tools like ChatGPT, which, since its launch in late 2022, has become a focal point for public and institutional engagement with AI. For clarity, this study defines GenAI as artificial intelligence capable of producing novel content such as text, images, code, and audio in response to user input. The term traffic metrics refers to web-based indicators of platform usage, including visit counts, user sessions, traffic share, and normalized usage per internet user or working-age population. These metrics provide measurable insights into the scale and distribution of AI engagement across geographic regions.

ChatGPT, developed by OpenAI, is one of the most widely adopted public-facing generative AI tools, known for its conversational text interface powered by large language models (LLMs). Compared to other publicly available platforms such as Gemini (Google), Claude (Anthropic), and Copilot (Microsoft), ChatGPT stands out for its early release, user-friendly interface, multilingual capabilities, and wide range of applications, spanning education, coding, business support, and everyday information retrieval (Shafik, 2023). As of March 2024, the dominant free version of ChatGPT available to most users was based on GPT-3.5, while GPT-4 was primarily accessible to paid subscribers via ChatGPT Plus. The data analyzed in this study most likely reflects user activity on the free, GPT-3.5-powered version, which represents the largest share of global traffic. Given its scale, accessibility, and integration into professional and educational settings, ChatGPT was chosen as a representative case for examining global generative AI usage trends.

The worldwide dissemination of GenAI platforms presents an unprecedented opportunity to study the diffusion of revolutionary technologies across continents, cultures, and economies. In contrast with conventional software, which typically necessitates complicated setups or huge financial investments, the majority of GenAI tools are based on straightforward web interfaces, hence providing an almost seamless entry point for global users (Mahadevappa et al., 2025; Shilaskar et al., 2025). Underneath this superficially global scope, a complex and refined array of adoption patterns lies, exhibiting significant variation across continents. The narrative of how various regions of the globe receive, understand, and adopt GenAI technologies provides a fascinating study of the intersections of technology, society, and culture. Each continent has its particular narrative regarding AI adoption, fueled by individual combinations of technological capabilities, economic conditions, regulatory environments, and cultural inclinations towards AI (Liu & Wang, 2024).

The variation in GenAI adoption is particularly pronounced when considered against the backdrop of advances in web traffic patterns. These digital breadcrumbs reveal more than just user data because they unveil the underlying patterns of worldwide technology uptake throughout the 21st century. North America's early adoption and heavy use demonstrate not just technological hegemony, but even cultural openness to experimentation with AI (Chang & Kidman, 2023). The risk-averse approach evident in Europe illustrates the significant impact of regulatory frameworks and privacy considerations on technological adoption trends (Novelli et al., 2024). Conversely, Asia's remarkable surge in AI traffic exemplifies the effectiveness of extensive user populations and mobile-first internet strategies (Lartey & Law, 2025). Meanwhile, Africa's nascent engagement patterns suggest the potential for leapfrogging technology adoption in emerging markets. The infrastructure underlying internet coverage in various continents presents vastly disparate experiences for users of AI. While in South Korea, for instance, the high-quality fiber optic infrastructure enables smooth usage of bandwidth-hungry AI applications, elsewhere, the limited bandwidth makes users in other parts resort to data workarounds through tactical solutions (Lee et al., 2021). The prevailing infrastructural conditions affect not just the user demographics of GenAI but also the modalities and contexts of use, thus creating diverse usage patterns reflective of regional technological capabilities and constraints.

Economic considerations introduce yet another layer of complication into the discussion of worldwide AI adoption. Although the potential for democratized access to AI via open platforms would seem to herald a more even playing field, things are far more complicated. The provision of advanced features, computational constraints, and reliance on subscription-based systems pose complex hindrances that could restrict economically disadvantaged groups from accessing advanced AI technologies (Jo, 2024; Kaur & Gupta, 2025). The variations in purchasing power between continents result in vast differences in relative spending on identical AI services and therefore could lead to new digital disparities. Cultural perceptions of AI vary widely across the globe based on different regions as affected by factors such as science fiction representations of AI in popular media and public statements about AI research (Emon & Khan, 2025). AI is perceived by some societies as a collaborative partner and productivity-increasing tool, leading to creative and diverse applications. There are cultures that are also apprehensive about AI, as it echoes underlying fears related to job loss and privacy issues (Rawashdeh, 2023; Solove, 2025). These cultural sensitivities are revealed in perceptive manners through traffic logs, user engagement patterns, and preferences for various platforms.

The legal framework for AI technologies has evolved as a fragmented set of differing approaches on various continents. The far-reaching AI Act adopted in Europe represents one extreme of the range marked by active regulation, while other continents prefer a more hands-off approach (Novelli et al., 2024). This regulatory divergence creates differing operating environments for AI platforms, influencing elements that stretch from data-collection practices to the range of functionalities offered. Users' awareness of these regulatory shields or barriers may influence their readiness to utilize AI platforms, leading to continent-based usage patterns.

Language barriers pose both threats and opportunities in the global AI ecosystem. Where primarily English AI platforms used to rule the roost, multilingual capabilities have opened the door to accessing completely new demographics and applications (Zenni & Andrew, 2023). The buzz created by the launch of local language support in AI platforms usually leads to phenomenal growth in local adoption, clearly indicating the significance of linguistic inclusiveness for technology adoption globally. The use of GenAI tools in professional and educational environments varies greatly on different continents, reflecting differing approaches to the implementation of technology in both formal and informal educational settings (Law, 2024). While certain settings welcome AI as a pedagogical tool, thus creating particular patterns of use within academic years, others continue to practice more conventional education and professional training methods. Differences among industries contribute to the overall patterns of traffic on the continent and reveal possibilities for insight into the embedding of AI into the texture of everyday life. Issues of security and privacy manifest in various guises around the world, depending upon cultural attitudes towards data sharing, historical experience with surveillance states, and varying degrees of trust in technology companies (Saeed et al., 2024). These concerns result in differing degrees of willingness to engage with AI sites, affecting such factors as the information shared at registration and the types of content that users are willing to contribute to and share.

At this juncture of technological advancement, trends in the adoption of GenAI across the globe have implications that reach far beyond academic curiosity; they provide important insight into the future landscape of international technology integration. The data on AI platforms provide a direct metric of human concordance with and accommodation of AI, and as such reflect the possible opportunities and challenges of our increasingly AI-pervaded existence.

This side-by-side comparison of GenAI traffic visits by continent provides a foundation for addressing key questions about technology, equity, culture, and human adaptation. In what ways do cultures address disruptive technologies differently? What are the factors facilitating or slowing their adoption? How can we ensure that the potential benefits of AI are experienced by all humans, regardless of geographical location or economic condition? The responses are embedded in traffic patterns, user interest data, and adoption trends that collectively help retell our shared narrative in the age of AI.

While the GenAI ecosystem comprises a growing array of tools, including Gemini, Claude, Copilot, and Midjourney, this study focuses specifically on ChatGPT due to its prominence, accessibility, and broad user base. It is essential to clarify that ChatGPT is not used here as a blanket representation of all public-facing GenAI tools, but rather as a proxy for examining large-scale patterns of user engagement with one leading platform. As such, the findings are indicative of ChatGPT-specific usage trends and should not be generalized to the entire GenAI landscape.

Overall Objective

To compare GenAI traffic visits across continents

Specific Objectives

i. To compare the number of visits across AI generative domains in March 2024.
ii. To compare the number of ChatGPT visits across different continents in March 2024.
iii. To compare the share in global ChatGPT traffic across different continents in March 2024.
iv. IV. To compare the ChatGPT traffic per internet user across different continents in March 2024.
v. To compare the ChatGPT traffic per person aged 15-64 across different continents in March 2024.

Chapter roadmap

The rest of this chapter is organized as follows: After the Introduction section 1, Section 2 details the related works. Methods are detailed in section 3. Section 4 presents the results and findings. The discussion is given in Section 5, the conclusion in Section 6, and the recommendations in Section 7.

RELATED WORKS

The increasing ubiquity of GenAI systems, particularly following the release of ChatGPT in late 2022, has fueled a surge of scholarly attention into how these technologies are being adopted and used globally. A particularly compelling angle in recent research has been the comparative analysis of traffic patterns, user demographics, and cross-continental adoption disparities. As the world witnesses an unprecedented acceleration in AI-driven transformations, understanding how different regions interact with GenAI technologies has become a foundational concern for both academia and policy. The broad enthusiasm for GenAI is reflected in a systematic literature review and bibliometric analysis that captured the global and interdisciplinary research footprint of ChatGPT (Khan et al., 2024). The research illustrated that the advent of ChatGPT, along with other GenAI platforms, has gained extensive traction in diverse domains like computer science, healthcare, education, and research, with considerable presence in 80 countries. Notably, the United States, China, and India have emerged as prominent producers of both academic and applied literature, with intensive involvement in GenAI in regions of high population density and technological advancement. However, this distribution also highlighted a significant underrepresentation of much of Africa and regions of South America, indicating the prevalence of infrastructural, as well as epistemological, gaps within the landscape of AI research and applications.

In addition to these results, the World Bank's empirical analysis leveraged unconventional real-time data sources like web analytics and Google Trends to examine the worldwide diffusion of GenAI tools (Liu & Wang, 2024). The results indicate that the top 40 GenAI platforms receive a total of around 3 billion visits each month, with ChatGPT accounting for more than 82 percent of the traffic. Despite its ubiquity, international traffic to ChatGPT constitutes a minority share compared to other sites, such as Google, simultaneously attesting to both its broad popularity and potential for expansion. Most of all, scientists noted middle-income nations as disproportionately major generators of GenAI traffic compared to their economic scale, now driving more than 50 percent of worldwide GenAI consumption. In contrast, low-income nations account for less than 1 percent, reflecting extensive disparities in line with broader trends of digital exclusion. Such differences extend beyond mere quantitative distinctions and encapsulate deep-seated structural obstacles pertaining to digital infrastructure, educational systems, and policy preparedness. The analysis of the determinants facilitating or inhibiting the adoption of GenAI technologies across various nations was strengthened by scholars in a subsequent study (Ali et al., 2025), which performed a negative binomial regression analysis on data from 136 nations. Their work placed in perspective the complexity of adoption: in addition to the paramount roles of technological infrastructure and research and development expenditures, organizational preparedness and policy stability are also significant determinants.

Countries with steady national AI policies, strongly secured cloud computing infrastructure, and AI industry-specific incentive mechanisms recorded significantly higher adoption levels. They successfully bridge micro-level organizational variables and macro-level governance by demonstrating how state policies and firm behavior jointly shape AI adoption in territories. Though macro-level data give us valuable information, individual-level reasons to utilize GenAI also reflect regional inequalities. Another study employed a survey-based approach to examine the experiential and psychological antecedents of ChatGPT utilization in Vietnam (Van Huy et al., 2024). The authors found that although pragmatic considerations such as performance expectancy and facilitating conditions impact early adoption, emotional and psychological factors, namely curiosity and hedonic motivation, significantly influence users' intentions to continue using the platform and recommend it to others. Social influence was observed to play comparatively less of a role, indicating that intrinsic motivation and personal initiative might adopt more central roles in technology adoption in certain cultural contexts, a result that would presumably vary quite significantly in socially interdependent or collectivist societies. Scholars in a subsequent study (Kim et al., 2024) provided an in-depth exploration of user psychology by integrating the Technology Acceptance Model (TAM) with Uses and Gratifications Theory (U&G), thereby creating an expanded framework for understanding users. Their analysis of a 400-user sample demonstrated that personalization and anthropomorphism features greatly enhanced trust in GenAI tools, with privacy issues reducing this trust. Trust plays a central role in user motivations to use both free and paid-for services. This interaction is brought into even sharper focus when considering the adoption of GenAI in regions with varying cultural responses to privacy and technology, such as Western Europe compared to East Asia or sub-Saharan Africa. In these models, the dynamics between the technology's nature and user perceptions are at the root of explaining regional take-up patterns.

The United States offers a unique lens for understanding the speed of GenAI uptake, as shown through the work of (Bick et al., 2024), who conducted nationally representative surveys across the nation. By the end of 2024, about 40 percent of Americans aged between 18 and 64 years had used GenAI tools, while 9 percent used them daily for work purposes. Compared to past experience with the introduction of the personal computer and the internet, the spread of GenAI has happened faster, particularly among more highly educated and technologically skilled workers. Their findings suggest that work-related adoption of GenAI is accelerating and already yielding measurable productivity gains. Notably, the rapid adoption of these technologies is driven by not only their innovative nature but also economic incentives, such as time saving and improved operational efficiency, hence asserting that utilitarian utility plays a role in motivating engagement as much as curiosity or newness. Additionally, the importance of governance and policy has been asserted in recent scholarly discourse. One study (Cheng & Wu, 2024) collated institutional guidelines on GenAI use from academic, governmental, and editorial groups. Scholars have highlighted how dominant actors in North America, Europe, and East Asia have developed infrastructure around transparency, authorship, and data provenance. In contrast, such processes are in their infancy in the Global South, where concerns regarding ethical parity and algorithmic harm are raised. This difference is mirrored in the comparative analysis of (Beltran et al., 2024), which compared five nations' soft laws and guidelines, including Australia, Canada, and South Korea. The authors do identify an increasing acknowledgment of data privacy threats, information leakage, and deteriorating public trust; however, they observed considerable variation in how these issues are embedded within policy frameworks. This divergence is mirrored in wider continental inequalities, with regions of greater economic development not just implementing GenAI at quicker speeds but also more dynamically participating in its governance. Governance issues are particularly acute in the African situation. A study by (Wakunuma & Eke, 2024) explored the political, ethical, and practical implications of GenAI in the African context. Their review highlighted the continent's contradictory position of being vulnerable to exploitation on one hand while well-placed to leverage the opportunities of AI in areas such as agriculture, health, and education. Yet, the absence of harmonized governance, coupled with systemic challenges such as the digital divide and under-resourced education systems, profoundly frustrates the region's capacity to make meaningful use of GenAI. The authors urge inclusive AI governance frameworks that reflect African contexts instead of adopting regulatory frameworks from the Global North. This speaks to the more general requirement for a refreshed, more plural, and more context-sensitive vision for global AI governance.

The public sector both presents opportunities and faces certain unique challenges for GenAI adoption. As noted in an earlier study (Beltran et al., 2024), governments are exploring GenAI for citizen engagement, policy-making, and administrative efficiency. However, in the absence of strong regulatory institutions, public application of GenAI risks exacerbating inequalities and undermining government institutions' trust. These dangers are more acute where institutional capacity is low and legal enforcement is weak, where GenAI can be misused or fail to bring expected gains. Taken together, the studies paint a complex yet coherent picture: Not only is GenAI a worldwide phenomenon, but it is also a geographically uneven, socially contingent, and policy-mediated one. While use patterns demonstrate rapid diffusion, particularly among middle- and high-income countries, there are significant variations across infrastructure, policy readiness, and user ability. Individual motivations, trust relationships, and cultural norms also condition the use and interpretation of GenAI technologies. This literature review makes it evident that an understanding of GenAI's world trajectory is about more than just tracking web traffic or installation rates; it is a matter of a subtle excavation of social, economic, psychological, and political processes that vary radically across continents. It's only by combining these viewpoints that we can start to get to grips with the long-term implications of GenAI's sudden emergence and make sure its benefits are shared fairly.

Although many early studies focused on the capabilities and risks of GenAI, more recent work has shifted toward analyzing global adoption and usage patterns. For instance, (McKinsey, 2023) explored sectoral adoption trends, while (Liu & Wang, 2024) provided continental traffic insights across major GenAI platforms. These studies inform the current work's decision to focus on ChatGPT, a tool with dominant visibility and accessibility, as a case for understanding regional disparities in generative AI interaction. Table 1 provides a summary of the regional highlights of GenAI Adoption from the reviewed studies.

Table 1. Key Regional Highlights of GenAI Adoption

Region	Key Trends & Features	Challenges
Africa	Mobile-first use, leapfrogging infrastructure gaps	Low compute power, digital literacy
Asia	Rapid commercial adoption, language localization	Regulation, data privacy
Europe	Regulated experimentation, strong policy frameworks	Compliance burdens, legal ambiguity
North America	Startup ecosystem, enterprise integration	Trust issues, misinformation spread

METHODOLOGY

This study employed a descriptive research design guided by a quantitative approach to systematically explore and compare GenAI traffic metrics, specifically ChatGPT, across continents. A descriptive quantitative design was selected because it enables the analysis of observable, empirical patterns, specifically web traffic data, across categorical groupings such as continents. The use of one-way ANOVA is appropriate for comparing means across more than two independent groups, allowing the study to determine whether statistically significant differences in ChatGPT usage exist between regions. This design aligns with the study's aim to provide a structured, comparative overview of generative AI tool engagement at a macro level. The design was selected to enable objective measurement and statistical testing of global usage variations in GenAI technologies, with a particular focus on ChatGPT.

Data Source and Collection

Secondary data were sourced from the World Bank's Policy Research Working Paper (Liu & Wang, 2024), titled Who on Earth is Using Generative AI?. The dataset compiled from web analytics and Google Trends includes global traffic statistics from March 2024 for over 40 leading GenAI platforms, including domain-specific traffic counts, global shares of ChatGPT usage, and normalized traffic figures per internet user and per person aged 15–64 across continents.

Hypotheses

The study was anchored around five hypotheses intended to statistically assess regional differences across multiple GenAI usage dimensions:

i. There is a significant difference in the number of visits across AI generative domains in March 2024.
ii. There is a significant difference in the number of ChatGPT visits across different continents in March 2024.
iii. There is a significant difference in the share in global ChatGPT traffic across different continents in March 2024.
iv. There is a significant difference in the ChatGPT traffic per internet user across different continents in March 2024.
v. There is a significant difference in the ChatGPT traffic per person aged 15–64 across different continents in March 2024.

Data Preparation and Cleaning

Raw data from the report were reviewed and cleaned. Non-applicable variables were excluded, and the dataset was formatted into analyzable variables categorized by continent. Each hypothesis corresponded to one specific dependent variable: total visits, ChatGPT visit count, share of global traffic, traffic per internet user, and traffic per working-age person, enabling structured hypothesis testing.

Descriptive Statistical Analysis

Descriptive statistics including mean, median, standard deviation, and range were computed for each metric across continents. These statistics provided an initial sense of central tendency and dispersion, highlighting variations in GenAI usage volume and intensity across global regions. To ensure appropriate application of parametric statistical tests, normality of the data distributions was assessed using both the Shapiro–Wilk test and the Kolmogorov–Smirnov test. These tests were conducted on each dependent variable within each continental group. A non-significant result ($p > 0.05$) indicated that the data were not significantly different from a normal distribution, supporting the use of parametric techniques.

One-Way ANOVA Analysis

In this study, the independent variable (IV) is continent, treated as a categorical variable with five levels: Africa, Asia, Europe, North America, and Oceania. The dependent variables (DVs) are quantitative measures of ChatGPT usage, including (i) total visits to ChatGPT, (ii) share in global ChatGPT traffic, (iii) visits per internet user, and (iv) visits per person aged 15–64. The ANOVA procedure was applied to examine whether significant differences in these usage metrics exist across the different continental groupings. The hypotheses were tested using One-Way

Analysis of Variance (ANOVA), as the data involved comparisons of means across multiple independent groups (i.e., continents). Each hypothesis corresponds to a distinct dependent variable, and the independent variable in all cases is the continent of origin. The null hypotheses in each case posited no significant difference among the group means, while the alternative hypotheses suggested at least one significant inter-group variation.

Ethical Considerations

As the study used publicly available secondary data from a reputable source (World Bank), it did not involve human participants and therefore required no ethical review. Nevertheless, all research was conducted in compliance with data ethics and academic citation standards.

RESULTS AND FINDINGS

This section presents an in-depth interpretation of the statistical analyses undertaken to understand global usage trends of GenAI tools, particularly ChatGPT, in March 2024. The results are structured around five key research objectives and are based on descriptive statistics, normality tests, and one-way ANOVA procedures.

Table 2. Descriptive Statistics and Normality Test

Variables	N	Mean	Std. Dev	Normality
Traffic visits across GenAI tools in March 2024 (million visits)	40	70.987	369.1649	0.000
ChatGPT traffic visits in march 2024 across different countries (million visits)	30	65.7073	93.23641	0.000
Share in global ChatGPT traffic aacross different countries (%)	30	2.8047	3.97779	0.000
ChatGPT traffic per internet user in march 2024	30	1.4017	0.46396	0.000
ChatGPT traffic per person aged 15-64 in march 2024	29	2.1248	0.65004	0.001

Source: World Bank (2024)

The normality test results, summarized in Table 1, revealed that all variables significantly deviated from a normal distribution at the 5% significance level, as indicated by the Kolmogorov–Smirnov test ($p < 0.05$ across all variables). This justified cautious application of ANOVA, supplemented by post-hoc analysis where applicable.

Comparing the number of visits across AI generative domains in march 2024

The first objective explored the variation in user traffic across three main types of GenAI domains: Chatbot-based, Image-based, and Video-based platforms. Although the One-Way ANOVA (F(2, 37) = 0.523, p = 0.597) indicated that there was no statistically significant difference in the number of visits among these domains, the descriptive data suggest otherwise in terms of practical relevance.

Table 3. One-Way ANOVA Comparing the number of visits across AI generative domains in march 2024

	Sum of Squares	df	Mean Square	F	Sig.
Between Groups	146156.658	2	73078.329	.523	.597
Within Groups	5168869.366	37	139699.172		
Total	5315026.024	39			

The results in Table 2 show no significant difference in the number of visits across AI generative domains in March 2024 (F(2, 37) = 0.523, p = 0.597). Table 3 indicates that the traffic visits were similar across the Chatbot, Image, and Video domains in March 2024. As illustrated in Figure 1, Chatbot-based tools (e.g., ChatGPT) overwhelmingly dominated user engagement with an average of 128 million visits, far exceeding visits to Image-based tools (9.9 million) and Video-based tools (3.4 million). This pattern, although not statistically significant, possibly due to high within-group variance, suggests a clear user preference or functional utility advantage in favor of chatbot platforms during the reporting period.

Figure 1. Comparing the number of visits across AI generative domains in March 2024

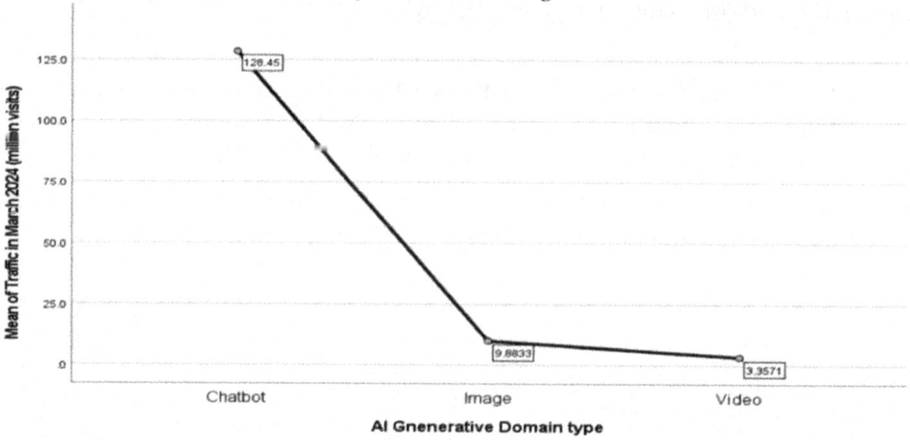

Comparing the number of ChatGPT visits across different continents in march 2024

The second analysis compared the total number of ChatGPT visits across four continents: North America, Asia, South America, and Europe. The ANOVA results **(F(3, 24) = 2.859, p = 0.058) fell just short of the conventional 5% threshold, indicating no statistically significant differences among the group means.**

Table 4. One-way ANOVA Comparing the number of ChatGPT visits across different continents in march 2024

	Sum of Squares	df	Mean Square	F	Sig.
Between Groups	64891.627	3	21630.542	2.859	.058
Within Groups	181599.113	24	7566.630		
Total	246490.739	27			

Nonetheless, the descriptive findings from Figure 2 provide insightful geographical trends. North America had the highest average ChatGPT traffic (203.8 million visits), suggesting early and deep integration of GenAI in professional, educational, and personal routines. Asia followed with 67 million visits, while South America and Europe recorded 54 million and 39 million visits, respectively. While not statistically conclusive, this trend reflects technological maturity, digital infrastructure, and user familiarity with AI tools in developed economies.

Figure 2. Comparing the number of ChatGPT visits across different continents in march 2024

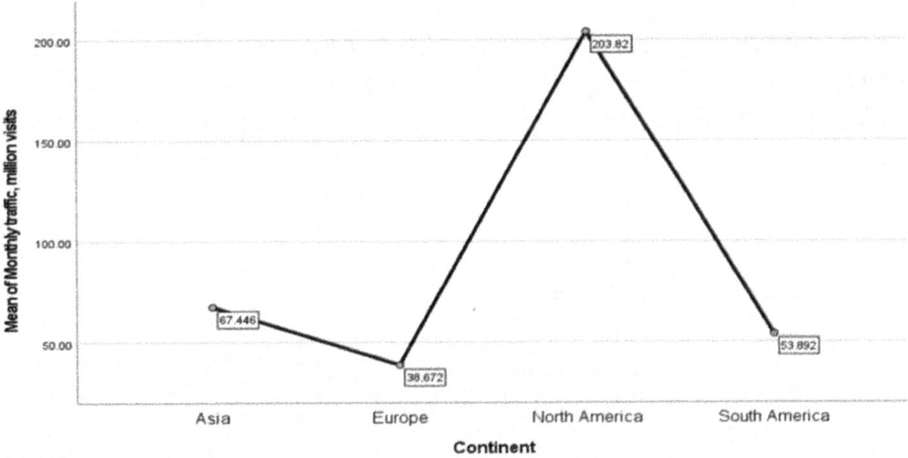

Comparing the share in global ChatGPT traffic across different continents in March 2024

The third objective examined whether the continental share of total global ChatGPT traffic varied significantly. The ANOVA again yielded a non-significant result ($F(3, 24) = 2.859$, $p = 0.058$), implying that no substantial disparities exist in relative traffic contributions when scaled to global totals. The findings are presented in Table 5.

Table 5. One-way ANOVA Comparing the share in global ChatGPT traffic across different continents in march 2024

	Sum of Squares	df	Mean Square	F	Sig.
Between Groups	118.118	3	39.373	2.859	.058
Within Groups	330.517	24	13.772		
Total	448.634	27			

Despite the statistical result, this analysis is critical in highlighting potential shifts toward democratized AI access. The relatively balanced shares suggest that GenAI tools are gaining traction across both developed and emerging regions, possibly indicating effective cross-border dissemination and the growing ubiquity of ChatGPT across socioeconomic contexts.

Comparing the ChatGPT traffic per internet user across different continents in march 2024

The fourth analysis sought to normalize ChatGPT usage by comparing visits per internet user across five continents. The results of the ANOVA (F(4, 25) = 1.386, p = 0.267) confirmed the absence of statistically significant differences, suggesting that internet users across continents exhibit relatively similar levels of engagement with ChatGPT. The findings are presented in Table 6.

Table 6. One-Way ANOVA Comparing the ChatGPT traffic per internet user across different continents in march 2024

	Sum of Squares	df	Mean Square	F	Sig.
Between Groups	1.133	4	.283	1.386	.267
Within Groups	5.109	25	.204		
Total	6.242	29			

This finding underscores a key insight: when adjusted for internet accessibility, user behavior with respect to GenAI tools appears to be globally convergent. In practical terms, it implies that once users are digitally connected, their likelihood of interacting with platforms like ChatGPT is comparable across regions, reflecting the tool's broad utility and appeal.

Figure 3. Comparing the ChatGPT traffic per internet user across different continents in March 2024

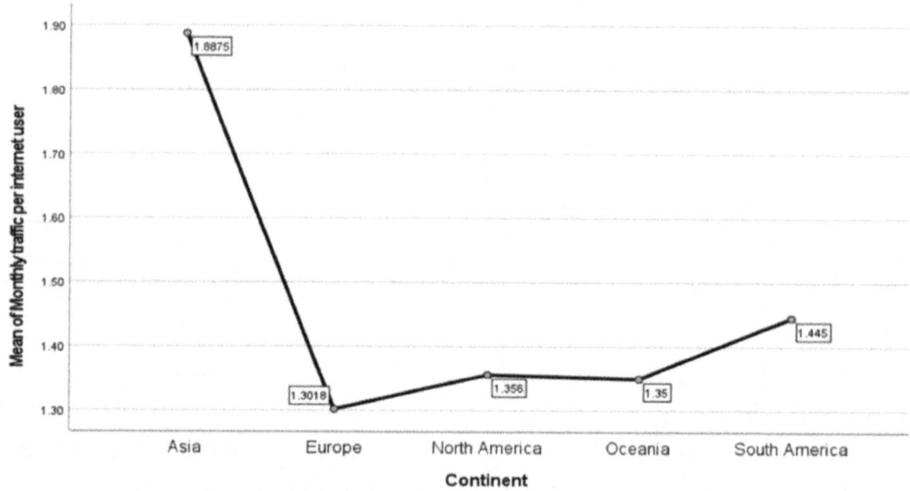

Comparing the ChatGPT traffic per person aged 15-64 across different continents in March 2024

Unlike previous dimensions, the fifth objective revealed a statistically significant difference in ChatGPT usage when normalized by the working-age population. The ANOVA ($F(3, 25) = 13.488$, $p = 0.000$) indicated that GenAI usage patterns vary substantially across continents when adjusted for the number of people aged 15–64. The findings are presented in Table 7.

Table 7. One-way ANOVA Findings Comparing the ChatGPT traffic per person aged 15-64 across different continents in march 2024

	Sum of Squares	df	Mean Square	F	Sig.
Between Groups	7.313	3	2.438	13.488	.000
Within Groups	4.518	25	.181		
Total	11.831	28			

Figure 4 depicts that Asia led in usage with an average of 4 visits per working-age person, followed by North America (3 visits), Oceania (2 visits), and Europe (1.8 visits). This pronounced disparity may stem from several region-specific factors. In Asia, rapid mobile-first digital adoption, high youth engagement with AI, and widespread use in education and content creation could explain elevated traffic. North America's high average is likely attributed to enterprise and academic use, whereas Europe's relatively lower average may reflect cultural or regulatory reservations about AI integration.

The results suggest that demographic density, digital literacy, cultural openness to automation, and economic digitization significantly influence normalized AI engagement. Unlike crude visit counts, this measure exposes regional behavioral patterns that align with broader technological and socioeconomic development trends.

Figure 4. Comparing the ChatGPT traffic per person aged 15-64 across different continents in march 2024

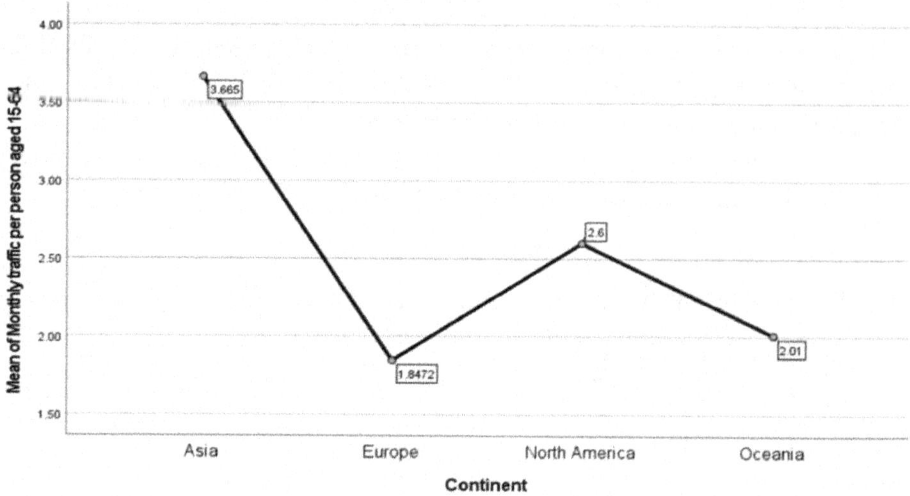

In summary, while several ANOVA tests returned non-significant results, the descriptive statistics and traffic visualizations reveal meaningful usage patterns that are not immediately evident through statistical significance alone. The only statistically confirmed hypothesis (H_5) demonstrates that regional disparities exist in ChatGPT usage per working-age individual, which could have implications for future AI education, marketing, and infrastructure planning. Meanwhile, the near-significant results in other tests indicate a need for continued longitudinal observation as GenAI adoption continues to evolve globally.

DISCUSSION

This study sought to explore and compare GenAI usage patterns on different continents based on traffic data from World Bank's March 2024 dataset. The findings, although largely not statistically significant on different dimensions, reveal important regional and behavioral trends that both affirm and counter literature on global adoption of GenAI. This discussion situates these results within the context of earlier studies, highlighting points of convergence and divergence, as well as the potential implications for future research and policy development. To begin with, the lack of statistically significant variation in the numbers of visits across various GenAI domains, i.e., Chatbots, Image, and Video tools, is consistent with the results presented by Khan et al. (2024), who noted the broad multidisciplinary use of

GenAI tools throughout. Nevertheless, whereas the statistical test in this research indicated no significant differences (p = 0.597), the descriptive statistics showed that Chatbot-based tools, and more specifically ChatGPT, outpaced others by far in terms of interactions with users. Such dominance validates the observation by Bick et al. (2024) that ChatGPT enjoys a leading status in the GenAI industry from its user-friendly text-based interface, growing usefulness across various professional and academic environments, and language processing abilities highly suited for addressing user productivity demands.

The absence of statistically significant difference, then, can be accounted for by heterogeneity in the group sizes, rather than indicative of homogeneous user preference. In comparing the use of ChatGPT among different continents, this study revealed that there was no significant difference at the 5% level (p = 0.058), yet North America had the highest average utilization. This result is partially aligned with the investigation by Liu and Wang (2024) that stated that while North America is still a main origin of traffic, middle-income countries, particularly in Asia, now account for over 50% of the traffic connected with GenAI. The proximity of the p-value to the established cut-off indicates that a larger sample or finer regional segmentation could reveal significant differences. Moreover, the convergence of global ChatGPT traffic observed could be attributed to the democratization of digital access and the linguistic versatility of the platform, as established in Wakunuma and Eke's (2024) evaluation of AI accessibility in Africa and Asia. Likewise, in the analysis of the distribution of global ChatGPT traffic among various continents, there was no difference of statistical significance (p = 0.058). Nevertheless, the descriptive findings revealed a tangible predominance of Asia and North America. These results are consistent with those of Liu and Wang (2024), who noted that traffic distribution across the globe is progressively more balanced over time, particularly as more mobile-first markets in South Asia, Latin America, and Africa gain speed. The fact that this study's traffic share results replicate those in the literature suggests that platform diffusion is growing more geographically even, if not yet equally intense or consequential. The analysis of ChatGPT traffic per user throughout the internet did not show any noticeable differences between continents (p = 0.267). Such consistency supports the findings by Kim et al. (2024), who claim that after gaining internet access, users' engagement with GenAI tools does not differ significantly across geographic regions. Their model, assuming trust, utility, and personalization as essential drivers of usage implies that these traits overcome cultural boundaries and linguistic varieties and appeal to global digital behavior. The implication is that variations in GenAI adoption have more to do with internet access and digital literacy, structural matters identified in Ali et al.'s (2025) global adoption study, than user choice.

Nevertheless, the most revealing finding of the current study is revealed in the analysis of ChatGPT traffic normalized for the 15–64-year-old population. This was the sole measure to indicate a statistically significant difference across continents ($p = 0.000$), with Asia leading in the highest interaction levels, followed by North America, Oceania, and Europe. This measure provides an essential context for interpreting broader patterns of usage. It would imply that the active population utilizing AI consists mainly of those in the working-age demographic, a group probably occupied with education, work, and content creation. The increased activity in Asia may mirror the extensive integration of ChatGPT into learning models, independent work frameworks, and computerized efficiency tools, a pattern previously observed by Huy et al. (2024) and anecdotally observed in regional news outlets. This finding is also consistent with Kim et al.'s (2024) argument that personalization and trust play a crucial role in the adoption of GenAI. In Asia, the integration of AI into everyday life, facilitated by super-app ecosystems, tech-driven education policies, and digital literacy among young people, may potentially catalyze adoption among the working population.

On the other hand, Europe's relatively lower uptake may be due to stricter regulatory regimes, data privacy concerns, and a cautious stance on the adoption of AI, as posited in Beltran et al. (2024). This regulatory climate, despite being for the higher purpose of ethics and security, can in fact dampen natural adoption rates. Second, the research results affirm the critical observation presented by Cheng and Wu (2024): policy and governance matter. The regional variations that have been noted, and especially the traffic per working-age individual when normalized, all point to the dynamic interactions among digital access, user aptitude, cultural openness, and regulatory climate. That traffic per internet user was relatively uniform, yet traffic per working-age adult was not, suggests that mere internet access is no longer the primary distinguishing factor. Instead, a population group's readiness—technically, economically, and culturally, to engage with AI is emerging as the new frontier in GenAI adoption research. Another argument supported by this evidence is that of Wakunuma and Eke (2024), who argue that Africa's absence from the top ranks of GenAI figures is more a function of infrastructural and policy limitations than of disinterest. If the continent's youth population gains wider access to AI technologies and digital literacy programs, the continent's traffic market share can increase exponentially, as observed in trends already developing in large areas of Asia and Latin America.

Collectively, these results and consistency with earlier research suggest that the worldwide uptake of ChatGPT and other GenAI technologies is proceeding at a rate toward mass adoption, but still subject to deeper socioeconomic processes and policy considerations. The convergence in per-capita user traffic across various regions is evidence of the broad appeal and adaptability of ChatGPT, while the divergences in usage among working-age populations are indicative of the effects of local conditions and institutional environments.

While this study focuses on traffic data for ChatGPT, it is essential to recognize that multiple contextual factors influence the adoption of GenAI. The proliferation of alternative tools, such as Google's Gemini, Anthropic's Claude, or Microsoft's Copilot, may lead to substitution effects, where users choose platforms based on availability, pricing, or user experience. Moreover, ChatGPT is increasingly embedded into third-party applications such as Microsoft Office, Bing, and Slack. Traffic through these integrations may not be fully reflected in standalone web traffic data, potentially underestimating actual usage in certain regions. These dynamics highlight the evolving nature of GenAI engagement and suggest that traffic-based analysis, while informative, may only capture part of the broader usage ecosystem. Additionally, language diversity and multilingual model capability influence adoption, particularly in regions with linguistically heterogeneous populations. While ChatGPT supports multiple languages, access to high-quality responses in local or minority languages may still lag, potentially limiting user engagement in parts of Africa, Asia, and Latin America.

Mobile internet penetration also plays a key role. In regions like Sub-Saharan Africa and South Asia, where mobile is the primary mode of internet access, device compatibility and data affordability can shape usage behavior. This may explain why traffic volume in these areas is not proportionate to their large working-age populations. Furthermore, data privacy regulations such as the European Union's General Data Protection Regulation (GDPR) may contribute to more cautious engagement with generative AI tools in certain regions. Concerns over how platforms like ChatGPT handle user data could influence usage patterns, particularly in jurisdictions with stringent digital oversight. While this study did not directly assess the environmental or economic impacts of generative AI, future research could explore how increased usage correlates with energy consumption, infrastructure demand, and economic shifts in labor or productivity.

Risks and Misuse of Generative AI

Despite the accelerating global adoption of generative AI tools, concerns about their misuse and unintended consequences are increasingly shaping the trajectory of adoption, particularly in regions with heightened regulatory oversight. While this chapter primarily analyzes usage behavior, it is essential to contextualize these patterns within the emerging landscape of GenAI-related risks. A major concern is the proliferation of AI-generated disinformation, particularly in politically sensitive environments. For instance, deepfake technologies have been used to manipulate political speeches and impersonate public figures, raising questions about the role of GenAI in undermining democratic discourse (Ferrara, 2024). Such risks have led several governments, especially in Europe, to call for stricter governance frameworks and digital authentication mechanisms[1]. In academic and professional domains, tools like ChatGPT have been linked to plagiarism and academic dishonesty, prompting universities to adopt AI-detection tools and reassess assessment policies (Bin-Nashwan et al., 2023). The risk landscape also includes security vulnerabilities, where prompt injection attacks or unrestricted access to GenAI APIs may be exploited for malicious purposes, including phishing, spam, and fraud (Ibrar et al., 2025). These possibilities have fueled regulatory hesitancy in certain jurisdictions, with some institutions implementing partial bans or usage restrictions pending further study. While the prevalence and impact of these misuses vary across regions, they contribute to adoption hesitancy, especially in areas with strong data protection norms or heightened sensitivity to digital harm. As such, GenAI's success in different markets may increasingly depend on the development of trust frameworks, responsible use guidelines, and adaptive policy environments.

CONCLUSION

This study sought to explore and compare GenAI traffic on the continental level, namely ChatGPT usage, by employing descriptive and inferential statistical analyses of worldwide data in March 2024. Out of the exploration of five grand hypotheses, the study offers a qualified examination of the uptake and utilization of GenAI tools in various regional contexts. While none of the most widely applied statistical tests revealed significant differences, the descriptive trends revealed insightful observations. GenAI tools that leverage chatbots, specifically ChatGPT, continue to dominate the AI field in terms of user engagement, thereby echoing earlier research that highlights the broad popularity, usability, and usefulness of text-based AI tools. Although no statistically significant differences in overall ChatGPT visits, percentage of global traffic, or traffic per internet user by continent were observed, the closeness of these measurements is a reflection of a growing global convergence towards AI interaction, once internet availability is assured.

Yet, the research identified a vast difference in ChatGPT use per working-age populace (15–64), with Asia at the top, followed by North America and Oceania. This observation identifies deeper regional patterns in AI use, which capture variations in digital literacy, institutional adoption of AI, youth population size, and cultural openness to emerging technologies.

The results align with previous research (e.g., Liu & Wang, 2024; Kim et al., 2024; Wakunuma & Eke, 2024), highlighting the role of demography, policy setting changes over time, and the robustness of technological infrastructure in influencing global uptake of GenAI. The research also supports the notion that availability of technology by itself will not ensure equitable use but, rather, substantial use is contingent upon local preparedness, education systems, and governance systems. Briefly, the worldwide diffusion of GenAI tools, as exemplified by ChatGPT, seems both swift and lopsided. This research adds to the expanding literature by providing a comparative, empirical snapshot of AI usage on the different continents. As GenAI is still developing, future studies must foreground longitudinal investigations, nation-state case studies, and the incorporation of qualitative data to continue investigating the social, cultural, and ethical aspects of the worldwide adoption of AI.

LIMITATIONS AND FUTURE WORK OF THE STUDY

Although secondary data enabled large-scale comparisons, its use introduces certain limitations. The analysis is limited by the absence of disaggregated demographic variables (e.g., age, gender, languages) and qualitative indicators such as user sentiment, trust, or skill level. While traffic metrics offer valuable macro-level insight, they may not fully capture the complexities of adoption behavior, risk perception, or digital readiness. Additionally, platform-specific constraints, such as regional access restrictions, embedded usage (e.g., in Bing or apps), or network latency, were not captured. These factors may affect traffic volume but are not observable in this dataset. Finally, this chapter is limited by the fact that it was based on secondary data. Future qualitative studies, including interviews with policymakers, AI startups, and digital inclusion NGOs, will complement our findings and enrich the policy dimension.

RECOMMENDATIONS

From the findings and comparative study of GenAI usage on different continents, the recommendations that are being put forward to bridge regional disparities in use and to achieve a more balanced global interaction with models such as ChatGPT are as follows:

Increase Access in Underserved Communities

Targeted efforts should be directed towards improving access to GenAI technology in areas with comparatively low uptake, especially in Europe and Oceania. This initiative entails expanding internet coverage, reducing costs associated with data and digital hardware, and expanding the broadband network. Concerted interventions by government agencies, private sector entities, and international development organizations are required to break down technological barriers to user uptake in these environments.

Promote Digital Literacy and AI Education

Mass use of GenAI requires not just access but digital competence. Efforts to enhance digital literacy should be prioritized, especially in nations having medium to low traffic volumes. Public awareness campaigns, training sessions focused on AI, and the integration of digital competencies into school curricula can empower users to engage more confidently and critically with AI tools.

Develop AI Solutions Regionally Tailored

The creators of GenAI tools need to follow a localization strategy that tunes tools to address the cultural, linguistic, and functional requirements of various user segments. The integration of multilingual capability, culturally appropriate prompts, and localized knowledge bases, for example, would greatly improve user experience and contextual applicability. Such a degree of customization has the potential to promote higher adoption rates and continued use in multicultural areas.

Create awareness through localized promotion activities

In regions such as Europe, where ChatGPT adoption is surprisingly minimal despite high internet penetration, targeted promotion is required to demonstrate the concrete utility and benefit of GenAI. Messaging must highlight everyday life practical application, education, business, research, etc., to help shift perceptions and instill confidence in these tools.

Engage Youth and Working-Age Populations

As the young and working population are the most active users of GenAI, policies and programs should be crafted to promote uptake among this demographic. Partnerships with educational institutions, universities, and corporate training institutes can provide first exposure to AI tools. Facilitating academic and workplace use of GenAI not only raises adoption but also innovation and productivity.

Invest in Equitable Global Digital Infrastructure

To achieve widespread and sustained adoption of GenAI technologies, governments and global leaders need to invest in developing robust and equitable digital infrastructure. This encompasses physical connectivity along with regulatory frameworks that facilitate open access, privacy protection, and responsible use of AI. Equitable investment in infrastructure is especially crucial in emerging economies where technological disparities can be leveraged to perpetuate systemic inequalities.

Through the implementation of the aforementioned recommendations, stakeholders can alleviate current disparities in GenAI contribution and create a more stable, equitable, and culturally sensitive AI environment worldwide.

KEY TAKEAWAYS

Generative AI Adoption Varies Widely Across Continents

Asia shows the highest per-capita usage, driven by mobile-first ecosystems and tech innovation, while Europe demonstrates slower uptake due to stricter regulations and data privacy laws.

Access to GenAI Reflects Infrastructure And Policy Disparities

Regions with stronger broadband coverage, cloud infrastructure, and regulatory clarity (e.g., North America, parts of Asia) show higher engagement levels compared to under-resourced areas.

Regulatory Environments Significantly Influence Adoption

Europe's cautious approach under the GDPR and forthcoming AI Act contrasts with the more flexible policies seen in Africa and Latin America, affecting both user behavior and platform availability.

Language and Localization Still Pose Adoption Barriers

Although some platforms support multiple languages, English remains dominant. Multilingual adoption in Latin America and Africa is still emerging, suggesting a need for better language support in GenAI tools.

Usage Patterns Have Surged Post-ChatGPT

A notable spike in traffic occurred globally after late 2022, marking a turning point in public accessibility and awareness of generative AI tools.

Risks such as Misinformation And Academic Misuse Are Emerging

These concerns are shaping both public perception and regulatory responses, especially in education, media, and politics.

REFERENCES

Ali, H., Mustafa, A., & Aysan, A. F. (2025). Global adoption of generative AI: What matters most? *Journal of Economy and Technology, 3*, 166–176. DOI: 10.1016/j.ject.2024.10.002

Beltran, M. A., Mondragon, M. I. R., & Han, S. H. (2024). Comparative Analysis of Generative AI Risks in the Public Sector. *ACM International Conference Proceeding Series*, 610–617. DOI: 10.1145/3657054.3657125

Bick, A., Blandin, A., Deming, D. J., & Louis, S. (2024). *The Rapid Adoption of Generative AI*. DOI: 10.3386/w32966

Bin-Nashwan, S. A., Sadallah, M., & Bouteraa, M. (2023). Use of ChatGPT in academia: Academic integrity hangs in the balance. *Technology in Society, 75*, 102370. DOI: 10.1016/j.techsoc.2023.102370

Chang, C. H., & Kidman, G. (2023). The rise of generative artificial intelligence (AI) language models - challenges and opportunities for geographical and environmental education. *International Research in Geographical and Environmental Education, 32*(2), 85–89. DOI: 10.1080/10382046.2023.2194036

Cheng, K., & Wu, H. (2024). Policy framework for the utilization of generative AI. *Critical Care, 28*(1), 1–2. DOI: 10.1186/s13054-024-04917-z PMID: 38637898

Emon, M. M. H., & Khan, T. (2025). The mediating role of attitude towards the technology in shaping artificial intelligence usage among professionals. *Telematics and Informatics Reports, 17*, 100188. DOI: 10.1016/j.teler.2025.100188

Ferrara, E. (2024). GenAI against humanity: Nefarious applications of generative artificial intelligence and large language models. *Journal of Computational Social Science, 7*(1), 549–569. DOI: 10.1007/s42001-024-00250-1

Feuerriegel, S., Hartmann, J., Janiesch, C., & Zschech, P. (2024). Generative AI. *Business & Information Systems Engineering, 66*(1), 111–126. DOI: 10.1007/s12599-023-00834-7

Fui-Hoon Nah, F., Zheng, R., Cai, J., Siau, K., & Chen, L. (2023). Generative AI and ChatGPT: Applications, challenges, and AI-human collaboration. *Journal of Information Technology Case and Application Research, 25*(3), 277–304. DOI: 10.1080/15228053.2023.2233814

Gupta, V. (2024). An Empirical Evaluation of a Generative Artificial Intelligence Technology Adoption Model from Entrepreneurs' Perspectives. *Systems 2024, Vol. 12, Page 103, 12*(3), 103. DOI: 10.3390/systems12030103

Ibrar, W., Mahmood, D., Al-Shamayleh, A. S., Ahmed, G., Alharthi, S. Z., & Akhunzada, A. (2025). Generative AI: A double-edged sword in the cyber threat landscape. *Artificial Intelligence Review*, *58*(9), 1–27. DOI: 10.1007/s10462-025-11285-9

Jo, H. (2024). Uncovering the Reasons behind Willingness to Pay for ChatGPT-4 Premium. *International Journal of Human-Computer Interaction*, *41*(2), 994–1009. DOI: 10.1080/10447318.2024.2307692

Kalinaki, K., Acaru, S. F., Kugonza, J., & Nsubuga, R. (2024). Towards an Intelligent Tomorrow: Machine Learning Enabling Sustainable Development. In *Methodologies* (pp. 66–89). Frameworks, and Applications of Machine Learning., DOI: 10.4018/979-8-3693-1062-5.ch004

Kalinaki, K., Malik, O. A., & Lai, C. D. T. (2024). Forest Canopy Cover Mapping in High-Resolution Multisource Satellite Imagery Using Attention Recurrent Residual U-Net. *IEEE Region 10 Annual International Conference, Proceedings/TENCON*, 400–403. DOI: 10.1109/TENCON61640.2024.10902705

Kalinaki, K., Namuwaya, S., Mwamini, A., & Namuwaya, S. (2023). Scaling Up Customer Support Using Artificial Intelligence and Machine Learning Techniques. In *Contemporary Approaches of Digital Marketing and the Role of Machine Intelligence* (pp. 23–45). IGI Global., DOI: 10.4018/978-1-6684-7735-9.ch002

Kalinaki, K., Yahya, U., Malik, O. A., & Lai, D. T. C. (2024). A Review of Big Data Analytics and Artificial Intelligence in Industry 5.0 for Smart Decision-Making. In *Human-Centered Approaches in Industry 5.0* (pp. 24–47). Human-Machine Interaction, Virtual Reality Training, and Customer Sentiment Analysis., DOI: 10.4018/979-8-3693-2647-3.ch002

Kaur, P., & Gupta, D. (2025). Exploring the dimensions of generative AI. *Progressive Computational Intelligence, Information Technology and Networking*, 159–165. DOI: 10.1201/9781003650010-26

Khan, N., Khan, Z., Koubaa, A., Khan, M. K., & Salleh, R. (2024). Global insights and the impact of generative AI-ChatGPT on multidisciplinary: A systematic review and bibliometric analysis. *Connection Science*, *36*(1), 2353630. DOI: 10.1080/09540091.2024.2353630

Kim, D., Kim, S., Kim, S., & Lee, B. H. (2024). Generative AI Characteristics, User Motivations, and Usage Intention. *Journal of Computer Information Systems*, 1–16. Advance online publication. DOI: 10.1080/08874417.2024.2442438

Lartey, D., & Law, K. M. Y. (2025). Artificial intelligence adoption in urban planning governance: A systematic review of advancements in decision-making, and policy making. *Landscape and Urban Planning*, *258*, 105337. DOI: 10.1016/j.landurbplan.2025.105337

Law, L. (2024). Application of generative artificial intelligence (GenAI) in language teaching and learning: A scoping literature review. *Computers and Education Open*, *6*, 100174. DOI: 10.1016/j.caeo.2024.100174

Lee, H., Jeong, S., & Lee, K. (2021). Estimating the deployment costs of broadband universal service via fiber networks in Korea. *Telecommunications Policy*, *45*(4), 102105. DOI: 10.1016/j.telpol.2021.102105

Liu, Y., & Wang, H. (2024). *Who on Earth is using generative AI?* https://documents.worldbank.org/curated/en/099720008192430535

Mahadevappa, P., Muzammal, S. M., & Tayyab, M. (2025). Introduction to Generative AI in Web Engineering: Concepts and Applications. Https://Services.Igi-Global.Com/Resolvedoi/Resolve.Aspx?Doi=10.4018/979-8-3693-3703-5.Ch015, 297–330. DOI: 10.4018/979-8-3693-3703-5.ch015

McKinsey. (2023). *The state of AI in 2023: Generative AI's breakout year*. https://www.mckinsey.com/capabilities/quantumblack/our-insights/the-state-of-ai-in-2023-generative-ais-breakout-year

Novelli, C., Casolari, F., Hacker, P., Spedicato, G., & Floridi, L. (2024). Generative AI in EU law: Liability, privacy, intellectual property, and cybersecurity. *Computer Law & Security Review*, *55*, 106066. DOI: 10.1016/j.clsr.2024.106066

Rawashdeh, A. (2023). The consequences of artificial intelligence: An investigation into the impact of AI on job displacement in accounting. *Journal of Science and Technology Policy Management*, *16*(3), 506–535. DOI: 10.1108/JSTPM-02-2023-0030

Saeed, M. M. A., Saeed, R. A., & Ahmed, Z. E. (2024). *Data Security and Privacy in the Age of AI and Digital Twins*., DOI: 10.4018/979-8-3693-1818-8.ch008

Sedkaoui, S., & Benaichouba, R. (2024). Generative AI as a transformative force for innovation: a review of opportunities, applications and challenges. *European Journal of Innovation Management, ahead-of-print*(ahead-of-print). DOI: 10.1108/EJIM-02-2024-0129

Shafik, W. (2023). Introduction to ChatGPT. In *Advanced Applications of Generative AI and Natural Language Processing Models* (pp. 1–25). DOI: 10.4018/979-8-3693-0502-7.ch001

Shafik, W., Hidayatullah, A. F., Kalinaki, K., & Aslam, M. M. (2024). Artificial Intelligence (AI) Assisted Computer Vision (CV) in Healthcare Systems. *Computer Vision and AI-Integrated IoT Technologies in the Medical Ecosystem*, 17–36. DOI: 10.1201/9781003429609-2

Shilaskar, S., Chakole, S., Samarth, A., & Shejole, P. (2025). GenAI based Data Extraction with Query-based Insights. *2025 International Conference on Emerging Smart Computing and Informatics, ESCI 2025*. DOI: 10.1109/ESCI63694.2025.10987906

Solove, D. J. (2025). Artificial Intelligence and Privacy. *Florida Law Review*, *77*, •••. https://heinonline.org/HOL/Page?handle=hein.journals/uflr77&id=1&div=&collection=

Van Huy, L., & Nguyen, T. (2024). Generative AI, Why, How, and Outcomes: A User Adoption Study. *AIS Transactions on Human-Computer Interaction*, *16*(1), 1–27. DOI: 10.17705/1thci.00198

Wakunuma, K., & Eke, D. (2024). Africa, ChatGPT, and Generative AI Systems: Ethical Benefits, Concerns, and the Need for Governance. *Philosophies 2024, Vol. 9, Page 80*, *9*(3), 80. DOI: 10.3390/philosophies9030080

Zakari, R. Y., Lawal, Z. K., Kalinaki, K., & Maiwada, U. D. (2024). Time Series Forecasting Techniques for Climate Trend Prediction. *Reshaping Environmental Science Through Machine Learning and IoT*, 282–310. DOI: 10.4018/979-8-3693-2351-9.ch014

Zenni, R. D., & Andrew, N. R. (2023). Artificial Intelligence text generators for overcoming language barriers in ecological research communication. *Austral Ecology*, *48*(7), 1225–1229. DOI: 10.1111/aec.13417

ENDNOTES

1 https://digital-strategy.ec.europa.eu/en/policies/eudi-regulation

Chapter 16
Inside the Black Box:
How Generative AI Creates Digital Synthetics

Valisher O. Sapayev
https://orcid.org/0000-0002-6751-5864
Mamun University, Uzbekistan

Noor Hayat
https://orcid.org/0000-0003-2726-2847
University of Central Punjab, Pakistan

Zokir Mamadiyarov
Termez University of Economics and Service, Uzbekistan

Munir Ahmad
https://orcid.org/0000-0003-4836-6151
Survey of Pakistan, Pakistan

ABSTRACT

This chapter unpacks the inner workings and outer implications of generative AI systems that produce synthetic text, images, audio, and video. While these models demonstrate remarkable creativity and utility, they also embody a core paradox: we increasingly rely on tools whose decision-making processes we do not fully understand. Exploring architectures such as transformers and diffusion models, we examine how outputs emerge from complex, probabilistic systems. The chapter also addresses the challenges of opacity, emergent behavior, and limited interpretability, highlighting the need for ethical oversight, technical safeguards, and societal trust in an age where synthetic content is reshaping digital reality.

DOI: 10.4018/979-8-3373-2277-3.ch016

INTRODUCTION

Over the past few years, the digital world has experienced extraordinary change with the meteoric emergence of generative artificial intelligence (generative AI) systems, technology that is able to generate human-quality content in different types of media, such as realistic images, fluid text, immersive audio, and lifelike video. The result of these models is what is termed digital synthetics: alleged creations of artificial representations of language, images, and sound that have the appearance of being actually real but are increasingly becoming indistinguishable from the actual. In contrast to the traditional software that is programmed with strictly defined rules, generative AI systems train and learn based on huge amounts of data and then produce content that resembles the structure and patterns it detects in the real-world data. This capacity to generate novel material has facilitated various uses, such as AI art, virtual assistants, deepfake videos, voice clones, and algorithmically produced music that are transforming the activities of a number of industrial sectors and even cultural traditions (Ahmad et al., 2025; Chapelle, 2025; Chen et al., 2023; Lee et al., 2024; Owoseni et al., 2024; Ramdurai & Adhithya, 2023).

Nevertheless, regardless of their unparalleled performance, the inner nature of such systems is quite inaccessible to the broad community and even to some developers applying them. With generative AI finding more common ground in the tools of daily interaction, it is even more important to comprehend how such models truly operate and work so that not only non-technical users, but also policymakers, educators and creatives as well as citizens who have to face an information-abundant and trust-deficient digital environment are less inclined to place the source of trust in a de facto black box. The key to this revolution lies in these super-architectures, such as Generative Adversarial Networks (GANs), Variational Autoencoders (VAEs), diffusion models, as well as Transformer-based models such as GPT and DALL·E. All these architectures have different functions in determining the nature of content that we edit and watch on the Internet (Bengesi et al., 2024; Cao et al., 2024).

In this chapter, the onion skin is cut on such so-called black box systems to show the basic key principles that power their synthetic creativity. It defines the fundamental types of models used to train and run generative AI, how they do it, the problems associated with their complexity, and the difficulty in understanding them. By revealing more about these systems and better explaining their manner of operation, the chapter provides technical insights on how the generative AI technology can and is used to form the synthetic content that is quickly entering our digital world.

FOUNDATIONS OF GENERATIVE AI

The era of generative AI can be described as one of the most important achievements in contemporary computing, when machines are no longer exposed to the task of data analysis but are capable of creation. This transition to the generation is the result of a successive series of deep learning architectural innovations made sequentially and brought together. To know how a machine can now generate original images, texts, or even songs, we need to cover the history of the most critical architectures that are behind this generative revolution: GANs, VAEs, diffusion models, and transformer-based models.

This long story started in 2013 with the VAEs - one of the first methods that now facilitates learning in decompressed, abstract representations of input data, called the latent space, by machines. The two fundamental elements of a VAE are an encoder, which compresses unencoded information into reconstruction variables, and a decoder, which tries to restore the original information out of these variables. The meaning of this encoding-decoding loop is not only that the model is able to recognize the structure of data, but it is also able to generate new data of the same theory as it has been sampled in the shared latent space. Despite being one of the earliest neural architectures that could produce novel content, especially images, VAEs produced outputs that were generic in most cases or blurry and lacking fine details. Notwithstanding the described limitations, VAEs set important theoretical precedents, particularly the utilization of probabilistic latent variables, which have a lasting impact on modern generative models in all modalities (Asperti et al., 2021; Papadopoulos & Karalis, 2023; Wei et al., 2020; Wei & Mahmood, 2021).

One of the biggest steps was made in 2014 when GANs were introduced, which took an entirely new approach to synthetic generation. Unlike the reconstruction-only, GANs come with a competitive training scheme which employs two neural nets: a generator and a discriminator that tries to produce fake data in resemble of the real and determines whether the created data is genuine or not. Such a competitive arrangement triggers the game-theoretic activity where the generator trains to generate more convincing output in its attempts to deceive the discriminator. The GANs rapidly became an epitome of high-quality image generation (Alqahtani et al., 2021; Bengesi et al., 2024; Biswas et al., 2023; Pan et al., 2019; Saxena & Cao, 2021; Yoon et al., 2020). After introducing StyleGAN, GANs got to a completely new level, where it became possible to generate hyper-realistic human faces and artistic images that stretched the boundaries of how realistic images could be. Nevertheless, GANs do also have their disadvantages; they are known to be very hard to train, and they can be quite unstable sometimes, such as the case of mode collapse, where the generator can generate similar things instead of a variety of different offerings Generative Adversarial Networks.

The researchers responded to some of these training difficulties by venturing into a new family of generative models that were based upon a more iterative process: diffusion models. Instead of the adversarial games, diffusion models also operate by iteratively adding noise to the input data until it appears as totally random and then learn how to undo this process to produce sensible outputs. The resulting generative process can be controlled very finely with this noise-based learning because it gives results of very high detail. Diffusion models have also been very costly in terms of computational resources, since they have a large number of iterative steps, which has increased their stability and controllability and made them well-suited to a new generation of image synthesis tools. Such successful examples as Stable Diffusion or Imagen demonstrate how accurate and detailed diffusion models can be, producing a high-resolution, detailed image when using only a textual description or an image as the input (Cao et al., 2024; Croitoru et al., 2023; Yang et al., 2024).

As such, innovations were playing out in the space of visual generation, and a transformative architecture was also appearing in the space of language modeling: the transformer. Transformers were proposed in 2017 to overcome the limitations of contextually modeling sequential data with the prior models due to their self-attention mechanism. With this strategy, the model can dynamically place value on the significance of individual aspects within a sequence to acquire long-term characteristics and semantic connections exceptionally successfully. The transformer design quickly fueled a surge of massive language models (LLMs), one of which is Generative Pretrained Transformer (GPT), which can generate coherent and contextually relevant text that has never been observed before. What is more relevant, the module and scalability of the transformer made it possible to transcend text; thus, a new type of model emerged, the so-called multimodal ones (i.e., those that are capable of producing images on the basis of textual descriptions), including DALL·E. Transformers are now central to almost every state-of-the-art generative AI architecture on account of their flexibility, effectiveness, and multi-modal mix-and-match properties (Bengesi et al., 2024; Rosário, 2024; Yadav, 2024; Zhang et al., 2023). Figure 1 shows the evolution of generative AI architectures from early probabilistic models to powerful multimodal transformers. Each layer builds upon the previous, increasing the complexity, realism, and flexibility of synthetic content generation.

Figure 1. Evolution of Generative AI Architectures

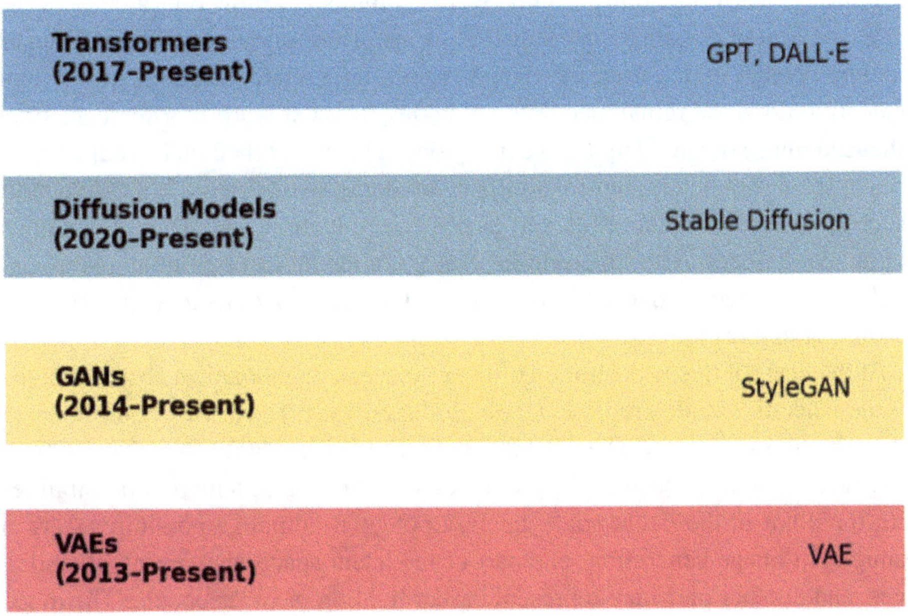

OPENING THE BLACK BOX

Generative AI may seem like that magic trick to a layperson: with only a prompt or an uploaded drawing, in a few seconds, a believable essay, piece of art, or song can be generated. However, behind this smooth interface is a very complex chain of events in which models can learn, represent, and generate content. Although math and engineering on such systems may be complicated, an intuitive understanding of the underlying workings of the system is possible. This section opens up the black box of generative AI to show the inner logics of how synthetic ingenuity can be possible, and examines the training of these models, how they manipulate information in latent space, and how they sample over learned representations to generate novel output.

The fundamental feature of all the generative models is learning based on exposure to big data (Gakhar et al., 2025; Mahmoud et al., 2025). To train models, millions of examples are fed to them, e.g., millions of images, documents, recordings of speech, or any combination. The model aims to identify the patterns, the regularities, and the statistical connections in this data. For instance, a model that learns to produce faces may learn to expect an average distance between eyes, symmetry of the face, or other changes created by different lighting. A language model, conversely, can be trained to learn such things as that words commonly follow one another or that sentences follow certain grammatical patterns. This training is normally supervised by oneself in directing the model is train on guessing sections of the input foundation upon the remainder- say, predicting the next word in a sentence or an enhanced image dreamed up from a noisy one. The model then makes adjustments to its internal parameters (typically billions of them), so that errors in its predictions match up with the real data as closely as possible, with time and in a process known as gradient descent. After the training process, a model can be fine-tuned using a smaller, more specific dataset in order to customize its outputs to a specific style, domain, or application.

At the heart of this is the notion of latent space, a mathematical abstraction that lets the generative models represent the high-dimensional data in a lower-dimensional representation (Asperti & Tonelli, 2023; Kaswan et al., 2023; Yee-King, 2022). Imagine the latent space as a playground of sorts or even as a map of possibilities, with the point in the latent space being a potential output (a possibility). As an example, in image generation, one part of the latent space may represent smiling faces, and another part may represent darker lighting or older people. The model that learns latent representations is essentially learning a method of extracting the essence of the data in a concise and workable form. This enables the possibility of rich applications, e.g., interpolating between two faces (a morph), editing features by moving coordinates (e.g., adding glasses, changing hair color), or driving outputs towards certain features. The abstract concepts, such as style, tone, or mood, are numerically encoded in the latent space where the model gains the possibility of combining the real and new.

When a model has attained some high-density latent representation based on the data, it has to produce outputs, and sampling methods come in at this point. Sampling denotes the act of taking values in the latent space or probability distribution over possible values. To put it simply, the model uses sampling to select what to generate. Random sampling is applied to some models, which enables them to be creative and variable, whereas others are used to manipulate the temperature controls of the model to determine how predictable or surprising the result is. As an example, in a text generator, a hot temperature would generate wacky, imaginative text, and a cold temperature would generate more conservative, steady-by-the-book writing. Diffusion models, in particular, perform their sampling in an iterative reverse-noise procedure that informs the model to approach a detailed, coherent image out of the true randomness presented at the start of the sampling procedure. Inference time is the point at which the process is usually done, and at this stage, the user notices, the model converts abstract representations into concrete contents.

When it comes to people that are not involved in machine learning, a useful analogy of a generative model as a great head chef (Alwahedi et al., 2024; Hla & Hindin, 2025; Rashidi et al., 2025). Training can be equated to that in the example, the chef has gone through thousands of different recipes, analyzing and trying different ingredients to learn the interaction that flavors and textures have. The pantry in the chef's mind is called latent space; we arrange ingredients and mix them up in new ways. The art of creating a new dish is known as sampling-how much to use of every ingredient, changing the flavoring, and putting it on a plate. The last meal, a dish, an image, or a text, is the result of a large number of previously unseen but trainable choices made out of experience and intuition that have been simulated into model parameters.

With time and demystification of these fundamental processes, training, latent space, and sampling, we can start to realize that generative AI is not merely generating random outputs, nor are they just repeating some historical data. Rather, it is the learned space of possibilities navigation as a combination of new material on the basis of statistical patterns and the directions of creativity encoded through training. It is a vital aspect to learn about this inner process because it is becoming increasingly autonomous and powerful in media, communication, and decision-making.

MODALITIES OF GENERATION OF TEXT, IMAGE, AUDIO, AND VIDEO

Synthetic media is the most prominent effect of generative AI: media that sounds realistic in any of these types: text, images, audio, and most recently, video. Although the fundamentals of generative modeling may still be the same, each modality has its own issues and needs different and unique architectures to provide natural and consistent outputs. This section decodes the production of various media types by leading-edge models, in addition to pointing out the shared abilities and special technical challenges that emerge in each area. Figure 2 presents a comparative complexity spectrum of generative modalities. Image and audio generation introduce spatial and temporal complexity, respectively, while video generation requires seamless coordination across both dimensions.

Figure 2. Complexity Spectrum of Generative Modalities

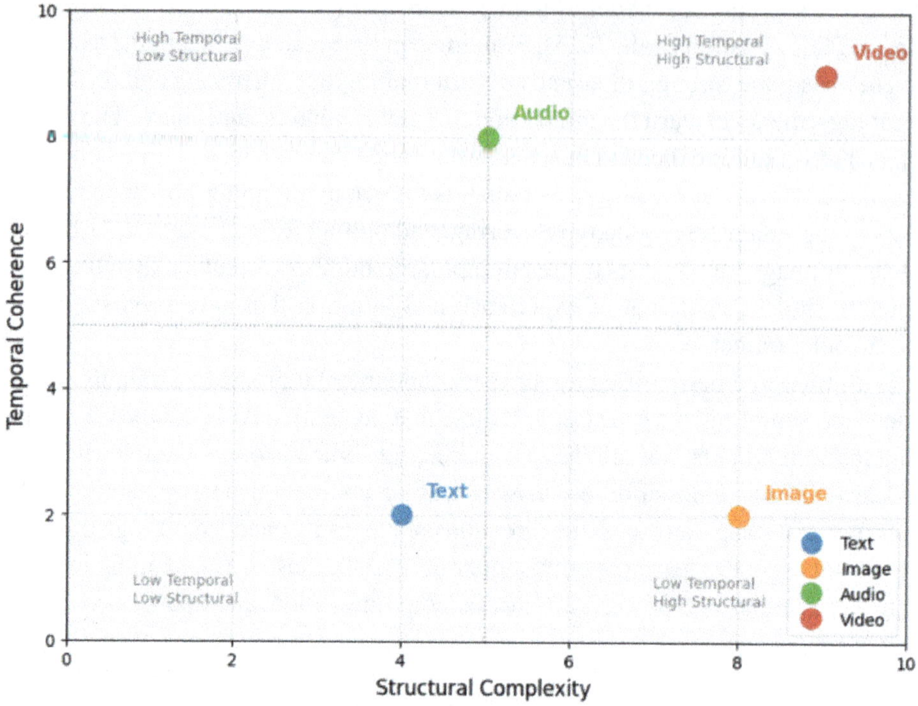

Text Generation

The text generation is the most developed, adapted to various tasks, and affects business in one way or another. Its wide usage is due to its urgency- everyone uses language and its fluidity- it can be made to perform many duties like composing emails, translating between languages, creating fiction, answering queries, tutoring, and even creating academic research. The technology behind this revolution is the transformer architecture that is the basis of some models like GPT (Generative Pretrained Transformer) by OpenAI, LLaMA (Large Language Model Meta AI) by Meta, and Claude by Anthropic (de Rosa & Papa, 2021; Lancaster, 2023).

Machine text generation is all about anticipating which word is next. They are trained on large-scale text corpora, that is, books, articles, websites, dialogues, and code repositories, to estimate the statistical probability that one word or phrase follows another. E.g., given the phrase "climate change is," the model will come to realize that words, such as the ones like it is real, a threat or an ongoing issue, are more likely than anything other. As this training continues, the model builds a probabilistic view of words and language, not just the words and grammar, but also shades of style, tone, and situational context, as well. The resulting model has this statistical learning capability, which allows it to replicate talk with an impressive fluency and flexibility in a way that is human-like.

The use of self-attention is what makes transformer-based models particularly effective because it enables them to weigh, prioritize words over long contexts. It implies that a model such as GPT-4 has access to the context of a word in a given paragraph to what a given reference made several lines prior can mean, allowing the model to create coherent and contextually relevant text that has cohesive themes. In contrast to other previous natural language models, which have difficulties when dealing with longer texts, it is quite easy to get single-sentence descriptions of multi-page documents with transformers. The uses of text generation are enormous. In artistic descriptions, such models have been applied to create poetry, screenplays, and couplets. To that end, they create legal summaries, draft reports, and automate customer service and responses in the professional setting. They help in writing code and debugging in programming. In learning, they can act as tutors, quiz authors, and tutoring characters. Text generation has proven to be diverse, as the models have been quickly incorporated into daily use, in search engines, writing assistants, virtual agents, etc.

Nevertheless, the strength of language models is the area of weakness. Since they can only produce what is likely to happen/be true and not necessarily what is true, they are faced with the odds of being in a state of hallucinations: acting with assurance and producing information that is not factual, of a false nature, or total fabrications. As an example, a model can make up references, misattributions of quotes, or wrong definitions, presenting them as fixed beyond any doubt. This poses threats in areas where precision and faith are paramount, like in healthcare, law, and education. The other restriction is long-term consistency. They may express themselves very well over short passages, but these models can quickly get lost when it comes to details in a narrative, or lose the thread of logical arguments in longer passages. The name of one of the characters can be altered halfway through the plot; the end might not agree with the arguments that have been presented earlier. These discrepancies are in line with the lack of a stable internal model of truth or memory, which differentiates generative models from what can truly be defined as intelligent agents. Even with these complications, text generators are becoming increasingly reliable and useful because of current improvements in retrieval augmented generation, fact-checking layers, and human-taught tuning.

Image Generation

Image making can be described as one of the most visually impressive capabilities of generative AI, which transforms text, sketches, or even any form of vague concept into a detailed, imaginative image. In its essence, this form of work could enable the machines to convert linguistic or symbolic information into encoded structures of pixels, or simply put, create on-demand art, illustrations, design, and photoreal text. Unlike previous techniques of computer graphics, which necessitated the intervention of human artists or procedural techniques, modern generative models automatically generate images that have a variety in terms of styles, semantic consistency, and flair (Bansal et al., 2024; Zhan et al., 2023).

Such models as DALL·E, Midjourney, and Stable Diffusion are taking the industry by storm because of the strength of combining diffusion models with transformer-based encoding. They do this by initially converting a text description (e.g., a cyberpunk city skyline at sunset) into a latent representation it one that semantically describes the text. This is, in turn, decoded to a visual structure with trained diffusion networks, which unfold a noise step by step to unveil an image. Basically, the model begins with visual noise and then develops recognizable forms, texture, and colors in the direction of the prompted input. What ensues is a unique picture that is visually attractive to signify the intended idea, even when the brief may be imaginative, surreal, or have no semblance to capture the photo through the camera.

Cross-modal alignment, which allows the model to bridge the gap between language and vision, is one of the breakthroughs that is possible with such systems. This is made possible by simultaneous training on very large collections of image-caption pairs that teach the model ways of associating words with visual representations. An example of these would be that it has become established that the term "skyline" would mean a horizontal arrangement of buildings, whereas the term "cyberpunk" would involve neon lighting, futuristic designs, and urban decay. Accuracy of such associations would be very important in the generation of work that fulfills human expectations and artistic intent. It is in this cross-modal translation that such models as Contrastive Language-Image Pretraining (CLIP) can facilitate positioning the visual output as semantically faithful to the prompt (Tu et al., 2023).

In addition to producing brand new images, these models are also shown to perform well at image manipulation tricks, such as inpainting (filling in missing portions of an image), style transfer (applying the aesthetics of one image to another), and image-to-image translation (altering the content of an image whilst maintaining its structure). This could be used, say, to supply a crudely drawn landscape and have the model turn it into something in Van Gogh style, or a damaged photo and reconstruct it with the missing parts, and so on. They provide new ways to create, in addition to professional editing software, putting equivalent tools in the hands of untrained graphic artists. Nevertheless, there are difficulties in image synthesis. The most basic one is upholding spatial coherence- those objects are made to look real, in terms of shapes, places, and proportions. A typical case of failure entails the misshaping of hands, disproportionate faces, or unbalanced faces, and light. Such artifacts may be the consequences of undermined learning or a lack of constraints at sampling. The other problem is to ensure high-resolution production with little fine detail and the absence of repetition and texture contamination. The ethical issues of image synthesis also include its use to create deepfakes and visually deceptive content or create other art without the permission of the original artist, in the style of a living artist.

Technically speaking, image generation requires huge databases and high-performance computing as well as carefully designed sampling procedures that incorporate stochasticity and determinism at different scales. The users would like to experiment with more original interpretations, and yet demand the visual clarity of results, as well as a certain degree of aesthetic finish. To find the right balance is necessary to fine-tune the variables, including prompt weighting, guidance scale, and traversing the latent space--as a method through which users and developers influence outputs of such powerful systems.

Audio Generation

Audio generation is a more sophisticated and technically complex area of development, even more so than text and image generation, as it is currently in the world of mainstream reality. All this is because of the way that human beings perceive sound. Our ears and brain are highly subject to passages in pitch, rhythm, tone, emotion, and above all, temporal continuity. The sound has to be natural and have no sudden emotion-impacting moments that would stop the flow of the sound to be comfortable over time. This is what makes audio generation one of the most demanding and at the same time one of the most powerful modalities in generative AI (Kumar et al., 2023; Mitra & Zualkernan, 2025).

The recent generative audio models can be divided into two groups: music and voice synthesis. When it comes to the sphere of music, the most developed tool is MusicLM, created by Google. This model is taught on a large amount of music data, gaining understanding not only of the form that melody, harmony, and rhythm take, but also how various genres and instruments are related. MusicLM can render new music compositions by text descriptions as an energetic band of jazz instrumentation that includes a saxophone solo or a movie-style orchestra with drums. It achieves this by using a top-down generative model that produces a low-level representation of the intended music, and iteratively improves it into a high-quality representation of high-quality audio waveforms. The models can create pieces of music several minutes long and make them sound amazingly coherent and stylistically aligned with the prompt- they sound as well done, and sometimes nearly as indistinguishable from a human creation.

Cousin to music synthesis is the field of voice cloning and speaking generation, where models such as VALL-E (by Microsoft) and Voicebox (by Meta) are hitting the news. Such systems are learned over corpora of human speech of various accents, languages as well and emotional states. Trained, they can perform a very good imitation of the voice of anybody given a few seconds of recordings to refer to; they can pick up not only the accent, the pitch, and the cadence but also the emotional intonation. The synthesized voices can be applied to virtual assistants, audiobooks, language translation, or content localization. However, their most controversial use is when they are used to create speech that the original speaker did not make. It is not only a technological wonder but raises increasingly troubling moral questions about what such capable synthesis of the most realistic speech can be able to generate.

The other fundamental issue that is particular to audio generation is to ensure fluidity and coherence across time. Where images can end up tolerating visual artifacts or neutrally appreciate them as a stylistic variation, even minimal jitter or artificial pause in audio can make the final product unpleasant or creepy to hear. This puts a lot of stress on sampling methods of the model, particularly those that record transitions between phonemes in a speech or beats in a piece of music on a microsecond scale. Also, modeling background noise, echo, breath sounds, and speaker intent requires very high-quality training data and modeling of the human auditory system. There is the emotional fidelity aspect of a layer of complexity. The speech of humans has a high reflection of affect, such as joy, sarcasm, hesitation, and anger, and using synthetic voices to convey these effects is still an open research problem. As an example, producing a phrase like I am okay with the right tones to reflect frustration or resignation would require not just an understanding of the syntax language but also the power to show emotions, something that the existing models are still learning but certainly cannot get right each time.

The development of audio production also changes rapidly, with its ethical implications being high. As realistic voice cloning technology is now readily available, criminals have easy access to such technology to impersonate a public figure, construct a false audio recording, or call someone under pretenses by employing audio deepfakes. Without proper management, such tools, in turn, pose a threat to individual privacy, online trust, and, indeed, national security. In this regard, it is important that we come up with watermarking technologies, verification systems of authenticity, and regulations to drive their responsible use. Nevertheless, the artistic possibilities of generative audio are massive. It can be used to democratize music production, provide a voice to those who have lost the ability to speak, create immersive music environments in virtual worlds, and be used to have real-time multilingual communication via expressive speech synthesis. With the advancing technology, it can be expected that the new models can not only speak but also listen, in the sense that they can comprehend what is happening around them and respond to noise, emotional intelligence, and social situations in natural ways.

Video Generation

Video generation is the most technically demanding Generative AI modality by far. It is not just the synthesis of one kind of content; it is the synthesis of sound, image, and type of orchestrated content combined and each layered on top of the other with time, movement, and narrative logic. To make an effective video, the model has to know how to make scenes move, how objects move in space, how the lighting and perspective change, and how to make up the continuity of the vision between shots. Such a combination of modalities stretches AI into new fields, demanding vision-language and temporal-dynamics innovation (Aldausari et al., 2023; Indira Priya et al., 2024; Masood et al., 2023; Yu et al., 2024).

Recent advances represent a turning point in the case of generative video. Cutting-edge models such as Sora by OpenAI, Gen-2, Pika, and Dream Machine by Luma depict the power of creating short video clips of a few seconds or less than a minute, and potentially even a few sketches or still photos, depending on the text translation or an image given. And imagine a user entering a sentence such as, "a cinematic drone shot of a sprawling mountain landscape during sunrise," and a few moments later, the model produces an entire moving scene, complete with sky animation, the shadows changing, and sun rays moving, the camera in motion as well. This new breed of video models is inspiring completely new modes of work across the content creator, advertiser, educator, and game development communities, and it makes possible a whole new visual production experience involving minimal equipment and actors and a small amount of software.

It is the presence of temporal coherence that is the key technical issue of video generation. Whereas image synthesis aims at depicting a single frame of visually appealing content, video generation needs to consider that each frame must lead to the next, not only visually but also in various ways, such as movement of objects, consistency, and realism. It implies that a dog traveling and moving across a frame should be the same size, form, and location in consecutive time steps, and the shadow should follow the traversing angle of the sun. The slightest flaws, such as sudden jumps in animation, weak motion blur, or jittery lines, can disrupt the sense of realism, which easily brings the viewers out of the synthetic world. Video models synthesize diffusion-based architectures with temporal-aware transformers. Diffusion layers are used to create high-resolution frames with progressive noise to coherent images, whereas the temporal models are used to track changes between the frames to create smooth visual and narrative flow. Other models, such as Sora, add the ability to understand 3D scenes; hence, they can better render camera movement and perspective change, which is essential in movie-like realism. The training of such systems, however, imposes the need to have remarkable amounts of video data, sometimes consisting of thousands of hours of annotated data across various settings, different lighting environments, and styles.

Although there is strong evidence that this field is making fundamental advancements, creating photorealistic and manipulable long-form video content as a class is still an open challenge in research. Even the most recent models are limited in that their videos exhibit artistic abstraction or stylization, and are not fine-grained to the extent of, say, controlling individual facial expressions, lip-sync, or the fine movements of objects. Such artifacts as object motion, which is jerky, discontinuous, repetitive, or sudden frame-to-frame inconsistency, are also prevalent, especially when there is high complexity in the scene or interactions between at least two actors. Compared to image generation, which solved the problem of photorealism to a large extent, video generation is still at its first steps of becoming photorealistic at scale.

In addition to technical difficulty, video synthesis also presents many ethical, legal, and societal issues. The same techniques that make it possible to offer creative storytelling and snappy animation are also used to create deepfake videos, imitations, or fake records that cannot be distinguished from authentic shots. This fact is why provenance tracking, content watermarking, and platform-level measures are being prioritized as the tools of video generation become available to the general populace. The only way to maintain the trustworthiness of digital media is to ensure that synthetic videos are also labeled, trackable, and applied responsibly. Nonetheless, the possible uses of generative video are numerous. It can speed up pre-vis in film and gaming as well as worldbuilding. In learning, it can dynamically show events or processes occurring in the past or under scientific procedures. It is used in training and simulation in that it can simulate a life-like virtual world to the pilot, surgeon, or emergency responders. The mixture between a real and a synthetic video content can become even more challenging as the models become more stable, longer, and manageable-changing the manner of the stories told, the way information is implied, and how the reality is mediated on digital screens. Table 1 synthesizes key distinctions, challenges, and applications across the four modalities discussed in this chapter. It serves as a reference point for understanding how each modality presents unique design and ethical considerations.

Table 1. Key characteristics, challenges, and use cases of generative AI across text, image, audio, and video modalities

Modality	Core Models & Tools	Input Type	Output Type	Key Challenges	Primary Use Cases
Text	GPT, Claude, LLaMA	Text prompt	Text	Factuality, consistency, hallucination	Writing, translation, coding, tutoring
Image	DALL·E, Midjourney, Stable Diffusion	Text/sketch	Image	Spatial realism, style control, and artifacts	Art, design, editing, and visualization
Audio	MusicLM, Voicebox, VALL-E	Text/audio seed	Music, voice	Temporal coherence, emotion, and voice fidelity	Music, dubbing, speech help, deepfakes
Video	Sora, Runway Gen-2, Pika	Text/image/ sketch	Short video	Motion continuity, photorealism, artifacts	Film, prototyping, education, simulation

However, modalities are becoming more and more similar and intertwined in spite of these differences. A large number of innovative models are multimedia, i.e., able to comprehend and generate in more than one medium. For example, DALL·E can interpret language and generate images; Sora can accept text commands that drive video generation; audio input and output are experimented with in GPT models, which implies real-time conversation agents. This convergence involves a movement away from the embodiment of single-task models to architecture unifications in which text, vision, and sound are seen as variant expressions of structured information. Nonetheless, every modality also presents distinctive research problems: language models have to deal with semantic and factual accuracy; image models have to learn appearance and aesthetics; audio models have to learn tone and emotion; and video models need to learn frame consistency and story consistency. Figure 3 presents core generation pipelines across modalities, whereas Table 2 presents representative failure types across generative AI modalities.

Figure 3. Input-to-Output Pipelines Across Modalities

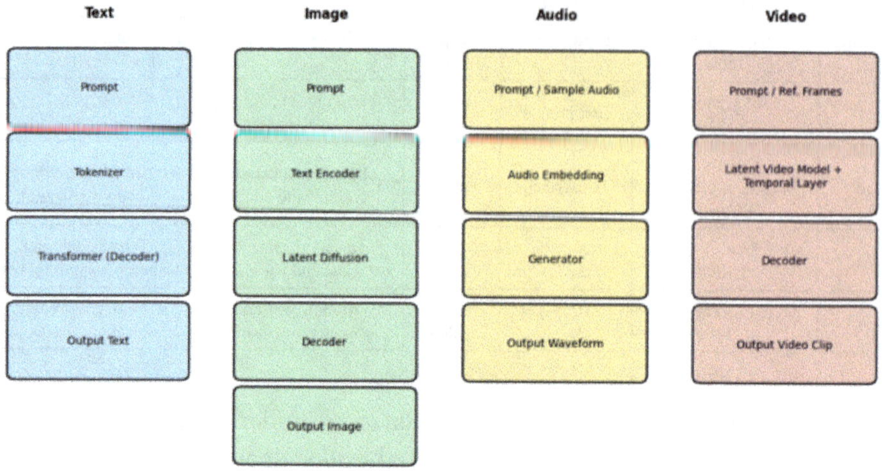

Table 2. Common Failure Modes Across Modalities

Modality	Failure Mode 1	Failure Mode 2	Failure Mode 3
Text	Factual hallucination	Repetition	Logical contradiction
Image	Distorted anatomy	Lighting inconsistencies	Unrealistic textures
Audio	Robotic intonation	Disjointed phonemes	Unnatural pauses
Video	Frame flickering	Inconsistent object motion	Unnatural transitions

EMERGENT COMPLEXITY: OPACITY, UNPREDICTABILITY, AND HUMAN OVERSIGHT

With more processes becoming centrally controlled by increasingly potent generative AI systems, they are becoming much more obscure, even to their designers. This black-box nature is a major cause of concern regarding transparency, accountability, and control. The crux of these issues is a disparity between the complexity of contemporary models and the depth of our capacity to interpret or anticipate their outputs and how or why they might arise. Scale brings with it emergent behavior, which is patterns, capabilities, or flaws that are never directly programmed and often never well anticipated.

The black-box characteristics of generative AI have to do with the fact that people cannot understand or explain how a model generates a certain outcome. GPT, DALL·E, and Stable Diffusion are made up of billions- check, or even trillion-scaled parameters, placed on an intricate framework of neural networks. Such parameters contain huge quantities of statistical information about the language, the pictures, or the other modalities, but are not easily explained. Generative models differ from traditional algorithms based on the fact that the latter have explicit steps that are logical, whereas models work by a probabilistic association, as well as the representations that are latent. Consequently, even when a given system generates a remarkably clever poem or a deceptive news story, there is frequently no easy way to come up with an explanatory reason that would explain this outcome. This makes such models hard to audit and understand what biases might exist, or otherwise work as intended.

To this intricacy has been added the phenomenon of emergent behavior. When models are trained on larger amounts of more diverse data, they might start to show characteristics that did not appear at all at smaller sizes, or were not observable at all. As an example, large language models have revealed the serendipitous capacity of performing mathematical computation, writing executable programs, translating between languages, or reasoning philosophically, even in cases where their training was not for these tasks. Such new abilities are positive, but they leave the future open-ended. The same model that has the ability to come up with a creative short story can also make up science, create a non-existent case law, or write prejudiced information with the given stimuli. These inconsistencies undermine the belief that there is greater reliability or control in a larger mass of data and data points.

In addition, safety and ethical issues that arise with generative models are unpredictable. In highly regulated fields like medicine, law, and education, the use of synthetic products can affect people in a severe manner. A hallucinated medical diagnosis, a fabricated legal reference, or a culturally insensitive generation of imagery can be used to deceive users, increase negative consequences, or undermine confidence in institutions. The fact that these models have a confident tone and well-polished outputs hides any issues that may be in the results and causes one to believe that there is an expert devising the output. It is not only dangerous in the figures that the models produce, but you also have to interpret and act on that information.

These systems also have a lot of challenges regarding human control and governance because they are mostly opaque and unpredictable. Software accountability is a field that already has established ways of holding software accountable by debugging code, running execution paths, or having them reviewed by looking through source logic, but all these methods fail in the case of AI systems that are based on learning large bodies of data and evolve during the process of training. The attempt to introduce transparency, including explainable AI methods or interpretable approximations, has only achieved partial success in actually unpacking the rule set of generative models. This discrepancy is the discrepancy between the increased influence of these systems and the tools that can be used to govern them or bend them towards human morals.

To do this, participants in the field of AI research and policy are considering new ethical and regulatory systems specific to generative AI. Some of them are documentation of datasets (model cards and data sheets), watermarking of outputs to indicate that it is generated synthetically, limitations on usage with usage policies, and frequent human-in-the-loop verification. Other developers are adding a layer of safety that limits damaging outputs or adding self-corrections made to a behavior depending on the response. Yet, the dynamism of these systems tends to be faster than the process of creating efficient safety measures, which makes the regulatory processes more reactive than proactive. Figure 4 shows overlapping domains of risk and oversight in generative AI.

Figure 4. Risk and Oversight Zones in Generative AI

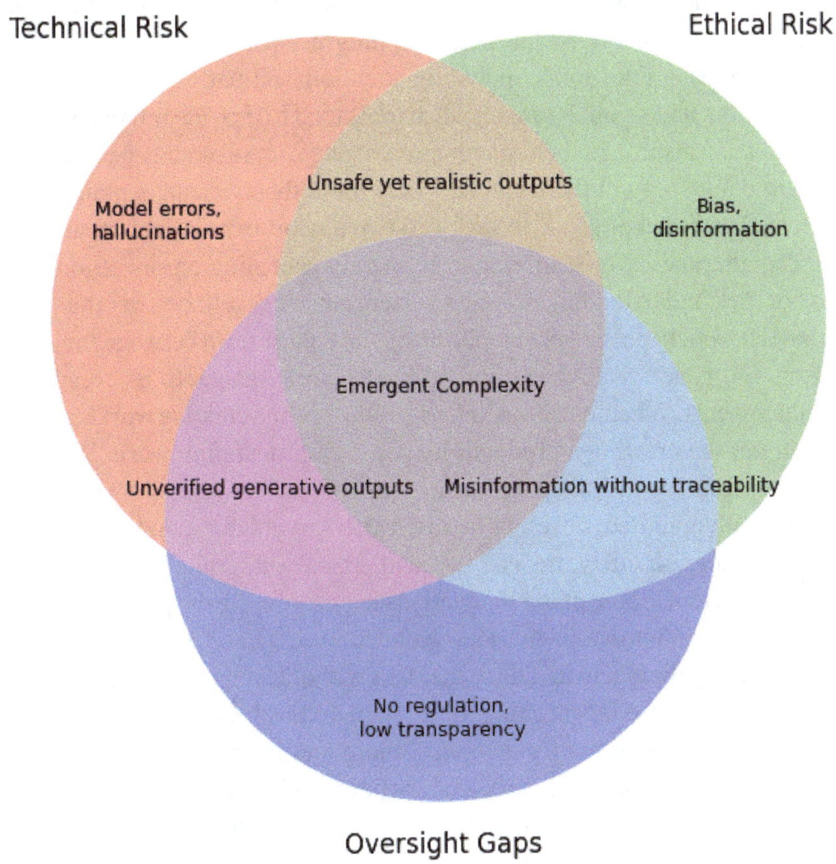

CONCLUSION

Because generative AI is still expanding and advertising itself as a piece of software that can write poems, develop architecture, create human-like voices, and even create cinematic video, we live in a strange relationship regarding its inner mechanisms. Such systems are based on huge neural networks and probabilistic training algorithms, and can be very hard to describe. We tend to know how generative models are trained and which mechanisms they follow, be it attention, latent spaces, or diffusion process, but we are, in most cases, still in the dark as to why they produce a particular output or when they acquire certain behaviors. Such an epistemic divide is what underpins the central paradox of modern AI, that we are learning to use tools we do not fully comprehend.

The issue of whether we have the potential to ever open the black box is a question on its own. Explainable AI has begun to progress with the creation of partial or lightweight transparency tools, including attention heatmaps or saliency scores, feature attribution, and so on. However, such tools tend to provide rather superficial insights as to the causal logic behind generative models. As models become large, the emergent complexity in them can be vastly more than we can explain, casting doubt on whether it makes sense to seek full explainability as a goal of such large systems. The alternative, instead, might be merely accepting an operational trust that does not seek understanding of how a system can work reliably, in other words.

This indecisiveness puts a new emphasis on the equilibrium between innovation and control. The generative AI is impossible to ignore, thanks to its creative and commercial impetus. Whether it is an industry redesign or a creative working model, these paradigms are creating a productivity and experimentation wave. With great power comes great responsibility. To ensure that generative systems are safe, traceable, and ethically guarded, developers, researchers, regulators, and users need to collaborate to integrate safety, traceability, and ethical guardrails into the lifecycle of a generative system. Some of the innovations in overcoming the problems include output watermarking, human-in-the-loop architectures, dataset auditing, and value alignment strategies that seem to reduce the risks without hamstringing innovations.

The consequences are far-reaching at the societal level. Generative AI systems are not only reflections of reality anymore; more and more, they are themselves shaping it because of what people see, hear, believe, and share. Trust in this type of landscape becomes a currency. Once the media, the credibility of digital content, or the integrity of the AI systems is no longer trusted, there is a risk that people can lose trust in digital communication as a whole. This not only makes governance and transparency an aspiration, but a civic imperative. Setting up structures that render accountability without stalling on forward-looking aspects is the thing what characterizes our time. Knowing generative AI might not imply total control, but instead a dynamic association of knowledgeable confidence, acts of responsible design, and flexible management. The way we have been going, the problem is no longer to open up the black box, but to make sure that what we open up there is good, it serves our objectives, it enhances rather than disrupts our digital future together.

REFERENCES

Ahmad, M., Ho, W. Y., Goyes Robalino, A. P., Maqsood, M., Ahmad, M., Ho, W. Y., Goyes Robalino, A. P., & Maqsood, M. (2025). Theoretical Implications of Generative AI for Content Generation in Geoinformatics Training. In A. Anna A. (Ed.), *Integration Strategies of Generative AI in Higher Education* (pp. 101–130). IGI Global Scientific Publishing. DOI: 10.4018/979-8-3693-5518-3.ch005

Aldausari, N., Sowmya, A., Marcus, N., & Mohammadi, G. (2023). Video Generative Adversarial Networks: A Review. In *ACM Computing Surveys* (Vol. 55, Issue 2). DOI: 10.1145/3487891

Alqahtani, H., Kavakli-Thorne, M., & Kumar, G. (2021). Applications of Generative Adversarial Networks (GANs): An Updated Review. *Archives of Computational Methods in Engineering*, *28*(2), 525–552. Advance online publication. DOI: 10.1007/s11831-019-09388-y

Alwahedi, F., Aldhaheri, A., Ferrag, M. A., Battah, A., & Tihanyi, N. (2024). Machine learning techniques for IoT security: Current research and future vision with generative AI and large language models. *Internet of Things and Cyber-Physical Systems*, *4*, 167–185. DOI: 10.1016/j.iotcps.2023.12.003

Asperti, A., Evangelista, D., & Loli Piccolomini, E. (2021). A Survey on Variational Autoencoders from a Green AI Perspective. *SN Computer Science*, *2*(4), 301. Advance online publication. DOI: 10.1007/s42979-021-00702-9

Asperti, A., & Tonelli, V. (2023). Comparing the latent space of generative models. *Neural Computing & Applications*, *35*(4), 3155–3172. DOI: 10.1007/s00521-022-07890-2

Bansal, G., Nawal, A., Chamola, V., & Herencsar, N. (2024). Revolutionizing Visuals: The Role of Generative AI in Modern Image Generation. *ACM Transactions on Multimedia Computing Communications and Applications*, *20*(11), 1–22. Advance online publication. DOI: 10.1145/3689641

Bengesi, S., El-Sayed, H., Sarker, M. K., Houkpati, Y., Irungu, J., & Oladunni, T. (2024). *Advancements in Generative AI: A Comprehensive Review of GANs, GPT, Autoencoders, Diffusion Model, and Transformers*. IEEE., DOI: 10.1109/ACCESS.2024.3397775

Biswas, A., Md Abdullah Al, N., Imran, A., Sejuty, A. T., Fairooz, F., Puppala, S., & Talukder, S. (2023). Generative Adversarial Networks for Data Augmentation. In *Data Driven Approaches on Medical Imaging*. DOI: 10.1007/978-3-031-47772-0_8

Cao, H., Tan, C., Gao, Z., Xu, Y., Chen, G., Heng, P. A., & Li, S. Z. (2024). A Survey on Generative Diffusion Models. *IEEE Transactions on Knowledge and Data Engineering*, *36*(7), 2814–2830. Advance online publication. DOI: 10.1109/TKDE.2024.3361474

Chapelle, C. A. (2025). Generative AI as game changer: Implications for language education. *System*, *132*, 103672. DOI: 10.1016/j.system.2025.103672

Chen, B., Zhu, X., & Díaz del Castillo, H. (2023). Integrating generative AI in knowledge building. *Computers and Education: Artificial Intelligence*, *5*, 100184. Advance online publication. DOI: 10.1016/j.caeai.2023.100184

Croitoru, F. A., Hondru, V., Ionescu, R. T., & Shah, M. (2023). Diffusion Models in Vision: A Survey. *IEEE Transactions on Pattern Analysis and Machine Intelligence*, *45*(9), 10850–10869. Advance online publication. DOI: 10.1109/TPAMI.2023.3261988 PMID: 37030794

de Rosa, G. H., & Papa, J. P. (2021). A survey on text generation using generative adversarial networks. *Pattern Recognition*, *119*, 108098. Advance online publication. DOI: 10.1016/j.patcog.2021.108098

Gakhar, S., Kondoju, V. P., Chauhan, S. S., Kumar, A., Kulkarni, S. S., & Goel, P. (2025). Generative AI for Real-Time Data Augmentation in Big Data Pipelines. *1st International Conference on Advances in Computer Science, Electrical, Electronics, and Communication Technologies, CE2CT 2025*, 1386–1391. DOI: 10.1109/CE2CT64011.2025.10941295

Hla, D. A., & Hindin, D. I. (2025). Generative AI & machine learning in surgical education. *Current Problems in Surgery*, *63*, 101701. DOI: 10.1016/j.cpsurg.2024.101701 PMID: 39922636

Indira Priya, P., Manju, M., Anitha, R., Harsavardhini, R., & Sri Harini, S. (2024). Transforming Text to Video: Leveraging Advanced Generative AI Techniques. *2024 Asian Conference on Intelligent Technologies, ACOIT 2024*. DOI: 10.1109/ACOIT62457.2024.10941483

Kaswan, K. S., Dhatterwal, J. S., Malik, K., & Baliyan, A. (2023). Generative AI: A Review on Models and Applications. *2023 International Conference on Communication, Security and Artificial Intelligence, ICCSAI 2023*, 699–704. DOI: 10.1109/ICCSAI59793.2023.10421601

Kumar, S., Musharaf, D., Musharaf, S., & Sagar, A. K. (2023). A Comprehensive Review of the Latest Advancements in Large Generative AI Models. *Communications in Computer and Information Science, 1920*, 90–103. Advance online publication. DOI: 10.1007/978-3-031-45121-8_9

Lancaster, T. (2023). Artificial intelligence, text generation tools and ChatGPT – does digital watermarking offer a solution? *International Journal for Educational Integrity, 19*(1), 10. Advance online publication. DOI: 10.1007/s40979-023-00131-6

Lee, G. H., Lee, K. J., Jeong, B., & Kim, T. (2024). Developing Personalized Marketing Service Using Generative AI. *IEEE Access : Practical Innovations, Open Solutions, 12*, 22394–22402. DOI: 10.1109/ACCESS.2024.3361946

Mahmoud, A. B., Kumar, V., & Spyropoulou, S. (2025). Identifying the Public's Beliefs About Generative Artificial Intelligence: A Big Data Approach. *IEEE Transactions on Engineering Management, 72*, 827–841. DOI: 10.1109/TEM.2025.3534088

Masood, M., Nawaz, M., Malik, K. M., Javed, A., Irtaza, A., & Malik, H. (2023). Deepfakes generation and detection: State-of-the-art, open challenges, countermeasures, and way forward. *Applied Intelligence, 53*(4), 3974–4026. Advance online publication. DOI: 10.1007/s10489-022-03766-z

Mitra, R., & Zualkernan, I. (2025). Music Generation Using Deep Learning and Generative AI: A Systematic Review. *IEEE Access : Practical Innovations, Open Solutions, 13*, 18079–18106. DOI: 10.1109/ACCESS.2025.3531798

Owoseni, A., Kolade, O., & Egbetokun, A. (2024). Enhancing Personalised Learning and Student Engagement Using Generative AI. *Generative AI in Higher Education*, 123–150. DOI: 10.1007/978-3-031-60179-8_5

Pan, Z., Yu, W., Yi, X., Khan, A., Yuan, F., & Zheng, Y. (2019). Recent Progress on Generative Adversarial Networks (GANs): A Survey. *IEEE Access : Practical Innovations, Open Solutions, 7*, 36322–36333. Advance online publication. DOI: 10.1109/ACCESS.2019.2905015

Papadopoulos, D., & Karalis, V. D. (2023). Variational Autoencoders for Data Augmentation in Clinical Studies. *Applied Sciences (Basel, Switzerland), 13*(15), 8793. Advance online publication. DOI: 10.3390/app13158793

Ramdurai, B., & Adhithya, P. (2023). The Impact, Advancements and Applications of Generative AI. *International Journal on Computer Science and Engineering, 10*(6), 1–8. Advance online publication. DOI: 10.14445/23488387/IJCSE-V10I6P101

Rashidi, H. H., Hu, B., Pantanowitz, J., Tran, N., Liu, S., Chamanzar, A., Gur, M., Chang, C. C. H., Wang, Y., Tafti, A., Pantanowitz, L., & Hanna, M. G. (2025). Statistics of Generative Artificial Intelligence and Nongenerative Predictive Analytics Machine Learning in Medicine. *Modern Pathology*, *38*(3), 100663. DOI: 10.1016/j.modpat.2024.100663 PMID: 39579984

Rosário, A. T. (2024). *Generative AI and Generative Pre-Trained Transformer Applications*. 45–71. : S Patil, S., Shelke, S., Velankar, M., Koul, T., & Rani, V. (2023). A Systematic Survey of Approaches Used in Computer Music Generation. *Journal of Innovations in Data Science and Big Data Management, 2*(1). https://doi.org/DOI: 10.4018/979-8-3693-1950-5.ch003

Saxena, D., & Cao, J. (2021). Generative Adversarial Networks (GANs). In *ACM Computing Surveys* (Vol. 54, Issue 3). DOI: 10.1145/3446374

Tu, W., Deng, W., & Gedeon, T. (2023). A Closer Look at the Robustness of Contrastive Language-Image Pre-Training (CLIP). *Advances in Neural Information Processing Systems*, *36*, 13678–13691.

Wei, R., Garcia, C., El-Sayed, A., Peterson, V., & Mahmood, A. (2020). Variations in Variational Autoencoders - A Comparative Evaluation. *IEEE Access : Practical Innovations, Open Solutions*, *8*, 153651–153670. Advance online publication. DOI: 10.1109/ACCESS.2020.3018151

Wei, R., & Mahmood, A. (2021). Recent Advances in Variational Autoencoders with Representation Learning for Biomedical Informatics: A Survey. *IEEE Access : Practical Innovations, Open Solutions*, *9*, 4939–4956. Advance online publication. DOI: 10.1109/ACCESS.2020.3048309

Yadav, A. B. (2024). Generative AI in the Era of Transformers: Revolutionizing Natural Language Processing with LLMs. *Journal of Image Processing and Intelligent Remote Sensing*, *42*(42), 54–61. Advance online publication. DOI: 10.55529/jipirs.42.54.61

Yang, L., Zhang, Z., Song, Y., Hong, S., Xu, R., Zhao, Y., Zhang, W., Cui, B., & Yang, M. H. (2024). Diffusion Models: A Comprehensive Survey of Methods and Applications. *ACM Computing Surveys*, *56*(4), 1–39. Advance online publication. DOI: 10.1145/3626235

Yee-King, M. (2022). Latent Spaces: A Creative Approach. *Springer Series on Cultural Computing*, 137–154. DOI: 10.1007/978-3-031-10960-7_8

Yoon, J., Drumright, L. N., & Van Der Schaar, M. (2020). Anonymization through data synthesis using generative adversarial networks (ADS-GAN). *IEEE Journal of Biomedical and Health Informatics*, *24*(8), 2378–2388. Advance online publication. DOI: 10.1109/JBHI.2020.2980262 PMID: 32167919

Yu, T., Yang, W., Xu, J., & Pan, Y. (2024). Barriers to Industry Adoption of AI Video Generation Tools: A Study Based on the Perspectives of Video Production Professionals in China. *Applied Sciences 2024, Vol. 14, Page 5770*, *14*(13), 5770. DOI: 10.3390/app14135770

Zhan, F., Yu, Y., Wu, R., Zhang, J., Lu, S., Liu, L., Kortylewski, A., Theobalt, C., & Xing, E. (2023). Multimodal Image Synthesis and Editing: The Generative AI Era. *IEEE Transactions on Pattern Analysis and Machine Intelligence*, *45*(12), 15098–15119. DOI: 10.1109/TPAMI.2023.3305243 PMID: 37624713

Zhang, E. Y., Cheok, A. D., Pan, Z., Cai, J., & Yan, Y. (2023). From Turing to Transformers: A Comprehensive Review and Tutorial on the Evolution and Applications of Generative Transformer Models. In *Sci* (Vol. 5, Issue 4). DOI: 10.3390/sci5040046

Conclusion

Digital synthetic data are a boon in so many applications. They help in terms of research and data analysis, data sharing (with privacy protections), machine learning, augmented human experiences, human expression, and other applications. Many are working to master the tools and to find new applications. Drawing out a topic from the background can help in terms of new thinking and new applications.

The potential for the use of synthetic data is not fully realized.

FUTURE SYNTHETIC DATA

The future will likely involve more nuanced, layered, raw, finished, and refined digital contents in various modalities and various applications.

- What of synthetic animations? Synthetic generated video shorts and full-length ones? Synthetic realia? Synthetic imaginaries? Synthetic emulated physics? Non-physics? Plausibilities and implausibilities? How will these enhance motion design?
- What will generated prompts look like? In various complexities? For what outcomes?
- What combinatorial mixes of synthetic datas may be used together in harmony? Disharmony?
- How will synthetic generation enhance character designs?

New raw synthetic data will likely emerge in new ways, too. Perhaps discernment of work origins will be harder in the future, if not impossible, especially given that synthetic data may be used in so many embedded parts of processes.

- What as-yet unthought-of data types can be created and used?

There are backlash effects, too, with special appreciation for human-created contents, with idiosyncrasies and uniquenesses appreciated that much more. The human hand can create works with less efficiency and more effort but also more heart, more empathy perhaps, and more hard-earned skills.

Over time, each method will be justified based on future needs and costs.

- How can layered synthetic datas be exploited together to answer real-world questions? Solve real-world problems? Enhance designs (Of products? Of services? Of characters?) Entertain? Entrance?

AN ASIDE

As this book was going to press, the publisher's development editor found several works that had hallucinated citations by generative AI, which indicated artificial contributions to the chapters. One was rejected, and the other passed muster after the authoring team's revisions and corrections and the publisher's re-review of the revised work. This anecdote shows something of synthetic data and its incursions in human endeavors. Ironically, the rejected chapter involved ethics and integrity, neither of which the authoring team demonstrated in their submission.

THANKS!

Thanks go to the contributors to this text. Each one offers unique expertise, insights, and perspectives.

Shalin Hai-Jew

2025

Compilation of References

Abay, N. C., Zhou, Y., Kantarcioglu, M., Thuraisingham, B., & Sweeney, L. (2019). Privacy preserving synthetic data release using deep learning. In *Machine Learning and Knowledge Discovery in Databases: European Conference, ECML PKDD 2018, Dublin, Ireland, September 10–14, 2018* [Springer International Publishing.]. *Proceedings, 18*(Part I), 510–526.

Abdullah, F., & Taveekitworachai, P. (2024) "The 1st ChatGPT4PCG Competition" *2024 IEEE Conference on Games (CoG)*. Doi: DOI: 10.1109/CoG57391.2024.00131

Abowd, J. M., & Vilhuber, L. (2008, September). How protective are synthetic data? In *International Conference on Privacy in Statistical Databases* (pp. 239-246). Berlin, Heidelberg: Springer Berlin Heidelberg.

Agarwal, A., Hunt, B. J., Stegemann, M., Rochwerg, B., Lamontagne, F., Siemieniuk, R. A., ... & Vandvik, P. O. (2020). A living WHO guideline on drugs for covid-19. *bmj, 370*.

Agchar I, Baumann I, Braun F, Pérez-Toro P A. (2024). "A Survey of Music Generation in the Context of Interaction". arXiv preprint arXiv.

Aggarwal, A., Mittal, M., & Battineni, G. (2021). Generative adversarial network: An overview of theory and applications. *International Journal of Information Management*. Advance online publication. DOI: 10.1016/j.ijinfomgt.2021.102363

Agrawal, G., Kaur, A., & Myneni, S. (2024). A review of generative models in generating synthetic attack data for cybersecurity. *Electronics (Basel), 13*(2), 322. DOI: 10.3390/electronics13020322

Ahmad, M., Ho, W. Y., Goyes Robalino, A. P., Maqsood, M., Ahmad, M., Ho, W. Y., Goyes Robalino, A. P., & Maqsood, M. (2025). Theoretical Implications of Generative AI for Content Generation in Geoinformatics Training. In A. Anna A. (Ed.), *Integration Strategies of Generative AI in Higher Education* (pp. 101–130). IGI Global Scientific Publishing. DOI: 10.4018/979-8-3693-5518-3.ch005

Ahmad, M., Shabbir, S., Roy, S. K., Hong, D., Wu, X., Yao, J., Khan, A. M., Mazzara, M., Distefano, S., & Chanussot, J. (2022). Hyperspectral Image Classification—Traditional to Deep Models: A Survey for Future Prospects. *IEEE Journal of Selected Topics in Applied Earth Observations and Remote Sensing*, *15*, 968–999. DOI: 10.1109/JSTARS.2021.3133021

Akkem, Y., Biswas, S. K., & Varanasi, A. (2024). A comprehensive review of synthetic data generation in smart farming by using variational autoencoder and generative adversarial network. *Engineering Applications of Artificial Intelligence*, *131*, 107881. Advance online publication. DOI: 10.1016/j.engappai.2024.107881

Aldausari, N., Sowmya, A., Marcus, N., & Mohammadi, G. (2023). Video Generative Adversarial Networks: A Review. In *ACM Computing Surveys* (Vol. 55, Issue 2). DOI: 10.1145/3487891

Alghushairy, O., & Ma, X. (2022). "Data storage". Encyclopedia of Big Data. Doi: DOI: 10.1007/978-3-319-32015-5_100047

Ali, H., Mustafa, A., & Aysan, A. F. (2025). Global adoption of generative AI: What matters most? *Journal of Economy and Technology*, *3*, 166–176. DOI: 10.1016/j.ject.2024.10.002

Alipour, P., & Gallegos, E. (2025). Leveraging generative AI synthetic and social media data for content generalizability to overcome data constraints in vision deep learning. *Artificial Intelligence Review*, *58*(5), 145. Advance online publication. DOI: 10.1007/s10462-025-11137-6

Alkhalifah, T., Wang, H., & Ovcharenko, O. (2021, October). MLReal: Bridging the gap between training on synthetic data and real data applications in machine learning. In *82nd EAGE Annual Conference & Exhibition* (Vol. 2021, No. 1, pp. 1-5). European Association of Geoscientists & Engineers. DOI: 10.3997/2214-4609.202113262

Allshire, A., Mittal, M., & Lodaya, V. (2022) "Transferring Dexterous Manipulation from GPU Simulation to a Remote Real-World Trifinger" 2022 IEEE/RSJ International Conference on Intelligent Robots and Systems (IROS). . Link: https://ieeexplore.ieee.org/document/9981776DOI: 10.1109/IROS47612.2022.9981776

Almirall, E., Callegaro, D., Bruins, P., Santamaría, M., Martínez, P., & Cortés, U. (2022). The use of Synthetic Data to solve the scalability and data availability problems in Smart City Digital Twins. *arXiv preprint arXiv:2207.02953*.

Alqahtani, H., Kavakli-Thorne, M., & Kumar, G. (2021). Applications of generative adversarial networks (GANs): An updated review. *Archives of Computational Methods in Engineering*, *28*(2), 525–552. Advance online publication. DOI: 10.1007/s11831-019-09388-y

Alvero, A. J., Lee, J., Regla-Vargas, A., Kizilcec, R. F., Joachims, T., & Antonio, A. L. (2024). Large language models, social demography, and hegemony: Comparing authorship in human and synthetic text. *Journal of Big Data*, *11*(1), 138. DOI: 10.1186/s40537-024-00986-7

Alwahedi, F., Aldhaheri, A., Ferrag, M. A., Battah, A., & Tihanyi, N. (2024). Machine learning techniques for IoT security: Current research and future vision with generative AI and large language models. *Internet of Things and Cyber-Physical Systems*, *4*, 167–185. DOI: 10.1016/j.iotcps.2023.12.003

Anderson, J. W., Kennedy, K. E., Ngo, L. B., Luckow, A., & Apon, A. W. (2014, October). Synthetic data generation for the Internet of Things. In *2014 IEEE International Conference on Big Data (Big Data)* (pp. 171-176). IEEE. DOI: 10.1109/BigData.2014.7004228

Andreini, P., Bonechi, S., Ciano, G., & Graziani, C. (2022). *Multi-stage synthetic image generation for the semantic segmentation of medical images*. Machine Learning and Knowledge Extraction., DOI: 10.3390/make4020028

Anjaneyulu, B. P., & Reddy, C. P. K. (2024). "DeepFake Detection using Convolutional Neural Networks (CNN) and Recurrent Neural Network (RNN)". *International Conference on Data Science and Applications*. Doi: DOI: 10.1109/ICDICI62993.2024.10810970

Ankitha, B., Srikanth, C., Venkatesh, D., Badrinath, D., Aditya, G., & Akila Agnes, S. (2023). Enhancing the resolution of Brain MRI images using Generative Adversarial Networks (GANs). *Proceedings of International Conference on Computational Intelligence and Sustainable Engineering Solution, CISES 2023*. DOI: 10.1109/CISES58720.2023.10183400

An, S., & Jeon, J. J. (2023). Distributional learning of variational autoencoder: Application to synthetic data generation. *Advances in Neural Information Processing Systems*.

Antonucci, D., Conselvan, F., Mascherbauer, P., Harringer, D., & Pozza, C. (2024). Synthetic data on buildings. In Learning and analytics in intelligent systems (pp. 203–226). https://doi.org/DOI: 10.1007/978-3-031-47909-0_7

Aragani, V. M., Anumolu, V. R., & Selvakumar, P. (2025). Democratization in the Age of Algorithms: Navigating Opportunities and Challenges. In Wongmahesak, K., Wekke, I., Seftyono, C., & Nurdin, N. (Eds.), *Democracy and Democratization in the Age of AI* (pp. 39–56). IGI Global Scientific Publishing., DOI: 10.4018/979-8-3693-8749-8.ch003

Aranjuelo, N., García, S., Loyo, E., Unzueta, L., & Otaegui, O. (2021). Key strategies for synthetic data generation for training intelligent systems based on people detection from omnidirectional cameras. *Computers & Electrical Engineering*, *92*, 107105. Advance online publication. DOI: 10.1016/j.compeleceng.2021.107105

Arnold, C., & Neunhoeffer, M. (2020). Really Useful Synthetic Data—A Framework to Evaluate the Quality of Differentially Private Synthetic Data. *arXiv Preprint arXiv:2004.07740.*

Arora, A., & Arora, A. (2022). Generative adversarial networks and synthetic patient data: Current challenges and future perspectives. *Future Healthcare Journal*, *9*(2), 190–193. DOI: 10.7861/fhj.2022-0013 PMID: 35928184

Arshad, M. H. (2024). *Comparing and Contrasting the Use of Reinforcement Learning to Drive an Autonomous Vehicle Around a Racetrack in Unity and Unreal Engine 5*. ProQuest.

Asperti, A., Evangelista, D., & Loli Piccolomini, E. (2021). A Survey on Variational Autoencoders from a Green AI Perspective. *SN Computer Science*, *2*(4), 301. Advance online publication. DOI: 10.1007/s42979-021-00702-9

Asperti, A., & Tonelli, V. (2023). Comparing the latent space of generative models. *Neural Computing & Applications*, *35*(4), 3155–3172. DOI: 10.1007/s00521-022-07890-2

Assefa, S. A., Dervovic, D., Mahfouz, M., Tillman, R. E., Reddy, P., & Veloso, M. (2020, October). Generating synthetic data in finance: opportunities, challenges and pitfalls. In *Proceedings of the First ACM International Conference on AI in Finance* (pp. 1-8). DOI: 10.1145/3383455.3422554

Avatavului, C. D., Ifrim, R. C., & Voncilă, M. (2023). "Can Neural Networks Enhance Physics Simulations?" BRAIN. *Broad Research in Artificial Intelligence and Neuroscience*, *14*(2), 76–92. Advance online publication. DOI: 10.18662/brain/14.2/445

Azizi, S., Kornblith, S., Saharia, C., Norouzi, M., & Fleet, D. J. (2023). Synthetic data from diffusion models improves imagenet classification. *arXiv Preprint arXiv:2304.08466.*

Azizi, Z., Zheng, C., Mosquera, L., Pilote, L., & El Emam, K. (2021). Can synthetic data be a proxy for real clinical trial data? A validation study. *BMJ Open*, *11*(4), e043497. DOI: 10.1136/bmjopen-2020-043497 PMID: 33863713

Bakhshi, H., Downing, J. M., Osborne, M. A., & Schneider, P. (2017). *The future of skills: Employment in 2030*. Pearson and Nesta.

Bandi, A., Adapa, P. V. S. R., & Kuchi, Y. E. V. P. K. (2023). The power of Generative AI: A review of requirements, models, Input–Output formats, evaluation metrics, and challenges. *Future Internet*, *15*(8), 260. DOI: 10.3390/fi15080260

Bansal, G., Nawal, A., Chamola, V., & Herencsar, N. (2024). Revolutionizing Visuals: The Role of Generative AI in Modern Image Generation. *ACM Transactions on Multimedia Computing Communications and Applications*, *20*(11), 1–22. Advance online publication. DOI: 10.1145/3689641

Bao, J., Wang, R., Wang, Y., Sun, A., Li, Y., Mi, F., & Xu, R. (2023, July). A synthetic data generation framework for grounded dialogues. In *Proceedings of the 61st Annual Meeting of the Association for Computational Linguistics (*Volume 1*: Long Papers)* (pp. 10866-10882). DOI: 10.18653/v1/2023.acl-long.608

Barbierato, E., Vedova, M. L. D., Tessera, D., Toti, D., & Vanoli, N. (2022). A methodology for controlling bias and fairness in synthetic data generation. *Applied Sciences (Basel, Switzerland)*, *12*(9), 4619. DOI: 10.3390/app12094619

Bauer A, Trapp S, Stenger M, Leppich R. (2024). "Comprehensive exploration of synthetic data generation A survey". arXiv preprint arXiv.

Bauer, A., Trapp, S., Stenger, M., Leppich, R., Kounev, S., Leznik, M., . . . Foster, I. (2024). Comprehensive exploration of synthetic data generation: A survey. *arXiv Preprint arXiv:2401.02524*.

Baumann, J., Castelnovo, A., Crupi, R., Inverardi, N., & Regoli, D. (2023, June). Bias on demand: a modelling framework that generates synthetic data with bias. In *Proceedings of the 2023 ACM Conference on Fairness, Accountability, and Transparency* (pp. 1002-1013). DOI: 10.1145/3593013.3594058

Behl, H. S., Baydin, A. G., Gal, R., Torr, P. H., & Vineet, V. (2020, August). Autosimulate:(quickly) learning synthetic data generation. In *European Conference on Computer Vision* (pp. 255-271). Cham: Springer International Publishing.

Behrouz, A., & Hashemi, F. "CS-MLGCN: Multiplex Graph Convolutional Networks for Community Search in Multiplex Networks," in *Proceedings of the 31st ACM International Conference on Information and Knowledge Management (CIKM '22)*, Atlanta, GA, USA, Oct. 2022, pp. 1-5. DOI: 10.1145/3511808.3557572

Bellovin, S. M., Dutta, P. K., & Reitinger, N. (2019). Privacy and synthetic datasets. *Stan. Tech. L. Rev.*, *22*, 1.

Beltran, M. A., Mondragon, M. I. R., & Han, S. H. (2024). Comparative Analysis of Generative AI Risks in the Public Sector. *ACM International Conference Proceeding Series*, 610–617. DOI: 10.1145/3657054.3657125

Bengesi, S., El-Sayed, H., Sarker, M. K., Houkpati, Y., Irungu, J., & Oladunni, T. (2024). *Advancements in Generative AI: A Comprehensive Review of GANs, GPT, Autoencoders, Diffusion Model, and Transformers*. IEEE., DOI: 10.1109/ACCESS.2024.3397775

Bernatovych, A. A., & Stetsenko, I. V. (2023). *Module-Based Scientific Physics Engine Architecture*. Problems in Programming., DOI: 10.15407/pp2023.03.030

Bernhard, M., Kakooee, R., Bedarf, P., & Dillenburger, B. (2021). TopoGAN - Topology Optimization with Generative Adversarial Networks. *AAG - Advances in Architectural Geometry, unpublished.*

Bessadok, A., Mahjoub, M. A., & Rekik, I. (2023, May 1). Graph Neural Networks in Network Neuroscience. *IEEE Transactions on Pattern Analysis and Machine Intelligence*, *45*(5), 5833–5848. DOI: 10.1109/TPAMI.2022.3209686 PMID: 36155474

Bhattarai, B., Baek, S., Bodur, R., & Kim, T. K. (2020, May). Sampling strategies for GAN synthetic data. In *ICASSP 2020-2020 IEEE International Conference on Acoustics, Speech and Signal Processing (ICASSP)* (pp. 2303-2307). IEEE. DOI: 10.1109/ICASSP40776.2020.9054677

Bhatt, S., & Sharma, S. (2025). Generative Artificial Intelligence Based Biomedical Applications for Pharmaceutical Industry. In *2025 International Conference on Computational, Communication and Information Technology (ICCCIT)*, 1–6. https://doi.org/DOI: 10.1109/ICCCIT62592.2025.10927855

Bick, A., Blandin, A., Deming, D. J., & Louis, S. (2024). *The Rapid Adoption of Generative AI*. DOI: 10.3386/w32966

Bin-Nashwan, S. A., Sadallah, M., & Bouteraa, M. (2023). Use of ChatGPT in academia: Academic integrity hangs in the balance. *Technology in Society*, *75*, 102370. DOI: 10.1016/j.techsoc.2023.102370

Bird, J. J., & Lotfi, A. (2024). Cifake: Image classification and explainable identification of ai-generated synthetic images. *IEEE Access : Practical Innovations, Open Solutions*, *12*, 15642–15650. DOI: 10.1109/ACCESS.2024.3356122

Biswas, A., Md Abdullah Al, N., Imran, A., Sejuty, A. T., Fairooz, F., Puppala, S., & Talukder, S. (2023). Generative Adversarial Networks for Data Augmentation. In *Data Driven Approaches on Medical Imaging*. DOI: 10.1007/978-3-031-47772-0_8

Blauth, T. F., Gstrein, O. J., & Zwitter, A. (2022). Artificial intelligence crime: An overview of malicious use and abuse of AI. *IEEE Access : Practical Innovations, Open Solutions, 10*, 77110–77122. DOI: 10.1109/ACCESS.2022.3191790

Blythe, J., & Tregubov, A. (2024). Synthetic Data Generation for Machine Learning Models with Cognitive Agent Simulations. In Lecture notes in computer science (pp. 73–83). https://doi.org/DOI: 10.1007/978-3-031-70415-4_7

Boell, S. K., & Cecez-Kecmanovic, D. (2015). On being 'systematic'in literature reviews in IS. *Journal of Information Technology, 30*(2), 161–173. DOI: 10.1057/jit.2014.26

Boikov, A., Payor, V., Savelev, R., & Kolesnikov, A. (2021). Synthetic data generation for steel defect detection and classification using deep learning. *Symmetry, 13*(7), 1176. DOI: 10.3390/sym13071176

Bolón-Canedo, V., Sánchez-Maroño, N., & Alonso-Betanzos, A. (2013). A review of feature selection methods on synthetic data. *Knowledge and Information Systems, 34*(3), 483–519. DOI: 10.1007/s10115-012-0487-8

Bommasani, R., Hudson, D. A., Adeli, E., ... (2022). On the opportunities and risks of foundation models. arXiv preprint arXiv:2203.02155.

Bordas, A., Masson, P. L., Thomas, M., & Weil, B. (2024). What is generative in generative artificial intelligence? A design-based perspective. *Research in Engineering Design, 35*(4), 427–443. DOI: 10.1007/s00163-024-00441-x

Borkman, S., Crespi, A., Dhakad, S., Ganguly, S., Hogins, J., Jhang, Y. C., . . . Yadav, N. (2021). Unity perception: Generate synthetic data for computer vision. *arXiv Preprint arXiv:2107.04259.*

Boutros, F., Struc, V., Fierrez, J., & Damer, N. (2023). Synthetic data for face recognition: Current state and future prospects. *Image and Vision Computing, 135*, 104688. DOI: 10.1016/j.imavis.2023.104688

Bozkus, T., & Mitra, U. (2024). Leveraging Digital Cousins for Ensemble Q-Learning in Large-Scale Wireless Networks. *IEEE Transactions on Signal Processing, 72*, 1114–1129. Advance online publication. DOI: 10.1109/TSP.2024.3366434

Braun, V., & Clarke, V. (2006). Using thematic analysis in psychology. *Qualitative Research in Psychology, 3*(2), 77–101. DOI: 10.1191/1478088706qp063oa

Buggineni, V., Chen, C., & Camelio, J. (2024). Enhancing manufacturing operations with synthetic data: A systematic framework for data generation, accuracy, and utility. *Frontiers in Manufacturing Technology*, *4*, 1320166. DOI: 10.3389/fmtec.2024.1320166

Buolamwini, J., & Gebru, T. (2018). Gender shades: Intersectional accuracy disparities in commercial gender classification. *Proceedings of Machine Learning Research*, *81*, 1–15.

Burgard, J. P., Kolb, J. P., Merkle, H., & Münnich, R. (2017). Synthetic data for open and reproducible methodological research in social sciences and official statistics. *AStA Wirtschafts-und Sozialstatistisches Archiv*, *11*(3-4), 233–244. DOI: 10.1007/s11943-017-0214-8

Cao, H., Tan, C., Gao, Z., Xu, Y., Chen, G., Heng, P. A., & Li, S. Z. (2024). A Survey on Generative Diffusion Models. *IEEE Transactions on Knowledge and Data Engineering*, *36*(7), 2814–2830. Advance online publication. DOI: 10.1109/TKDE.2024.3361474

Cardenuto, J. P., Yang, J., & Padilha, R. (2023). *The age of synthetic realities: Challenges and opportunities*. Foundations and Trends in Signal Processing., DOI: 10.1561/2000000113

Cardenuto, J. P., Yang, J., Padilha, R., Wan, R., Moreira, D., Li, H., Wang, S., Andaló, F., Marcel, S., & Rocha, A. (2023). The Age of Synthetic Realities: Challenges and opportunities. *APSIPA Transactions on Signal and Information Processing*, *12*(1). Advance online publication. DOI: 10.1561/116.00000138

Carlini, N., Tramer, F., Wallace, E., Jagielski, M., Herbert-Voss, A., Lee, K., . . . Erlingsson, Ú. (2021). Extracting training data from large language models. arXiv preprint arXiv:2012.07805.

Carlson, A., Skinner, K. A., Vasudevan, R., & Johnson-Roberson, M. (2018). Modeling camera effects to improve visual learning from synthetic data. In *Proceedings of the European Conference on Computer Vision (ECCV) Workshops* (pp. 0-0).

Carlucci, F. M., Russo, P., & Caputo, B. (2017, May). A deep representation for depth images from synthetic data. In *2017 IEEE International Conference on Robotics and Automation (ICRA)* (pp. 1362-1369). IEEE. DOI: 10.1109/ICRA.2017.7989162

Cascante-Bonilla, P., Shehada, K., Smith, J. S., Doveh, S., Kim, D., Panda, R., & Karlinsky, L. (2023). Going beyond nouns with vision & language models using synthetic data. In *Proceedings of the IEEE/CVF International Conference on Computer Vision* (pp. 20155-20165). DOI: 10.1109/ICCV51070.2023.01844

Casula, C., Leonardelli, E., & Tonelli, S. (2024, August). Don't augment, Rewrite? Assessing abusive language detection with synthetic data. In *Findings of the Association for Computational Linguistics ACL 2024* (pp. 11240-11247).

Chakradhar, B. S., Vijey, M., & Tunk, S. (2024). Synthetic data generation for marketing insights. In Chapman and Hall/CRC eBooks (pp. 195–215). https://doi.org/DOI: 10.1201/9781003472544-17

Chang, C. H., & Kidman, G. (2023). The rise of generative artificial intelligence (AI) language models - challenges and opportunities for geographical and environmental education. *International Research in Geographical and Environmental Education, 32*(2), 85–89. DOI: 10.1080/10382046.2023.2194036

Chapelle, C. A. (2025). Generative AI as game changer: Implications for language education. *System, 132*, 103672. DOI: 10.1016/j.system.2025.103672

Chauhan, C. (2024). The impact of generative artificial intelligence in scientific content synthesis for authors. *American Journal of Pathology, 194*(8), 1406–1408. DOI: 10.1016/j.ajpath.2024.06.002 PMID: 38879080

Chen, H., Waheed, A., Li, X., Wang, Y., Wang, J., Raj, B., & Abdin, M. I. (2024). On the diversity of synthetic data and its impact on training large language models. *arXiv preprint arXiv:2410.15226.*

Chen, B., Zhu, X., & Díaz del Castillo, H. (2023). Integrating generative AI in knowledge building. *Computers and Education: Artificial Intelligence, 5*, 100184. Advance online publication. DOI: 10.1016/j.caeai.2023.100184

Cheng, K., & Wu, H. (2024). Policy framework for the utilization of generative AI. *Critical Care, 28*(1), 1–2. DOI: 10.1186/s13054-024-04917-z PMID: 38637898

Chen, J., Chun, D., Patel, M., Chiang, E., & James, J. (2019). The validity of synthetic clinical data: A validation study of a leading synthetic data generator (Synthea) using clinical quality measures. *BMC Medical Informatics and Decision Making, 19*(1), 1–9. DOI: 10.1186/s12911-019-0793-0 PMID: 30871520

Chen, J., & Little, J. J. (2019). Sports camera calibration via synthetic data. In *Proceedings of the IEEE/CVF Conference on Computer Vision and Pattern Recognition Workshops* (pp. 0-0).

Chen, M., & Zhang, Y. (2014). Big data analytics: A survey. *Journal of Big Data, 1*(1), 1–32.

Chen, Y., Li, W., Chen, X., & Gool, L. V. (2019). Learning semantic segmentation from synthetic data: A geometrically guided input-output adaptation approach. In *Proceedings of the IEEE/CVF Conference on Computer Vision and Pattern Recognition* (pp. 1841-1850). DOI: 10.1109/CVPR.2019.00194

Chesney, B., & Citron, D. (2019). Deep fakes: A looming challenge for privacy, democracy, and national security. *California Law Review, 107*, 1753.

Cheuque, C., Querales, M., León, R., Salas, R., & Torres, R. (2022). An efficient multi-level convolutional neural network approach for white blood cells classification. *Diagnostics (Basel), 12*(2), 248. DOI: 10.3390/diagnostics12020248 PMID: 35204339

Chia, A. (2022). The Metaverse, but Not the Way You Think: Game Engines and Automation Beyond Game Development. *Critical Studies in Media Communication, 39*(3), 191–200. Advance online publication. DOI: 10.1080/15295036.2022.2080850

Chiodo, S. (2024). What AI "art" can teach us about art. *Journal of Aesthetics & Culture, 16*(1), 2395511. DOI: 10.1080/20004214.2024.2395511

Choi, Y., & Yoon, S. (2025). Enhancing in-situ model accuracy in operational building systems with augmentation-based synthetic operational data. *Journal of Building Engineering, 101*, 111623. DOI: 10.1016/j.jobe.2024.111623

Christou, P. A. (2023). How to use artificial intelligence (AI) as a resource, methodological and analysis tool in qualitative research? *Qualitative Report, 28*(7).

Civit, M., Civit-Masot, J., Cuadrado, F., & Escalona, M. J. (2022). A systematic review of artificial intelligence-based music generation: Scope, applications, and future trends. *Expert Systems with Applications, 209*, 118190. Advance online publication. DOI: 10.1016/j.eswa.2022.118190

Clausen, C. S. B., Ma, Z. G., & Jørgensen, B. N. (2022). Can We Benefit from Game Engines to Develop Digital Twins for Planning the Deployment of Photovoltaics? *Energy Informatics, 5*(S4), 42. Advance online publication. DOI: 10.1186/s42162-022-00222-7

Consalvo, M., & Staines, D. (2021). Reading Ren'Py: Game Engine Affordances and Design Possibilities. *Games and Culture, 16*(6), 762–778. Advance online publication. DOI: 10.1177/1555412020973823

Cortes, D., Bermejo, B., & Juiz, C. (2024). The use of CNNs in VR/AR/MR/XR: A systematic literature review. *Virtual Reality (Waltham Cross), 28*(154), 154. Advance online publication. DOI: 10.1007/s10055-024-01044-6

Cowap, A., Graham, Y., & Foster, J. (2023). Do Stochastic Parrots have Feelings Too? Improving Neural Detection of Synthetic Text via Emotion Recognition. *Findings of the Association for Computational Linguistics: EMNLP, 2023*, 9928–9946. Advance online publication. DOI: 10.18653/v1/2023.findings-emnlp.665

Croitoru, F. A., Hondru, V., Ionescu, R. T., & Shah, M. (2023). Diffusion Models in Vision: A Survey. *IEEE Transactions on Pattern Analysis and Machine Intelligence, 45*(9), 10850–10869. Advance online publication. DOI: 10.1109/TPAMI.2023.3261988 PMID: 37030794

Crosby, M., Pattanayak, P., Verma, S., & Kalyanaraman, V. (2016). Blockchain technology: Beyond Bitcoin. *Applied Innovation Review, 2*, 6–10.

Dahmen, J., & Cook, D. (2019). SynSys: A synthetic data generation system for healthcare applications. *Sensors (Basel), 19*(5), 1181. DOI: 10.3390/s19051181 PMID: 30857130

Dahmen, T., Trampert, P., Boughorbel, F., Sprenger, J., Klusch, M., Fischer, K., Kübel, C., & Slusallek, P. (2019). Digital reality: A model-based approach to supervised learning from synthetic data. *AI Perspectives, 1*(1), 1–12. DOI: 10.1186/s42467-019-0002-0

Dalal, R. (2025). *Core Concepts of Mechanics and Thermodynamics*. Educohack Press.

Dangi, S., Ghanshala, K., & Sharma, S. (2025). LIFT: Lightweight incremental and federated techniques for live memory forensics and proactive malware detection. *International Journal of Advanced Computer Science and Applications, 16*(4). Advance online publication. DOI: 10.14569/IJACSA.2025.0160445

Dankar, F. K., & Ibrahim, M. (2021). Fake it till you make it: Guidelines for effective synthetic data generation. *Applied Sciences (Basel, Switzerland), 11*(5), 2158. DOI: 10.3390/app11052158

Dankar, F. K., Ibrahim, M. K., & Ismail, L. (2022). A multi-dimensional evaluation of synthetic data generators. *IEEE Access : Practical Innovations, Open Solutions, 10*, 11147–11158. DOI: 10.1109/ACCESS.2022.3144765

Dash, A., Ye, J., & Wang, G. (2023). *A review of generative adversarial networks (GANs) and its applications in a wide variety of disciplines: from medical to remote sensing*. IEEE., DOI: 10.1109/ACCESS.2023.3234567

Davis, N., Liu, S., & Lovato, S. (2021). Creativity and AI in digital media: Implications for labour. *ACM Transactions on Multimedia Computing Communications and Applications, 17*(3s), 1–22.

De Melo, C. M., Torralba, A., Guibas, L., DiCarlo, J., Chellappa, R., & Hodgins, J. (2022). Next-generation deep learning based on simulators and synthetic data. *Trends in Cognitive Sciences*, *26*(2), 174–187. Advance online publication. DOI: 10.1016/j.tics.2021.11.008 PMID: 34955426

de Rosa, G. H., & Papa, J. P. (2021). A survey on text generation using generative adversarial networks. *Pattern Recognition*, *119*, 108098. Advance online publication. DOI: 10.1016/j.patcog.2021.108098

DeAndres-Tame, I., Tolosana, R., Melzi, P., Vera-Rodriguez, R., Kim, M., Rathgeb, C., Liu, X., Gomez, L. F., Morales, A., Fierrez, J., Ortega-Garcia, J., Zhong, Z., Huang, Y., Mi, Y., Ding, S., Zhou, S., He, S., Fu, L., Cong, H., & Menotti, D. (2025). Second frcsyn-ongoing: Winning solutions and post-challenge analysis to improve face recognition with synthetic data. *Information Fusion*, *120*, 103099. DOI: 10.1016/j.inffus.2025.103099

Debnath, P., Deshmukh, S. M., Laroia, M., Selvakumar, P., Manjunath, T. C., & Bhattacharya, S. (2025). Challenges Faced by Countries in Renewable Energy Adoption. In Özsungur, F., Chaychi Semsari, M., & Küçük Bayraktar, H. (Eds.), *Geopolitical Landscapes of Renewable Energy and Urban Growth* (pp. 295–320). IGI Global Scientific Publishing., DOI: 10.4018/979-8-3693-8814-3.ch010

Delp, S. L., Anderson, F. C., Arnold, A. S., Loan, P., Habib, A., John, C. T., Guendelman, E., & Thelen, D. G. (2007). OpenSim: Open-source software to create and analyse dynamic simulations of movement. *IEEE Transactions on Biomedical Engineering*, *54*(11), 1940–1950. DOI: 10.1109/TBME.2007.901024 PMID: 18018689

Delussu, R., Putzu, L., & Fumera, G. (2024). Synthetic Data for Video Surveillance Applications of Computer Vision: A Review. *International Journal of Computer Vision*, *132*(10), 1–37. DOI: 10.1007/s11263-024-02102-x

De, S., & Mitra, P. (2024). A novel technique of synthetic data generation for asset administration Shells in industry 4.0 scenarios. *IEEE Transactions on Artificial Intelligence*, *5*(10), 1–10. DOI: 10.1109/TAI.2024.3409516

Devgan, K., Sharma, S., Jangra, D., & Sharma, S. (2025b). Artificial Intelligence-Enabled Apple Disease Monitoring System for environmental protection and resource conservation. In Advances in computational intelligence and robotics book series (pp. 271–300). https://doi.org/DOI: 10.4018/979-8-3373-3246-8.ch011

Devliyal, S., Goyal, H. R., & Sharma, S. (2025b). ICDRET: A dynamic cyber risk estimation technique for intelligent Cyber-Physical Systems in PCS. *The Open Bioinformatics Journal*, *18*(1), e18750362380269. Advance online publication. DOI: 10.2174/0118750362380269250502062754

Devliyal, S., Sharma, S., & Goyal, H. R. (2024). EIAIMSPS: Edge inspired artificial intelligence-based multi stakeholders Personalized Security Mechanism in ICPS for PCS. *International Journal of Advanced Computer Science and Applications, 15*(8). Advance online publication. DOI: 10.14569/IJACSA.2024.01508117

Dina, A. S., Siddique, A. B., & Manivannan, D. (2022). Effect of balancing data using synthetic data on the performance of machine learning classifiers for intrusion detection in computer networks. *IEEE Access : Practical Innovations, Open Solutions, 10*, 96731–96747. DOI: 10.1109/ACCESS.2022.3205337

Dobrokvashina, A., Lavrenov, R., & Bai, Y. (2022) "Sensors Modelling for Servosila Engineer Crawler Robot in Webots Simulator" *2022 International Conference on Artificial Life and Robotics (ICAROB)*. DOI: 10.1109/MWENT55238.2022.9802400

Dolhansky, B., Bitton, J., Pflaum, B., Lu, J., Howes, R., Wang, M., & Ferrer, C. C. (2020). The deepfake detection challenge (dfdc) dataset. *arXiv preprint arXiv:2006.07397*.

Doron, G., Genway, S., Roberts, M., & Jasti, S. (2025). Generative AI: Driving productivity and scientific breakthroughs in pharmaceutical R&D. *Drug Discovery Today, 30*(1), 104272. DOI: 10.1016/j.drudis.2024.104272 PMID: 39675517

Dos Santos, R., & Aguilar, J. (2024). *A synthetic data generation system based on the variational-autoencoder technique and the linked data paradigm*. Progress in Artificial Intelligence., DOI: 10.1007/s13748-024-00328-x

Dosovitskiy, A., Ros, G., Codevilla, F., López, A., & Koltun, V. (2017). CARLA: An open urban driving simulator. *Conference on Robot Learning*, 1–16.

Dou, B., Zhu, Z., Merkurjev, E., Ke, L., & Chen, L. (2023). Machine learning methods for small data challenges in molecular science. *Chemical Science (Cambridge)*. Advance online publication. DOI: 10.1039/D2SC06141A PMID: 37384816

Dugan, L., Ippolito, D., Kirubarajan, A., Shi, S., & Callison-Burch, C. (2023, June). Real or fake text?: Investigating human ability to detect boundaries between human-written and machine-generated text. *Proceedings of the AAAI Conference on Artificial Intelligence, 37*(11), 12763–12771. DOI: 10.1609/aaai.v37i11.26501

Du, L., Zhang, R., & Wang, X. (2020). Overview of two-stage object detection algorithms. *Journal of Physics: Conference Series, 1544*(1), 012033. DOI: 10.1088/1742-6596/1544/1/012033

Dwivedi, P., & Islam, B. (2024). Generative Adversarial Networks Based Framework for Music Genre Classification. *SN Computer Science, 5*(8), 1149. Advance online publication. DOI: 10.1007/s42979-024-03531-8

Eigenschink, P., Reutterer, T., Vamosi, S., Vamosi, R., Sun, C., & Kalcher, K. (2023). Deep Generative Models for Synthetic Data: A survey. *IEEE Access : Practical Innovations, Open Solutions, 11*, 47304–47320. DOI: 10.1109/ACCESS.2023.3275134

El Emam, K. (2020). Seven ways to evaluate the utility of synthetic data. *IEEE Security and Privacy, 18*(4), 56–59. DOI: 10.1109/MSEC.2020.2992821

El Emam, K., Mosquera, L., Fang, X., & El-Hussuna, A. (2022). Utility metrics for evaluating synthetic health data generation methods: Validation study. *JMIR Medical Informatics, 10*(4), e35734. Advance online publication. DOI: 10.2196/35734 PMID: 35389366

El Fathi, T. E., Saad, A., Larhzil, H., Lamri, D., & Ibrahmi, E. M. A. (2025d). Integrating generative AI into STEM education: Enhancing conceptual understanding, addressing misconceptions, and assessing student acceptance. *Disciplinary and Interdisciplinary Science Education Research, 7*(1), 6. Advance online publication. DOI: 10.1186/s43031-025-00125-z

Emon, M. M. H., & Khan, T. (2025). The mediating role of attitude towards the technology in shaping artificial intelligence usage among professionals. *Telematics and Informatics Reports, 17*, 100188. DOI: 10.1016/j.teler.2025.100188

Endres, M., Venugopal, A. M., & Tran, T. S. (2022). Synthetic Data Generation: A Comparative study. In *IDEAS '22: Proceedings of the 26th International Database Engineered Applications Symposium (IDEAS'22)* (p. 94-102). ACM Digital Library. DOI: 10.1145/3548785.3548793

Eno, J., & Thompson, C. W. (2008). Generating synthetic data to match data mining patterns. *IEEE Internet Computing, 12*(3), 78–82. DOI: 10.1109/MIC.2008.55

European Commission. (2021). *Proposal for a regulation laying down harmonised rules on artificial intelligence.* AI Act.

Fabbri, M., Brasó, G., Maugeri, G., Cetintas, O., Gasparini, R., Ošep, A., & Cucchiara, R. (2021). Motsynth: How can synthetic data help pedestrian detection and tracking? In *Proceedings of the IEEE/CVF International Conference on Computer Vision* (pp. 10849-10859). DOI: 10.1109/ICCV48922.2021.01067

Farhadi, M., Feger, R., Fink, J., Wagner, T., & Stelzer, A. (2024). Combining MIMO DBF with Automotive Synthetic Aperture Radar Imaging and Phase Error Correction. *IEEE Access : Practical Innovations, Open Solutions, 12*, 31944–31959. Advance online publication. DOI: 10.1109/ACCESS.2024.3370231

Farid, H. (2022). Creating, using, misusing, and detecting deep fakes. *Journal of Online Trust & Safety, 1*(4). Advance online publication. DOI: 10.54501/jots.v1i4.56

Faulkner, C. A., Jankowski, D. S., Castellini, J. E.Jr, Zuo, W., Epple, P., Sohn, M. D., Kasgari, A. T. Z., & Saad, W. (2023). Fast prediction of indoor airflow distribution inspired by synthetic image generation artificial intelligence. *Building Simulation, 16*(7), 1219–1238. Advance online publication. DOI: 10.1007/s12273-023-0989-1 PMID: 37359832

Fazel, A., Yang, W., Liu, Y., Barra-Chicote, R., Meng, Y., Maas, R., & Droppo, J. (2021). Synthasr: Unlocking synthetic data for speech recognition. *arXiv Preprint arXiv:2106.07803*. DOI: 10.21437/Interspeech.2021-1882

Ferrara, E. (2024). GenAI against humanity: Nefarious applications of generative artificial intelligence and large language models. *Journal of Computational Social Science, 7*(1), 549–569. DOI: 10.1007/s42001-024-00250-1

Ferrari, F., & McKelvey, F. (2023). Hyperproduction: a social theory of deep generative models. *Distinktion: Journal of Social Theory*, 338–360. DOI: 10.1080/1600910X.2022.2137546

Feuerriegel, S., Hartmann, J., Janiesch, C., & Zschech, P. (2024). Generative AI. *Business & Information Systems Engineering, 66*(1), 111–126. DOI: 10.1007/s12599-023-00834-7

Figueira, A., & Vaz, B. (2022). Survey on synthetic data generation, evaluation methods and GANs. *Mathematics, 10*(15), 2733. DOI: 10.3390/math10152733

Fischer, J., & Mahon, B. Z. (2021). What Tool Representation, Intuitive Physics, and Action Have in Common: The Brain's First-Person Physics Engine. *Cognitive Neuropsychology, 38*(7-8), 455–467. Advance online publication. DOI: 10.1080/02643294.2022.2106126 PMID: 35994054

Fisher, S. A., Howard, J. W., & Kira, B. (2024). Moderating Synthetic Content: The Challenge of Generative AI. *Philosophy & Technology, 37*(4), 133. Advance online publication. DOI: 10.1007/s13347-024-00818-9 PMID: 39555113

Fonseca, J., & Bacao, F. (2023). Tabular and latent space synthetic data generation: A literature review. *Journal of Big Data, 10*(1), 115. DOI: 10.1186/s40537-023-00792-7

Foti, C., Anttila, D., Maniscalco, S., & Chiofalo, M. L. (2021). *Quantum Physics Literacy Aimed at K12 and the General Public*. Universe., DOI: 10.3390/ECU2021-09322

Fragkouli, S., Solanki, D., Castro, L. J., Psomopoulos, F. E., Queralt-Rosinach, N., Cirillo, D., & Crossman, L. C. (2024). Synthetic data: How could it be used in infectious disease research? *Future Microbiology, 19*(17), 1439–1444. DOI: 10.1080/17460913.2024.2400853 PMID: 39345126

Fredheim, R., & Pamment, J. (2024). Assessing the risks and opportunities posed by AI-enhanced influence operations on social media. *Place Branding and Public Diplomacy*. Advance online publication. DOI: 10.1057/s41254-023-00322-5

Freeman, C. D., Frey, E., Raichuk, A., & Girgin, S. (2021) "Brax—a Differentiable Physics Engine for Large Scale Rigid Body Simulation" arXiv preprint arXiv:2112.08241. doi: /arXiv.2112.08241.DOI: 10.48550

Frid-Adar, M., Klang, E., Amitai, M., Goldberger, J., & Greenspan, H. (2018). Synthetic data augmentation using GAN for improved liver lesion classification. *IEEE Transactions on Medical Imaging*, *38*(3), 915–925.

Frid-Adar, M., Klang, E., Amitai, M., Goldberger, J., & Greenspan, H. (2018). Synthetic data augmentation using GAN for improved liver lesion classification. *Neurocomputing*, *321*, 321–331. DOI: 10.1016/j.neucom.2018.09.013

Fridrich, J., & Goljan, M. (2009). Digital image forensics and the detection of image forgery. *IEEE Signal Processing Magazine*, *26*(2), 26–37. DOI: 10.1109/MSP.2008.931078

Fu, S., Tamir, N., Sundaram, S., Chai, L., Zhang, R., Dekel, T., & Isola, P. (2023). Dreamsim: Learning new dimensions of human visual similarity using synthetic data. *arXiv Preprint arXiv:2306.09344*.

Fui-Hoon Nah, F., Zheng, R., Cai, J., Siau, K., & Chen, L. (2023). Generative AI and ChatGPT: Applications, challenges, and AI-human collaboration. *Journal of Information Technology Case and Application Research*, *25*(3), 277–304. DOI: 10.1080/15228053.2023.2233814

Gakhar, S., Kondoju, V. P., Chauhan, S. S., Kumar, A., Kulkarni, S. S., & Goel, P. (2025). Generative AI for Real-Time Data Augmentation in Big Data Pipelines. *1st International Conference on Advances in Computer Science, Electrical, Electronics, and Communication Technologies, CE2CT 2025*, 1386–1391. DOI: 10.1109/CE2CT64011.2025.10941295

Gal, M. S., & Lynskey, O. (2023). Synthetic data: legal implications of the data-generation revolution. *IoWa L. rev., 109*, 1087.

Gal, M. S., & Lynskey, O. (2023). Synthetic data: Legal implications of the data-generation revolution. *SSRN*, *109*, 1087. DOI: 10.2139/ssrn.4414385

Gal, M. S., & Lynskey, O. (2024). Synthetic data: Legal implications of the data-generation revolution. *Iowa Law Review*, *109*, 1087.

Gandomi, A., & Haider, M. (2015). Beyond the hype: Big data concepts, methods, and analytics. *International Journal of Information Management, 35*(2), 137–144. DOI: 10.1016/j.ijinfomgt.2014.10.007

Gao, C., Killeen, B. D., Hu, Y., Grupp, R. B., Taylor, R. H., Armand, M., & Unberath, M. (2023). Synthetic data accelerates the development of generalizable learning-based algorithms for X-ray image analysis. *Nature Machine Intelligence, 5*(3), 294–308. DOI: 10.1038/s42256-023-00629-1 PMID: 38523605

Ge, T., Chan, X., Wang, X., Yu, D., Mi, H., & Yu, D. (2024). Scaling synthetic data creation with 1,000,000,000 personas. *arXiv preprint arXiv:2406.20094.*

Ge, T., Chan, X., Wang, X., Yu, D., Mi, H., & Yu, D. (2024). Scaling synthetic data creation with 1,000,000,000 personas. *arXiv Preprint arXiv:2406.20094.*

Georgiou, Y., Tsivitanidou, O., & Ioannou, A. (2021). *Learning Experience Design with Immersive Virtual Reality in Physics Education.* Springer., DOI: 10.1007/s11423-021-10055-y

Gholami, S., & Omar, M. (2023). Does synthetic data make Large Language Models more efficient? *arXiv Preprint arXiv:2310.07830.*

Gholamiankhah F, Mostafapour S, Arabi H. (2021). "Deep learning-based synthetic CT generation from MR images: comparison of generative adversarial and residual neural networks". arXiv preprint arXiv.

Giese, M. A., & Poggio, T. (2003). Neural mechanisms for the recognition of biological movements. *Nature Reviews. Neuroscience, 4*(3), 179–192. DOI: 10.1038/nrn1057 PMID: 12612631

Gillberg, J., Bergdahl, J., & Sestini, A. (2023) "Technical Challenges of Deploying Reinforcement Learning Agents for Game Testing in AAA Games" *2023 IEEE Conference on Games (CoG).* Doi: DOI: 10.1109/CoG57401.2023.10333194

Gill, S. S., Xu, M., Ottaviani, C., Patros, P., Bahsoon, R., Shaghaghi, A., Golec, M., Stankovski, V., Wu, H., Abraham, A., Singh, M., Mehta, H., Ghosh, S. K., Baker, T., Parlikad, A. K., Lutfiyya, H., Kanhere, S. S., Sakellariou, R., Dustdar, S., & Uhlig, S. (2022). AI for next generation computing: Emerging trends and future directions. *Internet of Things : Engineering Cyber Physical Human Systems, 19,* 100514. Advance online publication. DOI: 10.1016/j.iot.2022.100514

Giomi, M., Boenisch, F., Wehmeyer, C., & Tasnádi, B. (2022). A unified framework for quantifying privacy risk in synthetic data. *arXiv Preprint arXiv:2211.10459.*

Giuffrè, M., & Shung, D. L. (2023). Harnessing the power of synthetic data in healthcare: Innovation, application, and privacy. *NPJ Digital Medicine*, *6*(1), 186. DOI: 10.1038/s41746-023-00927-3 PMID: 37813960

Göbel, T., Baier, H., & Breitinger, F. (2023). Data for digital forensics: Why a discussion on "how realistic is synthetic data" is dispensable. *Digital Threats : Research and Practice*, *4*(3), 1–18. DOI: 10.1145/3609863

Goel, A., Goel, A. K., & Kumar, A. (2023). The role of artificial neural network and machine learning in utilizing spatial information. *Spatial Information Research*, *31*(3), 275–285. DOI: 10.1007/s41324-022-00494-x

Goetz, J., & Tewari, A. (2020). Federated learning via synthetic data. *arXiv Preprint arXiv:2008.04489.*

Goncalves, A., Ray, P., Soper, B., Stevens, J., Coyle, L., & Sales, A. P. (2020). Generation and evaluation of synthetic patient data. *BMC Medical Research Methodology*, *20*(1), 1–40. DOI: 10.1186/s12874-020-00977-1 PMID: 32381039

Gonzales, A., Guruswamy, G., & Smith, S. R. (2023). Synthetic data in health care: A narrative review. *PLOS Digital Health*, *2*(1), e0000082. DOI: 10.1371/journal.pdig.0000082 PMID: 36812604

Goodfellow, I., Pouget-Abadie, J., Mirza, M., Xu, B., Warde-Farley, D., Ozair, S., & Bengio, Y. (2014). Generative adversarial nets. *Advances in Neural Information Processing Systems*, *27*, 2672–2680.

Goyal, M., & Mahmoud, Q. H. (2024). A Systematic review of Synthetic data generation techniques using Generative AI. *Electronics (Basel)*, *13*(17), 3509. DOI: 10.3390/electronics13173509

Guan, J., Li, R., Yu, S., & Zhang, X. (2018, December). Generation of synthetic electronic medical record text. In *2018 IEEE International Conference on Bioinformatics and Biomedicine (BIBM)* (pp. 374-380). IEEE. DOI: 10.1109/BIBM.2018.8621223

Guan, S., Xu, C., Lin, M., & Greene, D. (2024, November). Effective synthetic data and test-time adaptation for OCR correction. In *Proceedings of the 2024 Conference on Empirical Methods in Natural Language Processing* (pp. 15412-15425). DOI: 10.18653/v1/2024.emnlp-main.862

Gui, J., Sun, Z., Wen, Y., Tao, D., & Ye, J. (2021). A review on generative adversarial networks: Algorithms, theory, and applications. *IEEE Transactions on Neural Networks and Learning Systems*. Advance online publication. DOI: 10.1109/TNNLS.2021.3088515

Gupta, A., Bhatt, D., & Pandey, A. (2021). Transitioning from real to synthetic data: Quantifying the bias in model. *arXiv Preprint arXiv:2105.04144.*

Gupta, V. (2024). An Empirical Evaluation of a Generative Artificial Intelligence Technology Adoption Model from Entrepreneurs' Perspectives. *Systems 2024, Vol. 12, Page 103, 12*(3), 103. DOI: 10.3390/systems12030103

Gupta, A., Mittal, A., & Jain, S. (2019). Synthetic data generation for machine learning. *International Journal of Computer Applications, 178*(7), 12–18. DOI: 10.5120/ijca2016910248

Gupta, A., Vedaldi, A., & Zisserman, A. (2016). Synthetic data for text localisation in natural images. In *Proceedings of the IEEE Conference on Computer Vision and Pattern Recognition* (pp. 2315-2324). DOI: 10.1109/CVPR.2016.254

Habibie, I., Tzionas, D., Black, M. J., & Theobalt, C. (2022). Learning human dynamics from sparse motion sensors using attention. *IEEE Transactions on Pattern Analysis and Machine Intelligence.*

Hai-Jew, S. (2022). *Common Visual Art in a Social Digital Age.* Nova Science Publishers. DOI: 10.52305/BKOT1869

Halliday, D., Walker, J., & Resnick, R. (2023). *Principles of physics.* John Wiley & Sons.

Hämäläinen, P., Tavast, M., & Kunnari, A. (2023, April). Evaluating large language models in generating synthetic HCI research data: a case study. In *Proceedings of the 2023 CHI Conference on Human Factors in Computing Systems* (pp. 1-19).

Handa, A., Patraucean, V., Badrinarayanan, V., Stent, S., & Cipolla, R. (2016). Understanding real world indoor scenes with synthetic data. In *Proceedings of the IEEE Conference on Computer Vision and Pattern Recognition* (pp. 4077-4085).

Han, S., Mao, H., & Dally, W. J. (2016). Deep compression: Compressing deep neural networks with pruning, trained quantization and Huffman coding. *International Conference on Learning Representations (ICLR).*

Hao, S., Han, W., Jiang, T., Li, Y., Wu, H., Zhong, C., . . . Tang, H. (2024). Synthetic data in AI: Challenges, applications, and ethical implications. *arXiv Preprint arXiv:2401.01629.*

Hartmann, J., Exner, Y., & Domdey, S. (2023). The power of generative marketing: Can generative AI reach human-level visual marketing content? SSRN *Electronic Journal.* DOI: 10.2139/ssrn.4597899

He, R., Sun, S., Yu, X., Xue, C., Zhang, W., Torr, P., . . . Qi, X. (2022). Is synthetic data from generative models ready for image recognition? *arXiv preprint arXiv:2210.07574.*

Heidari, F., Mahdavinejad, M., Werner, L. C., Roohabadi, M., & Sarmadi, H. (2021). Biocomputational Architecture Based on Particle Physics. *Frontiers in Energy Research, 9*, 620127. Advance online publication. DOI: 10.3389/fenrg.2021.620127

Hernandez, M., Epelde, G., Alberdi, A., Cilla, R., & Rankin, D. (2022). Synthetic data generation for tabular health records: A systematic review. *Neurocomputing, 493*, 28–45. DOI: 10.1016/j.neucom.2022.04.053

hicks, B. L. (2016). Beyond this or that: Challenging the limits of binary language in elementary education through poetry, Word Art, and creative bookmaking. *Teaching, Affirming, and Recognizing Trans and Gender Creative Youth: A Queer Literacy Framework*, 81-106.

Hilmes, B., & Rossenbach, N. (2024). On the effect of purely synthetic training data for different automatic speech recognition architectures. *arXiv preprint arXiv:2407.17997*. DOI: 10.21437/SynData4GenAI.2024-10

Hittmeir, M., Ekelhart, A., & Mayer, R. (2019, August). On the utility of synthetic data: An empirical evaluation on machine learning tasks. In *Proceedings of the 14th International Conference on Availability, Reliability and Security* (pp. 1-6). DOI: 10.1145/3339252.3339281

Hla, D. A., & Hindin, D. I. (2025). Generative AI & machine learning in surgical education. *Current Problems in Surgery, 63*, 101701. DOI: 10.1016/j.cpsurg.2024.101701 PMID: 39922636

Ho, J., Jain, A., & Abbeel, P. (2020). Denoising diffusion probabilistic models. arXiv preprint arXiv:2006.11239.

Hossain, M. Z., Uz Zaman, F., & Islam, M. R. (2023). Advancing AI-Generated Image Detection: Enhanced Accuracy through CNN and Vision Transformer Models with Explainable AI Insights. *2023 26th International Conference on Computer and Information Technology, ICCIT 2023*. DOI: 10.1109/ICCIT60459.2023.10440990

Howard, A. G.. (2019). Searching for MobileNetV3. *IEEE/CVF International Conference on Computer Vision (ICCV)*, 1314–1324.

Hu, T. Y., Armandpour, M., Shrivastava, A., Chang, J. H. R., Koppula, H., & Tuzel, O. (2022, May). Synt++: Utilizing imperfect synthetic data to improve speech recognition. In *ICASSP 2022-2022 IEEE International Conference on Acoustics, Speech and Signal Processing (ICASSP)* (pp. 7682-7686). IEEE.

Huang, Y. F., Lv, S. J., Tseng, K. K., Tseng, P. J., Xie, X., & Lin, R. F.-Y. (2023). Recent Advances in Artificial Intelligence for Video Production System. *Enterprise Information Systems, 17*(11), 2246188. Advance online publication. DOI: 10.1080/17517575.2023.2246188

Huang, Y., Guo, F., Zhang, H., & Yang, Z. (2022). An Efficient Computational Framework for Generating Realistic 3D Mesoscale Concrete Models Using Micro X-ray Computed Tomography Images and Dynamic Physics .... *Cement and Concrete Composites, 126*, 104347. Advance online publication. DOI: 10.1016/j.cemconcomp.2021.104347

Huang, Z., Li, X., Wang, H., & Duan, X. (2022). Detecting synthetic human motion using temporal consistency and motion statistics. *IEEE Transactions on Multimedia, 24*, 3472–3485.

Hwang, S. W., & Sugiyama, J. (2021). Computer vision-based wood identification and its expansion and contribution potentials in wood science: A review. *Plant Methods, 17*(47), 47. Advance online publication. DOI: 10.1186/s13007-021-00746-1 PMID: 33910606

Iantovics, L. B., & Enăchescu, C. (2022). Method for data quality assessment of synthetic industrial data. *Sensors (Basel), 22*(4), 1608. Advance online publication. DOI: 10.3390/s22041608 PMID: 35214509

Ibrahim, M., Al Khalil, Y., Amirrajab, S., Sun, C., Breeuwer, M., Pluim, J., & Dumontier, M. (2025). Generative AI for synthetic data across multiple medical modalities: A systematic review of recent developments and challenges. *Computers in Biology and Medicine, 189*, 109834. DOI: 10.1016/j.compbiomed.2025.109834 PMID: 40023073

Ibrar, W., Mahmood, D., Al-Shamayleh, A. S., Ahmed, G., Alharthi, S. Z., & Akhunzada, A. (2025). Generative AI: A double-edged sword in the cyber threat landscape. *Artificial Intelligence Review, 58*(9), 1–27. DOI: 10.1007/s10462-025-11285-9

Indira Priya, P., Manju, M., Anitha, R., Harsavardhini, R., & Sri Harini, S. (2024). Transforming Text to Video: Leveraging Advanced Generative AI Techniques. *2024 Asian Conference on Intelligent Technologies, ACOIT 2024*. DOI: 10.1109/ACOIT62457.2024.10941483

Israel, M. J., & Amer, A. (2023). Rethinking data infrastructure and its ethical implications in the face of automated digital content generation. *AI and Ethics, 3*(2), 427–439. Advance online publication. DOI: 10.1007/s43681-022-00169-1

Jacobsen, B. N. (2023). Machine learning and the politics of synthetic data. *Big Data & Society, 10*(1), 20539517221145372. DOI: 10.1177/20539517221145372

Jacobs, F., D'Amico, S., Benvenuti, C., Gaudio, M., Saltalamacchia, G., Miggiano, C., De Sanctis, R., Della Porta, M. G., Santoro, A., & Zambelli, A. (2023). Opportunities and challenges of synthetic data generation in oncology. *JCO Clinical Cancer Informatics, 7*(7), e2300045. DOI: 10.1200/CCI.23.00045 PMID: 37535875

Jaderberg, M., Simonyan, K., Vedaldi, A., & Zisserman, A. (2014). Synthetic data and artificial neural networks for natural scene text recognition. *arXiv Preprint arXiv:1406.2227.*

Jaderberg, M., Simonyan, K., Vedaldi, A., & Zisserman, A. (2016). Reading text in the wild with convolutional neural networks. *International Journal of Computer Vision, 116*(1), 1–20. DOI: 10.1007/s11263-015-0823-z

Jaipuria, N., Zhang, X., Bhasin, R., Arafa, M., Chakravarty, P., Shrivastava, S., & Murali, V. N. (2020). Deflating dataset bias using synthetic data augmentation. In *Proceedings of the IEEE/CVF Conference on Computer Vision and Pattern Recognition Workshops* (pp. 772-773). DOI: 10.1109/CVPRW50498.2020.00394

James, S., Harbron, C., Branson, J., & Sundler, M. (2021). Synthetic data use: Exploring use cases to optimise data utility. *Discover Artificial Intelligence, 1*(1), 15. DOI: 10.1007/s44163-021-00016-y

Jawahar, M., Ravi, V., Prassanna, J., Jasmine, S. G., Manikandan, R., Sekaran, R., & Kannan, S. (2022). Covmnet–deep learning model for classifying coronavirus (covid-19). *Health and Technology, 12*(5), 1009–1024. DOI: 10.1007/s12553-022-00688-1 PMID: 35966170

Jensen, R. I. T., Ferwerda, J., Jørgensen, K. S., Jensen, E. R., Borg, M., Krogh, M. P., Jensen, J. B., & Iosifidis, A. (2023). A synthetic data set to benchmark anti-money laundering methods. *Scientific Data, 10*(1), 661. DOI: 10.1038/s41597-023-02569-2 PMID: 37770445

Jerald, J. (2015). *The VR Book: Human-centred design for virtual reality.* ACM Books. DOI: 10.1145/2792790

Jo, H. (2024). Uncovering the Reasons behind Willingness to Pay for ChatGPT-4 Premium. *International Journal of Human-Computer Interaction, 41*(2), 994–1009. DOI: 10.1080/10447318.2024.2307692

Johansson, G. (1973). Visual perception of biological motion and a model for its analysis. *Perception & Psychophysics, 14*(2), 201–211. DOI: 10.3758/BF03212378

Johnson-Bey, S., Nelson, M. J., & Mateas, M. (2022). *Exploring the Design Space of Social Physics Engines in Games*. Springer., DOI: 10.1007/978-3-031-22298-6_36

Johnston, R. J., Boyle, K. J., Adamowicz, W., Bennett, J., Brouwer, R., Cameron, T. A., Hanemann, W. M., Hanley, N., Ryan, M., Scarpa, R., Tourangeau, R., & Vossler, C. A. (2017). Contemporary guidance for stated preference studies. *Journal of the Association of Environmental and Resource Economists*, *4*(2), 319–405. DOI: 10.1086/691697

Jordan, M. I., & Mitchell, T. M. (2015). Machine learning: Trends, perspectives, and prospects. *Science*, *349*(6245), 255–260. DOI: 10.1126/science.aaa8415 PMID: 26185243

Jordon, J., Szpruch, L., Houssiau, F., Bottarelli, M., Cherubin, G., Maple, C., . . . Weller, A. (2022). Synthetic data—what, why and how? *arXiv preprint arXiv:2205.03257.*

Jordon, J., Szpruch, L., Houssiau, F., Bottarelli, M., Cherubin, G., Maple, C., . . . Weller, A. (2022). Synthetic Data—what, why and how? *arXiv Preprint arXiv:2205.03257.*

Jordon, J., Yoon, J., & Van Der Schaar, M. (2018, September). PATE-GAN: Generating synthetic data with differential privacy guarantees. In *International Conference on Learning Representations*.

Joshi, I., Grimmer, M., Rathgeb, C., Busch, C., Bremond, F., & Dantcheva, A. (2024). Synthetic data in human analysis: A survey. *IEEE Transactions on Pattern Analysis and Machine Intelligence*, *46*(7), 4957–4976. DOI: 10.1109/TPAMI.2024.3362821 PMID: 38319772

Jouppi, N. P., Yoon, D. H., Kurian, G., Li, S., Patil, N., Laudon, J., Young, C., & Patterson, D. (2021). A domain-specific supercomputer for training deep neural networks. *Communications of the ACM*, *64*(7), 67–75. DOI: 10.1145/3360307

Juliani, A., ... (2020). Unity: A general platform for intelligent agents. arXiv preprint arXiv:1809.02627.

Kalinaki, K., Malik, O. A., & Lai, C. D. T. (2024). Forest Canopy Cover Mapping in High-Resolution Multisource Satellite Imagery Using Attention Recurrent Residual U-Net. *IEEE Region 10 Annual International Conference, Proceedings/TENCON*, 400–403. DOI: 10.1109/TENCON61640.2024.10902705

Kalinaki, K., Acaru, S. F., Kugonza, J., & Nsubuga, R. (2024). Towards an Intelligent Tomorrow: Machine Learning Enabling Sustainable Development. In *Methodologies* (pp. 66–89). Frameworks, and Applications of Machine Learning., DOI: 10.4018/979-8-3693-1062-5.ch004

Kalinaki, K., Namuwaya, S., Mwamini, A., & Namuwaya, S. (2023). Scaling Up Customer Support Using Artificial Intelligence and Machine Learning Techniques. In *Contemporary Approaches of Digital Marketing and the Role of Machine Intelligence* (pp. 23–45). IGI Global., DOI: 10.4018/978-1-6684-7735-9.ch002

Kalinaki, K., Yahya, U., Malik, O. A., & Lai, D. T. C. (2024). A Review of Big Data Analytics and Artificial Intelligence in Industry 5.0 for Smart Decision-Making. In *Human-Centered Approaches in Industry 5.0* (pp. 24–47). Human-Machine Interaction, Virtual Reality Training, and Customer Sentiment Analysis., DOI: 10.4018/979-8-3693-2647-3.ch002

Kaplan, R. M., & Milstein, A. (2021). Influence of a COVID-19 vaccine's effectiveness and safety profile on vaccination acceptance. *Proceedings of the National Academy of Sciences of the United States of America, 118*(10), e2021726118. DOI: 10.1073/pnas.2021726118 PMID: 33619178

Kapoor A, Gulli A, Pal S, Chollet F. (2022). "Deep Learning with TensorFlow and Keras Build and deploy supervised unsupervised deep and reinforcement learning models". Doi: DOI: 10.1007/978-1-4842-8246-5

Karras, T., Laine, S., & Aila, T. (2019). A style-based generator architecture for generative adversarial networks. *IEEE/CVF Conference on Computer Vision and Pattern Recognition (CVPR)*. DOI: 10.1109/CVPR.2019.00453

Karras, T., Laine, S., & Aila, T. (2020). A style-based generator architecture for generative adversarial networks. *IEEE Transactions on Pattern Analysis and Machine Intelligence, 43*(12), 4217–4228. DOI: 10.1109/TPAMI.2020.2970919 PMID: 32012000

Karst, F. S., Li, M. M., & Leimeister, J. M. (2025). SynDEc: A synthetic data ecosystem. *Electronic Markets, 35*(1), 7. DOI: 10.1007/s12525-024-00746-8

Kaswan, K. S., Dhatterwal, J. S., Malik, K., & Baliyan, A. (2023). Generative AI: A Review on Models and Applications. *2023 International Conference on Communication, Security and Artificial Intelligence, ICCSAI 2023*, 699–704. DOI: 10.1109/ICCSAI59793.2023.10421601

Kaur, P., & Gupta, D. (2025). Exploring the dimensions of generative AI. *Progressive Computational Intelligence, Information Technology and Networking*, 159–165. DOI: 10.1201/9781003650010-26

Khalaf, L. I., Alhamadani, B., Ismael, O. A., & Radhi, A. A. (2024). "Deep learning-based anomaly detection in network traffic for cyber threat identification". *Proceedings of the ACM on Management of Data*. Doi: DOI: 10.1145/3660853.3660932

Khalil, M., Vadiee, F., Shakya, R., & Liu, Q. (2025). Creating artificial students that never existed: Leveraging large language models and CTGANs for synthetic data generation. *arXiv preprint arXiv:2501.01793*. DOI: 10.1145/3706468.3706523

Khan F, (2024) "Optimization Techniques in Applied Mathematics: From Physics Simulations to Real-World Problems" Frontiers in Applied Physics and Mathematics.

Khan, N., Khan, Z., Koubaa, A., Khan, M. K., & Salleh, R. (2024). Global insights and the impact of generative AI-ChatGPT on multidisciplinary: A systematic review and bibliometric analysis. *Connection Science*, *36*(1), 2353630. DOI: 10.1080/09540091.2024.2353630

Kiefer, B., Ott, D., & Zell, A. (2022, August). Leveraging synthetic data in object detection on unmanned aerial vehicles. In *2022 26th International Conference on Pattern Recognition (ICPR)* (pp. 3564-3571). IEEE. DOI: 10.1109/ICPR56361.2022.9956710

Kietzmann, J., Lee, L. W., McCarthy, I. P., & Kietzmann, T. C. (2020). Deepfakes: Trick or treat? *Business Horizons*, *63*(2), 135–146. DOI: 10.1016/j.bushor.2019.11.006

Kim, D., Kim, S., Kim, S., & Lee, B. H. (2024). Generative AI Characteristics, User Motivations, and Usage Intention. *Journal of Computer Information Systems*, 1–16. Advance online publication. DOI: 10.1080/08874417.2024.2442438

Kim, Y. W., Mishra, S., Jin, S., Panda, R., Kuehne, H., Karlinsky, L., & Feris, R. (2022). How transferable are video representations based on synthetic data? *Advances in Neural Information Processing Systems*, *35*, 35710–35723.

Kingma, D. P., & Welling, M. (2013). Auto-encoding variational Bayes. arXiv preprint arXiv:1312.6114.

Kiran, A., & Kumar, S. S. (2024). A methodology and an empirical analysis to determine the most suitable synthetic data generator. *IEEE Access : Practical Innovations, Open Solutions*, *12*, 12209–12228. DOI: 10.1109/ACCESS.2024.3354277

Kittur, J., & Islam, T. (2021) "Serious Games in Engineering: The Current State, Trends, and Future" *2021 ASEE Virtual Annual Conference Content* Access. Doi: DOI: 10.18260/1-2--37709

Kohli, P., & Sharma, S. (2025). EDGE Artificial Intelligence-Enabled 6G communication security system for connected vehicles in revolutionizing urban development and governance. In IGI Global eBooks (pp. 431–462). https://doi.org/DOI: 10.4018/979-8-3373-1375-7.ch015

Kollias, D. (2022, October). Abaw: learning from synthetic data & multi-task learning challenges. In *European Conference on Computer Vision* (pp. 157-172). Cham: Springer Nature Switzerland.

Kong, Z., Lee, S. G., Ghosal, D., Majumder, N., Mehrish, A., Valle, R., . . . Catanzaro, B. (2024). Improving text to audio models with synthetic captions. *arXiv preprint arXiv:2406.15487.* DOI: 10.21437/SynData4GenAI.2024-1

Kortylewski, A., Schneider, A., Gerig, T., Egger, B., Morel-Forster, A., & Vetter, T. (2018). Training deep face recognition systems with synthetic data. *arXiv preprint arXiv:1802.05891.*

Kortylewski, A., Egger, B., Schneider, A., Gerig, T., Morel-Forster, A., & Vetter, T. (2019). Analyzing and reducing the damage of dataset bias to face recognition with synthetic data. In *Proceedings of the IEEE/CVF Conference on Computer Vision and Pattern Recognition Workshops* (pp. 0-0). DOI: 10.1109/CVPRW.2019.00279

Kotsiantis, S., Kanellopoulos, D., & Pintelas, P. (2006). Data preprocessing for supervised learning. *International Journal of Computational Science, 1*(2), 111–117.

Kotsis, K. T. (2024). ChatGPT in Teaching Physics Hands-on Experiments in Primary School. *European Journal of Education Studies, 11*(10). Advance online publication. DOI: 10.46827/ejes.v11i10.5549

Kozlov, S. V., & Radionovich, V. V. (2024). Digital Beamforming Features in Subaperture Antenna Arrays of Highly Informative Space Radars with Aperture Synthesis. *Doklady BGUIR, 21*(6), 70–74. Advance online publication. DOI: 10.35596/1729-7648-2023-21-6-70-74

Krishnan, P., & Jawahar, C. V. (2016). Generating synthetic data for text recognition. *arXiv Preprint arXiv:1608.04224.*

Kshetri, N. (2024). Generative artificial intelligence in the financial services industry. *Computer, 57*(6), 102–108. DOI: 10.1109/MC.2024.3382452

Kumar, A. S., & Rank, S. (2021) "BEAPS: Integrating Volumetric Dynamics in Virtual Agent Prototyping" Advances in Visual Computing: 16th International Symposium, ISVC 2021, Virtual Event, October 4–6, 2021, Proceedings, Part II. Doi: DOI: 10.1007/978-3-030-90439-5_34

Kumar, S., Musharaf, D., Musharaf, S., & Sagar, A. K. (2023). A Comprehensive Review of the Latest Advancements in Large Generative AI Models. *Communications in Computer and Information Science, 1920*, 90–103. Advance online publication. DOI: 10.1007/978-3-031-45121-8_9

Kumar, V., Raza, M. Q., & Rawat, D. B. (2020). Real-time big data analytics: Applications and challenges. *Computer Communications*, *160*, 257–275.

Kumawat, B., Sridevi, P. R. S., & Mehra, A. (2024) "Investigation of Tank Combat Simulation in Video Games" 2024 International Conference on Advances in Computing, Communication, and Control (ICAC3). Doi: DOI: 10.1109/ICAC2N63387.2024.10895254

Kumi, S., Hilton, M., Snow, C., Lomotey, R. K., & Deters, R. (2023, July). sleepsynth: Evaluating the use of synthetic Data in Health Digital Twins. In *2023 IEEE International Conference on Digital Health (ICDH)* (pp. 121-130). IEEE. DOI: 10.1109/ICDH60066.2023.00027

Kwatra, S., & Torra, V. (2024). Empirical evaluation of synthetic data created by generative models via attribute inference attack. In *IFIP advances in information and communication technology* (pp. 282–291). DOI: 10.1007/978-3-031-57978-3_18

Kyem, B. A., Asamoah, J. K., Huang, Y., & Aboah, A. (2024). Weather-Adaptive synthetic data generation for enhanced power line inspection using StarGAN. *IEEE Access : Practical Innovations, Open Solutions*, *1*, 193882–193901. Advance online publication. DOI: 10.1109/ACCESS.2024.3520120

Lai, J. W., & Cheong, K. H. (2022). *Educational Opportunities and Challenges in Augmented Reality: Featuring Implementations in Physics Education*. IEEE., DOI: 10.1109/ACCESS.2022.3161096

Lancaster, T. (2023). Artificial intelligence, text generation tools and ChatGPT–does digital watermarking offer a solution? *International Journal for Educational Integrity*, *19*(1), 10. DOI: 10.1007/s40979-023-00131-6

Lartey, D., & Law, K. M. Y. (2025). Artificial intelligence adoption in urban planning governance: A systematic review of advancements in decision-making, and policy making. *Landscape and Urban Planning*, *258*, 105337. DOI: 10.1016/j.landurbplan.2025.105337

Laver, K., George, S., Thomas, S., Deutsch, J. E., & Crotty, M. (2017). Virtual reality for stroke rehabilitation. *Cochrane Database of Systematic Reviews*, 11. PMID: 29156493

Law, L. (2024). Application of generative artificial intelligence (GenAI) in language teaching and learning: A scoping literature review. *Computers and Education Open*, *6*, 100174. DOI: 10.1016/j.caeo.2024.100174

LeCun, Y., Bengio, Y., & Hinton, G. (2015). Deep learning. *Nature*, *521*(7553), 436–444. DOI: 10.1038/nature14539 PMID: 26017442

Lee, G. H., Lee, K. J., Jeong, B., & Kim, T. (2024). Developing Personalized Marketing Service Using Generative AI. *IEEE Access : Practical Innovations, Open Solutions*, *12*, 22394–22402. DOI: 10.1109/ACCESS.2024.3361946

Lee, H., Jeong, S., & Lee, K. (2021). Estimating the deployment costs of broadband universal service via fiber networks in Korea. *Telecommunications Policy*, *45*(4), 102105. DOI: 10.1016/j.telpol.2021.102105

Lee, T. (2024). *Unity Engine Dissection: Improvement Points in Comparison between Unity Engine and Open-Source Engines*. Asia-Pacific Journal of Convergent Research and Innovation.

Le, T. A., Baydin, A. G., Zinkov, R., & Wood, F. (2017, May). Using synthetic data to train neural networks is model-based reasoning. In *2017 International Joint Conference on Neural Networks (IJCNN)* (pp. 3514-3521). IEEE. DOI: 10.1109/IJCNN.2017.7966298

Li, C., Li, L., Jiang, H., Weng, K., Geng, Y., Li, L., Ke, Z., Li, Q., Cheng, M., Nie, W., Li, Y., Zhang, B., Liang, Y., Zhou, L., Xu, X., Chu, X., Wei, X., & Wei, X. "YOLOv6: A Single-Stage Object Detection Framework for Industrial Applications," arXiv preprint, arXiv:2209.02976, Sep. 2022. Available: https://arxiv.org/abs/2209.02976

Li, Y. "Research and Application of Deep Learning in Image Recognition," 2022 IEEE 2nd International Conference on Power, Electronics and Computer Applications (ICPECA), Shenyang, China, 2022, pp. 994-999, DOI: 10.1109/ICPECA53709.2022.9718847

Li, Z., Zhu, H., Lu, Z., & Yin, M. (2023). Synthetic data generation with large language models for text classification: Potential and limitations. *arXiv preprint arXiv:2310.07849*. DOI: 10.18653/v1/2023.emnlp-main.647

Liang, Z., Li, H., Yu, N., Sun, K., & Cheng, R. (2025). Bridging Evolutionary Multiobjective Optimization and GPU Acceleration via Tensorization. *IEEE Transactions on Evolutionary Computation*, 1. Advance online publication. DOI: 10.1109/TEVC.2025.3555605

Libes, D., Lechevalier, D., & Jain, S. (2017, December). Issues in synthetic data generation for advanced manufacturing. In 2017 IEEE International Conference on Big Data (Big Data) (pp. 1746-1754). IEEE. DOI: 10.1109/BigData.2017.8258117

Li, C., Wang, L., & Lee, Y. (2021). Avatar-based rehabilitation system for post-stroke patients using motion synthesis. *Healthcare Technology Letters*, *8*(5), 94–100.

Lin, S., Wang, K., Zeng, X., & Zhao, R. (2023). Explore the power of synthetic data on few-shot object detection. In *Proceedings of the IEEE/CVF Conference on Computer Vision and Pattern Recognition* (pp. 638-647). DOI: 10.1109/CVPRW59228.2023.00071

Li, P., & Gai, S. (2023). Single image deraining using multi-scales context information and attention network. *Journal of Visual Communication and Image Representation, 90*, 103695. Advance online publication. DOI: 10.1016/j.jvcir.2022.103695

Lippmann, P., Spaan, M. T., & Yang, J. (2024). Exploring LLMs as a source of targeted synthetic textual data to minimize high confidence misclassifications. *arXiv preprint arXiv:2403.17860.*

Liu, Q., Deho, O., Vadiee, F., Khalil, M., Joksimovic, S., & Siemens, G. (2025). Can synthetic data be fair and private? A comparative study of synthetic data generation and fairness algorithms. *arXiv Preprint arXiv:2501.01785.* DOI: 10.1145/3706468.3706546

Liu, R., Wei, J., Liu, F., Si, C., Zhang, Y., Rao, J., . . . Dai, A. M. (2024). Best practices and lessons learned on synthetic data for language models. *arXiv Preprint arXiv:2404.07503.*

Liu, Y., & Wang, H. (2024). *Who on Earth is using generative AI?* https://documents.worldbank.org/curated/en/099720008192430535

Liu, Z., Liang, H., Huang, X., Xiong, W., Yu, Q., Sun, L., . . . Zhang, W. (2024). Synthvlm: High-efficiency and high-quality synthetic data for vision language models. *arXiv preprint arXiv:2407.20756.*

Liu, C., Habibie, I., Tzionas, D., & Theobalt, C. (2021). Physics-aware motion validation for trustworthy avatar animation. *IEEE/CVF Conference on Computer Vision and Pattern Recognition (CVPR)*, 2371–2380

Liu, M. Y., Huang, X., Yu, J., & Wang, T. C. (2021). "Generative adversarial networks for image and video synthesis: Algorithms and applications". *Proceedings of the IEEE Conference on Computer Vision and Pattern Recognition.* Doi: DOI: 10.1109/JPROC.2021.3049196

Liu, T., Tang, J., Vietri, G., & Wu, S. (2023, July). Generating private synthetic data with genetic algorithms. In *International Conference on Machine Learning* (pp. 22009-22027). PMLR.

Liu, T., Vietri, G., & Wu, S. Z. (2021). Iterative methods for private synthetic data: Unifying framework and new methods. *Advances in Neural Information Processing Systems, 34*, 690–702.

Liu, Y., Wang, Z., Zhou, X., & Zheng, L. (2023). A study of using synthetic data for effective association knowledge learning. *Machine Intelligence Research, 20*(2), 194–206. DOI: 10.1007/s11633-022-1380-x

Liu, Y., Xu, T., Zhang, L., & Wang, Y. (2020). Digital human synthesis in cinema: A case study on Gemini Man. *The Visual Computer*, *36*(11), 2357–2367.

Lomurno, E., & Matteucci, M. (2025). Federated Knowledge Recycling: Privacy-preserving synthetic data sharing. *Pattern Recognition Letters*, *191*, 124–130. DOI: 10.1016/j.patrec.2025.02.030

Long, L., Wang, R., Xiao, R., Zhao, J., Ding, X., Chen, G., & Wang, H. (2024). On LLMs-driven synthetic data generation, curation, and evaluation: A survey. *arXiv Preprint arXiv:2406.15126*. DOI: 10.18653/v1/2024.findings-acl.658

Long, Y., Kroeger, S., Zaeh, M. F., & Brintrup, A. (2025). Leveraging synthetic data to tackle machine learning challenges in supply chains: Challenges, methods, applications, and research opportunities. *International Journal of Production Research*, •••, 1–22. DOI: 10.1080/00207543.2024.2447927

Lopes, P. V., Silveira, L., Guimaraes Aquino, R. D., Ribeiro, C. H., Skoogh, A., & Verri, F. A. N. (2024). Synthetic data generation for digital twins: Enabling production systems analysis in the absence of data. *International Journal of Computer Integrated Manufacturing*, *37*(10-11), 1252–1269. DOI: 10.1080/0951192X.2024.2322981

López, C. A. F. (2022, December). On the legal nature of synthetic data. In *NeurIPS 2022 Workshop on Synthetic Data for Empowering ML Research*.

Lu Y, Shen M, Wang H, Wang X, van Rechem C. (2023). "Machine learning for synthetic data generation a review". arXiv preprint arXiv.

Lu, Y., Shen, M., Wang, H., Wang, X., van Rechem, C., Fu, T., & Wei, W. (2023). Machine learning for synthetic data generation: a review. *arXiv preprint arXiv:2302.04062*.

Lu, Y., Shen, M., Wang, H., Wang, X., van Rechem, C., Fu, T., & Wei, W. (2023). Machine learning for synthetic data generation: A review. *arXiv Preprint arXiv:2302.04062*. Kollias, D. (2022, October). Abaw: Learning from synthetic data & multi-task learning challenges. In *European Conference on Computer Vision* (pp. 157-172). Cham: Springer Nature Switzerland.

Lu, Y., Song, J., Chang, X., Bian, H., Maiti, S., & Watanabe, S. (2024). Syneslm: A unified approach for audio-visual speech recognition and translation via language model and synthetic data. *arXiv preprint arXiv:2408.00624*. DOI: 10.21437/SynData4GenAI.2024-7

Lu, B., & Gai, S. (2022). Single image rain removal based on multi scale progressive residual network. *Journal of Image and Graphics*, *27*(5), 1537–1553. Advance online publication. DOI: 10.11834/jig.210472

Lu, B., Gai, S., Xiong, B., & Wu, J. (2022). Single image deraining with dual U-Net generative adversarial network. *Multidimensional Systems and Signal Processing*, *33*(2), 485–499. Advance online publication. DOI: 10.1007/s11045-021-00806-8

Lu, L., Meng, X., Mao, Z., & Karniadakis, G. E. (2021). DeepXDE: A deep learning library for solving differential equations. *SIAM Review*, *63*(1), 208–228. DOI: 10.1137/19M1274067

Luo, J. (2024). A critical review of GenAI policies in higher education assessment: A call to reconsider the "originality" of students' work. *Assessment & Evaluation in Higher Education*, *49*(5), 651–664. DOI: 10.1080/02602938.2024.2309963

Lupidi, A., Gemmell, C., Cancedda, N., Dwivedi-Yu, J., Weston, J., Foerster, J., . . . Lomeli, M. (2024). Source2synth: Synthetic data generation and curation grounded in real data sources. *arXiv preprint arXiv:2409.08239.*

Lv, Z. (2023). Generative artificial intelligence in the metaverse era. *Cognitive Robotics*, *3*, 208–217. DOI: 10.1016/j.cogr.2023.06.001

Lychev, A. V. (2023). Synthetic Data Generation for Data Envelopment Analysis. *Data*, *8*(10), 146. DOI: 10.3390/data8100146

Ma, H., Geng, M., Wang, F., Zheng, W., Ai, Y., & Zhang, W. (2024). Data Augmentation of a Corrosion Dataset for Defect Growth Prediction of Pipelines Using Conditional Tabular Generative Adversarial Networks. *Materials (Basel)*, *17*(5), 1142. Advance online publication. DOI: 10.3390/ma17051142 PMID: 38473613

Mahadev Madgule, N. (2023). Vinayaka, Yeshwant M. Sonkhaskar, Dhiren Ramanbhai Patel, R. Karthikeyan, P. Selvakumar, (2023). Mechanical properties and microstructure of activated TIG welded similar joints of Inconel alloys by desirability approaches. *Materials Today: Proceedings*, *77*, 528–533. DOI: 10.1016/j.matpr.2022.12.250

Mahadevappa, P., Muzammal, S. M., & Tayyab, M. (2025). Introduction to Generative AI in Web Engineering: Concepts and Applications. Https://Services.Igi-Global.Com/Resolvedoi/Resolve.Aspx?Doi=10.4018/979-8-3693-3703-5.Ch015, 297–330. DOI: 10.4018/979-8-3693-3703-5.ch015

Maheshwari, G., Ivanov, D., & Haddad, K. E. (2024). Efficacy of synthetic data as a benchmark. *arXiv preprint arXiv:2409.11968.*

Mahmoud, A. B., Kumar, V., & Spyropoulou, S. (2025). Identifying the Public's Beliefs About Generative Artificial Intelligence: A Big Data Approach. *IEEE Transactions on Engineering Management, 72*, 827–841. DOI: 10.1109/TEM.2025.3534088

Maiorino, A., Padgett, Z., Wang, C., Yakubovskiy, M., & Jiang, P. (2023). Application and Evaluation of Large Language Models for the Generation of Survey Questions. *International Conference on Information and Knowledge Management, Proceedings*. DOI: 10.1145/3583780.3615506

Ma, J., Jiang, X., Fan, A., Jiang, J., & Yan, J. (2021). Image Matching from Handcrafted to Deep Features: A Survey. *International Journal of Computer Vision, 129*(1), 23–79. DOI: 10.1007/s11263-020-01359-2

Majeed, A. (2023). Attribute-centric and synthetic data based privacy preserving methods: A systematic review. *Journal of Cybersecurity and Privacy, 3*(3), 638–661. DOI: 10.3390/jcp3030030

Malik, S., Tyagi, A. K., & Mahajan, S. (2022). *Architecture generative model and deep reinforcement learning for IoT applications Deep learning perspective*. Artificial Intelligence-based Internet of Things., DOI: 10.1007/978-3-030-87059-1_9

Manduchi L, Pandey K, Meister C, Bamler R. (2024). "On the challenges and opportunities in generative AI". arXiv preprint arXiv.

Man, K., & Chahl, J. (2022). A review of synthetic image data and its use in computer vision. *Journal of Imaging, 8*(11), 310. DOI: 10.3390/jimaging8110310 PMID: 36422059

Mannino, M., & Abouzied, A. (2019, October). Is this real? Generating synthetic data that looks real. In *Proceedings of the 32nd Annual ACM Symposium on User Interface Software and Technology* (pp. 549-561). DOI: 10.1145/3332165.3347866

Manteghi, M. (2024). Can text and data mining exceptions and synthetic data training mitigate copyright-related concerns in generative AI? *Law, Innovation and Technology, 16*(2), 663–686. DOI: 10.1080/17579961.2024.2392928

Martín-Albo, D., Plamondon, R., & Vidal, E. (2014, September). Training of on-line handwriting text recognizers with synthetic text generated using the kinematic theory of rapid human movements. In *2014 14th International Conference on Frontiers in Handwriting Recognition* (pp. 543-548). IEEE. DOI: 10.1109/ICFHR.2014.97

Marwala, T., Fournier-Tombs, E., & Stinckwich, S. (2023). The use of synthetic data to train ai models: Opportunities and risks for sustainable development. *arXiv Preprint arXiv:2309.00652*.

Marwala, T. (2024). *Synthetic Versus Authentic Data.* The Balancing Problem in the Governance of Artificial Intelligence., DOI: 10.1007/978-981-97-9251-1_7

Masood, M., Nawaz, M., Malik, K. M., Javed, A., & Irtaza, A. (2023). Deepfakes generation and detection: State-of-the-art, open challenges, countermeasures, and way forward. *Applied Sciences (Basel, Switzerland).* Advance online publication. DOI: 10.3390/app13126941

Masood, M., Nawaz, M., Malik, K. M., Javed, A., Irtaza, A., & Malik, H. (2023). Deepfakes generation and detection: State-of-the-art, open challenges, countermeasures, and way forward. *Applied Intelligence, 53*(4), 3974–4026. Advance online publication. DOI: 10.1007/s10489-022-03766-z

Matsui, T., Taki, M., Pham, T. Q., Chikazoe, J., & Jimura, K. (2022). Counterfactual Explanation of Brain Activity Classifiers Using Image-To-Image Transfer by Generative Adversarial Network. *Frontiers in Neuroinformatics, 15*, 802938. Advance online publication. DOI: 10.3389/fninf.2021.802938 PMID: 35369003

McKinsey. (2023). *The state of AI in 2023: Generative AI's breakout year.* https://www.mckinsey.com/capabilities/quantumblack/our-insights/the-state-of-ai-in-2023-generative-ais-breakout-year

McLachlan, S. (2017). *Realism in synthetic data generation.* Massey University.

Meclea, M. A., Alexandru Silviu, G. O. G. A., & Bo coianu, M. (2024). *Aspects regarding artificial intelligence use in military and engineering sciences aircraft propulsion.* Scientific Research & Education in the Air Force-AFASES.

Mehrabi, N., Morstatter, F., Saxena, N., Lerman, K., & Galstyan, A. (2021). A survey on bias and fairness in machine learning. *ACM Computing Surveys, 54*(6), 1–35. DOI: 10.1145/3457607

Meiser, M., & Zinnikus, I. (2024). A survey on the use of synthetic data for enhancing key aspects of trustworthy AI in the energy domain: Challenges and Opportunities. *Energies, 17*(9), 1992. DOI: 10.3390/en17091992

Melzi, P., Tolosana, R., Vera-Rodriguez, R., Kim, M., Rathgeb, C., Liu, X., DeAndres-Tame, I., Morales, A., Fierrez, J., Ortega-Garcia, J., Zhao, W., Zhu, X., Yan, Z., Zhang, X.-Y., Wu, J., Lei, Z., Tripathi, S., Kothari, M., Zama, M. H., & Marras, M. (2024). FRCSyn-onGoing: Benchmarking and comprehensive evaluation of real and synthetic data to improve face recognition systems. *Information Fusion, 107*, 102322. DOI: 10.1016/j.inffus.2024.102322

Memarian B, Doleck T. (2023). “Fairness, Accountability, Transparency, and Ethics (FATE) in Artificial Intelligence (AI) and higher education: A systematic review”. Computers and Education: Artificial Intelligence. Doi: DOI: 10.1016/j.caeai.2023.100141

Merel, J., ... (2019). Learning human behaviours from motion capture by adversarial imitation. arXiv preprint arXiv:1707.02201.

Meyer, D., & Nagler, T. (2021). Synthia: Multidimensional synthetic data generation in Python. *Journal of Open Source Software*, *6*(65), 2863. DOI: 10.21105/joss.02863

Mildenhall, B.. (2020). NeRF: Representing scenes as neural radiance fields for view synthesis. *European Conference on Computer Vision (ECCV)*, 405–421. DOI: 10.1007/978-3-030-58452-8_24

Miletic, M., & Sariyar, M. (2024). Challenges of using synthetic data generation methods for tabular microdata. *Applied Sciences (Basel, Switzerland)*, *14*(14), 5975. DOI: 10.3390/app14145975

Mirsky, Y., & Lee, W. (2021). The creation and detection of deepfakes: A survey. *ACM Computing Surveys*, *54*(1), 1–41. DOI: 10.1145/3425780

Mirzayeva G, (2023) “The Role of Digital Educational Technologies in Teaching Physics” Science and Innovation. Doi: .DOI: 10.37872/2414-3087-2023.2.33357

Mishra, S., Panda, R., Phoo, C. P., Chen, C. F. R., Karlinsky, L., Saenko, K., & Feris, R. S. (2022). Task2sim: Towards effective pre-training and transfer from synthetic data. In *Proceedings of the IEEE/CVF Conference on Computer Vision and Pattern Recognition* (pp. 9194-9204). DOI: 10.1109/CVPR52688.2022.00898

Mitra, R., & Zualkernan, I. (2025). Music Generation Using Deep Learning and Generative AI: A Systematic Review. *IEEE Access : Practical Innovations, Open Solutions*, *13*, 18079–18106. DOI: 10.1109/ACCESS.2025.3531798

Moeslund, T. B., Hilton, A., & Krüger, V. (2006). A survey of advances in vision-based human motion capture and analysis. *Computer Vision and Image Understanding*, *104*(2–3), 90–126. DOI: 10.1016/j.cviu.2006.08.002

Moga, M., & Radu Moga, D. (2024). Character Development: Bridging Digital and Traditional Drawing Techniques. *Intelligent Human Systems Integration (IHSI 2024). Integrating People and Intelligent Systems*, *119*. Advance online publication. DOI: 10.54941/ahfe1004527

Mohammadabadi, S. M. S., Entezami, M., Moghaddam, A. K., Orangian, M., & Nejadshamsi, S. (2024). Generative artificial intelligence for distributed learning to enhance smart grid communication. *International Journal of Intelligent Networks*, *5*, 267–274. DOI: 10.1016/j.ijin.2024.05.007

Montanez, A. (2018). *SDV: an open source library for synthetic data generation* (Doctoral dissertation, Massachusetts Institute of Technology).

Morales, E. F., & Escalante, H. J. (2022). *A brief introduction to supervised unsupervised and reinforcement learning*. Computational Learning and Intelligence., DOI: 10.1016/B978-0-12-820125-1.00017-8

Morales-García, J., Bueno-Crespo, A., Terroso-Sáenz, F., Arcas-Túnez, F., Martínez-España, R., & Cecilia, J. M. (2023). Evaluation of synthetic data generation for intelligent climate control in greenhouses. *Applied Intelligence*, *53*(21), 24765–24781. DOI: 10.1007/s10489-023-04783-2

Moreira, D., Marcel, S., & Rocha, A. (2024). Synthetic Realities and Artificial Intelligence-Generated Contents. *IEEE Security and Privacy*, *22*(3), 7–10. DOI: 10.1109/MSEC.2024.3388244

Moreno, Y., Montero, A., Hidrobo, F., & Infante, S. (2023). "Synthetic Data Generator for an E-Learning Platform in a Big Data Environment". *International Conference on Technology and Education*. Doi: DOI: 10.1007/978-981-99-5414-8_39

Morse, C. (2021). *Gaming Engines: Unity, Unreal, and Interactive 3D Spaces*. Taylor & Francis.

Mosca, P., Bourhrara, L., & Calloo, A. (2024). APOLLO3: Overview of the New Code Capabilities for Reactor Physics Analysis. *Nuclear Science and Engineering*. Advance online publication. DOI: 10.1080/00295639.2024.2182831

Munappy, A. R., Bosch, J., Olsson, H. H., Arpteg, A., & Brinne, B. (2022). Data management for production quality deep learning models: Challenges and solutions. *Journal of Systems and Software*, *191*, 111359. Advance online publication. DOI: 10.1016/j.jss.2022.111359

Murtaza, H., Ahmed, M., Khan, N. F., Murtaza, G., Zafar, S., & Bano, A. (2023). Synthetic data generation State of the art in health care domain. *Computer Science Review*, *48*, 100546. Advance online publication. DOI: 10.1016/j.cosrev.2023.100546

Musalamadugu, T. S., & Kannan, H. (2023). *Generative AI for medical imaging analysis and applications*. Future Medicine AI. DOI: 10.2217/fmai-2023-0004

Mustonen, M. (2023). *Web-Based Game Engine Design*. LUT University.

Mylonas, G., Kalogeras, A., & Kalogeras, G. (2021). Digital Twins from Smart Manufacturing to Smart Cities: A Survey. *IEEE Transactions on Industrial Informatics*. Advance online publication. DOI: 10.1109/TII.2021.3084853

Naswir A F, (2024) "The Future of Artificial Intelligence in 2D Game Development" Journal of Data Science, Technology, and Computer Science.

Nelson, D. (2022). *What Is Synthetic Data?* Unite.AI.

Nerantzaki, M., Husser, C., Ryckelynck, M., & Lutz, J. F. (2024). Exchanging and Releasing Information in Synthetic Digital Polymers Using a Strand-Displacement Strategy. *Journal of the American Chemical Society*, *146*(10), 6456–6460. Advance online publication. DOI: 10.1021/jacs.3c13953 PMID: 38286022

Neuhausen, M., Herbers, P., & König, M. (2020). Using synthetic data to improve and evaluate the tracking performance of construction workers on site. *Applied Sciences (Basel, Switzerland)*, *10*(14), 4948. DOI: 10.3390/app10144948

Nichol, A. Q., & Dhariwal, P. (2021). Improved denoising diffusion probabilistic models. arXiv preprint arXiv:2102.09672.

Nikolenko, S. I. (2021). *Synthetic data for deep learning* (Vol. 174). Springer Nature. DOI: 10.1007/978-3-030-75178-4

NIST. (2023). *Artificial Intelligence Risk Management Framework (AI RMF 1.0)*. National Institute of Standards and Technology.

Novelli, C., Casolari, F., Hacker, P., Spedicato, G., & Floridi, L. (2024). Generative AI in EU law: Liability, privacy, intellectual property, and cybersecurity. *Computer Law & Security Review*, *55*, 106066. DOI: 10.1016/j.clsr.2024.106066

Nowok, B., Raab, G. M., & Dibben, C. (2016). synthpop: Bespoke creation of synthetic data in R. *Journal of Statistical Software*, *74*(11), 1–26. DOI: 10.18637/jss.v074.i11

Ock, J., No, H., & Kim, S. (2023). Poster: Exploring Synthetic Data Generation for Anomaly Detection in the 5G NWDAF Architecture. *Proceedings - International Conference on Distributed Computing Systems, 2023-July*. DOI: 10.1109/ICDCS57875.2023.00129

Offenhuber, D. (2024). Shapes and frictions of synthetic data. *Big Data & Society*, *11*(2), 20539517241249390. DOI: 10.1177/20539517241249390

Osuala, R., Skorupko, G., Lazrak, N., Garrucho, L., García, E., Joshi, S., Jouide, S., Rutherford, M., Prior, F., Kushibar, K., Díaz, O., & Lekadir, K. (2023). Medigan: A Python library of pretrained generative models for medical image synthesis. *Journal of Medical Imaging (Bellingham, Wash.)*, *10*(6). Advance online publication. DOI: 10.1117/1.JMI.10.6.061403 PMID: 36814939

Owoseni, A., Kolade, O., & Egbetokun, A. (2024). Enhancing Personalised Learning and Student Engagement Using Generative AI. *Generative AI in Higher Education*, 123–150. DOI: 10.1007/978-3-031-60179-8_5

Paganetti, H., Beltran, C., Both, S., Dong, L., Flanz, J., Furutani, K., Grassberger, C., Grosshans, D. R., Knopf, A.-C., Langendijk, J. A., Nystrom, H., Parodi, K., Raaymakers, B. W., Richter, C., Sawakuchi, G. O., Schippers, M., Shaitelman, S. F., Teo, B. K. K., Unkelbach, J., & Lomax, T. (2021). Roadmap: Proton Therapy Physics and Biology. *Physics in Medicine and Biology*, *66*(5), 05RM01. Advance online publication. DOI: 10.1088/1361-6560/abcd16 PMID: 33227715

Pakina, A. K., Sharma, A., & Kejriwal, D. (2025). AI-driven disinformation campaigns: Detecting synthetic propaganda in encrypted messaging via graph neural networks. *International Journal Science and Technology*, *4*(1), 12–24. DOI: 10.56127/ijst.v4i1.1960

Pandey, C., Sharma, S., & Bhadula, S. (2024). Artificial Intelligence Technique in Next Generation Healthcare Systems. In *2024 IEEE International Conference on Communication, Computing and Signal Processing (IICCCS)*, 1–6. https://doi.org/ DOI: 10.1109/IICCCS61609.2024.10763570

Panerati, J., Zheng, H., Zhou, S. Q., & Xu, J. (2021) "Learning to Fly—a Gym Environment with Pybullet Physics for Reinforcement Learning of Multi-Agent Quadcopter Control" *IEEE/RSJ International Conference on Intelligent Robots and Systems*. Doi: DOI: 10.1109/IROS51168.2021.9635857

Pantanowitz, J., Manko, C. D., Pantanowitz, L., & Rashidi, H. H. (2024). Synthetic data and its utility in pathology and laboratory medicine. *Laboratory Investigation*, *104*(8), 102095. DOI: 10.1016/j.labinv.2024.102095 PMID: 38925488

Pan, X., & Hamilton, A. F. de C. (2018). Why and how to use virtual reality to study human social interaction: The challenges of exploring a new research landscape. *British Journal of Psychology*, *109*(3), 395–417. DOI: 10.1111/bjop.12290 PMID: 29504117

Pan, Z., Yu, W., Yi, X., Khan, A., Yuan, F., & Zheng, Y. (2019). Recent Progress on Generative Adversarial Networks (GANs): A Survey. *IEEE Access : Practical Innovations, Open Solutions*, *7*, 36322–36333. Advance online publication. DOI: 10.1109/ACCESS.2019.2905015

Papadopoulos, D., & Karalis, V. D. (2023). Variational Autoencoders for Data Augmentation in Clinical Studies. *Applied Sciences (Basel, Switzerland)*, *13*(15), 8793. Advance online publication. DOI: 10.3390/app13158793

Park, K.. (2021). Nerfies: Deformable neural radiance fields. *IEEE/CVF International Conference on Computer Vision (ICCV)*, 5865–5874.

Parrot, M., Tajmouati, H., da Silva, V. B. R., Atwood, B. R., Fourcade, R., Gaston-Mathé, Y., Do Huu, N., & Perron, Q. (2023). Integrating synthetic accessibility with AI-based generative drug design. *Journal of Cheminformatics*, *15*(1), 83. Advance online publication. DOI: 10.1186/s13321-023-00742-8 PMID: 37726842

Partarakis, N., Zabulis, X., Zourarakis, D., Demeridou, I., Moreno, I., Dubois, A., Nikolaou, N., Fallahian, P., Arnaud, D., Crescenzo, N., Hee, P., & Stamou, A. (2025). Physics-Based Tool Usage Simulations in VR. *Multimodal Technologies and Interaction*, *9*(4), 29. Advance online publication. DOI: 10.3390/mti9040029

Pastuszak, J., & Węgierek, P. (2022). Photovoltaic cell generations and current research directions for their development. *Materials (Basel)*, *15*(16), 5542. Advance online publication. DOI: 10.3390/ma15165542 PMID: 36013679

Patel, P. (2024). Synthetic data. *Business Information Review*, *41*(2), 48–52. DOI: 10.1177/02663821241231101

Paulin, G., & Ivasic-Kos, M. (2023). Review and analysis of synthetic dataset generation methods and techniques for application in computer vision. *Artificial Intelligence Review*, *56*(9), 9221–9265. Advance online publication. DOI: 10.1007/s10462-022-10358-3

Pei, Y., & Zaïane, O. (2006). A synthetic data generator for clustering and outlier analysis.

Peng, X. B., ... (2018). Variational discriminator bottleneck: Improving imitation learning, inverse RL, and GANs by constraining information flow. arXiv preprint arXiv:1810.00821.

Peng, Y. H., Huq, F., Jiang, Y., Wu, J., Li, X. Y., Bigham, J. P., & Pavel, A. (2025). Dreamstruct: Understanding slides and user interfaces via synthetic data generation. In *European Conference on Computer Vision* (pp. 466-485). Springer, Cham. DOI: 10.1007/978-3-031-72691-0_26

Pérez-Porras, F. J., Triviño-Tarradas, P., Cima-Rodríguez, C., Meroño-de-Larriva, J. E., García-Ferrer, A., & Mesas-Carrascosa, F. J. (2021). Machine learning methods and synthetic data generation to predict large wildfires. *Sensors (Basel), 21*(11), 3694. DOI: 10.3390/s21113694 PMID: 34073312

Pezoulas, V. C., Zaridis, D. I., Mylona, E., Androutsos, C., Apostolidis, K., Tachos, N. S., & Fotiadis, D. I. (2024). Synthetic data generation methods in healthcare: A review on open-source tools and methods. *Computational and Structural Biotechnology Journal, 23*, 2892–2910. DOI: 10.1016/j.csbj.2024.07.005 PMID: 39108677

Piotrowski, W., Sher, Y., Grover, S., & Stern, R. (2023) "Heuristic Search for Physics-Based Problems: Angry Birds in PDDL+" *Proceedings of the International Conference on Automated Planning and Scheduling*. DOI: 10.1609/icaps.v33i1.27232

Planche, B., Wu, Z., Ma, K., Sun, S., Kluckner, S., Lehmann, O., & Ernst, J. (2017, October). Depthsynth: Real-time realistic synthetic data generation from cad models for 2.5 d recognition. In *2017 International Conference on 3d Vision (3DV)* (pp. 1-10). IEEE. DOI: 10.1109/3DV.2017.00011

Potluru, V. K., Borrajo, D., Coletta, A., Dalmasso, N., El-Laham, Y., Fons, E., . . . Balch, T. (2023). Synthetic data applications in finance. *arXiv Preprint arXiv:2401.00081.*

Pozzi, M., Noei, S., Robbi, E., Cima, L., Moroni, M., Munari, E., Torresani, E., & Jurman, G. (2024). Generating and evaluating synthetic data in digital pathology through diffusion models. *Scientific Reports, 14*(1), 28435. Advance online publication. DOI: 10.1038/s41598-024-79602-w PMID: 39557989

Prabhu Kavin, B., Karki, S., Hemalatha, S., Singh, D., Vijayalakshmi, R., Thangamani, M., Haleem, S. L. A., Jose, D., Tirth, V., Kshirsagar, P. R., & Adigo, A. G. (2022). Machine learning-based secure data acquisition for fake accounts detection in future mobile communication networks. *Wireless Communications and Mobile Computing, 2022*(1), 6356152. DOI: 10.1155/2022/6356152

Pradhyumna, P., & Shreya, G. P. (2021, August). Graph neural network (GNN) in image and video understanding using deep learning for computer vision applications. In *2021 Second International Conference on Electronics and Sustainable Communication Systems (ICESC)* (pp. 1183-1189). IEEE.

Prakash, A., Boochoon, S., Brophy, M., Acuna, D., Cameracci, E., State, G., & Birchfield, S. (2019, May). Structured domain randomization: Bridging the reality gap by context-aware synthetic data. In *2019 International Conference on Robotics and Automation (ICRA)* (pp. 7249-7255). IEEE. DOI: 10.1109/ICRA.2019.8794443

Prakash, S. R., & Singh, P. N. (2021). Object detection through region proposal based techniques. *Materials Today: Proceedings*, *46*, 3997–4002. DOI: 10.1016/j.matpr.2021.02.533

Pratapa, A., Bhat, G., Choudhury, M., Sitaram, S., Dandapat, S., & Bali, K. (2018, July). Language modeling for code-mixing: The role of linguistic theory based synthetic data. In *Proceedings of the 56th Annual Meeting of the Association for Computational Linguistics (*Volume 1*: Long Papers)* (pp. 1543-1553). DOI: 10.18653/v1/P18-1143

Puri, R., Spring, R., Patwary, M., Shoeybi, M., & Catanzaro, B. (2020). Training question answering models from synthetic data. *arXiv Preprint arXiv:2002.09599*. DOI: 10.18653/v1/2020.emnlp-main.468

Qais, M. H., Hasanien, H. M., Alghuwainem, S., & Loo, K. H. (2023). Propagation Search Algorithm: A Physics-Based Optimizer for Engineering Applications. *Mathematics*, *11*(20), 4224. Advance online publication. DOI: 10.3390/math11204224

Qian, Z., Cebere, B. C., & van der Schaar, M. (2023). Synthcity: Facilitating innovative use cases of synthetic data in different data modalities. *arXiv Preprint arXiv:2301.07573*.

Qian, Z., Callender, T., Cebere, B., Janes, S. M., Navani, N., & van der Schaar, M. (2024). Synthetic data for privacy-preserving clinical risk prediction. *Scientific Reports*, *14*(1), 25676. DOI: 10.1038/s41598-024-72894-y PMID: 39463411

Qian, Z., Davis, R., & van der Schaar, M. (2024). Synthcity: A benchmark framework for diverse use cases of tabular synthetic data. *Advances in Neural Information Processing Systems*, •••, 36.

Qiao, H., Bei, Y. J., & Jang, D. (2024). Digital Image Art Style Transfer Based on Deep Short and Long Term Memory. *Journal of Network Intelligence*, *9*(1).

Qiu, H., Yu, B., Gong, D., Li, Z., Liu, W., & Tao, D. (2021). Synface: Face recognition with synthetic data. In *Proceedings of the IEEE/CVF International Conference on Computer Vision* (pp. 10880-10890).

Rahman, M. M., Terano, H. J., Rahman, M. N., Salamzadeh, A., & Rahaman, M. S. (2023). ChatGPT and academic research: A review and recommendations based on practical examples. *Rahman, M., Terano, HJR, Rahman, N., Salamzadeh, A., Rahaman, S.(2023). ChatGPT and Academic Research: A Review and Recommendations Based on Practical Examples. Journal of Education. Management and Development Studies*, *3*(1), 1–12.

Rajotte, J. F., Bergen, R., Buckeridge, D. L., El Emam, K., Ng, R., & Strome, E. (2022). Synthetic data as an enabler for machine learning applications in medicine. *iScience*, *25*(11), 105331. DOI: 10.1016/j.isci.2022.105331 PMID: 36325058

Ramakrishnan, T., Mohan Gift, M. D., Chitradevi, S., Jegan, R., & Subha Hency Jose, P. Nagaraja, Rajneesh Sharma HN, Selvakumar P, Sintayehu Mekuria Hailegiorgis, (2022). "Study of Numerous Resins Used in Polymer Matrix Composite Materials", Advances in Materials Science and Engineering, vol. 2022, Article ID 1088926, 8 pages, 2022. https://doi.org/DOI: 10.1155/2022/1088926

Ramdurai, B., & Adhithya, P. (2023). The Impact, Advancements and Applications of Generative AI. *International Journal on Computer Science and Engineering*, *10*(6), 1–8. Advance online publication. DOI: 10.14445/23488387/IJCSE-V10I6P101

Ramesh, A., Dhariwal, P., Nichol, A., ... (2022). Hierarchical text-conditional image generation with CLIP latents. arXiv preprint arXiv:2204.06125.

Ramzan, F., Sartori, C., Consoli, S., & Recupero, D. R. (2024). Generative Adversarial Networks for Synthetic data Generation in Finance: Evaluating statistical similarities and quality assessment. *AI*, *5*(2), 667–685. DOI: 10.3390/ai5020035

Rashidi, H. H., Hu, B., Pantanowitz, J., Tran, N., Liu, S., Chamanzar, A., Gur, M., Chang, C. C. H., Wang, Y., Tafti, A., Pantanowitz, L., & Hanna, M. G. (2025). Statistics of Generative Artificial Intelligence and Nongenerative Predictive Analytics Machine Learning in Medicine. *Modern Pathology*, *38*(3), 100663. DOI: 10.1016/j.modpat.2024.100663 PMID: 39579984

Rashidi, H. H., Khan, I. H., Dang, L. T., Albahra, S., Ratan, U., Chadderwala, N., To, W., Srinivas, P., Wajda, J., & Tran, N. K. (2022). Prediction of tuberculosis using an automated machine learning platform for models trained on synthetic data. *Journal of Pathology Informatics*, *13*, 100172. DOI: 10.4103/jpi.jpi_75_21 PMID: 35136677

Rathore, M. M., Shah, S. A., Shukla, D., & Bentafat, E. (2021). The role of AI machine learning and big data in digital twinning A systematic literature review challenges and opportunities. *IEEE Transactions on Industrial Informatics*. Advance online publication. DOI: 10.1109/TII.2021.3084315

Rawashdeh, A. (2023). The consequences of artificial intelligence: An investigation into the impact of AI on job displacement in accounting. *Journal of Science and Technology Policy Management*, *16*(3), 506–535. DOI: 10.1108/JSTPM-02-2023-0030

Rawat, D., Sharma, S., & Bhadula, S. (2025b). CT Imaging-Based Deep Learning System for Non-Small Cell Lung Cancer Detection and Classification. *International Journal of Advanced Computer Science and Applications, 16*(5). Advance online publication. DOI: 10.14569/IJACSA.2025.0160561

Raza S, Qureshi R, Zahid A, Fioresi J, Sadak F. (2025). "Who is responsible? The data, models, users or regulations? Responsible generative AI for a sustainable future". arXiv preprint arXiv.

Reda, D. (2025). *Physics-Based Character Controllers with Reinforcement Learning*. University of British Columbia.

Reiter, J. P. (2002). Satisfying disclosure restrictions with synthetic data sets. *Journal of Official Statistics-Stockholm, 18*(4), 531–544.

Reiter, J. P. (2023). Synthetic data: A look back and a look forward. *Transactions on Data Privacy, 16*(1), 15–24.

Ren, S., He, K., Girshick, R., & Sun, J. (2017, June). Faster R-CNN: Towards real-time object detection with region proposal networks. *IEEE Transactions on Pattern Analysis and Machine Intelligence, 39*(6), 1137–1149. DOI: 10.1109/TPAMI.2016.2577031 PMID: 27295650

Richardson, E., Sela, M., & Kimmel, R. (2016, October). 3D face reconstruction by learning from synthetic data. In *2016 Fourth International Conference on 3D Vision (3DV)* (pp. 460-469). IEEE. DOI: 10.1109/3DV.2016.56

Rizvi, S. M. H., Naseer, A., Rehman, S. U., Akram, S., & Gruhn, V. (2024). Revolutionizing Agriculture: Machine and Deep Learning Solutions for Enhanced Crop Quality and Weed Control. *IEEE Access : Practical Innovations, Open Solutions, 12*, 11865–11878. Advance online publication. DOI: 10.1109/ACCESS.2024.3355017

Romanyuk, O. N., Shenshin, O. O., & Bobko, O. L. (2025). *GPU Stream Processing*. Innovative Technology and Scientific Progress.

Rombach, R., Blattmann, A., Lorenz, D., Esser, P., & Ommer, B. (2022). High-resolution image synthesis with latent diffusion models. *Proceedings of the IEEE/CVF Conference on Computer Vision and Pattern Recognition*. DOI: 10.1109/CVPR52688.2022.01042

Rosário, A. T. (2024). *Generative AI and Generative Pre-Trained Transformer Applications*. 45–71. : S Patil, S., Shelke, S., Velankar, M., Koul, T., & Rani, V. (2023). A Systematic Survey of Approaches Used in Computer Music Generation. *Journal of Innovations in Data Science and Big Data Management, 2*(1). https://doi.org/DOI: 10.4018/979-8-3693-1950-5.ch003

Rosenblatt, L., Liu, X., Pouyanfar, S., de Leon, E., Desai, A., & Allen, J. (2020). Differentially private synthetic data: Applied evaluations and enhancements. *arXiv Preprint arXiv:2011.05537.*

Rossenbach, N., Zeyer, A., Schlüter, R., & Ney, H. (2020, May). Generating synthetic audio data for attention-based speech recognition systems. In *ICASSP 2020-2020 IEEE International Conference on Acoustics, Speech and Signal Processing (ICASSP)* (pp. 7069-7073). IEEE. DOI: 10.1109/ICASSP40776.2020.9053008

Rujas, M., Del Moral Herranz, R. M. G., Fico, G., & Merino-Barbancho, B. (2024). Synthetic Data Generation in Healthcare: A Scoping Review of reviews on domains, motivations, and future applications. medRxiv *(Cold Spring Harbor Laboratory).* DOI: 10.1101/2024.08.09.24311338

Rujas, M., del Moral Herranz, R. M. G., Fico, G., & Merino-Barbancho, B. (2025). Synthetic data generation in healthcare: A scoping review of reviews on domains, motivations, and future applications. *International Journal of Medical Informatics, 195*, 105763. DOI: 10.1016/j.ijmedinf.2024.105763 PMID: 39719743

Rusanov, B., Hassan, G. M., Reynolds, M., Sabet, M., Rowshanfarzad, P., Bucknell, N., Gill, S., Dass, J., & Ebert, M. (2024). Transformer CycleGAN with uncertainty estimation for CBCT based synthetic CT in adaptive radiotherapy. *Physics in Medicine and Biology*, *69*(3), 035014. Advance online publication. DOI: 10.1088/1361-6560/ad1cfc PMID: 38198726

Saeed, M. M. A., Saeed, R. A., & Ahmed, Z. E. (2024). *Data Security and Privacy in the Age of AI and Digital Twins.*, DOI: 10.4018/979-8-3693-1818-8.ch008

Saharia, C.. (2022). *Photorealistic text-to-image diffusion models with deep language understanding*. Google Research.

Sakaridis, C., Dai, D., & Van Gool, L. (2018). Semantic foggy scene understanding with synthetic data. *International Journal of Computer Vision, 126*(9), 973–992. DOI: 10.1007/s11263-018-1072-8

Sakirin, T., & Kusuma, S. (2023c). A survey of Generative Artificial intelligence techniques. *Babylonian Journal of Artificial Intelligence, 10–14*, 10–14. Advance online publication. DOI: 10.58496/BJAI/2023/003

Saldanha, D. M. F., Dias, C. N., & Guillaumon, S. (2022). Transparency and accountability in digital public services Learning from the Brazilian cases. *Government Information Quarterly*, *39*(2), 101680. Advance online publication. DOI: 10.1016/j.giq.2022.101680

Salimans, T.. (2016). Improved techniques for training GANs. *NeurIPS, 29*, 2234–2242.

Sankaranarayanan, S., Balaji, Y., Jain, A., Lim, S. N., & Chellappa, R. (2018). Learning from synthetic data: Addressing domain shift for semantic segmentation. In *Proceedings of the IEEE Conference on Computer Vision and Pattern Recognition* (pp. 3752-3761). DOI: 10.1109/CVPR.2018.00395

Sannigrahi, S., Fraga-Silva, T., Oualil, Y., & Van Gysel, C. (2024, July). Synthetic query generation using large language models for virtual assistants. In *Proceedings of the 47th International ACM SIGIR Conference on Research and Development in Information Retrieval* (pp. 2837-2841). DOI: 10.1145/3626772.3661355

Sardar T H, Pandey B K. (2024). "Big data computing advances in technologies methodologies and applications". Doi: DOI: 10.1007/978-3-031-28647-1

Sarker, I. H. (2021). Deep learning a comprehensive overview on techniques taxonomy applications and research directions. *SN Computer Science*, *2*(6), 420. Advance online publication. DOI: 10.1007/s42979-021-00815-1 PMID: 34426802

Satava, R. M. (2002). Virtual reality surgical simulator: The first steps. *Surgical Endoscopy*, *17*(3), 451–455. PMID: 8503081

Saxena, D., & Cao, J. (2021). Generative Adversarial Networks (GANs). In *ACM Computing Surveys* (Vol. 54, Issue 3). DOI: 10.1145/3446374

Schieber, H., Demir, K. C., Kleinbeck, C., Yang, S. H., & Roth, D. (2024). Indoor synthetic data generation: A systematic review. *Computer Vision and Image Understanding*, *240*, 103907. DOI: 10.1016/j.cviu.2023.103907

Schwartz, D. (2021). *Making Games to Teach Physics and Mechanics*. RIT Scholar Works. DOI: 10.18260/1-2--36308

Sedkaoui, S., & Benaichouba, R. (2024). Generative AI as a transformative force for innovation: a review of opportunities, applications and challenges. *European Journal of Innovation Management, ahead-of-print*(ahead-of-print). DOI: 10.1108/EJIM-02-2024-0129

Segerstedt, J. (2021). *Animation for the Apex Game Engine using Neural Networks—Implementing Immersive Character Animation in an Industry-Proven Game Engine by Applying Machine Learning Techniques*. Chalmers University of Technology.

Seib, V., Lange, B., & Wirtz, S. (2020). Mixing real and synthetic data to enhance neural network training—A review of current approaches. *arXiv Preprint arXiv:2007.08781*.

Sengar, S. S., Hasan, A. B., Kumar, S., & Carroll, F. (2024). Generative artificial intelligence: A systematic review and applications. *Multimedia Tools and Applications, 84*(21), 23661–23700. Advance online publication. DOI: 10.1007/s11042-024-20016-1

Shafik, W. (2023). Introduction to ChatGPT. In *Advanced Applications of Generative AI and Natural Language Processing Models*. DOI: 10.4018/979-8-3693-0502-7.ch001

Shafik, W. (2024a). Artificial intelligence and Blockchain technology enabling cybersecurity in telehealth systems. In *Artificial Intelligence and Blockchain Technology in Modern Telehealth Systems*. DOI: 10.1049/PBHE061E_ch11

Shafik, W. (2024h). Dissecting the Role of Women in Cybersecurity and Information Technology: A Medical Perspective. In *Next-Generation Cybersecurity: AI, ML, and Blockchain* (pp. 325–350). Springer. DOI: 10.1007/978-981-97-1249-6_15

Shafik, W. (2024k). Generative AI for Social Good and Sustainable Development. In *Generative AI: Current Trends and Applications* (pp. 185–217). Springer. DOI: 10.1007/978-981-97-8460-8_10

Shafik, W. (2024l). Incorporating Artificial Intelligence for Urban and Smart Cities' Sustainability. In *Maintaining a Sustainable World in the Nexus of Environmental Science and AI* (pp. 23–58). IGI Global.

Shafik, W. (2024m). Shaping the Next Generation Smart City Ecosystem: An Investigation on the Requirements, Applications, Architecture, Security and Privacy, and Open Research Questions. In *Smart Cities: Innovations, Challenges and Future Perspectives* (pp. 3–52). Springer. DOI: 10.1007/978-3-031-59846-3_1

Shafik, W. (2024o). Sustainable development goal 14: explainable AI (XAI) for ocean health. In *Artificial Intelligence and Edge Computing for Sustainable Ocean Health* (pp. 167–198). Springer.

Shafik, W. (2024p). The Role of Artificial Intelligence in the Emerging Digital Economy Era. *Artificial Intelligence Enabled Management: An Emerging Economy Perspective*, 33.

Shafik, W. (2024q). The Role of Generative Artificial Intelligence in E-Commerce Fraud Detection and Prevention. In *Strategies for E-Commerce Data Security: Cloud, Blockchain, AI, and Machine Learning* (pp. 430–469). IGI Global. DOI: 10.4018/979-8-3693-6557-1.ch018

Shafik, W. (2025). Quantum Computing and Generative Adversarial Networks (GANs): Ethics, Privacy, and Security. In *Quantum AI and its Applications in Blockchain Technology* (pp. 111–156). IGI Global Scientific Publishing.

Shafik, W., Hidayatullah, A. F., Kalinaki, K., & Aslam, M. M. (2024). Artificial Intelligence (AI)-Assisted Computer Vision (CV) in Healthcare Systems. *Computer Vision and AI-Integrated IoT Technologies in the Medical Ecosystem*, 17–36. DOI: 10.1201/9781003429609-2

Shafik, W. (2024b). Artificial Intelligence and Internet of Things Roles in Sustainable Next-Generation Manufacturing. In *Sustainable Innovation for Industry 6.0* (pp. 207–239). IGI Global., DOI: 10.4018/979-8-3693-3140-8.ch012

Shafik, W. (2024e). Artificial Intelligence and the Medical Tourism. In *Examining Tourist Behaviors and Community Involvement in Destination Rejuvenation* (pp. 207–233). IGI Global. DOI: 10.4018/979-8-3693-6819-0.ch016

Shafik, W. (2024f). Artificial Intelligence Models to Prevent Forest Fires. In *AI and IoT for Proactive Disaster Management* (pp. 78–106). IGI Global., DOI: 10.4018/979-8-3693-3896-4.ch005

Shafik, W. (2024g). Deep Learning Impacts in the Field of Artificial Intelligence. In *Deep Learning Concepts in Operations Research* (pp. 9–26). Auerbach Publications., DOI: 10.1201/9781003433309-2

Shafik, W. (2024i). Environmental Conservation and Sustainable Development. In *Developing Digital Inclusion Through Globalization and Digitalization* (pp. 142–176). IGI Global., DOI: 10.4018/979-8-3693-4111-7.ch008

Shafik, W. (2024j). Ethical Use of Machine Learning Techniques in Smart Cities. In *Ethical Artificial Intelligence in Power Electronics* (pp. 21–47). CRC Press. DOI: 10.1201/9781032648323-3

Shafik, W. (2024n). Sustainable Development. In Viana Hassan, A. S. S. B. (Ed.), *Building Community Resiliency and Sustainability With Tourism Development* (1st ed., Vol. 1, pp. 1–30). IGI Global., DOI: 10.4018/979-8-3693-5405-6.ch001

Shahreza, H. O., Ecabert, C., George, A., Unnervik, A., Marcel, S., Di Domenico, N., . . . Fierrez, J. (2024, May). SDFR: Synthetic data for face recognition competition. In *2024 IEEE 18th International Conference on Automatic Face and Gesture Recognition (FG)* (pp. 1-9). IEEE.

Shahul Hameed, M. A., Qureshi, A. M., & Kaushik, A. (2024). Bias mitigation via synthetic data generation: A review. *Electronics (Basel)*, *13*(19), 3909. DOI: 10.3390/electronics13193909

Sharma, A., Khanna, V., & Gigras, Y. (2025). Data analytics and synthetic data generation for an IoT-based smart home. In Elsevier eBooks (pp. 29–45). https://doi.org/DOI: 10.1016/B978-0-443-26474-0.00013-0

Sharma, S., Tyagi, R., & Mohan, S. (2024). Generative Artificial Intelligence-Enabled climate change and Deep-Sea food webs. In Advances in environmental engineering and green technologies book series (pp. 199–224). https://doi.org/DOI: 10.4018/979-8-3693-6670-7.ch009

Sharma, S., Tyagi, R., & Mohan, S. (2025). Artificial intelligence and optimized computational modelling against climate crisis. In Elsevier eBooks (pp. 1–14). https://doi.org/DOI: 10.1016/B978-0-443-21905-4.00006-7

Sharma, S., Tyagi, R., & Mohan, S. (2025b). Blockchain-Based Security and Privacy in Digital Healthcare Recommender Systems. In In Healthcare Recommender Systems (pp. 221–237). https://doi.org/DOI: 10.1007/978-3-031-80056-6_10

Sharma, H., Juyal, A., & Das, M. (2023) "Future of Animation with Artificial Intelligence" *International Conference on Emerging Trends in Design & Arts.*

Sharma, S., Gupta, A., & Tyagi, R. (2024). Artificial Intelligence, Cyber Security and Vedic Scripture Integrated Sustainable Global Youth Leadership Development. In *2024 IEEE International Conference on Communication, Computing and Signal Processing (IICCCS)*, 1–6. https://doi.org/DOI: 10.1109/IICCCS61609.2024.10763618

Shende, A. D., Tekade, S. A., Deshmukh, A. A., Tembhurkar, S. P., & Selvakumar, P. (2025). Water Quality Monitoring and Control With AI. In Ruiz-Vanoye, J., & Díaz-Parra, O. (Eds.), *Smart Water Technology for Sustainable Management in Modern Cities* (pp. 51–80). IGI Global Scientific Publishing., DOI: 10.4018/979-8-3693-8074-1.ch003

Shermeyer, J., Hossler, T., Van Etten, A., Hogan, D., Lewis, R., & Kim, D. (2021). Rareplanes: Synthetic data takes flight. In *Proceedings of the IEEE/CVF Winter Conference on Applications of Computer Vision* (pp. 207-217).

Shi, Z. (2021). "A survey on audio synthesis and audio-visual multimodal processing". arXiv preprint arXiv:2108.00443.

Shilaskar, S., Chakole, S., Samarth, A., & Shejole, P. (2025). GenAI based Data Extraction with Query-based Insights. *2025 International Conference on Emerging Smart Computing and Informatics, ESCI 2025*. DOI: 10.1109/ESCI63694.2025.10987906

Shin, D., Perlmutter, D. D., Lim, J. S., & Jitkajornwanich, K. (2025). Fact-checking as epistemic infrastructure: From journalistic gatekeeping to platform governance. *Emerging Media, 3*(2), 183–202. DOI: 10.1177/27523543251344972

Shin, H. C., Tenenholtz, N. A., Rogers, J. K., Schwarz, C. G., Senjem, M. L., Gunter, J. L., & Andriole, K. P. (2018). Medical image synthesis for data augmentation and anonymization using generative adversarial networks. *International Workshop on Simulation and Synthesis in Medical Imaging*. DOI: 10.1007/978-3-030-00536-8_1

Shi, W., Cao, J., Zhang, Q., Li, Y., & Xu, L. (2020). Edge computing: Vision and challenges. *IEEE Internet of Things Journal*.

Siarohin, A.. (2018). Animating arbitrary objects via deep motion transfer. *IEEE/CVF Conference on Computer Vision and Pattern Recognition (CVPR)*, 2377–2386.

Siemuri, A., Selvan, K., Kuusniemi, H., Valisuo, P., & Elmusrati, M. S. (2022, December). A Systematic Review of Machine Learning Techniques for GNSS Use Cases. *IEEE Transactions on Aerospace and Electronic Systems*, *58*(6), 5043–5077. DOI: 10.1109/TAES.2022.3219366

Sindiramutty, S. R., & Prabagaran, K. R. V. (2025). *Ethics and Transparency in Secure Web Model Generation*. Generative AI for Secure Web Applications., DOI: 10.4018/978-1-6684-9573-9.ch005

Snoke, J., Raab, G. M., Nowok, B., Dibben, C., & Slavkovic, A. (2018). General and specific utility measures for synthetic data. *Journal of the Royal Statistical Society. Series A, (Statistics in Society)*, *181*(3), 663–688. DOI: 10.1111/rssa.12358

Sobota B, Pietriková E, (2023) "The Role of Game Engines in Game Development and Teaching" .DOI: 10.5772/intechopen.100922

Solove, D. J. (2025). Artificial Intelligence and Privacy. *Florida Law Review*, *77*, •••. https://heinonline.org/HOL/Page?handle=hein.journals/uflr77&id=1&div=&collection=

Soltana, G., Sabetzadeh, M., & Briand, L. C. (2017, October). Synthetic data generation for statistical testing. In *2017 32nd IEEE/ACM International Conference on Automated Software Engineering (ASE)* (pp. 872-882). IEEE. DOI: 10.1109/ASE.2017.8115698

Song, Y., & Ermon, S. (2021). Score-based generative modelling through stochastic differential equations. arXiv preprint arXiv:2011.13456.

Sreenivasulu, A., & Manimozhi, T. Satheesh kumar P, Selvakumar P. and Singh P, (2022). "The Model of Smart Sensing Device For Sensitive Nanoclusters Modification in Sensing Properties," 2022 2nd International Conference on Advance Computing and Innovative Technologies in Engineering (ICACITE), Greater Noida, India, 2022, pp. 1043-1047, DOI: 10.1109/ICACITE53722.2022.9823846

Stadler, T., Oprisanu, B., & Troncoso, C. (2020). Synthetic data-A privacy mirage. *arXiv Preprint arXiv:2011.07018.*

Stadler, T., Oprisanu, B., & Troncoso, C. (2022). Synthetic data–anonymisation groundhog day. In *31st USENIX Security Symposium (USENIX Security 22)* (pp. 1451-1468).

Strubell, E., Ganesh, A., & McCallum, A. (2019). Energy and policy considerations for deep learning in NLP. arXiv preprint arXiv:1906.02243. DOI: 10.18653/v1/P19-1355

Sugumaran K M D, Pandiyarajan V Y B S, Selvakumar P. and M. S, (2024). "Analysis of Electromyography Signals for Control of Mechanical Prosthesis using Machine Learning Techniques," 2024 2nd International Conference on Intelligent Data Communication Technologies and Internet of Things (IDCIoT), Bengaluru, India, 2024, pp. 1059-1067, .DOI: 10.1109/IDCIoT59759.2024.10467707

Summerville, A., Snodgrass, S., Mateas, M., & Ontañón, S. (2018). Procedural content generation via machine learning. *IEEE Transactions on Games, 10*(3), 257–270. DOI: 10.1109/TG.2018.2846639

Sundarreson, P., & Kumarapathirage, S. (2024). SentiGEN: Synthetic data generator for sentiment analysis. *Journal of Computing Theories and Applications, 1*(4), 461–477. DOI: 10.62411/jcta.10480

Sushko, N. (2024). *PAN 2024 multilingual TextDetox: exploring different regimes for synthetic data training for multilingual text detoxification.* Working Notes of CLEF.

Susser, D., & Seeman, J. (2024). Critical provocations for synthetic data. *Surveillance & Society, 22*(4), 453–459. DOI: 10.24908/ss.v22i4.18335

Tafazoli, D., Chirimbu, S. C., & Dejica-Cartis, A. (2014). Web 2.0 in English language teaching: Using word clouds. *Professional Communication and Translation Studies, 7*(1), 167–172. DOI: 10.59168/XQDJ2288

Takale, D. G., Mahalle, P. N., Sule, B., & Journals, M. A. T. (2024). Advancements and applications of generative artificial intelligence. *Journal of Information Technology and Sciences, 10*(1), 20–27.

Tamura, R., Tsuda, K., & Matsuda, S. (2023). NIMS-OS: An Automation Software to Implement a Closed Loop between Artificial Intelligence and Robotic Experiments in Materials Science. *Science and Technology of Advanced Materials.* Advance online publication. DOI: 10.1080/14686996.2023.2171319

Tan, C., Jian, H., & Margetts, B. (2019). *64 Generating synthetic electronic patient records*. DOI: 10.1136/archdischild-2019-gosh.64

Tang, R., Han, X., Jiang, X., & Hu, X. (2023). Does synthetic data generation of LLMs help clinical text mining? *arXiv preprint arXiv:2303.04360*.

Tao, C., Fan, X., & Yang, Y. (2024). Harnessing LLMs for API interactions: A framework for classification and synthetic data generation. *arXiv Preprint arXiv:2409.11703*. DOI: 10.1109/CAIT64506.2024.10962957

Taub, J., & Elliot, M. (2019). *The synthetic data challenge*. Joint UNECE/Eurostat Work Session on Statistical Data Confidentiality.

Taub, J., Elliot, M., & Sakshaug, J. W. (2020). The impact of synthetic data generation on data utility with application to the 1991 UK samples of anonymised records. *Transactions on Data Privacy*, *13*(1), 1–23.

Templet, R. M. (2021). *Game Physics: An Analysis of Physics Engines for First-Time Physics Developers*. ScholarWorks.

Tevet, G., ... (2022). Human motion diffusion model. arXiv preprint arXiv:2209.14916.

Thambawita, V., Salehi, P., Sheshkal, S. A., Hicks, S. A., Hammer, H. L., Parasa, S., Lange, T., Halvorsen, P., & Riegler, M. A. (2022). SinGAN-Seg: Synthetic training data generation for medical image segmentation. *PLoS One*, *17*(5), e0267976. Advance online publication. DOI: 10.1371/journal.pone.0267976 PMID: 35500005

Thomas, F., & Johnston, O. (1995). *The illusion of life: Disney animation* (Rev. ed.). Disney Editions.

Tian, Y., Fan, L., Isola, P., Chang, H., & Krishnan, D. (2024). Stablerep: Synthetic images from text-to-image models make strong visual representation learners. *Advances in Neural Information Processing Systems*, ●●●, 36.

Tiwald, P., Ebert, A., & Soukup, D. T. (2021). Representative & fair synthetic data. *arXiv Preprint arXiv:2104.03007*.

Tobin, J., ... (2017). Domain randomisation for transferring deep neural networks from simulation to the real world. arXiv preprint arXiv:1703.06907.

Tobin, J., Fong, R., Ray, A., Schneider, J., Zaremba, W., & Abbeel, P. (2017). Domain randomization for transferring deep neural networks from simulation to the real world. *IEEE/RSJ International Conference on Intelligent Robots and Systems (IROS)*. DOI: 10.1109/IROS.2017.8202133

Tolosana, R., Vera-Rodriguez, R., Fierrez, J., Morales, A., & Ortega-Garcia, J. (2020). Deepfakes and beyond: A survey of face manipulation and fake detection. *Information Fusion, 64*, 131–148. DOI: 10.1016/j.inffus.2020.06.014

Torkzadehmahani, R., Kairouz, P., & Paten, B. (2019). Dp-cgan: Differentially private synthetic data and label generation. In *Proceedings of the IEEE/CVF Conference on Computer Vision and Pattern Recognition Workshops* (pp. 0-0). DOI: 10.1109/CVPRW.2019.00018

Tremblay, J., Prakash, A., Acuna, D., Brophy, M., Jampani, V., Anil, C., & Birchfield, S. (2018). Training deep networks with synthetic data: Bridging the reality gap by domain randomization. In *Proceedings of the IEEE Conference on Computer Vision and Pattern Recognition Workshops* (pp. 969-977). DOI: 10.1109/CVPRW.2018.00143

Tripathi, S., Chandra, S., Agrawal, A., Tyagi, A., Rehg, J. M., & Chari, V. (2019). Learning to generate synthetic data via compositing. In *Proceedings of the IEEE/CVF Conference on Computer Vision and Pattern Recognition* (pp. 461-470). DOI: 10.1109/CVPR.2019.00055

Trocin, C., Hovland, I. V., Mikalef, P., & Dremel, C. (2021). How Artificial Intelligence affords digital innovation: A cross-case analysis of Scandinavian companies. *Technological Forecasting and Social Change, 173*, 121081. DOI: 10.1016/j.techfore.2021.121081

Tsai, M.-F., & Chen, C.-H. (2021). Spatial-Temporal Variation Graph Convolutional Networks (STV-GCN) for Skeleton-Based Emotional Action Recognition. *IEEE Access : Practical Innovations, Open Solutions*, *9*, 13870–13877. DOI: 10.1109/ACCESS.2021.3052246

Tulyakov, S.. (2018). MoCoGAN: Decomposing motion and content for video generation. *IEEE Conference on Computer Vision and Pattern Recognition (CVPR)*, 1526–1535. DOI: 10.1109/CVPR.2018.00165

Tu, W., Deng, W., & Gedeon, T. (2023). A Closer Look at the Robustness of Contrastive Language-Image Pre-Training (CLIP). *Advances in Neural Information Processing Systems, 36*, 13678–13691.

Tykoski S, (2022) "Mastering Game Design with Unity 2021: Immersive Workflows, Visual Scripting, Physics Engine, GameObjects, Player Progression, Publishing, and a Lot More"

Vaccari, C., & Chadwick, A. (2020). Deepfakes and disinformation: Exploring the impact of synthetic political video on deception, uncertainty, and trust in news. *Social Media + Society*, *6*(1), 2056305120903408. DOI: 10.1177/2056305120903408

Van Breugel, B., & van der Schaar, M. (2023). Beyond privacy: Navigating the opportunities and challenges of synthetic data. *arXiv Preprint arXiv:2304.03722*.

Van Breugel, B., Kyono, T., Berrevoets, J., & Van der Schaar, M. (2021). Decaf: Generating fair synthetic data using causally-aware generative networks. *Advances in Neural Information Processing Systems*, *34*, 22221–22233.

Van Breugel, B., Qian, Z., & Van Der Schaar, M. (2023, July). Synthetic data, real errors: how (not) to publish and use synthetic data. In *International Conference on Machine Learning* (pp. 34793-34808). PMLR.

Van Huy, L., & Nguyen, T. (2024). Generative AI, Why, How, and Outcomes: A User Adoption Study. *AIS Transactions on Human-Computer Interaction*, *16*(1), 1–27. DOI: 10.17705/1thci.00198

Varga, T., & Bunke, H. (2003, August). Generation of synthetic training data for an HMM-based handwriting recognition system. In *Seventh International Conference on Document Analysis and Recognition, 2003. Proceedings.* (pp. 618-622). IEEE. DOI: 10.1109/ICDAR.2003.1227736

Vdovenko D, (2021) "GU UNITY" VDU.

Verdoliva, L. (2020). Media forensics and deepfakes: An overview. *IEEE Journal of Selected Topics in Signal Processing*, *14*(5), 910–932. DOI: 10.1109/JSTSP.2020.3002101

Vijayakumar, G. (2024). Muniyandi Rajkumar, Rajiv Chandar N, Selvakumar P and Ramesh Duraisamy, (2024). Environmentally friendly TDS removal from wastewater by electrochemical ion exchange batch-type recirculation (EIR) technique. *Environmental Science. Water Research & Technology*, *10*, 826–835. DOI: 10.1039/D3EW00793F

Vinuesa, R., Brunton, S. L., & McKeon, B. J. (2023). The Transformative Potential of Machine Learning for Experiments in Fluid Mechanics. *Nature Reviews. Physics*, *5*(9), 536–545. Advance online publication. DOI: 10.1038/s42254-023-00622-y

Vohera, C., Chheda, H., & Chouhan, D. (2021) "Game Engine Architecture and Comparative Study of Different Game Engines" 2021 12th International Conference on Computing, Communication and Networking Technologies (ICCCNT). Doi: DOI: 10.1109/ICCCNT51525.2021.9579618

Wakunuma, K., & Eke, D. (2024). Africa, ChatGPT, and Generative AI Systems: Ethical Benefits, Concerns, and the Need for Governance. *Philosophies 2024, Vol. 9, Page 80, 9*(3), 80. DOI: 10.3390/philosophies9030080

Wang, K., Johnson, W. R., Lu, S., & Huang, X. (2023) "Real2Sim2Real Transfer for Control of Cable-Driven Robots via a Differentiable Physics Engine" *IEEE/RSJ International Conference on Intelligent Robots and Systems*. Doi: DOI: 10.1109/IROS55552.2023.10341811

Wang, P., Loignon, A. C., Shrestha, S., Banks, G. C., & Oswald, F. L. (2024). Advancing organizational science through synthetic data: A path to enhanced data sharing and collaboration. *Journal of Business and Psychology*, •••, 1–27.

Wang, Q., Gao, J., Lin, W., & Yuan, Y. (2019). Learning from synthetic data for crowd counting in the wild. In *Proceedings of the IEEE/CVF Conference on Computer Vision and Pattern Recognition* (pp. 8198-8207). DOI: 10.1109/CVPR.2019.00839

Wang, Q., Gao, J., Lin, W., & Yuan, Y. (2021). Pixel-wise crowd understanding via synthetic data. *International Journal of Computer Vision*, *129*(1), 225–245. DOI: 10.1007/s11263-020-01365-4

Wang, T., Zheng, X., Zhang, L., Cui, Z., & Xu, C. (2023). A graph-based interpretability method for deep neural networks. *Neurocomputing*, *555*, 126651. DOI: 10.1016/j.neucom.2023.126651

Wang, W. Y., Mazaitis, K., & Cohen, W. W. "Programming with Personalized PageRank: A Locally Groundable First-Order Probabilistic Logic," *Proc. Int. Conf. Mach. Learn. (ICML)*, 2013. DOI: 10.1145/2505515.2505573

Wang, W., Wang, A., Ai, Q., Liu, C., & Liu, J. (2019). AAGAN: Enhanced Single Image Dehazing with Attention-to-Attention Generative Adversarial Network. *IEEE Access : Practical Innovations, Open Solutions*, *7*, 173485–173498. Advance online publication. DOI: 10.1109/ACCESS.2019.2957057

Wang, W., Wong, R. C. W., & Wang, J. (2020). Generating and evaluating synthetic personalized data for recommender systems. [TIST]. *ACM Transactions on Intelligent Systems and Technology*, *11*(1), 1–23.

Wang, X., Xie, L., Dong, C., & Shan, Y. (2021). Real-ESRGAN: Training real-world blind super-resolution with pure synthetic data. In *Proceedings of the IEEE/CVF International Conference on Computer Vision* (pp. 1905-1914). DOI: 10.1109/ICCVW54120.2021.00217

Wardle, C., & Derakhshan, H. (2017). *Information disorder: Toward an interdisciplinary framework for research and policymaking* (Vol. 27). Council of Europe.

Waters, M. R., Inkman, M., Jayachandran, K., Kowalchuk, R. M., Robinson, C., Schwarz, J. K., Swamidass, S. J., Griffith, O. L., Szymanski, J. J., & Zhang, J. (2024). GAiN: An integrative tool utilizing generative adversarial neural networks for augmented gene expression analysis. *Patterns (New York, N.Y.)*, *5*(2), 100910. Advance online publication. DOI: 10.1016/j.patter.2023.100910 PMID: 38370125

Wei, J., Huang, D., Lu, Y., Zhou, D., & Le, Q. V. (2023). Simple synthetic data reduces sycophancy in large language models. *arXiv Preprint arXiv:2308.03958.*

Wei, R., Garcia, C., El-Sayed, A., Peterson, V., & Mahmood, A. (2020). Variations in Variational Autoencoders - A Comparative Evaluation. *IEEE Access : Practical Innovations, Open Solutions*, *8*, 153651–153670. Advance online publication. DOI: 10.1109/ACCESS.2020.3018151

Wei, R., & Mahmood, A. (2021). Recent Advances in Variational Autoencoders with Representation Learning for Biomedical Informatics: A Survey. *IEEE Access : Practical Innovations, Open Solutions*, *9*, 4939–4956. Advance online publication. DOI: 10.1109/ACCESS.2020.3048309

Wen-Mei WH, Kirk DB, El Hajj I, (2022) "Programming Massively Parallel Processors: A Hands-on Approach"

Wessel, M., Adam, M., Benlian, A., Majchrzak, A., & Thies, F. (2025). Generative AI and its Transformative Value for Digital Platforms. *Journal of Management Information Systems*. DOI: 10.1080/07421222.2025.2487315

Westerlund, M. (2019). The emergence of deepfake technology: A review. *Technology Innovation Management Review*, *9*(11), 39–52. DOI: 10.22215/timreview/1282

Whitney, C. D., & Norman, J. (2024, June). Real Risks of Fake Data: Synthetic Data, Diversity-Washing and Consent Circumvention. In *The 2024 ACM Conference on Fairness, Accountability, and Transparency* (pp. 1733-1744).

Wiehn, T. (2024). Synthetic data: From data scarcity to data pollution. *Surveillance & Society*, *22*(4), 472–476. DOI: 10.24908/ss.v22i4.18327

Williams, A., & Kovalerchuk, B. (2024, July). Synthetic data generation and automated multidimensional data labeling for AI/ML in general and circular coordinates. In *2024 28th International Conference Information Visualisation (IV)* (pp. 272-279). IEEE.

Williamson, S. M., & Prybutok, V. (2024). Balancing privacy and progress: A review of privacy challenges, systemic oversight, and patient perceptions in AI-driven healthcare. *Applied Sciences (Basel, Switzerland)*, *14*(2), 675. Advance online publication. DOI: 10.3390/app14020675

Wood, E., Baltrušaitis, T., Hewitt, C., Dziadzio, S., Cashman, T. J., & Shotton, J. (2021). Fake it till you make it: face analysis in the wild using synthetic data alone. In *Proceedings of the IEEE/CVF International Conference on Computer Vision* (pp. 3681-3691). DOI: 10.1109/ICCV48922.2021.00366

Wu, Y., Wang, Z., Zeng, D., Shi, Y., & Hu, J. (2023, June). Synthetic data can also teach: Synthesizing effective data for unsupervised visual representation learning. *Proceedings of the AAAI Conference on Artificial Intelligence, 37*(3), 2866–2874. DOI: 10.1609/aaai.v37i3.25388

Wu, Z., Wang, L., Wang, W., Shi, T., Chen, C., Hao, A., & Li, S. (2022, October). Synthetic data supervised salient object detection. In *Proceedings of the 30th ACM International Conference on Multimedia* (pp. 5557-5565). DOI: 10.1145/3503161.3547930

Xin, Y. L., Ge, G. P., Du, W., Wu, H., & Zhao, Y. (2024). Design of an Optical Physics Virtual Simulation System Based on Unreal Engine 5. *Applied Sciences (Basel, Switzerland), 14*(3), 955. Advance online publication. DOI: 10.3390/app14030955

Xiong, J., Yu, D., Liu, S., Shu, L., Wang, X., & Liu, Z. (2021). A review of plant phenotypic image recognition technology based on deep learning. *Electronics (Basel), 10*(1), 81. DOI: 10.3390/electronics10010081

Xu, J., Ruan, Y., Bi, W., Huang, G., Shi, S., Chen, L., & Liu, L. (2023). On synthetic data for back translation. *arXiv Preprint arXiv:2310.13675.*

Xue, B., & Zou, L. (2022). Knowledge graph quality management: A comprehensive survey. *IEEE Transactions on Knowledge and Data Engineering*, 1. Advance online publication. DOI: 10.1109/TKDE.2022.3150080

Yadav, A. B. (2024). Generative AI in the Era of Transformers: Revolutionizing Natural Language Processing with LLMs. *Journal of Image Processing and Intelligent Remote Sensing, 42*(42), 54–61. Advance online publication. DOI: 10.55529/jipirs.42.54.61

Yang, D., Xu, Z., Mo, W., Chen, Q., Huang, S., & Liu, Y. (2024). 3D vision and language pretraining with large-scale synthetic data. *arXiv preprint arXiv:2407.06084.*

Yang, Z., Band, N., Li, S., Candes, E., & Hashimoto, T. (2024). Synthetic continued pretraining. *arXiv preprint arXiv:2409.07431.*

Yang, L., Zhang, Z., Song, Y., Hong, S., Xu, R., Zhao, Y., Zhang, W., Cui, B., & Yang, M. H. (2024). Diffusion Models: A Comprehensive Survey of Methods and Applications. *ACM Computing Surveys, 56*(4), 1–39. Advance online publication. DOI: 10.1145/3626235

Yang, X., Fan, X., Wang, J., & Lee, K. (2022). Image translation based synthetic data generation for industrial object detection and pose estimation. *IEEE Robotics and Automation Letters*, *7*(3), 7201–7208. Advance online publication. DOI: 10.1109/LRA.2022.3180403

Ye-Bin, M., Hyeon-Woo, N., Choi, W., Kim, N., Kwak, S., & Oh, T. H. (2023). Exploiting synthetic data for data imbalance problems: Baselines from a data perspective. *arXiv Preprint arXiv:2308.00994*.

Yee-King, M. (2022). Latent Spaces: A Creative Approach. *Springer Series on Cultural Computing*, 137–154. DOI: 10.1007/978-3-031-10960-7_8

Yen, S. W., Yeh, C. K., Morace, C. C., Chen, S. Y., Lin, S. S., Chen, C. H., & Lee, T. Y. (2016). Content enhanced word art with depth perception. In *SIGGRAPH ASIA 2016 Posters* (pp. 1-2). DOI: 10.1145/3005274.3005280

Yim, M., Kim, Y., Cho, H. C., & Park, S. (2021, September). Synthtiger: Synthetic text image generator towards better text recognition models. In *International Conference on Document Analysis and Recognition* (pp. 109-124). Cham: Springer International Publishing. DOI: 10.1007/978-3-030-86337-1_8

Yoon, J. G., Yoon, J., Song, J. Y., Yoon, S. Y., Lim, C. S., Seong, H., & Kim, W. J. (2020). Clinical significance of a high SARS-CoV-2 viral load in the saliva. *Journal of Korean Medical Science*, *35*(20).

Yoon, J., Drumright, L. N., & Van Der Schaar, M. (2020). Anonymization through data synthesis using generative adversarial networks (ADS-GAN). *IEEE Journal of Biomedical and Health Informatics*, *24*(8), 2378–2388. Advance online publication. DOI: 10.1109/JBHI.2020.2980262 PMID: 32167919

Young, J., Graham, P., & Penny, R. (2009). Using Bayesian networks to create synthetic data. *Journal of Official Statistics*, *25*(4), 549–567.

Yu, T., Yang, W., Xu, J., & Pan, Y. (2024). Barriers to Industry Adoption of AI Video Generation Tools: A Study Based on the Perspectives of Video Production Professionals in China. *Applied Sciences 2024, Vol. 14, Page 5770, 14*(13), 5770. DOI: 10.3390/app14135770

Yue, X., Inan, H. A., Li, X., Kumar, G., McAnallen, J., Shajari, H., . . . Sim, R. (2022). Synthetic text generation with differential privacy: A simple and practical recipe. *arXiv preprint arXiv:2210.14348*.

Zakari, R. Y., Lawal, Z. K., Kalinaki, K., & Maiwada, U. D. (2024). Time Series Forecasting Techniques for Climate Trend Prediction. *Reshaping Environmental Science Through Machine Learning and IoT*, 282–310. DOI: 10.4018/979-8-3693-2351-9.ch014

Zeng, F., Li, W., Gao, W., & Pang, Y. (2024). Multimodal misinformation detection by learning from synthetic data with multimodal LLMs. *arXiv preprint arXiv:2409.19656*. DOI: 10.18653/v1/2024.findings-emnlp.613

Zenni, R. D., & Andrew, N. R. (2023). Artificial Intelligence text generators for overcoming language barriers in ecological research communication. *Austral Ecology*, *48*(7), 1225–1229. DOI: 10.1111/aec.13417

Zhan, F., Yu, Y., Wu, R., Zhang, J., Lu, S., Liu, L., Kortylewski, A., Theobalt, C., & Xing, E. (2023). Multimodal Image Synthesis and Editing: The Generative AI Era. *IEEE Transactions on Pattern Analysis and Machine Intelligence*, *45*(12), 15098–15119. DOI: 10.1109/TPAMI.2023.3305243 PMID: 37624713

Zhang, E. Y., Cheok, A. D., Pan, Z., Cai, J., & Yan, Y. (2023). From Turing to Transformers: A Comprehensive Review and Tutorial on the Evolution and Applications of Generative Transformer Models. In *Sci* (Vol. 5, Issue 4). DOI: 10.3390/sci5040046

Zhang, T. (2022). "Data Synthesis". Encyclopedia of Big Data. Doi: DOI: 10.1007/978-3-319-32010-6_503

Zhang, Y., Wu, J., Li, W., Li, B., Ma, Z., Liu, Z., & Li, C. (2024). Video instruction tuning with synthetic data. *arXiv Preprint arXiv:2410.02713*.

Zhang, H., Grimmer, M., Ramachandra, R., Raja, K., & Busch, C. (2021, May). On the applicability of synthetic data for face recognition. In *2021 IEEE International Workshop on Biometrics and Forensics (IWBF)* (pp. 1-6). IEEE. DOI: 10.1109/IWBF50991.2021.9465085

Zhang, H., Wu, C., Zhang, Z., Zhu, Y., Lin, H., Zhang, Z., & Smola, A. (2022). Resnest: Split-attention networks. In *Proceedings of the IEEE/CVF conference on computer vision and pattern recognition* (pp. 2736-2746).

Zhang, J., Zhang, L., & Zhang, Y. (2020). Face reenactment with GANs for deageing in film production. *Multimedia Tools and Applications*, *79*(39), 29319–29338.

Zhang, L., Gonzalez-Garcia, A., Van De Weijer, J., Danelljan, M., & Khan, F. S. (2018). Synthetic data generation for end-to-end thermal infrared tracking. *IEEE Transactions on Image Processing*, *28*(4), 1837–1850. DOI: 10.1109/TIP.2018.2879249 PMID: 30403630

Zhang, Y., Wang, J. Z., Li, K., Shi, M. M., Wen, Z., Jiao, M. G., & Bao, D. (2022). Regulation of the electronic structure of perovskites to improve the electrocatalytic performance for the nitrogen-reduction reaction. *Journal of Materials Chemistry. A, Materials for Energy and Sustainability, 10*(6), 2819–2825. Advance online publication. DOI: 10.1039/D1TA10534E

Zhang, Y., Zhang, R., & Lee, B. G. I. (2025). From Abstract Concepts to Engaged Learning: An IVR Game-Based Framework for Physics Education. *International Journal of Online Pedagogy and Course Design.*

Zhang, Z., Huang, S., Li, Y., Li, H., & Hao, H. (2022). Image detection of insulator defects based on morphological processing and deep learning. *Energies, 15*(7), 2465. DOI: 10.3390/en15072465

Zhavoronkov, A., Ivanenkov, Y. A., Aliper, A., Veselov, M., Aladinskiy, V., Aladinskaya, A., Terentiev, V. A., Polykovskiy, D. A., Kuznetsov, M. D., Asadulaev, A., Volkov, Y., Zholus, A., Shayakhmetov, R. R., Zhebrak, A., Minaeva, L. I., Zagribelnyy, B. A., Lee, L. H., Soll, R., Madge, D., & Aspuru-Guzik, A. (2019). Deep learning enables rapid identification of potent DDR1 kinase inhibitors. *Nature Biotechnology, 37*(9), 1038–1040. DOI: 10.1038/s41587-019-0224-x PMID: 31477924

Zheng, X., Liu, Y., Gunceler, D., & Willett, D. (2021, June). Using synthetic audio to improve the recognition of out-of-vocabulary words in end-to-end ASR systems. In *ICASSP 2021-2021 IEEE International Conference on Acoustics, Speech and Signal Processing (ICASSP)* (pp. 5674-5678). IEEE. DOI: 10.1109/ICASSP39728.2021.9414778

Zhou, J., Zhang, Y., Luo, Q., Parker, A. G., & De Choudhury, M. (2023). Synthetic Lies: Understanding AI-Generated Misinformation and Evaluating Algorithmic and Human Solutions. *Conference on Human Factors in Computing Systems - Proceedings*. DOI: 10.1145/3544548.3581318

Zhou, K.. (2021). PoseGAN: A pose-guided adversarial network for generating facial expressions. *Information Sciences, 545*, 283–296.

Zhou, Y.. (2020). Actor-centric motion modelling and control. *Computer Graphics Forum, 39*(2), 409–420.

Zhou, Z., Chen, X., & Li, L. (2021). Edge intelligence: Paving the last mile of artificial intelligence with edge computing. *Proceedings of the IEEE.*

Zhu, J. (2024). Synthetic data generation by diffusion models. *National Science Review, 11*(8), nwae276. DOI: 10.1093/nsr/nwae276 PMID: 39262927

Ziabari, A., Dubey, A., Venkatakrishnan, S., Frederick, C., Bingham, P., Dehoff, R., & Paquit, V. (2021). High Resolution X-Ray CT Reconstruction of Additively Manufactured Metal Parts using Generative Adversarial Network-based Domain Adaptation in AI-CT. *Microscopy and Microanalysis*, *27*(S1), 2940–2942. Advance online publication. DOI: 10.1017/S1431927621010254

Zielinski, C., Winker, M., Aggarwal, R., Ferris, L., Heinemann, M., Lapeña, J. F., Pai, S., Ing, E., Citrome, L., Alam, M., Voight, M., & Habibzadeh, F. (2023). Chatbots, Generative AI, and Scholarly Manuscripts: WAME Recommendations on Chatbots and Generative Artificial Intelligence in Relation to Scholarly Publications Revised May 31, 2023. *Philippine Journal of Otolaryngology Head and Neck Surgery*, *38*(1), 7. Advance online publication. DOI: 10.32412/pjohns.v38i1.2135

About the Contributors

Shalin Hai-Jew currently works as a grant writer at Hutchinson Community College. She co-edits the *HutchCC Grant Navigator* newsletter (a monthly) for the college community and beyond. Dr. Hai-Jew has worked in higher education in the U.S. and stateside for decades. She has worked as tenured faculty, newspaper advisor, course developer and instructional designer, grant writer, and researcher. She has B.A.s in English and psychology, an MA in English (from the University of Washington in Seattle), and an Ed.D. in Educational Leadership with a focus on public administration and online learning (from Seattle University). She has received several scholarships from multiple institutions of higher education. She has edited over 25 nonfiction educational books and authored a dozen.

***

Munir Ahmad is a seasoned professional in the realm of Spatial Data Infrastructure (SDI), Geo-Information Productions, Information Systems, and Information Governance, boasting over 25 years of dedicated experience in the field. With a PhD in Computer Science, Dr. Ahmad's expertise spans Spatial Data Production, Management, Processing, Analysis, Visualization, and Quality Control. Throughout his career, Dr. Ahmad has been deeply involved in the development and deployment of SDI systems specially in the context of Pakistan, leveraging his proficiency in Spatial Database Design, Web, Mobile & Desktop GIS, and Geo Web Services Architecture. His contributions to Volunteered Geographic Information (VGI) and Open Source Geoportal & Metadata Portal have significantly enriched the geospatial community. As a trainer and researcher, Dr. Ahmad has authored over 50 publications, advancing the industry's knowledge base and fostering innovation in Geo-Tech, Data Governance, and Information Infrastructure, and Emerging Technologies. His commitment to Research and Development (R&D) is evident in his role as a dedicated educator and mentor in the field.

Nilesh Anute is an Associate Professor at Sri Balaji University, Pune. With a strong academic background and years of teaching experience, Dr. Anute specializes in the fields of management and business studies. He is known for his engaging teaching style, research contributions, and dedication to student development. Dr. Anute has actively participated in various academic conferences and has published research papers in reputed journals. His commitment to excellence in education and research continues to inspire students and peers alike.

Rahul Bir is an academic professional with a strong foundation in computer science, currently serving as an Assistant Professor at Brainware University since July 2023. He holds an M.Sc. in Computer Science from the University of North Bengal (2020), following a B.Sc. in the same field from the same institution. Rahul is a UGC NET (July 2020) and WBSET (2022) qualified scholar, showcasing his academic competence and dedication to higher education. His teaching journey includes one year as a TGT Math teacher at Kendriya Vidyalaya Sangathan, Bagdogra, and two years as a guest faculty member at the Centre for Innovative Studies, Techno India Group. With expertise in programming languages such as C, C++, and Python, and a keen interest in data structures, operating systems, and computer networks, Rahul brings a rich mix of knowledge and practical insights to his students. Fluent in Bengali, English, and Hindi, Rahul hails from Barasat, West Bengal. He actively participates in academic events, including international webinars on computational technologies.

A Prabhu Chakkaravarthy was born on 18th February 1985 in Tamilnadu, India. He received his Ph.D. in Computer Science and Engineering from Sathyabama Institute of Science and Technology, Chennai, India. He is graduated from Anna University, Chennai in 2007 with the Bachelor of Technology degree in Computer Science and Engineering. He received the Master of Technology degree in Computer Science and Engineering from Sathyabama University, Chennai in 2012 with Distinction. He is currently working as an Assistant Professor in Networking and Communications, SRM Institute of Science and Technology, Kattankulathur, Chennai, India. He has published more than 70 research papers in national, international conferences, journals including 8 in Science Citation Indexed journals with 120 citations. He has published 10 patents, 2 patents granted. His research interests include IoT, Artificial Intelligence, Deep Learning, Data Science, Signal Processing and Image Processing.

Saurabh Chandra is an accomplished legal scholar and currently serves as an Associate Professor & Assistant Dean at the School of Law, Bennett University, India. With a robust academic foundation, he earned his Bachelor of Arts and

Bachelor of Laws (B.A. LL.B.) from the prestigious Aligarh Muslim University (A.M.U.). He further specialized in Business Laws, completing his Master of Laws (LL.M.) at the renowned National Law School of India University (NLSIU), Bangalore. Dr. Chandra's academic journey culminated in a Doctorate in Law (Ph.D.) from the esteemed National Law University (NLU) Delhi. In addition to his impressive legal education from some of India's most prestigious institutions, Dr. Chandra also holds a master's degree in management with specialization in Human Resources in Management and another in Journalism and Mass Communication. With almost two decades of teaching experience, he has numerous publications to his credit. His expertise is complemented by a robust portfolio of scholarly publications. By bridging the gap between theoretical insights and real-world applications, Dr. Chandra's work serves as a valuable resource for academics, legal practitioners, and industry professionals alike. Dr. Chandra brings a wealth of experience in Academic Administration, having successfully held various key administrative positions. His diverse qualifications and extensive experience make him a distinguished figure in the academic and legal communities.

Kashvi Chaturvedi, is studying B.Tech, specialisation in Artificial Intelligence and Data Science at Ajeenkya DY Patil University. She started solidifying her programming skills by completing various certification courses like Data Structures in C, Tableau functions and Operating Systems along with doing few projects using languages like Python, C, C++ and creating Web Page using HTML, CSS and Java Script. She had also attended an Arduino-Based Automation course at Dolphin Labs to enhance her software knowledge. There, she learned how to work with sensors, microcontrollers, and circuit design. She has also written multiple review papers on topics such as Geographic Information Systems, Comparative Analysis of BJT Configurations, Swiss Bank vs. European Central Bank: Data Analysis. She is also researching on and learning many topics such as Small Modular Reactors, Bibliometric Analysis, UNSDGs etc. Because of her interest in latest technologies, she also participated in Global Hyperloop Competition hosted by IIT Madras representing her college as team Vac-Vectoor where she learnt more about various subtopics from the mentors.

Susanta Das is a seasoned educator with a proven record of teaching and mentoring diverse students, working at multiple institutions, and providing service to the various initiatives of organizations. He received his Ph.D. and M.A. degrees from Western Michigan University (WMU), USA, and M.Sc. degree from Banaras Hindu University (BHU), India all in Physics. He continued his research as a Marie-Curie post-doctoral fellow at Stockholm University, Stockholm, Sweden on beam diagnostics for the DESIREE (Double ElectroStatic Ion Ring ExpEriment) in the project DITANET (Diagnostic Techniques for particle Accelerators – a Marie-Curie initial training NETwork), at the Indian Institute of Science Education and Research-Kolkata, India on high-pressure physics as a Project Scientist-B, and at the University of Electro-Communications, Tokyo, Japan on ion-surface interactions as a Post-doctoral fellow. Throughout his career, Dr. Das worked and collaborated with many researchers, Ph.D., master, and visiting students from different countries. He further visited many countries to discuss research and an international conference participant (UK, Italy, Germany, Greece, Belgium, Bulgaria, Romania, Brazil etc.). He co-authored several research articles in WoS/Scopus indexed international journals, conference proceedings, and scholarly book chapters published by renowned international publishing houses. Beyond his research endeavors, Dr. Das has a storied history of service to academic and administrative committees, exhibiting his commitment to the growth and development of educational institutions. His experience includes tenure at Central University South Bihar, Sri Sri University, and P.K. University before assuming his current role at Ajeenkya DY Patil University. He received the Marie-Curie post-doctoral fellowship in Sweden, Science Academies' Summer Research Fellowship in India, Gwen Frostic Doctoral Fellowship, Department Graduate Research and Creative Scholar Award by WMU, and Leo R. Parpart Doctoral Fellowship by Dept. of Physics, WMU, among many others, throughout his academic journey. At present, Dr. Das continues to delve into cutting-edge fields, with a keen interest in nanotechnology, quantum computing, and data science. His multifaceted contributions, spanning teaching, research, and administrative leadership, showcase a dedicated professional who is instrumental in advancing the vision and mission of the institutions he serves.

Noor Hayat is working as Assistant Professor at University of Central Punjab. He completed his doctoral degree from Universiti Sains Malaysia. His research interests are political and international communication, artificial intelligence, media framing and agenda setting and climate change

S. Ida Evangeline, presently working as an Assistant Professor in the Department of Electrical and Electronics Engineering at the Government College of Engineering, Tirunelveli, Tamilnadu, India. She obtained her Bachelor's in Electrical and Electronics Engineering from Manonmaniam Sundaranar University in 2003, and Master's in Power Systems Engineering from Anna University, Chennai, Tamilnadu, India in 2005. She completed her PhD degree in 2023 at Anna University, Chennai, Tamilnadu, India. She has published six research papers in international journals. Her current research interests include optimal power flow, renewable energy, soft computing, machine learning, and deep learning.

J. Dhanalakshmi received the B.E. degree in Computer Science and Engineering and the M.E. degree in Computer Science and Engineering from Anna University, Tamil Nadu, India, in 2015 and 2017, respectively. She was ranked 31st at Anna University for her Master's degree in Computer Science and Engineering. She was a software developer at Fermtronics Solutions until 2018. She completed her Ph.D. degree at the School of Computer Science and Engineering, B.S. Abdur Rahman Crescent Institute of Science and Technology, located in Chennai, India. She is currently pursuing a postdoctoral fellowship at the National Kaohsiung University of Science and Technology, which is located in Kaohsiung, Taiwan. She is currently an assistant professor at the SRM Institute of Science and Technology, KTR Campus.

Kassim Kalinaki (Member, IEEE) is a computer science researcher, technologist, and educator with over a decade of experience bridging academic research and industry expertise. He holds a Diploma in Computer Engineering from Kyambogo University, a Diploma in IT, as well as BSc and MSc degrees in Computer Science and Engineering from the Islamic University of Technology, Bangladesh. He earned his PhD in Computer Science from the School of Digital Science, Universiti Brunei Darussalam (UBD). In addition to his academic qualifications, Dr. Kalinaki has obtained several industry-recognized certifications in cybersecurity and Big Data Analytics, solidifying his reputation as a trusted advisor to organizations striving for digital transformation and resilience. His research areas include: Ecological Informatics, Healthcare Informatics, Data Analytics, Computer Vision, Machine Learning (ML), Deep Learning (DL), Digital Image Processing, Cybersecurity, Internet of Things (IoT), Remote Sensing, and Educational Technologies. He has authored and co-authored over 60 peer-reviewed articles published by leading publishers such as Springer, Elsevier, Taylor & Francis, IGI Global, Emerald, Wiley, IEEE, and MDPI, advancing state-of-the-art research and innovation. He also serves as a reviewer for numerous international journals, contributing to the quality and integrity of scholarly output in his fields. Since 2014, Dr. Kalinaki has lectured at the Islamic University in Uganda (IUIU), where he served as Head of the Computer Science Department from 2019 to 2022. Passionate about mentorship and knowledge-sharing, he is committed to shaping future-ready talent and fostering technological innovations that contribute to building a secure, intelligent, and sustainable digital world.

Samrat Kundu is an Assistant Professor in the CSS Department at Brainware University, with over three years of experience in teaching and technical roles. He holds an MCA from the University of North Bengal and a BCA from MAKAUT, W.B. Skilled in programming (Java, Python, C/C++), video editing, and database management, he is passionate about teaching and student development. Samrat is fluent in English, Hindi, and Bengali, and enjoys games, music, and tech innovation.

M.Jegadeeswaran is currently working as a Assistant Professor, Department Of Agricultural Engineering at Agricultural Engineering at Rathinam Technical Campus, Coimbatore, Tamilnadu, India. He Completed B.E (Agriculture Engineering). in the year of MAY 2020, at Paavai Engineering College (Autonomous), Namakkal and M.E (Environmental Engineering) at Gnanamani College of Technology (Anna University), Namakkal,.

Zokir Mamadiyarov is a Professor of Economics at Mamun University, Khiva, Uzbekistan, and also serves as Professor at the Department of Finance and Tourism at Termez University of Economics and Service, Termez, Uzbekistan. He earned his academic qualifications in economics and finance and has over two decades of teaching and research experience in higher education. Prof. Mamadiyarov is a member of several national and regional academic initiatives and actively participates in conferences and journal publications. His primary research interests cover economic policy, financial systems, tourism management, and sustainable regional development.

T.C. Manjunath was born on 6th February 1967. He did his BE from RV College of Engg. in 1989, ME in Robotics & Automation from LD College of Engg. in 1992 with First Rank and PhD in Systems & Control Engg. from IIT Bombay in the year 2007. He is currently working as the Dean Research R D and as Professor in the Dept. of Computer Science & Engineering with specialization in IoT, Cyber Security & Blockchain Technology in Rajarajeswari College of Engineering, Bengaluru, Karnataka. He has completed B.E. Degree from R.V. College of Engg., M.E. Govt.'s LD College of Engg., Gujarat Univ., Ahmedabad in First Rank and Ph.D. in Systems and Control Engineering from the Indian Institute of Technology Bombay IIT Bombay He has got a teaching academic, research and administrative experience of more than 35+ years 10 years as Principal, 8 years as HoD rest as faculty and has published more than 1200+ research papers in various national & international journals and conferences, with 175+ patents, 25 book-chapters, 25 awards and also published 25 text books in various field of science, engineering and technology, completed projects worth Rs. 9 Crores till date apart from guiding 10 PhD scholars. He is a member of 30 Professional Society Chapters and is a fellow of IE, IETE, AMIEE and a chartered engineer. He got the best research scholar award in 2006 in IIT Bombay amongst all the research scholars of the institute. He has got 10 awards in academics and research activities.

Moses Matovu is an IT professional and academician with more than 20 years experience. His research interests include AI, Data structures and algorithms.

Nabukeera Madinah hold a Ph.D in Public Administration (University Malaya), Masters in Public Administration -Local Government (IUIU), Diploma in Law (IUIU), Postgraduate Diploma in Education management -PGDMATHE (IUIU), Postgraduate Diploma in Public Administration (UMI), Postgraduate Diploma in Project Planning (UMI) and Bachelors' of Public Administration (IUIU) . A lecturer, Faculty of Social Sciences – Department of Political Science and Public Administration Kyambogo University (KyU). She has vast experience in teaching at the university level. She is passionate about research and he has contributed many articles in international peer-reviewed journals. She has presented papers in both local and international conferences. Areas of expertise and interest are; leadership-administrative management, governance, public policy, Gender/Gender Based Violence and Politics and Public Administrationn. Besides academics, Dr Nabukeera served in different administrative positions at University Level. Much of Nabukeera's recent research also focused on political accountability and equity, Performance of NIRA, health governance systems accountability, students' mothers, Local government efficiency, governance of solid waste and service delivery, electoral democracy and election violence, power & individual merit in multiparty politics, governance of road accidents and Kasese bloodbath. In this respect, the scholar is recently working on a number of publications a book chapter with Palgrave Handbook on Global Social Change i.e., demographic, population and social change in Uganda.

P. Selvakumar is currently an Associate Professor in the Department of Science and Humanities at Nehru Institute of Technology, Coimbatore, Tamil Nadu, India. Prior to joining Nehru Institute of Technology, he held positions as an Assistant Professor in the Department of Science and Humanities various engineering colleges in Coimbatore, Tamil Nadu, India, spanning from 2011 to 2023. Dr.P. Selvakumar's academic and research interests lie primarily in the field of phytochemistry. He has made significant contributions to his field, with an impressive record that includes 30 journal publications, 12 patents, and 20 conference papers. Additionally, he has organized 5 seminars and conferences, further demonstrating his commitment to academic and scientific exchange. Throughout his career, Dr. P.Selvakumar has been recognized for his dedication and excellence in teaching, having been honored with several awards, including the Best Teacher Awards. He holds a prominent position in the academic community, contributing actively to research.

Deepanshu Pachauri, is an Assistant Professor in the Sharda School of Law, Sharda University, Greater Noida. He has publications in scopus indexed journals. He has attended national and international faculty development programmes. He has also participated in national and international conferences and seminars. His teaching and research experience is of 7 years.

Sandip Pal is currently serving as an Assistant Professor at Brainware University since July 2023, with prior experience as a Guest Lecturer at Midnapore College (Autonomous) in both BCA and M.Sc. Computer Science programs from August 2022 to June 2023. He holds an M.Sc. in Computer Science with a CGPA of 8.67 and a B.Sc. (Hons.) in the same field with 72.62%, both from Midnapore College. Sandip has qualified for the UGC NET and WB SET in Computer Science and has a keen research interest in machine learning, deep learning, natural language processing, and image processing. He has completed several academic projects including image segmentation using Rough K-Means clustering and an online bus reservation system. His technical skills include C, C++, Python, Java, SQL, and web technologies, and he has taught subjects such as DBMS, Operating Systems, Computer Networks, and Artificial Intelligence.

Guru Basava Aradhya S is a seasoned academician with 17 years of experience in management education. He currently serves as the Director & Professor at Chandigarh School of Business, Mohali. He holds a Ph.D. in Management and has mentored five Ph.D. scholars. His expertise includes HR, marketing, and business analytics. A prolific researcher, he has published in SCOPUS, ABDC, and UGC CARE journals. He has organized numerous international conferences and contributed to NAAC and NBA accreditations. His industry collaborations include faculty exchange programs and consultancy projects. Passionate about academia, he actively fosters innovation, research, and student development.

Saravana Kumar S is a dedicated academic professional in the Department of Mathematics at Vinayaka Mission's Kirupananda Variyar Engineering College, a constituent institution of Vinayaka Mission's Research Foundation (Deemed to be University), located in Salem, Tamil Nadu. With a strong commitment to teaching and research, he contributes significantly to the academic development of students and the growth of the institution. His expertise in mathematics plays a crucial role in shaping analytical and problem-solving skills among engineering students.

S.Seenivasan is currently working as a Associate Professor Department of Mechanical Engineering at Rathinam Technical Campus, Coimbatore,TamilNadu, India. He completed BE (Mechanical Engineering) in the year of May 2009 at Kumaraguru College of Technology, Coimbatore and ME (Engineering Design) in the year May 2012 at Anna University Regional campus,Coimbatore. Ph.D (Mechanical Engineering) in the Year 2021 at Anna University, Chennai, in the area of Composite Materials,machine learning,deep learning . He has 12 years of teaching & 5 Years of Research Experience.

Sunil Sankathala is pursuing B. Tech Computer Science and Engineering at Ajeenkya DY Patil University, with a diverse range of interests that bridge technology, sustainability, and innovation. His academic focus lies in computer science fundamentals like programming, data structures, algorithms, and software development. He actively supplements this with hands-on experience in web development and robotics. Beyond the core curriculum, He has a strong research inclination in emerging and impactful fields. His current research interests include Artificial Intelligence in Healthcare, bioplastics, nano bioplastics, e-waste management, and solid waste management—areas where he believes technology can make a meaningful difference in solving real-world problems. He has been publishing research articles and book chapters in reputed national and international journals and books. He is also part of an ambitious Hyperloop project, which involves interdisciplinary collaboration and cutting-edge engineering aimed at revolutionizing transportation. Working on this project has enhanced his skills in teamwork, innovation, and systems thinking. In the tech domain, he has developed proficiency in web development using technologies such as HTML, CSS, JavaScript, and React. His interests in robotics further reflect building smart, responsive systems that interact with the physical world. Throughout his course, he has helped strengthen problem-solving and collaborative skills. He is proficient in languages such as Python, C, C++, MySQL, Ubuntu, Machine Learning and familiar with frameworks and tools like React, Git.

Valisher Sapayev is a lecturer at the Department of General Professional Subjects at Mamun University, Khiva, Uzbekistan. He holds an academic background in pedagogical sciences and has been actively involved in higher education teaching and research. He regularly contributes to academic projects at Mamun University, with a focus on modern teaching methodologies, student engagement, and the integration of innovative pedagogical technologies in the learning process. His research interests include pedagogy, educational psychology, curriculum design, and the role of digital and interactive methods in university education.

Wasswa Shafik (Member, IEEE) received a Bachelor of Science degree in information technology from Ndejje University, Uganda, a Master of Engineering degree in Information Technology (Computer and Communication Networks) from Yazd University, Iran, and a PhD in Computer Science with the School of

Digital Science, Universiti Brunei Darussalam, Brunei Darussalam. He is also the Founder and Research Director of the Dig Connectivity Research Laboratory (DCRLab) after serving as a Research Associate at the Network Interconnectivity Research Laboratory at Yazd University. Prior to this, he worked as a Community Data Analyst at Population Services International (PSI-Uganda), a Community Data Officer at the Programme for Accessible Health Communication (PACE-Uganda), a Research Assistant at the Socio-Economic Data Centre (SEDC-Uganda), Uganda, an Assistant Data Officer at TechnoServe, Kampala, Uganda, and Asmaah Charity Organization as Cheif Executive Officer. He has hundreds of publications with renowned publishers. His research interests include Artificial Intelligence, Smart Agriculture, Digital Health, Green Computing, Security and Privacy.

Rizwan Shaikh is a distinguished academician, researcher, and software consultant with over 30 years of rich experience. He holds a Ph.D. in Computer Applications from Pune University, India. His doctoral work focused on the evaluation and analysis of e-Governance systems. Dr. Rizwan currently serves as Professor (Computer Applications) in prestigious D Y Patil Group of Institutes, Pune. He was Director (R&D, Science and Technology) at Anam Green Revolution (AGR) IT Solutions LLP. As a Board of Studies (BoS) member and syllabus designer, he contributed to curriculum reforms at RTM University and SPPU. He has authored and co-authored over 25 research papers on e-Governance, CRM, semantic web, and cloud computing. He has received research grants from BCUD, University of Pune, and holds a patent on electronic assessment systems. He has presented research at international conferences organized by IIM Bangalore, CSIR-IPR, and London Business School. Dr. Rizwan is an active member of professional bodies including INETA, CSI, ISTE, and the National Entrepreneurship Network (NEN). He played a pioneering role in establishing EDP cells in collaboration with IIM Bangalore and IIT Mumbai. He has served as coordinator for MBA online exams across 153 institutions affiliated with SPPU since 2013. As an educator, he has taught Computer Science, Management, Physics, and Engineering subjects. His contributions span academic leadership, accreditation processes (NAAC, NBA), and institutional development. He is a passionate advocate for value-based education and digital transformation in governance. With his cross-disciplinary expertise, he mentors students, faculty, and professionals in research and innovation. His vision is to create impactful digital ecosystems through governance, education, and technology.

Mohit Sharma is a research scholar at the Institute of Management Studies and Research, Maharshi Dayanand University, Rohtak, Haryana, India. His research focuses on the dynamic relationship between brand image and purchase intentions, particularly in the context of brand extensions within the FMCG sector. He has co-authored several scholarly works and continues to contribute to the academic community. Mohit can be reached at mohitattri.rs.imsar@mdurohtak.ac.in

Sachin Sharma has worked as a Professor at Amity School of Engineering and Technology, Amity University, Punjab, Mohali, India since January 2025. He also worked as Associate Dean of International Affairs and Professor, Department of Computer Science and Engineering at Graphic Era Deemed to be University, Dehradun, Uttarakhand, India from February, 2018 to January 2025. He has experience of 15+ years in industry and academia. He is also Co-founder and Chief Technology officer (CTO) of IntelliNexus LLC, Arkansas, USA based company. He also worked as a Senior Systems Engineer at Belkin International, Inc., Irvine, California, USA. He received his Philosophy of Doctorate (Ph.D.) degree in Engineering Science and Systems specialization in Systems Engineering from University of Arkansas at Little Rock, USA with 4.0 out 4.0 GPA and M.S. degree in Systems engineering from University of Arkansas at Little Rock with 4.0 out 4.0 GPA. Dr. Sharma has authored/coauthored over 270+ publications in the form of books, patents, and papers in refereed journals and conference proceedings. He holds 70+ US, Indian and international patents in the area of IoT, wireless communication.

Vatsala Sharma is a passionate Assistant Professor in Fine Arts, currently teaching at CGC Jhanjeri in the B.Sc. Multimedia program. With an MFA from Apeejay College of Fine Arts, she blends academic instruction with real-world experience in design and media aesthetics. Her expertise spans art theory, media production, and design communication. Vatsala is adept at using tools like Photoshop, Illustrator, and Figma, and has led numerous student activities, workshops, and departmental projects. Her portfolio includes international freelance work and internships at prestigious institutions like the National Gallery of Modern Art. She continues to mentor and inspire through creative teaching.

Arunima Shastri is a legal academic with a focus on the intersections of law, technology, and social justice. She holds an LL.M. in Corporate and Commercial Law from NLU Odisha and is currently completing her Ph.D. at Gujarat National Law University, where her research critically explores the interplay between intellectual property rights, visual art, deepfakes, and digital expression. Her teaching and research interests include intellectual property law, data privacy, blockchain regulation, and human rights law, with a particular emphasis on LGBTQ+ rights. Through her work,she bridges doctrinal legal analysis with contemporary technological and cultural developments, contributing to scholarship and curriculum development in emerging areas of law.

Aditya Shrivastav is an aspiring technologist currently pursuing a Bachelor of Technology (B. Tech) in Computer Science and Engineering at Ajeenkya D. Y. Patil University. He holds solid programming expertise in C, C++, and Python, alongside strong command over data and business tools such as Microsoft Excel and Tally software. Aditya is a member of Team Vac-Vectoor Hyperloop and participated with the team in the Global Hyperloop Competition organized by IIT Madras, contributing to cutting-edge transportation solutions. He actively engages in hackathons and various technical events to expand his skills and contribute to innovative problem-solving. He has authored and co-authored book chapters in the domains of solid waste management, bioplastics, nanotechnology applications, and artificial intelligence, showcasing his multidisciplinary approach and research acumen. Aditya is also working on an innovative project on predictive e-waste management. The project is targeted toward IT departments, schools, and colleges, aiming to support sustainable practices and resource optimization. In parallel with his academic journey, Aditya is serving as a faculty intern at Disha Computer Institute, where he teaches C, C++, basic Python, basic Java, Excel formulas, and Tally software. His dedication and teaching proficiency earned him a Letter of Appreciation and the opportunity to conduct a workshop on Advanced Excel at G. H. Raisoni College of Arts, Commerce and Science. With a strong foundation in technology, a passion for sustainability, and hands-on experience in both teaching and project development, Aditya Shrivastav is committed to driving impact through research, education, and innovation.

Bhupinder Singh working as Professor at Sharda University, India. His areas of publications as Smart Healthcare, Medicines, fuzzy logics, artificial intelligence, robotics, machine learning, deep learning, federated learning, IoT, PV Glasses, metaverse and many more. He has 3 books, 139 paper publications, 163 paper presentations in international/national conferences and seminars, participated in more than 40 workshops/FDP's/QIP's, 25 courses from international universities of repute, organized more than 59 events with international and national academicians and industry people's, editor-in-chief and co-editor in journals, developed new courses. He has given talks at international universities, resource person in international conferences such as in Nanyang Technological University Singapore, Tashkent State University of Law Uzbekistan; KIMEP University Kazakhstan, All'ah meh Tabatabi University Iran, the Iranian Association of International Criminal law, Iran and Hague Center for International Law and Investment, The Netherlands, Northumbria University Newcastle UK,

Ajay Upadhyaya is presently working as an Professor in the Department of Computer Engineering at SAL Engineering & Technical Institute, SAL Education,

Ahmedabad, Gujarat. He has completed his PhD in Computer Engineering and his M.Tech. in Computer Science and Engineering. He has 18 years of teaching experience. His areas of specialisation include Mobile and Vehicular Ad-hoc Networks, Security in Networks, Machine Learning, Data Science, and IOT. In addition to his academic endeavours, he participates actively in faculty development programmes, short-term initiatives, and seminars. He conducted more than 120 workshops and training programmes on various technical topics at different institutes for students and faculty members. He has more than 22 publications in various refereed national and international conferences and journals.

Aryan Vasa is pursuing his B.Tech. in computer science at the School of Engineering of Ajeenkya D Y Patil University, Pune, Maharashtra, India. his research interests are AI, ethics in AI, and applications of AI in nanotechnology.

Index

Symbols

100-Word Novelas 205

A

Agency 298
AI-Generated Content 46, 66, 67, 68, 70, 71, 76, 77, 78, 139, 337, 345, 356, 357, 358, 360, 361, 362, 363, 364, 365, 367, 368, 369, 370, 371, 373, 374
AI Governance 78, 79, 387
Algorithm 17, 58, 156, 166, 177, 184, 192, 195, 213, 224, 230, 231, 235, 262, 263, 279, 305, 349
Alphanumeric Images 283, 285
Artificial Intelligence 2, 4, 5, 7, 9, 10, 14, 22, 25, 26, 27, 28, 32, 35, 38, 52, 54, 56, 57, 58, 60, 61, 62, 63, 74, 75, 76, 77, 80, 81, 83, 84, 87, 88, 90, 92, 95, 97, 106, 113, 116, 118, 120, 121, 122, 124, 139, 146, 148, 150, 152, 183, 186, 199, 206, 221, 222, 223, 227, 228, 229, 232, 237, 239, 240, 246, 248, 249, 250, 251, 267, 268, 275, 277, 278, 279, 280, 281, 284, 296, 297, 299, 307, 310, 315, 335, 342, 350, 353, 354, 356, 357, 359, 362, 372, 380, 405, 406, 407, 408, 410, 430, 431, 432
Artmaking Generative Ai 285, 292
Authentication 73, 123, 140, 355, 356, 357, 358, 359, 360, 361, 362, 363, 364, 366, 372, 375, 400

B

Big Data 31, 32, 33, 38, 39, 40, 44, 52, 54, 55, 59, 60, 61, 64, 65, 76, 77, 78, 80, 83, 84, 85, 86, 198, 200, 206, 217, 322, 327, 328, 331, 406, 413, 430, 431, 432
black box 349, 409, 410, 413, 428

C

Content Creation 8, 16, 32, 40, 47, 58, 59, 60, 61, 62, 63, 68, 70, 73, 76, 136, 139, 356, 360, 371, 374, 375, 379, 395, 398
Content Generation 14, 32, 37, 38, 40, 42, 43, 56, 58, 59, 60, 61, 62, 63, 64, 65, 71, 72, 77, 82, 111, 227, 234, 365, 412, 429

D

Data Augmentation 24, 36, 54, 56, 65, 74, 75, 107, 123, 133, 142, 177, 202, 225, 242, 247, 305, 310, 327, 429, 430, 431
Data Generation 2, 8, 22, 26, 41, 54, 61, 69, 70, 76, 80, 81, 82, 83, 84, 85, 86, 89, 90, 93, 107, 113, 115, 120, 121, 122, 198, 199, 200, 201, 202, 207, 217, 218, 219, 220, 230, 231, 248, 297, 298, 299, 309, 313, 315, 322, 323, 324, 325, 326, 328, 329, 330, 331, 333, 334, 336
Deepfake Detection 4, 18, 66, 80, 342, 358, 363, 366, 371, 376
Deepfakes 15, 19, 34, 45, 46, 49, 56, 58, 67, 68, 71, 72, 83, 91, 115, 137, 145, 147, 223, 245, 337, 338, 340, 341, 342, 345, 346, 347, 349, 350, 360, 364, 367, 368, 372, 373, 375, 376, 377, 419, 420, 423, 431
Deep Learning 2, 4, 6, 7, 14, 19, 22, 25, 28, 29, 32, 35, 42, 52, 55, 56, 57, 58, 59, 60, 62, 70, 71, 72, 73, 74, 75, 76, 77, 78, 79, 81, 82, 83, 84, 85, 94, 97, 112, 113, 122, 126, 130, 140, 151, 153, 158, 167, 168, 180, 181, 202, 203, 207, 218, 223, 228, 229, 233, 236, 248, 249, 299, 310, 322, 323, 324, 330, 340, 347, 356, 359, 361, 367, 372, 376, 377, 411, 431
Deep Learning Models 6, 7, 14, 57, 58, 71, 72, 75, 84, 94, 153, 158, 207, 223, 229, 356, 367, 372
Diffusion Models 2, 3, 5, 8, 10, 18, 19, 20, 26, 29, 35, 37, 50, 52, 55, 60, 62, 63, 66, 76, 78, 111, 124, 125, 126, 127,

129, 136, 142, 144, 201, 220, 322, 409, 410, 411, 412, 414, 418, 430, 432
Digital Synthetic Data 183, 184, 355
Digital Synthetic Generation 31, 32, 33, 34, 35, 38, 40, 42, 44, 47, 48, 49, 52, 53, 57, 58, 59, 60, 61, 62, 63, 64, 65, 66, 67, 68, 69, 70, 71, 72, 73, 74, 75, 76, 77, 78, 79, 226
Digital Synthetic Outputs 1, 6, 18, 22, 24, 28, 221, 222, 223, 224, 225, 226, 239, 240, 241, 243, 244, 337, 340, 355, 359, 362, 363
Digital Synthetics 351, 409, 410
Digital Synthetic Text 222

E

Emergent Behavior 409, 424, 425
Ethical AI 61, 77, 78, 79, 338, 357, 359
Explainable AI 16, 18, 20, 52, 246, 250, 348, 349, 352, 426, 428

F

Fiction 87, 205, 206, 210, 215, 339, 382, 416
Finance 2, 3, 4, 6, 7, 18, 20, 26, 40, 45, 49, 77, 114, 166, 183, 198, 331

G

Generative AI 1, 2, 3, 4, 5, 6, 7, 8, 9, 10, 11, 12, 13, 14, 15, 16, 18, 20, 21, 22, 23, 24, 25, 26, 27, 28, 49, 50, 79, 83, 84, 85, 90, 92, 117, 141, 198, 199, 205, 206, 209, 210, 216, 218, 220, 221, 222, 224, 227, 228, 229, 231, 232, 233, 234, 236, 237, 238, 239, 241, 242, 243, 245, 246, 248, 249, 251, 283, 285, 286, 287, 292, 293, 295, 297, 299, 316, 317, 318, 319, 320, 326, 355, 379, 380, 381, 387, 388, 399, 403, 404, 405, 406, 407, 408, 409, 410, 411, 412, 413, 415, 417, 419, 421, 423, 424, 425, 426, 427, 428, 429, 430, 431, 432, 433
Generative Models 1, 2, 3, 4, 5, 6, 8, 9, 14, 15, 17, 19, 21, 25, 26, 29, 32, 33, 36, 37, 40, 41, 43, 44, 45, 48, 49, 50, 52, 60, 69, 71, 75, 76, 78, 79, 84, 111, 114, 127, 133, 134, 136, 137, 138, 142, 195, 224, 225, 226, 227, 228, 229, 230, 233, 237, 242, 243, 244, 325, 326, 357, 411, 412, 413, 414, 417, 418, 425, 426, 427, 428, 429
Graph Neural Network 154, 159, 180

H

Healthcare 2, 3, 4, 6, 7, 10, 14, 15, 17, 18, 19, 20, 23, 25, 26, 27, 29, 31, 32, 33, 37, 40, 41, 45, 49, 50, 52, 68, 69, 70, 71, 72, 73, 74, 75, 76, 77, 78, 79, 86, 101, 103, 114, 118, 121, 122, 139, 143, 145, 147, 149, 154, 166, 174, 177, 183, 184, 190, 191, 196, 199, 201, 208, 221, 236, 241, 296, 299, 301, 324, 331, 353, 355, 356, 359, 360, 365, 369, 371, 373, 380, 384, 408, 417
Human-in-the-Loop Verification 140, 426

I

Interpretability 19, 29, 49, 52, 94, 141, 144, 180, 230, 409

M

Machine Learning 2, 8, 10, 17, 23, 31, 32, 33, 34, 35, 39, 40, 41, 52, 54, 55, 56, 58, 59, 60, 61, 62, 63, 67, 72, 75, 78, 79, 80, 81, 83, 84, 85, 89, 91, 96, 97, 104, 105, 109, 111, 120, 121, 124, 125, 130, 132, 136, 141, 142, 143, 144, 149, 154, 170, 180, 183, 186, 188, 200, 201, 202, 207, 208, 218, 221, 226, 228, 237, 242, 243, 249, 250, 257, 265, 267, 268, 270, 274, 279, 280, 285, 296, 297, 299, 304, 309, 310, 312, 322, 325, 327, 329, 331, 332, 334, 338, 342, 344, 353, 354, 355, 356, 359, 361, 363, 364, 365, 369, 372, 406, 408, 415, 429, 430, 432
Media Literacy 358, 365, 370, 372, 373, 374

N

Neural Networks 5, 28, 31, 32, 34, 35, 36, 42, 56, 60, 63, 70, 72, 73, 80, 82, 97, 98, 99, 106, 109, 125, 126, 129, 133, 141, 146, 147, 148, 150, 151, 152, 153, 154, 158, 164, 176, 179, 180, 227, 228, 229, 231, 250, 275, 279, 303, 308, 310, 327, 328, 342, 354, 361, 364, 373, 425, 427
Newsletter 210, 295, 297, 316, 317, 318, 319, 321
Nonfiction 205, 210
Numbers 228, 284, 287, 293, 317, 396

O

Object Classification 175, 176, 177
Object Detection 71, 72, 86, 99, 166, 167, 168, 169, 173, 174, 175, 176, 179, 180, 181, 259, 304, 308, 313, 314, 328, 335

P

Physics Engines In Game Design for Digital Play 253
Privacy Preservation 20, 118, 184, 202, 303, 311

R

Research 1, 2, 11, 17, 20, 21, 22, 23, 24, 25, 28, 32, 39, 40, 46, 48, 50, 52, 54, 55, 62, 64, 76, 77, 79, 84, 85, 86, 97, 98, 102, 103, 114, 115, 121, 126, 128, 129, 130, 133, 145, 148, 151, 152, 153, 158, 166, 172, 177, 179, 180, 183, 184, 188, 190, 191, 192, 195, 196, 199, 200, 205, 206, 207, 209, 212, 213, 216, 217, 219, 221, 222, 224, 226, 227, 228, 232, 236, 238, 241, 242, 243, 244, 249, 253, 275, 276, 278, 280, 285, 292, 295, 296, 297, 298, 299, 305, 306, 308, 311, 312, 315, 321, 323, 326, 329, 341, 353, 354, 358, 359, 362, 363, 365, 371, 374, 375, 376, 377, 382, 384, 385, 388, 390, 396, 397, 398, 399, 400, 401, 402, 405, 408, 416, 420, 422, 423, 426, 429
Robotic Training 123

S

Software 1, 11, 12, 13, 14, 84, 104, 107, 109, 112, 125, 133, 146, 154, 183, 184, 187, 188, 189, 191, 193, 218, 240, 241, 253, 254, 257, 258, 260, 267, 269, 270, 280, 285, 286, 309, 330, 333, 350, 381, 410, 418, 421, 425, 427
Spun Text 205, 210, 216, 220
Synthetic Data 1, 2, 3, 4, 5, 6, 7, 8, 9, 15, 16, 17, 18, 19, 20, 22, 23, 25, 26, 28, 33, 36, 37, 41, 42, 43, 44, 52, 54, 58, 59, 61, 63, 65, 67, 68, 69, 70, 75, 76, 77, 78, 79, 80, 81, 82, 83, 84, 86, 87, 88, 89, 90, 91, 92, 93, 94, 95, 100, 101, 102, 103, 107, 108, 109, 110, 112, 113, 114, 115, 116, 117, 118, 119, 120, 121, 122, 126, 132, 133, 149, 183, 184, 185, 186, 187, 188, 189, 190, 191, 192, 193, 194, 195, 196, 197, 198, 199, 200, 201, 202, 203, 205, 206, 207, 208, 209, 210, 216, 217, 218, 219, 220, 229, 231, 237, 242, 247, 248, 284, 293, 296, 297, 298, 299, 300, 301, 302, 303, 304, 305, 306, 307, 308, 309, 310, 311, 312, 313, 314, 315, 316, 321, 322, 323, 324, 325, 326, 327, 328, 329, 330, 331, 332, 333, 334, 335, 336, 353, 355, 360, 365, 366, 367, 368, 371
Synthetic Imagery 295, 308, 321
Synthetic Media 5, 14, 15, 16, 31, 36, 45, 47, 48, 55, 58, 60, 63, 71, 126, 149, 337, 338, 339, 340, 342, 345, 346, 347, 348, 350, 351, 356, 357, 358, 359, 360, 361, 362, 363, 364, 365, 366, 367, 368, 369, 370, 371, 372, 373, 374, 375, 415
System Prototyping 183, 188

T

Text Art 284, 293
Text Prompts 60, 316, 380
Text-to-Image Synthesis 11, 69, 70
Transformers 2, 5, 6, 8, 10, 11, 12, 20, 21, 22, 28, 29, 35, 70, 71, 72, 74, 166, 176, 177, 236, 299, 347, 409, 412, 417, 421, 429, 432, 433

U

User-Prompted Synthetic Data 284, 293, 297, 336

V

Validation 17, 66, 67, 68, 82, 102, 123, 140, 145, 147, 189, 192, 193, 206, 236, 322, 324, 355, 356, 358, 359, 360, 361, 362, 363, 365, 366, 371, 372
Video Games 58, 62, 70, 72, 123, 226, 232, 255, 256, 257, 258, 259, 265, 277

W

Word Art 283, 284, 285, 286, 287, 291, 292, 293, 294, 295, 317
Word Images 304

www.ingramcontent.com/pod-product-compliance
Lightning Source LLC
LaVergne TN
LVHW062337300926
849161LV00007B/1728